Algebraic Analysis

Volume I

PROFESSOR MIKIO SATO

Algebraic Analysis

Papers Dedicated to Professor Mikio Sato on the Occasion of His Sixtieth Birthday

Volume I

Edited by

Masaki Kashiwara
Takahiro Kawai

Research Institute for Mathematical Sciences
Kyoto University
Sakyo-ku, Kyoto
Japan

ACADEMIC PRESS, INC.

Harcourt Brace Jovanovich, Publishers

Boston   San Diego   New York
Berkeley   London   Sydney
Tokyo   Toronto

ACADEMIC PRESS, INC.
1250 Sixth Avenue, San Diego, CA 92101

United Kingdom Edition published by
ACADEMIC PRESS INC. (LONDON) LTD.
24–28 Oval Road, London NW1 7DX

Library of Congress Cataloging-in-Publication Data

Algebraic analysis: papers dedicated to Professor Mikio Sato on the
  occasion of his sixtieth birthday/edited by Masaki Kashiwara,
  Takahiro Kawai.
      p. cm.
    Includes bibliographies and index.
    ISBN 0-12-400465-2 (v. 1). ISBN 0-12-400466-0 (v. 2)
    1. Mathematical analysis. 2. Algebra. 3. Sato, Mikio, Date–
  I. Sato, Mikio, Date–   . II. Kashiwara, Masaki, Date–   .
  III. Kawai, Takahiro.
  QA300.A45   1988
  515—dc19   88-6331
                                                                CIP

Printed in the United States of America

89 90 91 92   9 8 7 6 5 4 3 2 1

Dedication

We and the contributors to this collection dedicate these articles on algebraic analysis and related topics to Professor Mikio Sato.

We would like to celebrate the birthday of Professor Sato, initiator of algebraic analysis in the twentieth century.

Congratulations on your sixtieth birthday, Professor Sato!

April 18, 1988

Editors
Michio Jimbo
Masaki Kashiwara
(co-editor-in-chief)
Takahiro Kawai
(co-editor-in-chief)
Hikosaburo Komatsu
Tetsuji Miwa
Mitsuo Morimoto

VOLUME EDITORS:

Michio Jimbo

Research Institute for Mathematical Sciences
Kyoto University
Sakyo-ku, Kyoto
Japan

Masaki Kashiwara

Research Institute for Mathematical Sciences
Kyoto University
Sakyo-ku, Kyoto
Japan

Takahiro Kawai

Research Institute for Mathematical Sciences
Kyoto University
Sakyo-ku, Kyoto
Japan

Hikosaburo Komatsu

Department of Mathematics
The University of Tokyo
Bunkyo-ku, Tokyo
Japan

Tetsuji Miwa

Research Institute for Mathematical Sciences
Kyoto University
Sakyo-ku, Kyoto
Japan

Mitsuo Morimoto

Department of Mathematics,
Sophia University,
Chiyoda-ku, Tokyo
Japan

Contents

Contents of Volume II

Contributors

Numbers in parentheses refer to the pages on which the authors' contributions begin. Numbers greater than 470 indicate contributions in Volume II.

Takashi Aoki (19), *Department of Mathematics and Physics, Kinki University, Kowakae 3-4-1, Higashi-Osaka-shi, Osaka 577, Japan*

K. Aomoto (25), *Department of Mathematics, Nagoya University, Furo-cho, Chikusa-ku, Nagoya 464, Japan*

Helen Au-Yang (29), *Institute for Theoretical Physics, State University of New York at Stony Brook, Stony Brook, New York 11794-3840*

Gunter Bengel (41), *Mathematisches Institut der Westfälischen Wilhelms Universität, Roxeler Strasse 64, 44 Münster, Federal Republic of Germany*

J. Bros (49), *Service de Physique Théorique, Institut de Recherche Fondamentale, CEA, CEN-Saclay, B.P. 2, 91191 Gif-sur-Yvette Cédex, France*

Etsuro Date (75), *Department of Mathematics, College of General Education, Kyoto University, Yoshida-Nihonmatsu-cho, Sakyo-ku, Kyoto 606, Japan*

Leon Ehrenpreis (85), *Department of Mathematics, Temple University, Broad St. and Montgomery Ave., Philadelphia, Pennsylvania 19122*

L. D. Faddeev (129), *Steklov Institute, Fontanka 27, Leningrad 191011, USSR*

H. Flaschka (141), *Department of Mathematics, University of Arizona, Tucson, Arizona 85721*

Victor Guillemin (155), *Department of Mathematics, Massachusetts Institute of Technology, 77 Massachusetts Avenue, Cambridge, Massachusetts 02139*

B. Helffer (171), *Department of Mathematics, Université de Nantes, 44072 Nantes Cédex, France, and Centre de Mathématiques, Ecole Polytechnique, 91128 Palaiseau Cédex, France*

C. Denson Hill (185), *Department of Mathematics, State University of New York at Stony Brook, Stony Brook, New York 11794-3840*

Ryogo Hirota (203), *Department of Applied Mathematics, Faculty of Engineering, Hiroshima University, Oaza-Shitami, Saijo-cho, Higashi-Hiroshima, Hiroshima 724, Japan*

D. Iagolnitzer (217), *Service de Physique Théorique, CEN-Saclay, B.P. 2, 91190 Gif-sur-Yvette Cédex, France*

Jun-ichi Igusa (231), *Department of Mathematics, Johns Hopkins University, Baltimore, Maryland 21218*

Mitsuru Ikawa (243), *Department of Mathematics, Osaka University, Machikaneyama-cho 1-1, Toyonaka, Osaka 560, Japan*

S. Iyanaga (9), *Otsuka 6-12-4, Bunkyo-ku, Tokyo 112, Japan*

Michio Jimbo (75, 253), *Research Institute for Mathematical Sciences, Kyoto University, Sakyo-ku, Kyoto 606, Japan*

Akiro Kaneko (267), *Department of Mathematics, University of Tokyo, Komaba 3-8-1, Meguro-ku, Tokyo 153, Japan*

Masaki Kashiwara (1, 277), *Research Institute for Mathematical Sciences, Kyoto University, Sakyo-ku, Kyoto 606, Japan*

Kiyomi Kataoka (287), *Department of Mathematics, Faculty of Science, Tokyo Metropolitan University, Fukazawa, Setagaya-ku, Tokyo 158, Japan*

Yukiyosi Kawada (305), *Nagasaki 6-37-17, Toshima, Tokyo 171, Japan*

Takahiro Kawai (1, 309, 331), *Research Institute for Mathematical Sciences, Kyoto University, Sakyo-ku, Kyoto 606, Japan*

Tatsuo Kimura (345), *Department of Mathematics, University of Tsukuba, Tennodai 1-1-1, Niihari-gun, Ibaraki 305, Japan*

Hikosaburo Komatsu (357), *Department of Mathematics, University of Tokyo, Hongo 7-3-1, Bunkyo-ku, Tokyo 113, Japan*

Michio Kuga (373), *Department of Mathematics, State University of New York at Stony Brook, Stony Brook, New York 11794-3840*

Yves Laurent (381), *Département de Mathématiques, Université de Paris-Sud, Bâtiment 425, 91405 Orsay Cédex, France*

J.-L. Lieutenant (393), *Department of Mathematics, place du 20 Août 7, 4000 Liège, Belgium*

B. Malgrange (403), *Institut Fourier, B.P. 74, 38402 St. Martin d'Hères Cédex, France*

Barry M. McCoy (29), *Institute for Theoretical Physics, State University of New York at Stony Brook, Stony Brook, New York 11794-3840*

Z. Mebkhout (413), *U.E.R. de Mathématiques, Tour 45-55 5ème étage, Université de Paris 7, 2 place Jussieu, 75251 Paris Cédex 05, France*

Tetsuji Miwa (75, 253), *Research Institute for Mathematical Sciences, Kyoto University, Sakyo-ku, Kyoto 606, Japan*

Masatake Mori (423), *Denshi-Joho-Kogaku, Institute of Information Sciences and Electronics, University of Tsukuba, Tennodai 1-1-1, Niihari-gun, Ibaraki 305, Japan*

Mitsuo Morimoto (439), *Department of Mathematics, Sophia University, Kioi-cho, Chiyoda-ku, Tokyo 102, Japan*

Yasuo Morita (457), *Mathematical Institute, Tohoku University, Aza-Aoba Aramaki, Sendai-shi, Miyagi 980, Japan*

Motohico Mulase (473), *Department of Mathematics, University of California, Los Angeles, California 90024* xviii

Masakazu Muro (493), *Department of Mathematics, Kochi University, Akebono-cho 2-5-1, Kochi 780, Japan*

Yoshimasa Nakamura (505), *Department of Mathematics, Gifu University, Yanagito 1-1, Gifu 501-11, Japan*

Noboru Nakanishi (517), *Research Institute for Mathematical Sciences, Kyoto University, Sakyo-ku, Kyoto 606, Japan*

Isao Naruki (527), *Research Institute for Mathematical Sciences, Kyoto University, Sakyo-ku, Kyoto 606, Japan*

Masatoshi Noumi (549), *Department of Mathematics, Sophia University, Kioi-cho 7-1, Chiyoda-ku, Tokyo 102, Japan*

Toshinori Ôaku (571), *Department of Mathematics, Yokohama City University, Seto 22-2, Kanazawa-ku, Yokohama 236, Japan*

Takayuki Oda (587), *Department of Mathematics, Faculty of Science, Niigata University, Ninomachi 8050, Ikarashi 2 no chou, Niigata 950-21, Japan*

Yasunori Okabe (601), *Department of Mathematics, Hokkaido University, Kita 13-jo, Nishi 6, Kita-ku, Sapporo 060, Japan*

Masato Okado (75, 253), *Research Institute for Mathematical Sciences, Kyoto University, Sakyo-ku, Kyoto 606, Japan*

Kazuo Okamoto (647), *Department of Mathematics, University of Tokyo, Komaba 3-8-1, Meguro-ku, Tokyo 153, Japan*

Takashi Ono (659), *Department of Mathematics, Johns Hopkins University, Baltimore, Maryland 21218*

Toshio Oshima (667, 681), *Department of Mathematics, University of Tokyo, Hongo 7-3-1, Bunkyo-ku, Tokyo 113, Japan*

Jacques H. H. Perk (29), *Institute for Theoretical Physics, State University of New York at Stony Brook, Stony Brook, New York 11794-3840*

F. Pham (699), *Department of Mathematics, Université de Nice, Parc Valrose, 06034 Nice Cédex, France*

Mario Rasetti (727), *Dipartimento di Fisica, Politecnico di Torino, 10100 Torino, Italy*

Tullio Regge (727), *Dipartimento di Fisica Teorica, Università di Torino, via Pietro Giuria 1, 10125 Torino, Italy*

N. Yu. Reshetikhin (129), *Steklov Institute, Fontanka 27, Leningrad 191011, USSR*

Gian-Carlo Rota (13), *Department of Mathematics, Massachusetts Institute of Technology, 77 Massachusetts Avenue, Cambridge, Massachusetts 02139*

Yutaka Saburi (681), *Chiba Tanki College, Kuonodai 1-3-1, Ichikawa 272, Japan*

Kyoji Saito (735), *Research Institute for Mathematical Sciences, Kyoto University, Sakyo-ku, Kyoto 606, Japan*

Fumihiro Sato (789), *Mathematisches Institut der Universität, Bunsenstrasse 3-5, 3400 Göttingen, Federal Republic of Germany, and Department of Mathematics, Rikkyo University, Nishi-Ikebukuro 3-34-1, Toshima-ku, Tokyo, Japan*

Pierre Schapira (809), *Centre Scientifique et Polytechnique, Département de Mathématiques, Université de Paris-Nord, Avenue J. B. Clément, 93430 Villetaneuse, France*

Jiro Sekiguchi (821), *Department of Mathematics, University of Electro-Communications, Dentsudai, Chofugaoka 1-5-1, Chofu 182, Tokyo, Japan*

J. Sjöstrand (171), *Department of Mathematics, Université de Paris-Sud, 91405 Orsay, France*

Henry P. Stapp (309), *Lawrence Laboratory, University of California, Berkeley, California 94720*

Hidetoshi Tahara (837), *Department of Mathematics, Sophia University, Kioi-cho 7-1, Chiyoda-ku, Tokyo 102, Japan*

Shinichi Tajima (849), *Department of Mathematics, Niigata University, Ninomachi 8050, Ikarashi, Niigata 950-21, Japan*

K. Takasaki (853), *Research Institute for Mathematical Sciences, Kyoto University, Kitashirakawa, Sakyo-ku, Kyoto 606, Japan*

Yoshitsugu Takei (331), *Department of Mathematics, Faculty of Science, Kyoto University, Sakyo-ku, Kyoto 606, Japan*

L. A. Takhtajan (129), *Steklov Institute, Fontanka 27, Leningrad 191011, USSR*

Shuang Tang (29), *Institute for Theoretical Physics, State University of New York at Stony Brook, Stony Brook, New York 11794-3840*

Nobuyuki Tose (837), *Department of Mathematics, Faculty of Science, University of Tokyo, 7-3-1 Hongo, Bunkyo, Tokyo 113, Japan*

Keisuke Uchikoshi (883), *Department of Mathematics, National Defense Academy, Boeidai, Hashirimizu 1-10-20, Yokosuka 239, Japan*

Kimio Ueno (893), *Department of Mathematics, Yokohama City University, 22-2 Seto, Kanazawa-ku, Yokohama 236, Japan*

J.-L. Verdier (901), *Ecole Normale Supérieure, 45 rue d'Ulm, 75230 Paris Cédex 05, France*

Ryoko Wada (439), *Department of Mathematics, Sophia University, Kioi-cho, Chiyoda-ku, Tokyo 102, Japan*

Masato Wakayama (681), *Sanzo 985, Higashimura-cho, Fukuyama 729–02, Japan*

Hirofumi Yamada (893), *Department of Mathematics, College of Science, University of the Ryukyus, Nishihara-cho, Okinawa 903-01, Japan*

Masao Yamazaki (911), *Mathematical Sciences Research Institute, 1000 Centennial Drive, Berkeley, California 94720, and Department of Mathematics, Faculty of Science, University of Tokyo, Hongo 7-3-1, Bunkyo-ku, Tokyo 113, Japan*

Tamaki Yano (927), *Department of Mathematics, Faculty of Science, Saitama University, Shimo-Okubo 255, Uruawa-shi, Saitama 338, Japan*

Kunio Yoshino (943), *Department of Mathematics, Jochi University, Kioi-cho 7-1, Chiyoda-ku, Tokyo 102, Japan*

Kosaku Yosida (17), *Kajiwara 3-24-4, Kamakura 247, Japan*

Introduction

Professor Mikio Sato and his Work

In 1952 Professor Mikio Sato graduated from the Department of Mathematics of the University of Tokyo, majoring in number theory under the direction of Professor S. Iyanaga. He then spent a few years of graduate study in physics, with Professor S. Tomonaga as his advisor, at Tokyo Bunrika University. In 1958 he became an assistant at the Department of Mathematics of the University of Tokyo. In 1963 he received the degree of Doctor of Science from the University of Tokyo.

In 1970 he became a professor at the Research Institute for Mathematical Sciences (RIMS), Kyoto University, after having held memberships at Tokyo Kyoiku University (the new name of Tokyo Bunrika University), the Institute for Advanced Study, Osaka University, Columbia University and the University of Tokyo. He has been the director of the Institute since 1987.

After settling in Kyoto, Professor Sato made numerous visits to the United States, Europe and Israel to give lectures and attend conferences. Those visits included a one-year stay at the University of Nice during the academic year 1972. He was one of the main speakers at the International Congress of Mathematicians at Nice (1970) and at Warsaw (1983). He was awarded the Asahi Prize in 1970 (with H. Komatsu as joint recipient), the Japan Academy Prize in 1976, and the Fujiwara Prize in 1987 for his outstanding research in mathematics. In 1984, he was elected as a Contributor to

Algebraic Analysis, Volume I

Copyright © 1988 by Academic Press, Inc.
All rights of reproduction in any form reserved.
ISBN 0-12-400465-2

Culture, one of the greatest honors that the Government of Japan offers to distinguished scholars and artists.

The research of Professor Sato ranges over vast areas of mathematics and physics, as can be observed in the bibliography below. Still, his whole research seems to aim at one thing, i.e., the renaissance of "Algebraic Analysis" of Euler; he always tries to find something enduring that cannot be affected by volatile mathematical fashions. Such mathematics is what Euler, Gauss, Abel, Jacobi, Riemann, and others pursued, but it has become less emphasized since the turn of the century. Professor Sato is really a mathematical scientist in its best sense, namely in the same sense that Archimedes, Newton, Euler, Gauss, Riemann, Dirac, and others are mathematical scientists. With audacity and thorough thinking he has dug out many remarkable structures from the depths of nature, and at the same time he has made substantial contributions to the understanding of several problems in physics, making full use of the mathematics he has developed.

Although it is impossible to give here even a modest description of all the achievements that Professor Sato has made and is making, we will list some of them in chronological order.

Professor Sato invented the theory of hyperfunctions in 1957 ([1], [2], [3], [4]). In formulating his new theory of generalized functions, he developed the theory of relative cohomology groups, independently of A. Grothendieck. Furthermore, he obtained the theory of something similar to derived categories in this connection, although the full account of the theory was not published in spite of his announcement.

As is mentioned in Iyanaga's contribution to this volume, around 1960 Professor Sato became discouraged by the rather cold reaction to the theory of hyperfunctions from the mathematical world in spite of the initial enthusiastic reaction. This seems to have driven him to concentrate on number theory and some related topics. Most of them, except for the theory of prehomogeneous vector spaces ([9], [20], [30]), were not published, but they include some conjectures on the distribution of zeros of congruence zeta functions and the study of the relation between the Ramanujan conjecture and the Weil conjecture ([5]). The last one gave an impetus to the study of Kuga-Shimura (*Ann. of Math.*, **82** (1965)) and Y. Ihara (*Ann. of Math.*, **85** (1967)), leading to the final affirmative answer to the Ramanujan conjecture due to P. Deligne (Sém. Bourbaki, Exposé 355, 1968-69, Lect. Notes in Math. No. 179, Springer).

Thus he spent the early 1960s in studying the algebraic aspect of algebraic analysis. Here is, however, one important thing that is of a different flavor and worth mentioning concerning his activities in those days: At a colloquium at the University of Tokyo (1960), he emphasized the importance of the cohomological treatment of (overdetermined) systems of linear differential equations (cf. [6], [8]) and introduced the primitive form of the notion of maximally overdetermined systems so that one might characterize and analyze "natural" functions by such systems. (The terminology "maximally overdetermined system" has now been superseded by a shorter one: holonomic system.) This talk may be regarded as a declaration of the opening of algebraic analysis of the twentieth century. Professor Sato once said ([9]) that one motivation for constructing the theory of pre-homogeneous vector spaces had been to make a laboratory where he could try to exemplify the idea of controlling functions by maximally overdetermined systems. As byproducts, this laboratory produced the notion of a-functions, b-functions and c-functions.

After his return to the University of Tokyo in 1968, the enthusiastic activity of Komatsu, M. Morimoto and the students of Komatsu called him back to the analytic aspect of algebraic analysis. Inspired by the celebrated work of F. John on the plane wave decomposition, he introduced the notion of microfunctions in 1969 ([7], [8], [9], [12]). Together with the discovery of Maslov and Egorov that one can find a transformation of pseudo-differential operators compatible with the contact transformation, this opened a new era in analysis; the monumental works of Sato–Kawai–Kashiwara ([17]) and (Duistermaat-) Hörmander (*Acta Math.*, **127** (1971), **128** (1972)) established "Microlocal Analysis," the local analysis on the cotangent bundle. One of the most important discoveries in [17], besides the sheaf of microfunctions, is the so-called structure theorem for general overdetermined systems of microdifferential equations. It asserts that, generically speaking, any system is microlocally a direct sum of very simple equations, i.e., partial de Rham systems, partial Cauchy–Riemann systems, and Lewy–Mizohata systems. This theorem gives decisive results on the propagation of singularities and the (non-)existence of solutions of general overdetermined systems.

His stay at Nice mentioned earlier gave him an opportunity to talk with F. Pham, D. Iagolnitzer and D. Olive. First he recognized that the analysis of the S-matrix done by Iagolnitzer–Stapp (*Commun. Math. Phys.*, **14** (1969)) could be regarded as a prototype of microfunction theory. Encouraged by this observation, he proposed to study Green's function in quantum field theory and related functions with the aid of microlocal analysis ([22], [27]). In 1976, in order to embody this proposal, Professor Sato began with

T. Miwa and M. Jimbo the study of the two-dimensional Ising model. Just around that time, Wu–McCoy–Tracy–Barouch (*Phys. Rev.* B **13** (1976)) made a surprising discovery, that the two-point correlation function for the scaling limit of the two-dimensional Ising model can be explicitly written down with the help of the Painlevé transcendent. Upon seeing this result, Professor Sato immediately noticed the close connection between the Ising model and the classical theory of Schlesinger on the monodromy-preserving deformation of linear ordinary differential equations (*J. Reine Angew. Math.*, **141** (1912)), and he, Miwa and Jimbo succeeded in the explicit calculation of general n-point functions by making use of solutions of nonlinear differential equations which arise as a condition to preserve the monodromy structure in deforming a system of linear differential equations ([34], [36]). Professor Sato was further interested in the relation between monodromy-preserving deformations and spectrum-preserving deformations studied by P. Lax (*Comm. Pure Appl. Math.*, **21** (1968)), and he began to study nonlinear differential equations of the soliton type. He was particularly interested in Hirota's intriguing method of constructing concrete solutions of soliton-type equations. Collaborating with Mrs. Sato, in 1981, he obtained the following beautiful result ([47], [48], [49]): a solution of Kodomtsev–Petriashivili equations corresponds to a point of an infinite-dimensional Grassmann manifold, and (almost) all hitherto-known solutions of soliton-type equations can be obtained as an orbit of a subgroup of the transformation group $GL(\infty)$ of the infinite-dimensional Grassmann manifold. Further, Hirota's bilinear equations can be explained as Plücker relations defining the Grassmann manifold as a submanifold of projective space.

Although Professor Sato has been exceptionally successful in very fundamental problems in analysis, he still seems not to have been completely satisfied with what he has done so far; he is trying to attack challenging problems such as finding some structure theorem for nonlinear differential equations. Since "Kanreki," a Chinese expression of the sixtieth birthday, means in Japan releasing a person from all social duties, we sincerely wish Professor Sato to use this privilege to develop his theory further. At the same time, we and all of his other students promise him to try to embody his ideas.

Masaki Kashiwara
Takahiro Kawai

Publications of Professor Mikio Sato

[1] On a generalization of the concept of functions, *Proc. Japan Acad.* **34** (1958) 126–130.

[2] On a generalization of the concept of functions, II, *Ibid.* **34** (1958) 604–608.

[3] Theory of hyperfunctions, *Sûgaku*, **10** (1958) 1–27 (in Japanese).

[4] Theory of hyperfunctions, I and II, *J. Fac. Sci. Univ. Tokyo, Sect. I*, **8** (1959/60) 139–193 and 387–437.

[5] Weil's conjecture and Ramanujan's conjecture, *Sûgaku-no-Ayumi*, **10**-2 (1963) 2–7 (notes by H. Hijikata, in Japanese).

[6] On systems of linear differential equations, *RIMS Kokyuroku*, No. 59 (1968) 225–237 (with S. Hitotumatu, in Japanese).

[7] Structure of hyperfunctions, *Report of Katata Conf.* (1969) 4-1–4-30 (notes by T. Kawai, in Japanese).

[8] Hyperfunctions and partial differential equations, *Proc. Internat. Conf. on Functional Analysis and Related Topics, 1969*, Univ. Tokyo Press, Tokyo, 1970, pp. 91–94.

[9] The Sato-volume of *Sûgaku-no-Ayumi*, **15**-1 (1970) (notes by M. Kashiwara, H. Enomoto and T. Shintani, in Japanese).

[10] Mathematical theory of Maya game, *RIMS Kokyuroku*, No. 98 (1970) 105–135 (notes by H. Enomoto, in Japanese).

[11] Regularity of hyperfunction solutions of partial differential equations, *RIMS Kokyuroku*, No. 114 (1971) 105–123.

[12] Regularity of hyperfunction solutions of partial differential equations, *Actes Congrès Internat. Math., 1970*, Gauthier-Villars, Paris, 1971, **2**, pp. 785–794.

[13] On hyperfunctions and the sheaf $\mathscr{C}$—An introduction to algebraic analysis, *RIMS Kokyuroku*, No. 126 (1971) 1–113 (notes by Y. Namikawa, in Japanese).

[14] On Pseudo-differential equations in hyperfunction theory, *RIMS Kokyuroku*, No. 162 (1972) 136–144 (with T. Kawai and M. Kashiwara).

[15] On the structure of single linear pseudo-differential equations, *Proc. Japan Acad.*, **48** (1972) 643–646 (with T. Kawai and M. Kashiwara).

[16] Microlocal structure of a single linear pseudo-differential equation, *Séminaire Goulaouic-Schwartz, 1972-1973*, Exposé 18.

[17] Microfunctions and pseudo-differential equations, *Lecture Notes in Math.*, No. 287, Springer, 1973, pp. 265-529 (with T. Kawai and M. Kashiwara).

[18] The theory of pseudo-differential equations in the framework of hyperfunctions, *Sûgaku*, **25** (1973) 213-238 (with T. Kawai and M. Kashiwara, in Japanese).

[19] Pseudo-differential equations and theta functions, *Astérisque* **2** et **3** (1973) 286-291.

[20] On zeta functions associated with prehomogeneous vector spaces, *Ann. of Math.*, **100** (1974) 131-170 (with T. Shintani).

[21] Pseudo-differential equations and theta functions, *RIMS Kokyuroku*, No. 201 (1974) 247-252 (reprinted from [19]).

[22] On the micro-analyticity of the *S*-matrix. I., *RIMS Kokyuroku*, No. 209 (1974) 11-25 (notes by T. Kawai, in Japanese).

[23] The determinant of matrices of pseudo-differential operators, *Proc. Japan Acad.*, **51** (1975) 17-19 (with M. Kashiwara).

[24] Reduced *b*-function, *RIMS Kokyuroku*, No. 225 (1975) 1-15 (notes by T. Yano, in Japanese).

[25] Microlocal calculus. I., *RIMS Kokyuroku*, No. 226 (1975) 114-167 (notes by M. Jimbo, in Japanese).

[26] Fourier transform with imaginary Lagrangian variety, *RIMS Kokyuroku*, No. 248 (1975) pp. 212-260 (with T. Miwa, M. Kashiwara and M. Muro, in Japanese).

[27] Recent development in hyperfunction theory and its applications to physics (Microlocal analysis of *S*-matrices and related quantities), *Lecture Notes in Physics*, No. 39, Springer, 1975, pp. 13-29.

[28] Microlocal study of infinite systems, *RIMS Kokyuroku*, No. 248 (1975) 261-292 (with T. Miwa and M. Kashiwara, in Japanese).

[29] Dimension formula for the Landau singularity, *RIMS Kokyuroku*, No. 266 (1976) 77-87 (with T. Miwa and M. Jimbo, in Japanese).

[30] A classification of irreducible prehomogeneous vector spaces and their relative invariants, *Nagoya Math. J.*, **65** (1977) 1-155 (with T. Kimura).

[31] Holonomy structure of Landau singularities and Feynman integrals, *Publ. RIMS, Kyoto Univ.*, **12** suppl. (1977) 387-438 (with T. Miwa, M. Jimbo and T. Oshima).

[32] Studies on holonomic quantum fields. I, *RIMS Kokyuroku*, No. 287 (1977) 42-48 (with T. Miwa and M. Jimbo).

[33] Studies on holonomic quantum fields, *RIMS Kokyuroku*, No. 295 (1977) 77-87 (with T. Miwa and M. Jimbo, in Japanese).

[34] Studies on holonomic quantum fields, I-VIII, X-XII, XV, XVI, *Proc. Japan Acad.*, **53** A (1977) 6-10, 147-152, 153-158, 183-185, 219-224, **54** A (1978) 1-5, 36-41, 221-225, 309-313, **55** A (1979) 6-9, 73-77, 267-272, **56** A (1980) 317-322 (with T. Miwa, M. Jimbo and partly with Y. Môri).

[35] Studies on holonomic quantum fields—a deformation theory of linear differential equations, *RIMS Kokyuroku*, No. 321 (1978) 132-160 (with T. Miwa and M. Jimbo, in Japanese).

[36] Holonomic quantum fields I-V, *Publ. RIMS, Kyoto Univ.*, **14** (1978) 223-267 (I), **15** (1979) 201-278 (II), 577-629 (III), 871-972 (IV), **16** (1980) 531-584 (V), 137-151 (IV Suppl), **17** (1981) (with T. Miwa and M. Jimbo).

[37] Higher dimensional holonomic quantum fields, *RIMS Kokyuroku*, No. 349 (1979) 45-50 (with T. Miwa and M. Jimbo, in Japanese).

[38] Federbush model, *RIMS Kokyuroku*, No. 349 (1979) 56-63 (with T. Miwa and M. Jimbo, in Japanese).

[39] On the density matrix for impenetrable boson, *RIMS Kokyuroku*, No. 361 (1979) 180–183 (with T. Miwa, M. Jimbo, in Japanese).

[40] H. Q. F. '79, *RIMS Kokyuroku*, No. 375 (1980) 6–12 (with T. Miwa, M. Jimbo and Y. Môri, in Japanese).

[41] Density matrix of impenetrable bose gas and the fifth Painlevé transcendent, *Physica* 1 D (1980) 80–158 (with M. Jimbo, T. Miwa and Y. Môri).

[42] Micro-local analysis of prehomogeneous vector spaces, *Inventiones Math.*, **62** (1980) 117–179 (with M. Kashiwara, T. Kimura and T. Oshima).

[43] Holonomic quantum fields—The unanticipated link between deformation theory of differential equations and quantum field theory, *Lecture Notes in Physics*, No. 116, Springer, 1980, 119–142 (with M. Jimbo, J. Miwa and Y. Môri).

[44] Aspects of holonomic quantum fields, *Lecture Notes in Physics*, No. 126, Springer, 1980, 429–491 (with T. Miwa and M. Jimbo).

[45] Problems on deformation theory, *RIMS Kokyuroku*, No. 388 (1980) 1–10 (in Japanese).

[46] On Hirota's bilinear equations, *RIMS Kokyuroku*, No. 388 (1980) 183–204 (with Y. Môri, in Japanese).

[47] On Hirota's bilinear equations, II, *RIMS Kokyuroku*, No. 414 (1981) 181–202 (with Y. Sato, in Japanese).

[48] Soliton equations as dynamical systems on infinite dimensional Grassmann manifolds, *RIMS Kokyuroku*, No. 439 (1981) 30–46.

[49] Soliton equations as dynamical systems on infinite dimensional Grassmann manifolds, *Lect. Notes in Num. Appl. Anal.* **5** (1982) 259–271 (with Y. Sato).

[50] Linear differential equations of infinite order and theta functions, *Adv. Math.*, **47** (1983) 300–325 (with M. Kashiwara and T. Kawai).

[51] M. Sato. Monodromy theory and holonomic quantum fields—a new link between mathematics and theoretical physics. (Talk at the Warsaw Congress, 1983. Unpublished).

[52] Microlocal analysis of theta functions, *Advanced Study in Pure Math.* **4** (1984) 267–289 (with M. Kashiwara and T. Kawai).

Three Personal Reminiscences

S. Iyanaga

12-4 Otsuka 6-chome
Bunkyo-ku
Tokyo, Japan

I have been acquainted with Mikio Sato since the early 1950s. I am not in a position to enumerate exactly how many times I met him, at which places, and on which occasions. He has left me, however, with innumerable memories, three of which left particularly strong impressions.

1.

I believe it was the year 1956, or perhaps in 1957. I met him in my study room at my home on a winter evening. I remember well that a gas heater was burning. Rather suddenly after some months of non-contact, he had called me over the phone to fix the date of this visit.

Sato graduated from the University of Tokyo in 1952. During his last year there in my seminar he, Shimura and others studied algebraic geometry using Weil's *Foundations*. He told me that he was obliged to teach at a high school to help family finances. I remember that he did well in my seminar, but after graduation he expressed his desire to turn to the study of theoretical

Algebraic Analysis, Volume I

physics under the direction of Professor S. Tomonaga at Tokyo Bunrika University. I could not but encourage him. He was admitted to the graduate school of that university and thus his wishes were realized. He then visited me only a few times a year to tell me what he was doing.

In 1955 we had an International Symposium on Algebraic Number Theory in Tokyo and Nikko. Weil came to this Symposium and was impressed by the contributions of Shimura and Taniyama. While understanding that Sato was making progress in the study of physics, I regretted then that he was not with us.

It was a year or so after this Symposium that I received a telephone call from Sato. I was happy to have another occasion to meet with him. I thought that he would bring some news of discovery in physics or some new idea of algebraic geometry perhaps related to algebraic number theory. The content of his talk was quite different: a new idea to define "generalized functions" using boundary values of analytic functions. The idea is quite different from that of Schwartz's distributions, and the defined objects are not entirely the same. But Sato's "hyperfunctions" are as useful as, and in some occasions even more useful than, Schwartz's distributions! I was fascinated by his talk and listened to him till late into the night. I even asked him if he did not wish to come back to mathematics, and I believe he answered yes!

As soon as I had an occasion to meet Yosida at the Department of Mathematics, I enthusiastically told him of Sato's idea and asked him if we would engage him as assistant. It was very fortunate that there was a vacant post at that time, and Yosida immediately agreed to my proposal. Of course we had to obtain Professor Tomonaga's agreement. Fortunately, we got his agreement, and Sato came to join our department in 1957.

In later years, the research of Sato and of his school came to have contact with physics. That may have its remote origin in the days he spent with Professor Tomonaga.

2.

Sato was invited to the Institute for Advanced Study in Princeton, New Jersey, in 1960. When he arrived at the Institute in early September, I happened to be there.

In early summer of that year, I was invited to the Five Hundred Year Festival of Basel University. After the Festival in Basel, I spent one month in Paris. I spent another month in Princeton before going to Chicago where I was invited in late September. My wife accompanied me. Sato came to Princeton while I was there.

Weil and Schwartz were also at the Institute. I was happy to introduce Sato to them. I had thought that they would immediately and enthusiastically recognize the value of Sato's theory. Unfortunately this expectation was not initially fulfilled.

I believe Sato had sent them reprints of his paper published in the Journal of the Faculty of Science of the University of Tokyo. His theory of hyperfunctions is expressed in this paper in terms of relative cohomology theory. Though this form of presentation is extremely beautiful, apparently it was not to the likings of Weil and Schwartz. Schwartz tried to interpret it in the framework of his own theory, which was certainly not easy. On the other hand, Sato was not yet used to speaking in English and could not demonstrate his excellent art in explaining his idea by verbal expressions.

It took time before the importance of Sato's theory became widely known to the international mathematical community. It became particularly well known to the French School, and Sato was invited with his colleagues Kawai and Kashiwara to Nice University in 1972 for several months.

It was certainly useful to Sato to stay at Princeton in 1960, despite his disappointment at the beginning. He got used to the circumstances outside Japan and it was there, I believe, that he got his idea of prehomogeneous vector spaces. I remember an enjoyable time at the Institute apartment complex where Sato lived. Sato invited my wife and me for supper. Unmarried, he cooked for himself. He treated us to a meal of delicious fried chicken.

3.

In 1987 Sato was one of the winners of the Fujiwara Prize. I was invited to the Awarding Ceremony of this Prize, and I was particularly happy to see him with his beautiful wife, who collaborates in his mathematical work.

Sato had already won several other important prizes, including the Japan Academy Prize and the Asahi Prize. He missed the Fields Medal because of its age restriction, but everyone recognizes the importance of his work and its influence. I have mentioned here only the Fujiwara Prize because the memory of its Awarding Ceremony remains most vivid to me.

Let me express my most sincere wishes for the further development of his scientific work, his school, and for the prosperity of his family!

Mikio Sato: Impressions

Gian-Carlo Rota

Department of Mathematics
Massachusetts Institute of Technology
Cambridge, Massachusetts

Spring, 1966, was the golden time of a year that we, the survivors of today, recognize to have been one of the golden ages of mathematics. On the lovely campus of Columbia University in New York, one felt as if one were in a secure citadel, far from the madding crowd. From the fifth floor of the mathematics building, one took pleasure in making out the distant noises of Broadway as they filtered in through the windows, muted by the new leaves; and one basked in the illusion that the forces of intellect would for once be holding in check the senseless babble without.

The weekly mathematics colloquium was well attended by both the local faculty and by New York mathematicians; the crowded dinners were followed by hours of chatter, well through the evening. At one of these dinners, I saw for the first time a young Japanese mathematician, sitting silently between Eilenberg and Lang, and watching his two neighbors alternately, as if they were in a tennis match and he the spectator, as they engaged in a loud exchange on the role of category theory (a hot subject at the time). Serge Lang introduced him without his usual casualness, adding in fact (what for Lang was more unique than unusual) a complimentary remark

Algebraic Analysis, Volume I

sotto voce. Since that meeting, I remember (probably mistakenly) Mikio Sato as having blue eyes (the color that matches the clarity of his thought).

As we became acquainted, we discovered an affinity of interests that we thought we would have ample time to develop (young as we were, we believed in infinite time). We took long walks in town. The estrangement of the canyons of midtown Manhattan made mathematical conversation all the more coveted and cozy. To me, these walks were a stroke of good luck: I was privileged to be one of the first to hear the outlines of Sato's theory of hyperfunctions, as well as to see glimmers of a great many sparkling ideas, which are now permanent assets in the patrimony of mathematics. I remember the instant when he first succeeded—after several attempts—in making me understand his main point. As we walked, the noise of the tumultuous street traffic seemed suddenly dampened by the brilliance of an idea.

I eventually found the courage to ask him to give a series of lectures on his theory of zeta functions to our minute group at Rockefeller University. I felt flattered when he accepted immediately.

His audience consisted of Hans Rademacher, at the time in his seventies, but still beaming with energy and wit; Larry Harper, fresh out of school and of California; and myself. It was one of very few occasions when I was able to write down a complete text on the spot, directly as Sato spoke. The theorems flowed with his words with graceful but slightly deceptive ease. As the speaker got carried away by his subject, his English became flawless too.

I have often reread those lecture notes, partly because I have found no better reference to Sato's theory, and partly to recall the setting of those memorable meetings. Beneath the huge square windows of the building on Sutton Place the traffic of the Queensborough Bridge flowed silently. The thick blue carpet in the office cut off all noise except the speaker's voice. It was as if all obstacles had been removed that stood in the way of the purest mathematical experience.

After his last lecture, Sato invited all members of his audience to dinner at a Japanese restaurant. We sat next to each other at a large square table, and watched the cook's acrobatics, as is the custom in Japanese steak houses. Sato sat next to Hans Rademacher. They both sensed it would be the last time they would meet. They discoursed far and wide on the history of number theory and on the future of zeta functions. Hans Rademacher had tears in his eyes when we all parted, late in the evening.

Twenty years have gone by since that day, and distance has stood in the way of our meeting again. Often, I have had the opportunity of listening to expositions of Sato's work. On these occasions, I fancy myself as the secret sharer of his voice, back in the blue office in Sutton Place. I am glad that time and the width of an ocean do not stand in the way now, as I have the honor of introducing this volume, where the contributions of students and friends bear witness to Mikio Sato's first sixty years of a splendid life in mathematics.

Sato, a Perfectionist

Kôsaku Yosida

Kajwara, Kamakura
Japan

Professor Mikio Sato graduated from the University of Tokyo in 1952, specializing in mathematics. Just after that he also studied and graduated from the University of Tokyo in 1954, specializing in physics. And between 1954 and 1958, Sato studied as a graduate student in the Tokyo Educational University.

At last in 1958 he was inspired with a happy idea of a generalization of functions. This idea was to "take boundary value of a holomorphic function as a hyperfunction." More precisely, let $\mathbf{R}^1$ be the field of real numbers, and let $\mathbf{C}^1$ be the field of complex numbers. Then, let $\mathcal{O}(\mathbf{C}^1 - \mathbf{R}^1)$ be the totality of one-variable holomorphic functions on $(\mathbf{C}^1 - \mathbf{R}^1)$, and let similarly $\mathcal{O}(\mathbf{C}^1)$ be the totality of one-variable holomorphic functions on $\mathbf{C}^1$. Now, "the residue class mod $\mathcal{O}(\mathbf{C}^1)$ containing $F(z) = F(x + iy) \in \mathcal{O}(\mathbf{C}^1 - \mathbf{R}^1)$" is called by Sato the hyperfunction $f(x)$ defined through $F(z)$.

Thus the hyperfunction defined through the derivative $F'(z)$ of $F(z)$ is the (hyperfunction) derivative $f'(x)$ of $f(x)$. Hence, Sato's hyperfunction is infinitely differentiable in the hyperfunction sense.

The hyperfunction of several variables is also defined by Sato through the fine use of homology theory in topology, so that Sato's hyperfunction is infinitely differentiable with respect to the several variables.

Algebraic Analysis, Volume I

Moreover, the totality of the hyperfunctions of Sato contains the totality of Laurent Schwartz's hyperfunctions (distributions) as a proper subset.

I thus encouraged Sato to finish writing the monograph of his theory of hyperfunctions in the *Journal of the Faculty of Science, University of Tokyo*, 1959. The editor of this journal complained to me often that Mr. Sato was very slow in proofreading his monograph. I replied to the editor that Mr. Sato is a perfectionist about publishing his papers and monographs, witness his study of mathematics and physics for many years.

Existence of Local Holomorphic Solutions of Differential Equations of Infinite Order

Takashi Aoki

Department of Mathematics and Physics
Kinki University, Higashi-Osaka
Osaka, Japan

A differential operator of infinite order is an infinite sum of differential operators

$$P = \sum_{\alpha} a_{\alpha}(x) D_x^{\alpha}$$

satisfying the following conditions:

(*i*)   Each a_{α} is holomorphic in an open set X in $\mathbf{C}^n$,
(*ii*)   For every compact set K in X, $\lim_{|\alpha| \to \infty} |\alpha| \sup_{x \in K} |a_{\alpha}(x)|^{1/|\alpha|} = 0$.

Here we use the familiar notation of multi-indices: $D_x^{\alpha} = D_1^{\alpha_1} \cdots D_n^{\alpha_n}$, $D_j = \partial/\partial x_j$, etc. If we set $P(x, \xi) = \sum_{\alpha} a_{\alpha}(x)\xi^{\alpha}$ for $(x, \xi) \in X \times \mathbf{C}^n$, (*i*) and (*ii*) are equivalent to the following:

$P(x, \xi)$ is holomorphic in $X \times \mathbf{C}^n$ and for each compact K in X and for every $h > 0$ there is $C_h > 0$ such that $|P(x, \xi)| \leq C_h \exp(h|\xi|)$.

Algebraic Analysis, Volume I

We call $P(x, \xi)$ the (total) symbol of P. Of course, every differential operator of finite order with holomorphic coefficients in X satisfies (i) and (ii). We are interested in the case where infinitely many a_α are not zero. Differential (or microdifferential) operators of infinite order are very important in studying the structure of partial differential equations of finite order ([10], [21], [22]). Moreover, differential equations of infinite order play a role in investigating theta functions ([14], [18], [19], [20]).

In this article, we discuss some results concerning the existence of local holomorphic solutions of a differential equation of infinite order $Pu = f$, f being a given holomorphic function. We say that P is locally solvable at $\mathring{x} \in X$ if $P : \mathcal{O}_{\mathring{x}} \to \mathcal{O}_{\mathring{x}}$ is surjective. Here $\mathcal{O}$ denotes the sheaf of holomorphic functions on X and $\mathcal{O}_{\mathring{x}}$ the stalk of $\mathcal{O}$ at $\mathring{x}$.

In the constant coefficient case, Martineau [17] obtained the following

Theorem 1. *If P is non-zero and with constant coefficients, then P is locally solvable at each point in $\mathbf{C}^n$.*

In fact, Martineau discussed the existence of global solutions (see [17]; also [8], [16]). We restrict our discussion to local solvability, for our main interest is in the case of variable coefficients.

Theorem 1 is proved using functional analysis. Let us remark that the stalk $\mathcal{O}_{\mathring{x}}$ is a DFS-space ([7], [15]) and its dual is isomorphic to $\mathscr{B}^\infty_{\{\mathring{x}\}|\mathbf{C}^n}$, the space of holomorphic hyperfunctions ([10], [18]). We will abbreviate $\mathscr{B}^\infty_{\{\mathring{x}\}|\mathbf{C}^n}$ to $\mathscr{B}^\infty$. By the Fourier–Borel transform ([17]), $\mathscr{B}^\infty$ is isomorphic to the space $E_0(\mathbf{C}^n)$ of entire functions of infra-exponential type (i.e., of order 1, minimal type). Here the topology of $E_0(\mathbf{C}^n)$ is given by the family of norms $\|\varphi\|_\varepsilon = \sup|\varphi(\xi)| \exp(-\varepsilon|\xi|)$ ($\varepsilon > 0$). To prove the surjectivity of $P : \mathcal{O}_{\mathring{x}} \to \mathcal{O}_{\mathring{x}}$, it suffices to show the following:

 (a) The image of $P : \mathcal{O}_{\mathring{x}} \to \mathcal{O}_{\mathring{x}}$ is dense.
 (b) The image of $P : \mathcal{O}_{\mathring{x}} \to \mathcal{O}_{\mathring{x}}$ is closed.

Statement (a) is equivalent to

 (a′) ${}^t P : \mathscr{B}^\infty \to \mathscr{B}^\infty$ is injective.

Here ${}^t P$ is the transposed operator of P. By the Fourier–Borel transform, (a′) is equivalent to the following trivial statement:

 (a″) ${}^t P(\xi)\varphi(\xi) = 0 \,(\varphi \in E_0(\mathbf{C}^n))$ implies $\varphi(\xi) = 0$.

Here ${}^t P(\xi)$ is the total symbol of ${}^t P$. By the closed range theorem, (b) is equivalent to

(b′) The image of $'P: \mathscr{B}^\infty \to \mathscr{B}^\infty$ is closed.

As in the preceding argument, (b′) is equivalent to

(b″) $'P(\xi)E_0(\mathbf{C}^n)$ is closed in $E_0(\mathbf{C}^n)$.

This follows from a division theorem for entire functions ([6]).

When the operator P has variable coefficients, the virtue of the Fourier–Borel transform decreases, for the Fourier–Borel transform of $'Pu$ is not so simple as in the constant coefficient case. Thus we cannot expect such a general result as Theorem 1. Under some natural conditions, however, we can prove the local solvability of P (Theorems 3, 4). These conditions are invariant under changes of local coordinates. We remark that Ishimura [9] has found some interesting conditions for local solvability. Under his conditions, the argument in the constant coefficient case is available. The conditions, however, depend on the choice of coordinates.

To state our theorem, we have to prepare the notion of a characteristic set. A point $x^* = (\overset{\circ}{x}, \overset{\circ}{\xi})$ in the cotangent space T^*X (identified with $X \times \mathbf{C}^n$) is said to be non-characteristic with respect to P if there exist a conic neighborhood Ω of x^* and a positive number r such that the total symbol $P(x, \xi)$ of P never vanishes in the set $\Omega \cap \{(x, \xi); |\xi| > r\}$. We denote by $\mathrm{Ch}(P)$ the complement of the set of all non-characteristic elements with respect to P. We call $\mathrm{Ch}(P)$ the characteristic set of P ([4], [12]). Although the total symbol depends on the choice of local coordinates, we have ([4], Theorem 2.1.2)

Theorem 2. $\mathrm{Ch}(P)$ *is a well-defined closed conic subset in the cotangent bundle* T^*X.

The following theorem gives a sufficient condition for local solvability of P in terms of $\mathrm{Ch}(P)$ ([4], Theorem 4.1.1).

Theorem 3. *Let $\overset{\circ}{x}$ be a point in X. If there exists $\overset{\circ}{\xi} \in T^*_{\overset{\circ}{x}}X - \{0\}$ such that the set $\{\lambda \in \mathbf{C}^*; (\overset{\circ}{x}, \lambda\overset{\circ}{\xi}) \in \mathrm{Ch}(P)\}$ is contained in the right half-plane* $\mathrm{Re}\,\lambda > 0$, *then P is locally solvable at $\overset{\circ}{x}$.*

In [4], we have constructed a solution of $Pu = f$ directly by using the symbol of inverse of P. Here we prove Theorem 3 as a corollary of the following theorem which has been suggested by Prof. Kashiwara.

Theorem 4. *Let $\overset{\circ}{x}$ be a point in X. If there exists $\overset{\circ}{\eta} \in T^*_{\overset{\circ}{x}} X - \{0\}$ such that P is invertible in $\mathscr{E}^{\mathbf{R}}$ at $(\overset{\circ}{x}, \lambda \overset{\circ}{\eta})$ for each λ satisfying $\operatorname{Re} \lambda \leq 0$, then P is locally solvable at $\overset{\circ}{x}$.*

Here $\mathscr{E}^{\mathbf{R}}$ denotes the sheaf of pseudodifferential operators ([3], [10], [11]).

Let us prove Theorem 4. By a suitable change of coordinates, we may assume $\overset{\circ}{x} = (0, \ldots, 0)$, $\overset{\circ}{\eta} = (1, 0, \ldots, 0)$. If $\alpha > 0$ and $\varepsilon > 0$ are sufficiently small, P is invertible in $\mathscr{E}^{\mathbf{R}}$ at $z^*(\alpha, \theta) = (\alpha, 0, \ldots, 0; e^{i\theta}, 0, \ldots, 0)$ for every θ such that

$$\theta \in \left(\frac{\pi}{2} - \varepsilon, \frac{3\pi}{2} + \varepsilon \right).$$

Let f be a holomorphic function defined in a neighborhood of $\overset{\circ}{x}$. We set $Y = \{x \in X; x_1 = \alpha\}$. If α is small, $(1/2\pi i) f \log(x_1 - \alpha)$ defines an element $\psi \in \mathscr{C}^{\mathbf{R}}_{Y|X, z^*(\alpha, \theta)}$ for every θ. Here $\mathscr{C}^{\mathbf{R}}_{Y|X}$ denotes the sheaf of holomorphic microfunctions ([10], [21]). There exists the inverse P^{-1} of P in $\mathscr{E}^{\mathbf{R}}_{z^*(\alpha, \theta)}$ for every θ such that $(\pi/2) - \varepsilon < \theta < (3\pi/2) + \varepsilon$. Since $\mathscr{E}^{\mathbf{R}}_{z^*(\alpha, \theta)}$ acts on $\mathscr{C}^{\mathbf{R}}_{Y|X, z^*(\alpha, \theta)}$, we can let P^{-1} operate on ψ and obtain an element $\varphi = P^{-1} \psi \in \mathscr{C}^{\mathbf{R}}_{Y|X, z^*(\alpha, \theta)}$ for every $\theta \in ((\pi/2) - \varepsilon, (3\pi/2) + \varepsilon)$. Let g be a defining function of φ. Then g can be holomorphically continued to $|\arg(x_1 - \alpha)| < \pi + \varepsilon$. If α is sufficiently small, we can assume g is holomorphic at $\overset{\circ}{x} = 0$. Let $g_\pm$ be two branches of g corresponding respectively to $\arg(x_1 - \alpha) \in (\pm\pi - \varepsilon, \pm\pi + \varepsilon)$ and we set $u = g_+ - g_-$. Since $P\varphi = P^{-1}(P\psi) = \psi$ in $\mathscr{C}^{\mathbf{R}}_{Y|X}$ and α is small, $Pg - (1/2\pi i) f \log(x_1 - \alpha)$ is holomorphic in an open neighborhood of $\overset{\circ}{x} = 0$ which intersects with Y. Therefore we have $Pu = Pg_+ - Pg_- = f$ and we obtain a holomorphic solution u of $Pu = f$ at $\overset{\circ}{x}$.

Combining Theorem 4 and the following theorem ([5], Theorem 1), we have Theorem 3.

Theorem 5. *If $x^* \notin \operatorname{Ch}(P)$, then P is invertible in $\mathscr{E}^{\mathbf{R}}$ at x^*.*

We remark that the condition of Theorem 3 is satisfied if P is of finite order and the principal symbol of P does not vanish identically at $\overset{\circ}{x}$, for the characteristic variety of P coincides with $\operatorname{Ch}(P)$ and it is invariant under the action of $\mathbf{C}^*$. The assumption of Theorem 3 is, however, much stronger than that of Theorem 1, the constant coefficient case. Unfortunately, we cannot weaken the assumption of Theorem 3 at the present stage. Let us remark that under a fairly good condition, we have the following, which corresponds to Statement (a) in the constant coefficient case.

Proposition 6. *Let $\overset{\circ}{x}$ be a point in X. If there exists $\overset{\circ}{\xi} \in T^*_{\overset{\circ}{x}} X$ such that $(\overset{\circ}{x}, \overset{\circ}{\xi}) \notin \mathrm{Ch}(P)$, then the image of $P : \mathcal{O}_{\overset{\circ}{x}} \to \mathcal{O}_{\overset{\circ}{x}}$ is dense.*

It suffices to show $^tP \colon \mathscr{B}^\infty \to \mathscr{B}^\infty$ is injective. If $^tPu = 0$ for $u \in \mathscr{B}^\infty$, we have $^tPu = 0$ in $\mathscr{C}^{\mathbf{R}}_{\{\overset{\circ}{x}\}|X}$. Since $\mathrm{Ch}(^tP) = \{(x, \xi);\ (x, -\xi) \in \mathrm{Ch}(P)\}$ ([4]), we have $(\overset{\circ}{x}, -\overset{\circ}{\xi}) \notin \mathrm{Ch}(^tP)$. By Theorem 5, ^tP is invertible in $\mathscr{E}^{\mathbf{R}}$ at $(\overset{\circ}{x}, -\overset{\circ}{\xi})$. Hence we have $u = (^tP)^{-1}\ ^tPu = 0$ in $\mathscr{C}^{\mathbf{R}}_{\{\overset{\circ}{x}\}|X}$ at $(\overset{\circ}{x}, -\overset{\circ}{\xi})$. This implies $u = 0$ in $\mathscr{B}^\infty$.

References

[1] T. Aoki, Calcul exponential des opérateurs microdifférentiels d'ordre infini, I. *Ann. Inst. Fourier, Grenoble* **33–4** (1983) 227–250.

[2] T. Aoki, Calcul exponentiel des opérateurs microdifférentiels d'ordre infini, II. *Ibid.* **36–2** (1986) 143–165.

[3] T. Aoki, Symbols and formal symbols of pseudodifferential operators, *Advanced Studies in Pure Math.* **4** (1984) 181–208.

[4] T. Aoki, Existence and continuation of holomorphic solutions of differential equations of infinite order, *Adv. in Math.* (to appear).

[5] T. Aoki, M. Kashiwara, and T. Kawai, On a class of linear differential operators of infinite order with finite index, *Adv. in Math.* **62** (1986) 155–168.

[6] V. Avanissian, Fonctions plurisousharmoniques, différences de deux fonctions plurisousharmoniques de type exponentiel, *C. R. Acad. Sc., Paris, t.* **252** (1961) 499–500.

[7] A. Grothendieck, Sur les espaces (F) et (DF), *Summs Brasil. Math.* **3** (1954) 57–122.

[8] R. Ishimura, Théorèmes d'existence et d'approximation pour les équations aux dérivées partielles lineaires d'ordre infini, *Publ. RIMS, Kyoto Univ.* **16** (1980) 393–415.

[9] R. Ishimura, Existence locale de solutions holomorphes pour les équations différentiels d'ordre infini, *Ann. Inst. Fourier, Grenoble* **35–3** (1985) 49–57.

[10] M. Kashiwara and T. Kawai, On holonomic systems of microdifferential equations, III. *Publ. RIMS, Kyoto Univ.*, **17** (1981) 813–979.

[11] M. Kashiwara and P. Schapira, Micro-hyperbolic systems, *Acta Mathematica* **142** (1979) 1–55.

[12] T. Kawai, On the theory of Fourier hyperfunctions and its applications to partial differential equations with constant coefficients, *J. Fac. Sci., Univ. Tokyo, Sect. IA* **17** (1970) 467–517.

[13] T. Kawai, An example of a complex of linear differential operators of infinite order, *Proc. Japan Acad. Ser. A* **59** (1983) 113–115.

[14] T. Kawai, The Fabry-Ehrenpreis gap theorem and linear differential equations of infinite order, *American J. Math.* **109** (1987) 57–64.

[15] H. Komatsu, Ultradistributions, I, Structure theorems and a characterization, *J. Fac. Sci. Univ. Tokyo, Sect IA* **20** (1973) 25–105.

[16] Ju. F. Korobeinik, The existence of an analytic solution of an infinite order differential equation and the nature of its domain of analyticity, *Math. USSR Sbornik* **9-1** (1969) 53-71.

[17] A. Martineau, Equations différentiels d'ordre infini, *Bull. Soc. Math. France* **95** (1967) 109-154.

[18] Sato, M. Pseudo-differential equations and theta functions, *Asterisque* **2** et **3** (1973) 286-291.

[19] M. Sato, M. Kashiwara, and T. Kawai, Linear differential equations of infinite order and theta functions, *Adv. in Math.* **47** (1983) 300-325.

[20] M. Sato, M. Kashiwara, and T. Kawai, Microlocal analysis of theta functions, *Advanced Studies in Pure Math.* **4** (1984) 267-289.

[21] M. Sato, T. Kawai, and M. Kashiwara, Microfunctions and pseudo-differential equations, *Lecture Notes in Math.*, No. 287, Springer, Berlin-Heidelberg-New York, 1973, pp. 265-529.

[22] K. Uchikoshi, Microlocal analysis of partial differential operators with irregular singularities, *J. Fac. Sci. Univ. Tokyo, Sect. IA* **30** (1983) 299-332.

A Note on Holonomic q-Difference Systems

K. Aomoto

Department of Mathematics
Nagoya University
Nagoya, Japan

1. Introduction

In [A3] the author has investigated integrals of power products of poly-nomials as solutions of maximally overdetermined difference systems with respect to variables of exponents. One may easily imagine that one can develop a similar theory of maximally overdetermined, namely holonomic, systems of q-difference equations. In recent years there has been consider-able interest in the q-analog of special functions such as q-basic series and their extensions (see [A2], [M1], [M2], etc). It seems that these are intimately related to holonomic q-difference systems. C. Praagman has investigated n commuting linear difference operators including (2) below, at least in formal series settings (see [P]). But his result doesn't seem to match the connection problems in which we are mainly interested. Under this context, in this note, following the almost same procedure as in [A3], we show an existence theorem of solutions in one direction at infinity for holonomic q-difference systems *with coefficients of rational functions*, where we make use of the

Algebraic Analysis, Volume I

classical existence theorem for q-difference systems in one variable due to R. D. Carmichael, G. D. Birkhoff and C. R. Adams (see [C], [B] and [A1]).

In a succeeding paper, we shall investigate associated connection problems among solutions having asymptotic behaviours in various directions at infinity for the Jackson integral, which is a q-analog of the Jordan–Pochhammer integral:

$$F(x_1, \ldots, x_n) = \int t^{\alpha_0 - 1} \prod_{j=1}^{n} \frac{(t/x_j)_\infty}{(tq^{\alpha_j}/x_j)_\infty} \, d_q t,$$

for $(x_1, \ldots, x_n) \in \mathbf{C}^n$, $q \in \mathbf{C}$, $|q| < 1$ and $\alpha_0, \alpha_1, \ldots, \alpha_n \in \mathbf{C}^{n+1}$.

2. Existence Theorem

The n commuting q-difference operators $T_1, \ldots, T_n$ are defined by

$$T_j f(x_1, \ldots, x_n) = f(x_1, \ldots, x_{j-1}, q x_j, x_{j+1}, \ldots, x_n) \tag{1}$$

for a function f on $(\mathbf{C}^*)^n$. Consider a system of n linear q-difference equations

$$T_j \mathbf{f} = \mathbf{f} A_j(x) \tag{2}$$

for a vector-valued function $\mathbf{f} = (f_1(x), \ldots, f_m(x))$ on $(\mathbf{C}^*)^n$, where $A_j(x)$ are m-th order matrix-valued rational functions on $\mathbf{C}^n$. We assume the following condition for $A_j(x)$:

$$A_k(x) \cdot T_k A_j(x) = A_j(x) \cdot T_j A_k(x) \tag{3}$$

for arbitrary j, k. Then (2) defines a holonomic q-difference system of equations of order m. As each $A_j(x)$ is rational, it has a Laurent expansion in x_1 as follows:

$$A_j(x) = x_1^{\mu_j}\{A_j^{(0)}(x_2, \ldots, x_n) + x_1^{-1} A_j^{(1)}(x_2, \ldots, x_n) + \cdots\}, \tag{4}$$

where each $A_j^{(\nu)}(x_2, \ldots, x_n)$ denotes a polynomial function of $x_2, \ldots, x_n$.

We make the following assumptions of generic property in the x_1 axis for each $A_j(x)$:

(i) $A_j^{(0)}(x)$ is constant and invertible. It will be denoted by $A_j^{(0)}$.

(ii) Let the eigenvalues of $A_1^{(0)}$ be $\alpha_1, \alpha_2, \ldots, \alpha_m$. Then $\alpha_i/\alpha_j \neq 1$, $q^{\pm 1}, q^{\pm 2}, \ldots$ for all i, j such that $i \neq j$.

The following lemma is an easy consequence of (3).

Lemma 1. (*i*) $\mu_j = 0$ *for $j > 1$.*
(*ii*) $A_1^{(0)} \cdot A_j^{(0)} = A_j^{(0)} \cdot A_1^{(0)}$ *for $j > 1$.*

Lemma 2. *There exists a unique matrix-valued meromorphic function $S(x)$ in $(\mathbf{C}^*)^n$, which has a Laurent expansion in x_1 for $|x_1| > R$ and $(x_2, \ldots, x_n) \in D_R$,*

$$S(x) = 1 + x_1^{-1} S^{(1)}(x_2, \ldots, x_n) + x_1^{-2} S^{(2)}(x_2, \ldots, x_n) + \cdots, \qquad (5)$$

such that

$$S \cdot A_1 = A_1^{(0)} \cdot T_1 S, \qquad (6)$$

where $S^{(\nu)}(x_2, \ldots, x_n)$ are polynomials in $x_2, \ldots, x_n$. R denotes a positive number and D_R denotes a domain in $\mathbf{C}^{n-1}$ such that $D_R \subset D_{R'}$ for $R < R'$ and $\bigcup_{R > 0} D_R = \mathbf{C}^{n-1}$. $S(x)$ is a meromorphic function of $x_1, x_2, \ldots, x_n$ in $\mathbf{C}^* \times \mathbf{C}^{n-1}$. As a result $\Phi_0(x) = q^{\mu_1(t^2 - t)/2} x_1^{\log A_1^{(0)}/\log q} S(x)$ is the unique solution of the ordinary q-difference equation

$$T_1 \Phi_0(x) = \Phi_0(x) A_1(x) \qquad (7)$$

having the asymptotic behaviour

$$\Phi_0(x) \sim q^{\mu_1(t^2 - t)/2} x_1^{\log A_1^{(0)}/\log q} \qquad (8)$$

for $x_1 \to \infty$, $0 < \arg t < 2\pi$, where t denotes $\log x_1 / \log q$. Note that $\log A_1^{(0)}$ is well defined because $A_1^{(0)}$ is semi-simple and invertible.

Proof. One has only to repeat the proof in [C], being careful of the behaviour of $S(x)$ with respect to $x_2, \ldots, x_n$.

We denote by $\bar{A}_j$ the matrix $S \cdot A_j \cdot (T_j S)^{-1}$. Then $\bar{A}_j$ turns out to be equal to $A_j^{(0)}$. This follows from the fact that $\bar{A}_j(x)$ themselves satisfy the compatibility condition (2.3) and that $\bar{A}_1 = A_1^{(0)}$. We denote by $\Lambda_j = \log A_j^{(0)}/\log q$. Remark that the Λ_j can be chosen such that they commute each other. Then we have

Theorem. *The equations (2) have the unique matrix solution*

$$\Phi = q^{\mu_1(t^2 - t)/2} \cdot x_1^{\Lambda_1} \cdot x_2^{\Lambda_2} \cdots x_n^{\Lambda_n} S(x) \qquad (9)$$

having the asymptotic behaviour

$$\Phi \sim q^{\mu_1(t^2 - t)/2} \cdot x_1^{\Lambda_1} \cdot x_2^{\Lambda_2} \cdots x_n^{\Lambda_n} \qquad (10)$$

for $x_1 \to \infty$, $0 < \arg t < 2\pi$.

Proof. One has only to prove that (9) satisfies the equation $T_j \Phi = \Phi \cdot A_j(x)$ for each $j > 1$. Consider the matrix $T_j \Phi \cdot A_j(x)^{-1}$. Then

$$T_1(T_j\Phi \cdot A_j(x)^{-1}) = T_1 T_j \Phi \cdot (T_1 A_j(x))^{-1}$$

$$= T_j T_1 \Phi \cdot (T_1 A_j(x))^{-1} = T_j(\Phi \cdot A_1(x)) \cdot (T_1 A_j(x))^{-1}$$

$$= T_j \Phi \cdot T_j A_1(x) \cdot (T_1 A_j(x))^{-1} = T_j \Phi \cdot (A_j(x))^{-1} \cdot A_1(x),$$

$$(11)$$

because of (3), i.e., $T_j \Phi \cdot A_j(x)^{-1}$ satisfies the same ordinary q-difference equation as Φ and has the same asymptotic behaviour as Φ when $x_1 \to \infty$, $0 < \arg t < 2\pi$, in view of the fact $q^{\Lambda_j} = A_j^{(0)}$ and $A_j \sim A_j^{(0)}$. Hence the former must coincide with the latter. The theorem has been proved.

References

[A1]   C. R. Adams, On the linear ordinary q-difference equation, *Annals of Math.* **30** (1929) 195–205.

[A2]   G. Andrews, Problems and prospects for basic hypergeometric functions, in *Theory and Application of Special Functions* (R. Askey, ed.), pp. 191–224, Academic Press, New York, 1975.

[A3]   K. Aomoto, Les équations aux differences linéaires et les integrales des fonctions multiformes, *J. Fac. Sci. Univ. Tokyo* **22** (1975) 271–297.

[B]   G. D. Birkhoff, The generalized Riemann problem for linear differential and the allied problems for linear difference and q-difference equations, *Proc. Am. Acad. Arts and Sci.* **49** (1914) 521–568.

[C]   R. D. Carmichael, The general theory of linear q-difference equations, *Amer. J. Math.* **34** (1912) 147–168.

[M1]   S. Milne, Hypergeometric series well-poised in $SU(n)$ and a generalization of Biedenharn's G functions, *Adv. in Math.* **36** (1980) 169–211.

[M2]   S. Milne, Basic hypergeometric series very well poised in $U(n)$, *J. Math. Anal. and Appl.* **118** (1986) 263–277.

[P]   C. Praagman, Formal decomposition of n commuting partial linear difference operators, *Duke Math. J.* **51** (1984) 331–353.

Solvable Models in Statistical Mechanics and Riemann Surfaces of Genus Greater than One

Helen Au-Yang, Barry M. McCoy,
Jacques H. H. Perk, and Shuang Tang

Institute for Theoretical Physics
State University of New York at Stony Brook
Stony Brook, New York

For over four decades the two-dimensional Ising model has been a profound source of inspiration for the physics of phase transitions. For almost two decades there has been a long search for other solvable models in which the Ising restriction of two states (spin up and spin down) per site can be generalized to N states per site, and by now there are many such solvable N-state models known. However, most of them have not proven to be as instructive as the Ising model because for none of them has anyone been able to calculate very much exact information about correlation functions.

Most recently, however, it was discovered [1, 2] that there is in fact an N-state generalization of the Ising model which seems to possess all of its nice properties. This model is the chiral Potts model (or Z_N symmetric model) defined by

Algebraic Analysis, Volume I

$$\mathscr{E} = -\sum_{j,k} \sum_{n=1}^{N-1} \{E_n^v(\sigma_{j,k}\sigma_{j+1,k}^*)^n + E_n^h(\sigma_{j,k}\sigma_{j,k+1}^*)^n\} \tag{1}$$

with

$$\sigma_{j,k}^N = 1. \tag{2}$$

This model is obviously the Ising model if $N = 2$. It is our purpose here to review what is known about this model for arbitrary N and to indicate in what ways the results of the Ising case can generalize.

A most natural place to commence the investigation of any two-dimensional statistical mechanical model is to study its transfer matrix. For the model (1), if we construct the transfer matrix T which goes diagonally through the square lattice we have

$$T = \sum_{j=1}^{N-1} l_j (Z_{-M}^\dagger)^j \prod_{k=-M}^{M-1} (L_k L_{k+1/2}) L_M (Z_M)^j, \tag{3}$$

where for each integer k

$$L_k = \sum_{n=0}^{N-1} \bar{l}_n X_k^n, \qquad L_{k+1/2} = \sum_{n=0}^{N-1} l_n (Z_k Z_{k+1}^\dagger)^n, \tag{4}$$

$$X_j = I_N \otimes \cdots \otimes \overset{j\text{th}}{X} \otimes \cdots \otimes I_N, \tag{5a}$$

$$Z_j = I_N \otimes \cdots \otimes \overset{j\text{th}}{Z} \otimes \cdots \otimes I_N, \tag{5b}$$

I_N is the $N \times N$ unit matrix, the elements of the $N \times N$ matrices X and Z are

$$Z_{j,m} = \delta_{j,m} \omega^{j-1}, \tag{6a}$$

$$X_{j,m} = \delta_{j,m+1 \pmod N}, \tag{6b}$$

$\omega = e^{2\pi i/N}$, and the relation between $l_n(\bar{l}_n)$ and $E_n^h(E_n^v)$ is (with $l_0 = \bar{l}_0 = 1$ chosen as a normalization convention)

$$l_n = \frac{\sum_{j=0}^{N-1} \omega^{-nj} \exp\left[(kT)^{-1} \sum_{m=1}^{N-1} E_m^h \omega^{mj}\right]}{\sum_{j=0}^{N-1} \exp\left[(kT)^{-1} \sum_{m=1}^{N-1} E_n^h \omega^{mj}\right]}, \tag{7a}$$

$$\bar{l}_n = \exp\left[(kT)^{-1} \sum_{m=1}^{N-1} E_m^v(\omega^{mn} - 1)\right]. \tag{7b}$$

Perhaps the most important question for which one can use the transfer matrix is to see whether or not the weights $\bar{l}_n$ and l_n can be chosen as functions of a single variable u so that the commutation relation of Baxter [3],

$$[T(u), T(u')] = 0, \tag{8}$$

holds, which extends to the quantum mechanical case the classical concept of complete integrability [4]. This question has been studied [1, 2] with the additional restriction that there be a value of u (chosen here for convenience to be zero) such that when $u \to 0$

$$l_n \to \alpha_n u \quad \text{and} \quad \bar{l}_n \to \bar{\alpha}_n u \quad \text{for } n \geq 1. \tag{9}$$

In this limit

$$T(u) = 1 + u\mathcal{H} + O(u^2), \tag{10}$$

where

$$\mathcal{H} = \sum_{j=-M}^{M} H_{j,j+1} = \sum_{j=-M}^{M} \sum_{n=1}^{N-1} \{\bar{\alpha}_n (X_j)^n + \alpha_n (Z_j Z_{j+1}^\dagger)^n\} \tag{11}$$

and (8) reduces to the "quantum Lax pair" equation [5]

$$[T(u), \mathcal{H}] = 0. \tag{12}$$

The following results have been obtained:

(1) For arbitrary $N \neq 4$ it is necessary that for (8) to hold either the model reduces to the critical scalar Potts model [6]

$$\alpha_m = \bar{\alpha}_m = \alpha \quad \text{for all } m \geq 1, \tag{13}$$

or [1, 2]

$$\alpha_m = \frac{e^{i(2m-N)\phi/N}}{\sin(\pi m/N)} \tag{14a}$$

and

$$\bar{\alpha}_m = \frac{\lambda\, e^{i(2m-N)\bar{\phi}/N}}{\sin(\pi m/N)} \tag{14b}$$

with

$$\cos \phi = \lambda \cos \bar{\phi}. \tag{15}$$

For $N = 4$ these solutions also exist but there are a few additional solutions for the α's [1] which do not give Riemann surfaces of genus greater than 1.

(2) For the case $\lambda = 1$, $l_n = \bar{l}_n$ and $N \geq 3$ the equation (8) will hold if

$$\frac{l_n}{l_0} = b^{2n} \prod_{k=1}^{n} \frac{[\omega^{-(k-1)/2}y - \omega^{(k-1)/2}z]}{[\omega^{-(N-k)/2}x - \omega^{(N-k)/2}z]}, \tag{16a}$$

where $b = e^{i\phi/N}$ and

$$b^{-N}(x^N - z^N) = b^N(y^N - z^N). \tag{16b}$$

This has been explicitly verified for $N = 3, 4$, [1, 2] and 5 and is certainly valid in general. The limit (9) is regained by choosing u such that as $u \to 0$

$$x = z[1 - 2ib^N u + O(u^2)],$$

$$y = z[1 - 2ib^{-N}u + O(u^2)]. \tag{17}$$

The genus of the Fermat curve (16b) is $(N-1)(N-2)/2$ if $b^{2N} \neq 0, 1, \infty$, and by a simple rescaling the moduli of the curve are obviously independent of b. When $b = 1$ the result (16) reduces to the genus zero result of Fateev and Zamolodchikov [7].

(3) For the case $N = 3$ equation (8) will hold if l_1 and l_2 lie on the curve [1]

$$-\left(\frac{\bar{\alpha}_1^3 - \bar{\alpha}_2^3}{\alpha_1^3 - \alpha_2^3}\right)\left(\frac{\alpha_1\alpha_2}{\bar{\alpha}_1\bar{\alpha}_2}\right) 3(\omega - \omega^{-1})l_0l_1l_2(l_0^2 - l_1l_2)(l_1^2 - l_2l_0)(l_2^2 - l_0l_1)$$

$$= (l_0^3 + l_1^3 + l_2^3 - 3l_0l_1l_2)(l_0^3l_1^3 + l_1^3l_2^3 + l_2^3l_0^3 - 3l_0^2l_1^2l_2^2)$$

$$-\left(\frac{\alpha_1^3 + \alpha_2^3}{\alpha_1^3 - \alpha_2^3}\right)(l_1^3 - l_2^3)(l_2^3 - l_0^3)(l_0^3 - l_1^3). \tag{18}$$

This ninth-order curve has six ordinary triple points at $(1, 0, 0)$, $(0, 1, 0)$, $(0, 0, 1)$, $(1, 1, 1)$, $(1, \omega, \omega^2)$, and $(1, \omega^2, \omega)$ and hence has genus 10.

The above results are obtained by Baxter's method [3, 8] of reducing the global condition (8) to a local equation by introducing an auxiliary matrix $M_\mu(u, u')$. Thus, (8) will hold if an $M_\mu(u, u')$ can be found such that for

μ an integer or half integer,

$$M_{\mu-1/2}(u, u')L_{\mu}(u)L_{\mu-1/2}(u') = L_{\mu}(u')L_{\mu-1/2}(u)M_{\mu}(u, u'), \qquad (19)$$

where M is of the form

$$M_k(u, u') = \sum_{n=0}^{N-1} \bar{x}_n(u, u')(X_k)^n, \qquad (20a)$$

$$M_{k+1/2}(u, u') = \sum_{n=0}^{N-1} x_n(u, u')(Z_k Z_{k+1}^{\dagger})^n. \qquad (20b)$$

The crucial ingredient to our results is that we do *not* make the assumption that u can be so chosen that $M_{\mu}(u, u')$ becomes a function of $u - u'$ alone. If this assumption is made then (19), which, following Onsager [9] and Baxter [3, 8], is called a "star-triangle" equation, reduces to the equations for factorizable S-matrices (sometimes called Yang–Baxter equations) first introduced by McGuire [10]. If this factorizable S-matrix restriction is made we may apply the criterion of Reshetikhin [11] which says that $H_{j,j+1}$ must obey

$$[H_{12} + H_{23}, [H_{12}, H_{23}]] = O_{12} - O_{23}, \qquad (21)$$

where $O_{j,j+1}$ is some nearest-neighbor operator. This shows that only the critical Potts case and the $b=1$ case of Fateev and Zamolodchikov [7] are obtained. Indeed, it may be shown in general that if the factorizable S-matrix assumption is made then only genus-zero and genus-one uniformizing curves can be obtained [12]. However, in general, we may use (20) in (19) to write

$$\sum_{k=0}^{N-1} x_{m-k}\bar{l}_n(u)l_k(u')\omega^{nk} = \sum_{k=0}^{N-1} \bar{l}_k(u')l_m(u)\bar{x}_{n-k}\omega^{mk}, \qquad (22)$$

then eliminate $\bar{x}_n$ and x_n from (22) altogether, and find that (8) will hold if

$$\frac{V_{ab}V_{00}}{V_{0b}V_{a0}} = \frac{\bar{V}_{ba}\bar{V}_{00}}{\bar{V}_{0a}\bar{V}_{b0}}, \qquad (23a)$$

where

$$V_{ab} = \sum_{m=0}^{N-1}\sum_{k=0}^{N-1} \omega^{am+bk+mk}l_m(u)\bar{l}_k(u') \qquad (24a)$$

and

$$\bar{V}_{ab} = \sum_{m=0}^{N-1}\sum_{k=0}^{N-1} \omega^{am+bk+mk}\bar{l}_m(u)l_k(u'). \qquad (24b)$$

Furthermore, from (23a) we may readily derive

$$\frac{V_{ab}V_{cd}}{V_{cb}V_{ad}} = \frac{\bar{V}_{dc}\bar{V}_{ba}}{\bar{V}_{da}\bar{V}_{bc}}. \tag{23b}$$

These equations have a useful symmetry which may be seen if we define the Fourier transform

$$S_m = \sum_{k=0}^{N-1} \omega^{mk} l_k, \tag{25a}$$

$$\bar{S}_m = \sum_{k=0}^{N-1} \omega^{mk} \bar{l}_k. \tag{25b}$$

Thus V_{ab} (24a) has the three equivalent expressions

$$V_{ab} = N^{-1}\omega^{-ab} \sum_{m=0}^{N-1}\sum_{k=0}^{N-1} \omega^{ak+bm-km} S_m(u)\bar{S}_k(u') \tag{26a}$$

$$= \sum_{k=0}^{N-1} \omega^{bk} \bar{l}_k(u') S_{a+k}(u) \tag{26b}$$

$$= \sum_{m=0}^{N-1} \omega^{am} l_m(u) \bar{S}_{b+m}(u'), \tag{26c}$$

with three similar expressions for $\bar{V}_{ab}$. Comparing (26a) with (23) and (24) we see that (23) is invariant if

$$\omega \to \omega^{-1}, \qquad l_k \to S_{N-k},$$

and

$$\bar{l}_k \to \bar{S}_{N-k}. \tag{27}$$

It is now straightforward to use (26b) in (23b) and let $u' \to 0$ to obtain

$$\sum_{m=1}^{N-1} \alpha_m(\omega^{am} - \omega^{cm})\left(\frac{\bar{S}_{b+m}}{\bar{S}_b} - \frac{\bar{S}_{d+m}}{\bar{S}_d}\right) = \sum_{m=1}^{N-1} \bar{\alpha}_m(\omega^{bm} - \omega^{dm})\left(\frac{S_{a+m}}{S_a} - \frac{S_{c+m}}{S_c}\right). \tag{28a}$$

Then, if we multiply (28a) by $\bar{S}_b\bar{S}_d S_a S_c \omega^{-(a\alpha+b\beta+c\gamma+d\delta)}$ and sum a, b, c, d from 0 to $N-1$ we may derive

$$\sum_{m=1}^{N-1} \bar{\alpha}_m(\omega^{\alpha m} - \omega^{\gamma m})\left(\frac{\bar{l}_{\beta-m}}{\bar{l}_\beta} - \frac{\bar{l}_{\delta-m}}{\bar{l}_\delta}\right) = \sum_{m=1}^{N-1} \alpha_m(\omega^{\beta m} - \omega^{\delta m})\left(\frac{l_{\alpha-m}}{l_\alpha} - \frac{l_{\gamma-m}}{l_\gamma}\right). \tag{28b}$$

Furthermore, putting $c = d = 0$ in (28a), multiplying by $N^{-1}\omega^{-ma-nb}$, and summing over a and b from 0 to $N-1$, we obtain, for $1 \le m, n \le N-1$,

$$\alpha_m \sum_{k=0}^{N-1} \frac{\bar{S}_{m+k}}{\bar{S}_k} \omega^{-nk} = \bar{\alpha}_n \sum_{k=0}^{N-1} \frac{S_{n+k}}{S_k} \omega^{-mk}, \tag{29a}$$

and, similarly,

$$\bar{\alpha}_{N-m} \sum_{k=0}^{N-1} \frac{\bar{l}_{m+k}}{\bar{l}_k} \omega^{nk} = \alpha_{N-n} \sum_{k=0}^{N-1} \frac{l_{n+k}}{l_k} \omega^{mk}. \tag{29b}$$

In addition we find there is a value of u (called u_0) such that as $u \to u_0$

$$S_k \to c\bar{\alpha}_{N-k}(u-u_0)S_0,$$
$$\bar{S}_k \to c\alpha_{N-k}(u-u_0)\bar{S}_0, \qquad 1 \le k \le N-1, \tag{30}$$

where c is a constant. For the self-dual case (16) u_0 is the place $x = \omega^{-1/2}b^2 z$, $y = \omega^{-1/2}b^{-2}z$, $(l_k \equiv 1)$. In this case $u=0$ and $u=u_0$ are transformed into each other by one of the $6N^2$ automorphisms of the Fermat curve (16b). The results (13)-(18) all follow from a study of equations (23)-(30).

The Ising model ($N=2$) holds its distinguished place amongst solvable models because of its relation to a free fermion field theory. This relation is what is ultimately behind all the various nonlinear difference equations, nonlinear differential equations, and deformation theory results which have been derived for the Ising model. The most detailed statement of the result is that the Ising model is related to a free fermion theory on a lattice. This connection was first shown by Kaufman [13] and has deeply influenced all subsequent discussions. It is widely believed that for the Z_N model with $b=1$ there is also some sort of free theory of "parafermions" underlying the model and in an appropriate continuum limit an impressive number of results have been derived using the conformal field theory approach [14, 15]. Unfortunately, no one has been able to replicate Kaufman's analysis on the lattice for $N \ge 3$, because, even though the basic definition of fermions may be generalized to parafermions [16], the resulting objects do not seem to obey linear equations on the lattice.

However, there is another procedure of solving the Ising model which does appear to generalize to $N \ge 3$. This is the original method of Onsager [9]. Onsager does not base his solution on either the commutation relation of Baxter (8) or the fermions of Kaufman but rather on the set of commutation relations

$$[A_k, A_m] = 4G_{k-m}, \tag{31a}$$

$$[G_m, A_k] = 2(A_{k+m} + A_{k-m}), \qquad (31b)$$

$$[G_m, G_k] = 0, \qquad (31c)$$

where, in our notation for $N = 2$,

$$A_1 = \bar{\alpha}_1 \sum_{j=-M}^{M} X_j, \qquad A_0 = -\alpha_1 \sum_{j=-M}^{M} Z_j Z_{j+1}. \qquad (32)$$

From these relations and the periodicity statement

$$A_{n+(2M+1)} = -CA_n = -A_n C, \qquad (33)$$

where C is the operator that reverses the sign of all spins, Onsager finds a free spectrum for the transfer matrix. It is surely of great importance that for the special case of the solvable chiral Potts model

$$\phi = \bar{\phi} = \pi/2. \qquad (34)$$

Von Gehlen and Rittenberg [16] showed that for $N \geq 3$

$$[A_0, [A_0, [A_0, A_1]]] = \text{const}[A_0, A_1] \qquad (35a)$$

and

$$[A_1, [A_1, [A_1, A_0]]] = \text{const}'[A_0, A_1], \qquad (35b)$$

which were posited by Dolan and Grady [17] as a condition for complete integrability. Indeed (31) may be derived from (35) but (33) remains to be generalized. Furthermore, numerical work of von Gehlen and Rittenberg [16] and of Howes, Kadanoff, and den Nijs [18] shows remarkable regularities which are reminiscent of those seen 43 years ago by Onsager. All of this suggests strongly that the correlation functions for the $N > 2$ chiral Potts models based on the higher-genus curves can be studied to a similar depth. Therefore, since the very title of the volume is *Algebraic Analysis* and since the deformation theory of the Ising model was pioneered by Prof. Sato [19] it is perhaps not inappropriate here to discuss the prospects for the algebraic analysis of the chiral Potts models.

We begin our discussion of these prospects by recalling a few well known results for the Ising model. By necessity the results must be on the lattice and not in the scaling limit. Indeed, for $N \geq 3$ the general picture of Howes, Kadanoff, and den Nijs [18] for the quantum spin chain (11) is that even in the self-dual case a phase factor such as $e^{i\theta(m-n)}$ appears in the correlation of operators at sites m and n and hence there is always some remnant of the lattice left in the final results even in an asymptotic limit.

All Ising correlations are expressed as finite determinants which have elements of the form [20]

$$\int_{-\pi}^{\pi} d\phi_1 \int_{-\pi}^{\pi} d\phi_2 \frac{P(\phi_1, \phi_2)}{\Delta(\phi_1, \phi_2)}, \tag{36}$$

where $P(\phi_1, \phi_2)$ is a polynomial in $e^{i\phi_1}$ and $e^{i\phi_2}$ and

$$\Delta(\phi_1, \phi_2) = \cosh(2E_1/kT) \cosh(2E_2/kT) - \sinh(2E_1/kT) \cos \phi_1$$
$$- \sinh(2E_2/kT) \cos \phi_2. \tag{37}$$

The fact which is constantly used is that the only region of ϕ_1, ϕ_2 space which contributes to integrals like (36) is where the restriction

$$\Delta(\phi_1, \phi_2) = 0 \tag{38}$$

holds. This, of course, is equivalent to the statement that the variables $x = e^{i\phi_1}$, $y = e^{i\phi_2}$ lie on the Riemann surface defined by the plane cubic curve

$$xy \cosh(2E_1/kT) \cosh(2E_2/kT) - 2y(x^2+1) \sinh(2E_1/kT)$$
$$- 2x(y^2+1) \sinh(2E_2/kT) = 0, \tag{39}$$

which is an elliptic curve (of genus 1) and modulus $k^{-1} = \sinh(2E_1/kT) \sinh(2E_2/kT)$. This is exactly the same curve and modulus which uniformized the star-triangle equation for $N = 2$ and strongly suggests that the uniformizing curve for the $N \geq 3$ models plays the same role as does $\Delta(\phi_1, \phi_2) = 0$ for the Ising model. This is a more precise lattice version of the assumption of free parafermions for the $b = 1$ model made in [15].

The simplest correlation function for the Ising model is the nearest diagonal correlation which for $T < T_c$ is [20]

$$\langle \sigma_{0,0} \sigma_{1,1} \rangle = \frac{2}{\pi} E(k), \tag{40}$$

where E is the complete elliptic integral of the second kind. There are clear candidates for generalization of this expression to higher-genus Riemann surfaces in the integrals around period loops of the normal Abelian differentials of the second kind. More generally, the diagonal correlation function for the Ising model is the $n \times n$ Toeplitz determinant

$$\langle \sigma_{0,0} \sigma_{n,n} \rangle = \begin{vmatrix} a_0 & \cdots & a_{-n+1} \\ \vdots & & \vdots \\ a_{n-1} & \cdots & a_0 \end{vmatrix}, \tag{41}$$

where

$$a_n = \frac{1}{2\pi} \int_0^{2\pi} d\theta \, e^{-in\theta} \left[\frac{1 - k \, e^{-i\theta}}{1 - k \, e^{i\theta}} \right]^{1/2}. \tag{42}$$

The integrands of (42) are clearly interpretable as Abelian differentials with a pole of order $2n + 2$ for $n \geq 0$ and $-2n$ for $n \leq -1$ so that again completely analogous expressions are available for Riemann surfaces of arbitrary genus. Finally, we note that if the spins of the Ising model are not on the diagonal the correlations involve elliptic integrals of the third kind, and that once again this has an immediate extension to Abelian integrals of the third kind for arbitrary Riemann surfaces.

If these suggested analogies can be shown to hold, then the entire deformation theory of Sato will have many new applications of direct physical relevance.

Finally we wish to make some closing remarks about other places to search for models with higher-genus uniformizing curves. First of all, just as two nearest-neighbor Ising models can be coupled together to form an eight-vertex model one should seek to couple two identical $\mathbf{Z}_N$ models together with three and four spin couplings. This will surely lead to new solvable models. Similarly, the higher-genus $\mathbf{Z}_N$ models should be studied in triangular lattices. This must almost surely lead to new solvable models. But these may very well not lead to any new Riemann surfaces. Instead, the situation may be like the genus-one case and will produce either the same free theory as the chiral Potts case or produce an interacting parafermion theory just as the eight-vertex model produces an interacting four-fermion theory. Indeed, the major unsolved conceptual problem at this stage is the following: can one expect to find a system of commuting transfer matrices for any Riemann surface, or are there special properties, such as a large group of automorphisms, which a Riemann surface must possess before it can produce a set of commuting transfer matrices? Until this question can be resolved the true place of Riemann surfaces and algebraic geometry in the theory of solvable statistical mechanical models cannot be said to be truly understood.

Acknowledgements

We are most pleased to acknowledge fruitful discussions with Prof. M. Kuga, Prof. C. H. Sah, Prof. M. L. Yan and Prof. A. B. Zamolodchikov.

This work has been supported by the National Science Foundation under Grant DMR-8505419.

References

[1] H. Au-Yang, B. M. McCoy, J. H. H. Perk, S. Tang, and M. L. Yan, *Phys. Lett.* **123A** (1987) 219.

[2] The case $N = 4$ is treated in B. M. McCoy, J. H. H. Perk, S. Tang, and C. H. Sah, *Phys. Lett.* **125A** (1987) 9.

[3] R. J. Baxter, *Phys. Rev. Lett.* **26** (1971) 832; *Ann. Phys.* **70** (1972) 193.

[4] See, for example, V. I. Arnol'd, *Usp. Mat. Nauk* **18**, No. 6 (1963) 91 [*Russian Mathematical Surveys* **18**, 6 (1963) 85].

[5] B. M. McCoy and T. T. Wu, *Nuovo Cimento* **56B** (1968) 311; B. Sutherland, *J. Math. Phys.* **11** (1970) 3183; L. Onsager, in *Critical Phenomena in Alloys, Magnets, and Superconductors* (R. E. Mills, E. Ascher, and R. I. Jaffee, eds.), McGraw-Hill, New York, 1971, p. 3.

[6] H. N. V. Temperley and E. H. Lieb, *Proc. Roy. Soc. London* **A322** (1971) 251; R. J. Baxter, *J. Phys.* **C6** (1973) L445; F. Y. Wu, *Rev. Mod. Phys.* **54** (1982) 235.

[7] V. A. Fateev and A. B. Zamolodchikov, *Phys. Lett.* **92A** (1982) 37.

[8] R. J. Baxter, *J. Stat. Phys.* **28** (1982) 1; *Exactly Solved Models in Statistical Mechanics*, Academic Press, London, 1982.

[9] L. Onsager, *Phys. Rev.* **65** (1944) 117; G. H. Wannier, *Rev. Mod. Phys.* **17** (1945) 50; M. E. Fisher, *Phys. Rev.* **113** (1959) 969; I. Syozi, in *Phase Transitions and Critical Phenomena*, Vol. 1 (C. Domb and M. S. Green, eds.), Academic Press, London, 1972, p. 269.

[10] J. B. McGuire, *J. Math. Phys.* **5** (1964) 622; C. N. Yang, *Phys. Rev.* **168** (1968) 1920.

[11] P. P. Kulish and E. K. Sklyanin, in *Integrable Quantum Field Theories* (J. Hietarinta and C. Montonen, eds.), Lecture Notes in Physics 151, Springer-Verlag, Berlin, 1982, p. 61.

[12] J. M. Maillard, *J. Math. Phys.* **27** (1986) 2776; J. B. McGuire and J. M. Freeman (Florida Atlantic Univ. preprint); J. B. McGuire and C. A. Hurst (Florida Atlantic Univ. preprint).

[13] B. Kaufman, *Phys. Rev.* **76** (1949) 1232.

[14] A. B. Zamolodchikov, in *Critical Phenomena, 1983 Braşov School Conference* (V. Ceauşescu, G. Costache and V. Georgescu, eds.), Birkhäuser, Boston, 1985, p. 402; A. A. Belavin, A. M. Polyakov, and A. B. Zamolodchikov, *J. Stat. Phys.* **34** (1984) 763; *Nucl. Phys.* **B241** (1984) 333.

[15] A. B. Zamolodchikov and V. A. Fateev, *Zh. Eksp. Teor. Fiz.* **89** (1985) 380 [*Sov. Phys. JETP* **62** (1985) 215].

[16] G. von Gehlen and V. Rittenberg, *Nucl. Phys.* **B257** [FS14] (1985) 351.

[17] L. Dolan and M. Grady, *Phys. Rev.* **D25** (1982) 1587.

[18] S. Howes, L. P. Kadanoff, and M. den Nijs, *Nucl. Phys.* **B215** [FS7] (1983) 169.

[19] M. Sato, T. Miwa, and M. Jimbo, *Publ. RIMS, Kyoto. Univ.* **14** (1978) 223; **15** (1979) 201, 577, 871.

[20] B. M. McCoy and T. T. Wu, *The Two-dimensional Ising Model* (Harvard Univ. Press, Cambridge, 1973).

Linearization and Singular Partial Differential Equations

Gunter Bengel

Mathematisches Institut der Westfälischen Wilhelms Universität
Münster, Federal Republic of Germany

1. Introduction

Consider the autonomous differential equation $x' = f(x)$ near a fixed point x_0 which we take to be 0; so $f(0) = 0$ and we write $f(x) = Ax + F(x)$ where $F(x) = O(|x|^2)$. We get the equation

$$x' = Ax + F(x), \tag{1}$$

and we look for a smooth change of coordinates $x = y + u(y)$ such that (1) is transformed into its linear part

$$y' = Ay. \tag{2}$$

It is easy to see that u has to satisfy the singular partial differential equation

$$\sum a_{jk} y_k \partial_j u_i = \sum a_{ik} u_k + F_i(y + u(y)), \tag{3}$$

where $A = (a_{jk})$.

Algebraic Analysis, Volume I

Let $(\lambda_1, \ldots, \lambda_n) = \mathrm{spec}(A)$ be the eigenvalues of A and set $\sigma(\lambda, m) = \lambda - (m_1\lambda_1 + \cdots + m_n\lambda_n)$ where m_j is entire.

Definition. A is said to satisfy the Sternberg condition of order k if $\sigma(\lambda, m) \neq 0$ for all $\lambda \in \mathrm{spec}(A)$ and $2 \leq |m| \leq k$.

Sternberg [7], [8] has shown that there exists a solution u of the conjugation problem of a certain smoothness class if F is smooth, A satisfies a Sternberg condition and Re $\lambda_j \neq 0$ for $j = 1, \ldots, n$. Recently G. Sell [6] has given improved smoothness results. In the analytic case results are due to H. Poincaré and C. L. Siegel (see [2], [3] for details).

Here we only treat the case of a source, so in the sequel we suppose Re $\lambda > 0$ for $\lambda \in \mathrm{spec}(A)$. Of course this solves also the case of a sink. To formulate our result, set $\mu = \min\{\mathrm{Re}\ \lambda_j\}$ and $\Lambda = \max\{\mathrm{Re}\ \lambda_j\}$ and $\rho = \Lambda/\mu$; ρ is called the *spectral spread* of A.

Theorem 1. *Let F be of class C^k near 0 and $\partial^\alpha F(0) = 0$ for $|\alpha| \leq k$, and suppose $k > \rho$. Set $s = [(k+\rho)/2]$ if $(k+\rho)/2$ is not entire and $s = (k+\rho)/2 - 1$ else. Then there exists a C^s-linearization $x = y + u(y)$ transforming (1) into (2).*

In [6] G. Sell proves this under the conditions $F \in C^{2k}$, Re $\sigma(\lambda, m) \neq 0$ for $\lambda \in \mathrm{spec}(A)$, $|m| = k$ with $s = [k/\rho]$. For small ρ near 1 and large k this is better than our value for s. For $\rho \geq 2$ our result gives a bigger s. Very precise results in the two-dimensional case are given by D. Stowe [9]. If the Sternberg condition of order k is satisfied there is a polynomial change of coordinates which transforms (1) into $y' = Ay + G(y)$ with $\partial^\alpha G(0) = 0$ for $|\alpha| \leq k$; for a proof see [4] or [6]. So Theorem 1 immediately gives

Theorem 2. *Let F be of class C^k in the neighborhood of 0, $F(0) = 0$, $DF(0) = 0$ and suppose A satisfies the Sternberg condition of order k with $k > \rho$. Then the conclusion of Theorem 1 holds.*

Remark. Suppose A satisfies the Sternberg condition of order k but not of order $k + 1$. Then there is m with $|m| = k + 1$ such that $\sum m_j\lambda_j = \lambda$ for some $\lambda \in \mathrm{spec}(A)$. Taking real parts gives Re $\lambda = \sum m_j$ Re λ_j and since $0 < \mu \leq \mathrm{Re}\ \lambda_j \leq \Lambda$ we get $(k+1)\mu \leq \sum m_j$ Re $\lambda_j = \mathrm{Re}\ \lambda \leq \Lambda$. So $k+1 \leq \Lambda/\mu = \rho$. Therefore the condition $k > \rho$ in Theorem 2 can only be satisfied if A satisfies the Sternberg condition of all orders, which is of course the generic case. Nevertheless, results in the case $k \leq \rho$ would be interesting. For the two-dimensional case cf. [9].

The proof of Theorem 1 proceeds along the lines of the paper [5] by Ise and Nagumo. We solve (3) by the method of characteristics and use the "auxiliary theorem" of Ise and Nagumo to estimate also the derivatives of u.

The difference between our treatment and that in [5] is that Ise and Nagumo consider the inverse transform $y = x + v(x)$, which gives a more complicated partial differential equation and does not so easily give estimates for the derivatives of u.

2. Preliminaries

Since we are interested only in local results we will suppose that F has compact support contained in the ball $\{x; |x| \leq \varepsilon\}$. This can be achieved by a multiplication with a suitable cut-off function. By a conjugation with a linear mapping we can bring A in a real normal form such that along the main diagonal we have the real parts of the eigenvalues, $a_{jj} = \mathrm{Re}\, \lambda_j$, and such that

$$\left| \sum_{j \neq k} a_{jk} x_j x_k \right| \leq \delta |x|^2,$$

where $\delta > 0$ can be chosen arbitrarily small. So we get

$$(\mu - \delta)|x|^2 \leq \sum a_{jk} x_j x_k \leq (\Lambda + \delta)|x|^2. \tag{4}$$

In the sequel we replace y by x in equation (3).

Consider now the characteristics of (3); these are the curves $x = x(t)$ given by the differential equation $x' = Ax$. The normal to the sphere $S = \{x; |x| = 2\varepsilon\}$ is proportional to x and, since $\langle x, x' \rangle = \langle Ax, x \rangle \geq (\mu - \delta)|x|^2 = (\mu - \delta)(2\varepsilon)^2 \neq 0$ on S if δ is small, S is not characteristic. So we can solve uniquely the Cauchy problem

$$\sum_{k,j} a_{jk} x_k \partial_j u_i = \sum_k a_{ik} u_k + F_i(x + u(x)), \qquad u = 0 \text{ on } S,$$

by solving the differential equations

$$u'_j = \sum a_{jk} u_k + F_j(x + u(x)).$$

Since the characteristics $x(t)$ cover the domain $D = \{x; 0 < |x| < 2\varepsilon\}$ simply we get a solution u in D which by the unicity of the Cauchy problem is zero for $|x| > \varepsilon$ and is C^k in D. To prove Theorem 1 we only have to prove that u is C^s at the origin. To do this we use the following "auxiliary theorem" of Ise and Nagumo.

Proposition. *Let φ be the solution of the Cauchy problem*

$$\sum a_{jk} x_k \partial_j \varphi_\nu = Q_\nu(x, \varphi), \qquad \varphi = 0 \text{ on } S \tag{5}$$

where Q is C^1 and let ω be C^1 in D with $\omega > 0$ on S. Suppose we have

$$\sum_{j,k} a_{jk} x_k \partial_j \omega(x) \ge \frac{1}{\omega(x)} \sum_\nu Q_\nu(x, \varphi) \varphi(x) \tag{6}$$

for all $x \in D$ such that $|\varphi(x)| = \omega(x)$. Then $|\varphi(x)| \le \omega(x)$ in D.

Proof. An easy calculation shows that the right-hand side of (6) is $\partial_t \omega(x(t))$ and the left-hand side is $\partial_t |\varphi(x(t))|$ at the points $x(t)$ where $|\varphi(x)| = \omega(x)$. So, since $\omega(x(0)) > 0 = \varphi(x(0))$ and $\partial_t \omega(x(t)) \ge \partial_t |u(x(t))|$ whenever $|u| = \omega$, we must have $|\varphi(x)| \le \omega(x)$ in D.

3. Proof of Theorem 1

Since $\partial^\alpha F(0) = 0$ for $|\alpha| \le k$ we have $\partial^\alpha F(\xi) = O(|\xi|^{k-|\alpha|})$ by Taylor's formula.

3.1. Estimates for u

In the proposition of part II take $u_\nu = u_i$ and Q the right-hand side of (3). Take $\omega(x) = C|x|^k$. Then $\omega(x) > 0$ on S and $\partial_j \omega = k \cdot C \cdot |x|^{k-2} \cdot x_j$. This gives

$$\sum a_{jk} x_k \partial_j \omega = k \cdot C \cdot |x|^{k-2} \sum a_{jk} x_j x_k \ge k \cdot C(\delta - \mu) \cdot |x|^k.$$

On the other hand

$$\left| \sum a_{ik} u_i u_k + \sum u_i F_i(x + u(x)) \right| \le (\Lambda + \delta)|u|^2 + |F(x+u)| \cdot |u|$$

$$\le (\Lambda + \delta)|u|^2 + |u| \cdot (C_0|x|^k + C_1 \varepsilon^{k-1} \cdot |u|),$$

where the last term comes from the estimate

$$|F(x+u)| \le |F(x)| + |DF(\xi)| \cdot |u| \le C_0|x|^k + C_1 \cdot \varepsilon^{k-1} \cdot |u|$$

since F has support in $|x| \le \varepsilon$, so $|\xi| \le \varepsilon$.

If $|u(x)| = \omega(x) = C \cdot |x|^k$ we get

$$\frac{1}{\omega}\left|\sum Q_i u_i\right| \le (\Lambda + \delta)|u| + C_0|x|^k + C_1\varepsilon^{k-1}|u|.$$

In order to get (6) we must have

$$\frac{\Lambda + \delta}{\mu - \delta} + \frac{C_0 + CC_1\varepsilon^{k-1}}{C(\mu - \delta)} \le k,$$

which is possible if we choose δ and ε small and C big enough.

3.2. Estimates for the First Derivative

To make things clearer we treat the first derivative separately. Take now $\varphi_\nu = (Du)_{ij} = \partial_j u_i = u_{ij}$ and $\omega(x) = C|x|^{k-1}$. Deriving (3) with respect to x_p we get

$$\sum_{j,k} a_{jk}x_k\partial_j u_{ip} = \sum_k a_{ik}u_{kp} - \sum_k a_{kp}u_{ik} + \partial_p(F_i(x + u(x)))$$

$$= Q_{ip}(x, Du).$$

The last term is $\partial_p(F_i(x + u(x))) = (\partial_p F_i)(x+u) + \sum(\partial_k F_i)(x+u) \cdot u_{kp}$, which is $O(|x+u|^k) + O(|x+u|^{k-1}) \cdot |Du|$ and, since $|u| \le C|x|^k \le C|x|$, we get $|x+u| = O(|x|)$. Combining all this we get at points where $|Du(x)| = \omega(x)$

$$\frac{1}{\omega}\left|\sum_{i,p} Q_{ip}u_{ip}\right| = \frac{1}{\omega}\left|\sum_{i,k,p} a_{ik}u_{kp}u_{ip} - \sum_{i,k,p} a_{kp}u_{ik}u_{ip} - \sum_{i,k,p} u_{ip}\partial_p F_i(x+u)\right|$$

$$\le (\Lambda - \mu + 2\delta) \cdot |Du| + C_1|x|^{k-1} + C_2|x|^{k-1} \cdot |Du|$$

$$\le (k-1)C(\mu - \delta) \cdot |x|^{k-1} \le \sum_{j,k} a_{jk}x_k\partial_j\omega(x),$$

if

$$\frac{\Lambda - \delta}{\mu - \delta} - 1 + \frac{C_1 + CC_2\varepsilon^{k-1}}{C(\mu - \delta)} \le k - 1,$$

which is possible since $k > \rho$ for small enough ε, δ and great C.

3.3. Higher-Order Derivatives

Let $|\alpha| = r \leq s$. By the choice of s we have $|\rho - r| < k - r$ for all such r, and s is the biggest integer for which this holds. Set $\partial^\alpha u_i = u_{i,\alpha} = \varphi_\nu$ and call the collection of all the $u_{i,\alpha}$ $D^r u$. Deriving (3) we get

$$\sum_{i,k} a_{jk} x_k \partial_j u_{i \cdot \alpha} = \sum_k a_{ik} u_{k,\alpha} - \sum_{j,k} a_{jk} \alpha_k u_{i,\alpha+(k,j)} + \partial^\alpha (F_i(x + u(x))),$$

where the symbol $\alpha + (k, j)$ means subtracting 1 from α at the k-th place and adding 1 at the j-th, especially $\alpha + (k, k) = \alpha$. The last term, if written down explicitly (cf. [1]), is a linear combination of several kinds of expressions. First there is $(\partial^\alpha F_i)(x + u) = O(|x|^{k-r})$. Then there are products of derivatives of F multiplied by lower-order derivatives of u which by induction have already been estimated by $\partial^\beta u = O(|x|^{k-|\beta|})$. And last there are terms containing a first-order derivative of F multiplied by an r-th order derivative of u; these terms are of order $O(|x|^{k-1}) \cdot |D^r u|$. Combining all this we get

$$\frac{1}{\omega} \left| \sum Q_\nu \varphi_\nu \right| \leq \frac{1}{\omega} \left| \sum a_{ik} u_{k,\alpha} u_{i,\alpha} - \sum a_{ik} \alpha_k u_{i,\alpha+(k,j)} u_{i,\alpha} \right| + \left| \sum u_{i,\alpha} \partial^\alpha (F_i(x + u)) \right|$$

$$\leq \left| (\Lambda + \delta) - r(\mu - \delta) \right| \cdot |Du| + C_1 |x|^{k-r} + C_2 \varepsilon^{k-1} |D^r u|$$

$$\leq C(\mu - \delta)(k - r) \cdot |x|^{k-r},$$

where we have set $\omega(x) = C|x|^{k-r}$ and considered the points x where $|D^r u(x)| = \omega(x)$. For the last inequality to hold it is sufficient that

$$\left| \frac{\Lambda + \delta}{\mu - \delta} - r \right| + \frac{C_1 + CC_2 \varepsilon}{C(\mu - \delta)} \leq k - r.$$

This is possible by the choice of $r \leq s$ if δ, ε are small and C big enough. By Proposition 13.1 in [1], u is of class C^s also at zero.

References

[1] R. Abraham and J. Robbins, *Transversal Mappings and Flows*. Benjamin, New York, 1967.
[2] G. Bengel, Convergence of formal singular partial differential equations, *Advances in Microlocal Analysis*, Nato ASI Series Vol. 169, 1–14, 1985.
[3] G. Bengel and R. Gérard, Formal and convergent solutions of singular partial differential equations, *Manuscripta Math.* **38** (1982) 343–373.
[4] P. Hartman, *Ordinary Differential Equations*, Wiley, New York, 1964.

[5] K. Ise and M. Nagumo, On the normal forms of differential equations in the neighborhood of an equilibrium point, *Osaka Math. J.* **9** (1957) 221-234.

[6] G. Sell, Smooth linearization near a fixed point, *Amer. J. Math.* **107** (1985) 1035-1091.

[7] S. Sternberg, Local contractions and a theorem of Poincaré, *Amer. J. Math.* **79** (1957) 809-824.

[8] S. Sternberg, On the structure of local homeomorphisms of Euclidean n-space, *Amer. J. Math.* **80** (1958) 623-631.

[9] D. Stowe, Linearization in two dimensions, *J. Diff. Equations* **63** (1986) 183-226.

On the Notions of Scattering State, Potential and Wave-Function in Quantum Field Theory: An Analytic-Functional Viewpoint

J. Bros

Service de Physique Théorique
Institut de Recherche Fondamentale
CEA, CEN-Saclay
Gif-sur-Yvette, France

1. Particles and Fields

The quantum theory of scattering phenomena involving particles in mutual short-range interaction has been developed in two mathematical frameworks which differ considerably by their concepts and by the way in which the Hilbert space of states is introduced. These are respectively the frameworks of wave-mechanics (W.M.) and of (massive) quantum-field theory (Q.F.T.). While the first theory, dominated by the concept of wave-function, stands in a certain L^2 space of space-time variables $(\mathbf{x}, t)$ (or of the Fourier conjugate momentum and energy variables $(\mathbf{p}, E)$) and applies to "low-energy" (i.e. non-relativistic) physics, the other one, based on the concept of field operator $A(f) = \int A(x)f(x)\, dx =$

Algebraic Analysis, Volume I

$\int A(-p)f(p)\,dp$ $(x=(\mathbf{x},t),p=(\mathbf{p},p^{(0)})$, with $\mathbf{x}$ and $\mathbf{p}$ in $\mathbf{R}^d)$ stands in an abstract Hilbert space of states $\mathcal{H}$, which is supposed to take into account all "high-energy" (i.e. relativistic) scattering phenomena (including creation and annihilation of particles) in a synthetic way. A basic discomfort is created by the conceptual gap between the two theories, since the former should in principle be re-obtained as a non-relativistic approximation of the latter. It is the purpose of this essay to try to reduce this discomfort by exhibiting a closely parallel presentation of the two theories for the (simplest) case of two-particle scattering states. It will be shown in particular that all the notions and algebraic results that pertain to the W.M. framework have their exact counterparts in the Q.F.T. framework, provided the "real analysis viewpoint" is replaced by an appropriate "complex analysis viewpoint." In particular the distribution character of the wave-function $\Psi(\mathbf{p};E)$ in energy-momentum space will have an analytic-functional counterpart in Q.F.T., provided a relevant space of test-functions $f(\mathbf{p},p^{(0)};E)$, holomorphic with respect to the complex energy variable $p^{(0)}$ is introduced. This modest example of the beautiful algebraic simplicity of complex analysis may in some sense be affiliated to the spirit of Professor M. Sato's research, when, at the glory time of dispersion relations in physics, he was led to imbed distribution theory in his general theory of hyperfunctions [1].

As a matter of fact, it is only in the ideal case of particles without mutual interaction that a simple connection exists between the two mathematical frameworks, namely between the free Schrödinger wave-equation and the free quantum field theory. In this case, the Q.F.T. Hilbert space $\mathcal{H}$ is indeed identical with a certain Fock space $\mathcal{F}=\bigoplus_{n\geq 0}\mathcal{F}_n$, where $\mathcal{F}_0=\{\lambda\Omega;\lambda\in\mathbf{C}\}$, Ω being the so-called "vacuum state," and where the subspace $\mathcal{F}_n$ of "n-field states," spanned by the set $\{A(f_1)\cdots A(f_n)\Omega;f_i\in S(\mathbf{R}^{d+1}),1\leq i\leq n\}$ is isomorphic to the L^2 space of symmetric functions $\hat{f}(p_1,p_2,\ldots,p_n)$ on the "n-particle mass shell manifold" $(H_m^+)^{\times n}$, defined by: $H_m^+=\{p=(\mathbf{p},p^{(0)})\in\mathbf{R}^{d+1},\ p^2\equiv p^{(0)2}-\mathbf{p}^2=m^2,\ p^{(0)}>0\}$ and equipped with a Lorentz invariant measure; this isomorphism is given by the simple rule: $\hat{f}=\text{sym}\,f_1\ldots f_n|(H_m^+)^{\times n}$ (where "sym" denotes symmetrization with respect to the n vectors $p_1,\ldots,p_n$). In particular, one can say that for each test-function f in $\mathcal{A}_1=S(\mathbf{R}_{(p)}^{d+1})$, the corresponding "one-field state" $A(f)\Omega$ is represented by the (energy-momentum) "one-particle wave-function" $\Psi=f\times\delta(s)$, where $s\equiv p^2-m^2$ and $\delta(s)$ is the corresponding Dirac measure with support H_m^+: Ψ is the Fourier-transform of a "positive-energy wave-function" $\psi(\mathbf{x},t)$ satisfying the relativistic version of the free Schrödinger equation, namely the Klein–Gordon equation

$$\left(-\frac{\partial^2}{\partial t^2}+\Delta_{\mathbf{x}}-m^2\right)\psi(\mathbf{x}, t)=0$$

(in units where the light velocity c is taken equal to 1). We note that the one-field state $\Phi^{(\dot{f})}=A(f)\Omega$ is actually associated with a *class* $\dot{f}$ of test-functions f, modulo the subspace $\mathscr{B}_1=\{f; f=s\times\chi, \chi\in\mathscr{A}_1\}$ and we call β_1 the isomorphism from $\dot{\mathscr{A}}_1=\mathscr{A}_1/\mathscr{B}_1$ into $\mathscr{H}$ defined by $\Phi^{(\dot{f})}=\beta_1(\dot{f})$; we also call α_1 the isomorphism from $\dot{\mathscr{A}}_1$ into $\mathscr{A}'_1=S'(\mathbf{R}^{d+1}_{(p)})$ defined by $\Psi^{(\dot{f})}=f\times\delta(s)=\alpha_1(\dot{f})$. As indicated above, similar isomorphisms exist for the n-field or n-particle states (with n arbitrary) as a consequence of the Fock structure of $\mathscr{H}$.

In the case of interacting particles, the previous simple connection between field states and particle states disappears, except however for the subspace of one-field or one-particle states, provided the energy-momentum spectrum of the field theory considered contains one (or several) discrete hyperboloid shell H^+_m: we shall assume in the following that this special spectral property of Q.F.T., denoted by (Sp), is satisfied.

The conceptual gap between the two theories results from the situation that we now review. In W.M., each state considered throughout its evolution in time (or "*sub specie aeternatis*," according to Heisenberg's viewpoint) is represented by a "wave-function" $\psi\{\mathbf{x}, t\}$ or equivalently by its Fourier transform $\Psi(\mathbf{p}, E)$ satisfying, in the sense of distributions, a given Schrödinger equation. For the case of two-particle states described in the space of the relative momentum and energy variables $(\mathbf{p}, E)$, the latter can be written (with a suitable choice of the mass parameter) as follows:

$$(E-\mathbf{p}^2)\Psi(\mathbf{p}, E)-\int \tilde{V}(\mathbf{p}-\mathbf{p}')\Psi(\mathbf{p}', E)\, d\mathbf{p}'=0, \tag{1}$$

where $\tilde{V}$ is the Fourier-transform of a given short-range potential $V(\mathbf{x})$, and is therefore analytic (at least in a tubular neighbourhood of real $\mathbf{p}$ space). The Hilbertian norm of Ψ is defined as the L^2 norm of $\psi(\cdot, t)$ (independent of t and reexpressed in terms of Ψ); it is interpreted physically as the total presence probability of the particles at time t, which must be preserved in time.

In Q.F.T., the basic states, which span the Hilbert space $\mathscr{H}$, are the n-field states $\Phi^{(f)}=A(f_1)\cdots A(f_n)\Omega$ (with $f=f_1\times\cdots\times f_n$ in $\mathscr{A}_1^{\otimes n}$), the vacuum state Ω being the unique eigenvector with eigenvalue zero for the set of energy-momentum operators. The space $\mathcal{H}$ and the field can alternatively be introduced by a G.N.S.-type construction, starting from a given positive linear functional $\mathcal{W}$ on the tensor $*$-algebra $\mathcal{A}$ of test-functions generated by $\mathcal{A}_1$. From this viewpoint (initiated by the Wightman reconstruction theorem [2], and deeply developed by H. Borchers [3]), $\mathcal{H}$ can be considered as isomorphic to the (closure of the) quotient-space $\mathcal{A}/\mathcal{B}$, if $\mathcal{B}$ denotes the null-ideal of $\mathcal{W}$. All dynamical properties of a given field theory, including those of general nature specified in the axiomatic approach of Q.F.T. [2], can be described either by conditions on the field $A(x)$ itself (i.e. on its vacuum expectation values or Green functions) or by conditions on the subspace $\mathcal{B}$ of the algebra $\mathcal{A}$; for example the "locality property," which states that $[A(f), A(g)] = 0$ for all couples (f, g) whose inverse Fourier-transforms $\tilde{f}, \tilde{g}$ have "acausal supports," may be expressed by saying that the corresponding elements $f \otimes g - g \otimes f$ of $\mathcal{A}$ belong to $\mathcal{B}$.

From this brief survey, the following questions emerge:

(1) How can particle scattering phenomena be described in the Q.F.T. approach and give rise to the same notion of scattering operator as in the W.M. approach?

(2) Is there a "test-function approach" of W.M. and if this is the case, what is the substitute to the isomorphism α_1 (depending now on the potential V), for re-obtaining the wave-function description?

(3) Is there a substitute to the notion of wave-function in the Q.F.T. approach, and if this is the case, how is it obtained from the G.N.S.-type test-function description of states? What Schrödinger-type equation should it satisfy and what would play the role of the potential in this equation?

A rather satisfactory answer to the first question has been given in the axiomatic approach of Q.F.T. by the Haag–Ruelle theory of scattering states [4, 5], which is valid under the condition that the field theory considered satisfies the spectral property (Sp) previously described. This theory constructs two isomorphisms i_{ex} (ex = in, out) from the Fock-space $\mathcal{F}$ onto subspaces $\mathcal{H}_{ex}$ of $\mathcal{H}$; each state in $\mathcal{H}_{ex}$ can be interpreted as a scattering state, i.e., a state that behaves like an n-particle state with well-defined "wave-packet" on $(H_m^+)^{\times n}$, when the time tends to $+\infty$ or $-\infty$. Technically, these states are constructed as limits of appropriate families of n-field states $\Phi(f, t)$ depending on a time-parameter t, namely $\Phi_{in}^{(\hat{f})} = \lim_{t \to -\infty} \Phi(f, t)$ and $\Phi_{out}^{(\hat{f})} = \lim_{t \to +\infty} \Phi(f, t)$, where the wave-packet $\hat{f}$ is defined by a formula similar to that of the free-field case, namely $\hat{f} = \operatorname{sym} f|_{(H_m^+)^{\times n}}$. The scattering operator S is then defined in the Hilbert space $\mathcal{H}$ of Q.F.T. by the isometry: $\Phi_{in}^{(\hat{f})} = S\Phi_{out}^{(\hat{f})}$ (i.e., $S = i_{in} \circ i_{out}^{-1}$).

In the W.M. framework, a similar theory of scattering states via Haag–Ruelle-type limits was performed by K. Hepp [5]; however, in contrast with Q.F.T. scattering theory, whose arguments are based on the axioms of field theory (in particular, on the locality and spectral axioms), the W.M. derivation of [5] is of course based on estimates involving the potential, and is therefore of very different nature. On the other hand, in the standard W.M. scattering theory [12], the initial and final wave-packets f_{in}, f_{out} associated with a given scattering state Ψ are directly computed in terms of the wave-function $\Psi(\mathbf{p}, E)$ of this state (rather than obtained by a limiting procedure) and the "scattering-matrix" ("or S-matrix") is then directly defined by $f_{out}(\mathbf{p}) = \int S(\mathbf{p}, \mathbf{p}')f_{in}(\mathbf{p}') \, d\mathbf{p}'$ (in the sense of distributions). In Q.F.T., there is also a direct way of computing the asymptotic manifestations, f_{in}, f_{out} of a given field-state $\Phi^{(f)}$: f_{in} and f_{out} are indeed computable in terms of a representative test-function f of the state $\Phi^{(f)}$, via the theory of "reduction formulas" [5, 6] (which contain the "Green's functions" of the field as basic kernels characterizing the theory). However, in contrast with the W.M. case, it is a "test-function" and not a "wave-function" that is involved in this Q.F.T. formalism. Moreover, an important discrepancy between the W.M. and Q.F.T. frameworks comes from the fact that in a given field theory, *there may exist field-states* $\Phi^{(f)}$ *which are not scattering states* (if $\mathcal{H}_{ex} \neq \mathcal{H}$). A field theory is said to satisfy the "*asymptotic completeness property*" if $\mathcal{H}_{ex} = \mathcal{H}$.

From this comparative survey of scattering theory in the W.M. and Q.F.T. frameworks, we conclude that the remaining obscurities in this parallel are precisely linked with the second and third questions that we have set above, and which we shall partly answer in the following.

In this article, we shall restrict ourselves to the case of two-particle scattering states and (in Q.F.T.) of two-field states, and we will try to give a synthetic viewpoint on the two Hilbertian representations (namely the "wave-function aspect" and the G.N.S. or "test-function aspect"), valid in both W.M. and Q.F.T. frameworks. More precisely, we will show that the diagram of Fig. 1 (in which the arrows represent vector-space homomorphisms) can be implemented with completely similar formulas in the W.M. and Q.F.T. frameworks; this will provide positive answers to our second and third questions for the case of two-particle and two-field states.

In this diagram, $\mathcal{A}$ will denote a relevant space of "test-functions" f with appropriate analyticity properties, and $\mathcal{A}'$ a corresponding space of distribu-

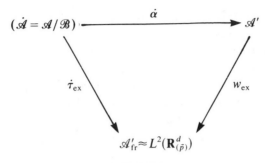

FIGURE 1

tions or analytic functionals Ψ (namely the "wave-functions") which will satisfy a certain Schrödinger-type equation $\sigma \cdot \Psi = 0$. The subspace $\mathscr{B}$ of $\mathscr{A}$ will correspondingly be defined as the set of functions f of the form $f = \sigma \cdot \chi$, with χ in $\mathscr{A}$. $\mathscr{A}'_{\text{fr}}$ (fr = "free") will denote a space of distributions Ψ_{ex} that satisfy the corresponding *free* Schrödinger-type equation $\sigma_{\text{fr}} \cdot \Psi_{\text{ex}} = 0$ (as the space $\mathscr{A}'_1$ above): $\mathscr{A}'_{\text{fr}}$ will contain the "asymptotic forms" (Ψ_{in} and Ψ_{out}) of a given state, represented either by a class $\dot{f}$ in $\dot{\mathscr{A}}$ or by the wave-function $\Psi^{(\dot{f})} = \alpha(\dot{f})$ in $\mathscr{A}'$; $\dot{\tau}_{\text{ex}}$ and w_{ex} will denote the corresponding "asymptotic mappings" (i.e., $\Psi_{\text{ex}} = \dot{\tau}_{\text{ex}}(\dot{f}) = w_{\text{ex}}(\Psi^{(\dot{f})})$, ex = in or out).

The kernels of the various mappings $\dot{\alpha}$, $\dot{\tau}_{\text{ex}}$, w_{ex} (and their inverses) and the Hilbertian norm in $\dot{\mathscr{A}}$ will be expressed in terms of two basic functions of the theory B and F, mutually related by a resolvent integral equation. The function B, which we define as $(-2i\pi)^{-1} \tilde{V}(\mathbf{p} - \mathbf{p}')$ in W.M., will be a Bethe–Salpeter-type kernel in the Q.F.T. axiomatic framework; we will call B the "irreducible interaction kernel." The function F, which we call "the amputated Green function" is linked to the resolvent of the Hamiltonian operator in W.M., and to the ("chronological") product of four field-operators in Q.F.T. The analytic background which concerns the properties of B and F and of special composition products $\circ_+$, $\circ_-$, $*$, is briefly presented in Section 2.

Sections 3 and 4 are devoted to the explicit construction of the diagram of Fig. 1, respectively in the W.M. and in the Q.F.T. frameworks. In the latter case, the realization of states in the Hilbert space $\mathscr{H}$ (in terms of appropriate field products) will require the implementation of another commutative diagram, represented in Fig. 2. In the latter, i_{ex} represents the Haag–Ruelle isomorphism from $\mathscr{A}'_{\text{fr}}$ onto the corresponding subspace of two-particle scattering states $\mathscr{H}^{(2)}_{\text{ex}}$ which is contained in a certain two-field state subspace $\mathscr{H}^{(2)}$ of $\mathscr{H}$. π_{ex} denotes the orthogonal projection of $\mathscr{H}^{(2)}$

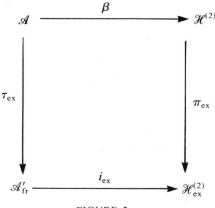

FIGURE 2

onto its subspace $\mathcal{H}_{ex}^{(2)}$, and β will be an appropriate substitute to the standard G.N.S.-type mapping mentioned above, for which the analyticity properties of the elements f of $\mathcal{A}$ will play a crucial role. The subspace $\mathcal{H}^{(2)}$ is said to satisfy the *"asymptotic completeness"* property if $\mathcal{H}^{(2)} = \mathcal{H}_{ex}^{(2)}$ (i.e., if π_{ex} is the identity). The validity or non-validity of the latter can be simply characterized in terms of analyticity properties of the function B, and the fact that the analogue of this asymptotic completeness property is always valid in the W.M. framework, admits a clear interpretation from this viewpoint.

2. Irreducible Interaction Kernels and the Analytic Background of Two-Particle Scattering

In each W.M. or Q.F.T. theory (of short-range interactions), the basic functions B and F are analytic functions of a "global energy variable" E and of two momentum (vector)-variables p and p'. It is convenient to consider them as E-dependent kernels acting in complex p-space and to denote them by $B(p, p'; E)$ and $F(p, p'; E)$, since they are mutually related by a resolvent integral equation of the following form:

$$F(p, p'; E) = B(p, p'; E)$$

$$+ \int_{\Gamma(E)} F(p, p''; E) B(p'', p'; E) s^{-1}(p'', E) \mu(p''; E). \qquad (2)$$

In the latter, $s(p, E)$ is a certain polynomial function, and the integration cycle $\Gamma(E)$, whose dimension is that of real p-space, is chosen such that it avoids the poles of the so-called "propagator function" s^{-1}. The kinematics and the precise prescriptions for making use of equation (2) are fixed by the W.M. or by the Q.F.T. framework; they can be described in a parallel way as follows.

2.1. Kinematics

In Q.F.T., the variables $p = (p^{(0)}, \mathbf{p})$, $p' = (p'^{(0)}, \mathbf{p}')$ vary in $\mathbf{C}^{d+1} = \mathbf{C} \times \mathbf{C}^d$ and are related to four "field-energy-momentum" variables p_1, p_2, p_1', p_2' such that $p_1 + p_2 = p_1' + p_2' = P$, by the formulas: $p = (p_1 - p_2)/2$, $p' = (p_1' - p_2')/2$. Moreover, a special Lorentz frame has been chosen in which the "total energy-momentum" vector P is of the form: $P = (2[E + 1]^{1/2}, \mathbf{0})$.

In $\mathbf{C}_{(p)}^{d+1} \times \mathbf{C}_{(E)}$, one considers the (reducible) complex manifold S with equation $s(p, E) \equiv s_+(p, E) \times s_-(p, E) = 0$, where the functions $s_\pm(p, E) = E - \mathbf{p}^2 + p^{(0)}[p^{(0)} \pm 2(E + 1)^{1/2}]$ correspond respectively to $s_+ = p_1^2 - m^2$, $s_- = p_2^2 - m^2$, if the mass m is chosen equal to 1. If we put $S = S_+ \cup S_-$, with $S_\pm = \{(p, E); s_\pm(p, E) = 0\}$, the complex "mass shell manifold" is defined by $M^{(c)} = S_+ \cap S_-$ and represented by the equations

$$M^{(c)}: \qquad p^{(0)} = 0, \qquad E - \mathbf{p}^2 = 0. \qquad (3)$$

The real mass-shell $M = M^{(c)} \cap \mathbf{R}_{(p, E)}^{d+2}$ is the set of physical energy-momentum configurations for two relativistic particles with individual variables p_1, p_2 in H_m^+.

In W.M., the vectors p, p' reduce to $p = (0, \mathbf{p}) \equiv \mathbf{p}$, $p' = (0, \mathbf{p}') \equiv \mathbf{p}'$, with $\mathbf{p}, \mathbf{p}'$ varying in $\mathbf{C}^d$. $\mathbf{p}$ and $\mathbf{p}'$ are associated respectively with the initial and final two-particle momentum configurations considered in the barycentric frame, namely $\mathbf{p} = \mathbf{p}_1 = -\mathbf{p}_2$, $\mathbf{p}' = \mathbf{p}_1' = -\mathbf{p}_2'$ (these relations being identical with the previous relativistic ones in the chosen Lorentz frame). In $\mathbf{C}_{(\mathbf{p})}^d \times \mathbf{C}_{(E)}$, one then defines the complex "energy shell" $S = M^{(c)}$ by the equation

$$s(p, E) \equiv E - \mathbf{p}^2 = 0. \qquad (4)$$

The real energy shell $M = S \cap \mathbf{R}_{(\mathbf{p}, E)}^{d+1}$ corresponds to the physical configurations for non-relativistic particles with equal masses $m = 1$ (i.e., reduced mass $m = \frac{1}{2}$ in the barycentric system).

2.2. Analyticity Domain of B and Integration Prescriptions for Equation (2)

In the "normal" case, interpreted below as the case when the asymptotic completeness property holds, the irreducible interaction kernel $B(p, p'; E)$

is analytic in a domain of the form $D \times D \times \Delta$, where, in W.M. $D = \underline{D} = \{\mathbf{p} \in \mathbf{C}^d, |\text{Im} \mathbf{p}| < \varepsilon\}$ and $\Delta = \mathbf{C}$ (note that $B \equiv (-2i\pi)^{-1} \tilde{V}(\mathbf{p} - \mathbf{p}')$ is in fact constant with respect to E), while, in Q.F.T. $D = \{p^{(0)} \in \mathbf{C}; |\text{Re} p^{(0)}| < \frac{1}{2}\} \times \underline{D}$ and Δ is a suitable neighbourhood of a given set $\delta = \{E \in \mathbf{R}; 0 \le E < E_1\}$, with $E_1 \le \frac{5}{4}$ ($E = \frac{5}{4}$ corresponds to the three-particle kinematical threshold).

In Eq. (2), $\mu(p; E)$ is equal to $-2i\pi\nu(\mathbf{p})$, with $\nu\{\mathbf{p}\} = dp^{(1)} \wedge \cdots \wedge dp^{(d)}$ in the W.M. case, and to $\rho(p, E) dp^{(0)} \wedge \nu(\mathbf{p})$ in the Q.F.T. case, ρ being an appropriate damping factor (holomorphic in D and equal to 1 for (p, E) in $M^{(c)}$). The integration cycle $\Gamma(E)$ is a $(d+1)$- (resp. d-) dimensional cycle in Q.F.T. (resp. in W.M.) whose class $\dot{\Gamma}(E)$ in $H^F(D\backslash\{p; (p, E) \in S\})$ "varies continuously with E" (in the sense of [10]) and is specified, for $E < 0$, by the following representative: $\Gamma(E) = \Gamma_0 \equiv \{p; p^{(0)} \in \mathbf{R}, \mathbf{p} \in \mathbf{R}^d\}$ in Q.F.T. (resp. $\{\mathbf{p} \in \mathbf{R}^d\}$ in W.M.). The set $\{\dot{\Gamma}(E); E \in \Delta\backslash\{0\}\}$ is shown to define a section of a homology bundle $\{(E, \dot{\Gamma}(E)); E \in \hat{\Delta}\}$ whose basis $\hat{\Delta}$ is the two-sheeted or universal covering of $\Delta\backslash\{0\}$, according to whether d is odd or even. The "physical sheet" of $\hat{\Delta}$, is defined over $\Delta\backslash\delta$ by the prescription $\dot{\Gamma}(E) = \dot{\Gamma}_0$ for $E < 0$. In this sheet, $\dot{\Gamma}(E)$ admits two limits, denoted by $\dot{\Gamma}_+(E)$ and $\dot{\Gamma}_-(E)$, on the border $\{E > 0\}$ from the respective sides $\text{Im } E > 0$ and $\text{Im } E < 0$. Near $E = 0$, the set $M_E = \{p; (p, E) \in M\}$ is (in view of equations (3), (4)) a vanishing sphere which produces a "simple quadratic pinching" situation for $\dot{\Gamma}(E)$ (in the sense of [7]) and the following homological relation holds (for $E > 0$): $\dot{\Gamma}_+(E) - \dot{\Gamma}_-(E) = \tilde{M}_E$, where $\tilde{M}_E$ is the "Leray's coboundary" [8] of M_E in $H(D\backslash\{p; (p, E) \in S\})$.

Let us now introduce the space $\mathscr{A}$ of functions $f(p, E)$ with support in $\{(p, E); E \in \delta\}$, holomorphic[1] with respect to p in D and μs^{-1}-integrable on $\dot{\Gamma}_\pm(E)$. We then denote by $\circ_+$ and $\circ_-$ the operations of μs^{-1}-integration on the respective classes of cycles $\Gamma_+(E)$ and $\Gamma_-(E)$ and by $*$ the operation of integration on the manifold M_E with the measure defined by the form $-2i\pi \cdot \mu/ds|_{M_E}$ in W.M. or $(2i\pi)^2 \mu/ds_+ \wedge ds_-|_{M_E}$ in Q.F.T. It then follows from the previous homological relation between $\dot{\Gamma}_+(E)$ and $\dot{\Gamma}_-(E)$ and from Leray's residue theorem [8] that one has the relation

$$\circ_+ - \circ_- = *, \tag{5}$$

which applies for instance to functions f, g in $\mathscr{A}$ or to μs^{-1}-integrable kernels as B (e.g., $f \circ_+ g - f \circ_- g = f * g$, $B \circ_+ f - B \circ_- f = B * f$ etc...). In particular, the limits of Eq. (2) on $\{E > 0\}$ from the sides $\text{Im } E > 0$, $\text{Im } E < 0$,

---

[1] The use of such a space of analytic functions in a similar context has already been introduced in [15] for the case $d = 1$ of Q.F.T. (namely, in the course of a proof of the two-particle asymptotic completeness property in the constructive (euclidean) approach of certain two-dimensional field models).

can be written respectively:

$$F_+ = B + F_+ \circ_+ B, \qquad F_- = B + F_- \circ_- B, \qquad (6)$$

if F_+, F_- denote the corresponding boundary values of F in the physical sheet of $\hat{\Delta}$.

Remark. In the W.M. case, formula (5) is just an extension to $\mathscr{A}$ of the distribution identity $(-2i\pi)^{-1}[(s^{-1})_+ - (s^{-1})_-] = \delta(s)$, since in this case $\dot{\Gamma}_+(E)$ and $\dot{\Gamma}_-(E)$ can be represented by $\mathbf{R}^d_{(\mathbf{p})}$, up to local $\pm i\varepsilon$-distortions around M_E. In the Q.F.T. case, one can still write formula (5) under the form $(2i\pi)^{-2}[(s^{-1})_+ - (s^{-1})_-] = \delta(s)$, but *only in the sense of analytic functionals* on $\mathscr{A}$, and in reference to the representatives of $\dot{\Gamma}_\pm(E)$ described in Section 4.2: the latter contain parts $\Gamma_0^\pm$ obtained from a real region Γ_0 (in $\mathbf{R}^{d+1}_{(\mathbf{p})}$) by "$\pm i\varepsilon$-distortions" that correspond in fact to the respective distribution-prescriptions $(s^{-1})_+ \equiv (s_+^{-1})_+ \times (s_-^{-1})_+$ and $(s^{-1})_- \equiv (s_+^{-1})_- \times (s_-^{-1})_-$.

2.3. Results Concerning the Analytic Structure of F

Let us assume that B enjoys the "normal" analyticity and μs^{-1} integrability properties described *above*, and moreover satisfies the "reality condition" $\overline{B(p, p'; E)} = -B(\bar{p}', \bar{p}; \bar{E})$; the latter (which expresses the real-valuedness of the potential in the W.M. case) corresponds to the Hermitian character of all operators representing "observable quantities" (field operators, Hamiltonian, etc).

Under these assumptions and by using the Fredholm theory[2] on $\Gamma(E)$ together with the relation (5), one obtains the following results [9, 10, 11] for F:

(*i*) F is meromorphic on $D \times D \times \hat{\Delta}$ and of the form $F(p, p'; E) = \mathscr{N}(p, p'; E)/\mathscr{D}(E)$ with $\mathscr{N}$(resp. $\mathscr{D}$) analytic in $D \times D \times \hat{\Delta}$(resp. $\hat{\Delta}$).

(*ii*) In the physical sheet, poles of F can only exist on the set $(E < 0)$ and are then interpreted as "bound states" (in W.M.) or "composite particles" (in Q.F.T.). In the unphysical sheet(s), poles can be produced at complex values of E and may then be interpreted as "resonances" or (in Q.F.T.) "unstable particles."

---

[2] In W.M. and for $d \geq 2$, a more refined application (or version) of Fredholm theory [12] is required for obtaining the meromorphic form of F described in (*i*). In Q.F.T., the Fredholm formulas themselves are applicable thanks to the introduction of the damping factor ρ in μ, and correspondingly of the regularized irreducible kernel $B(\equiv B_{(\rho)})$ satisfying equation (2): this presentation of the Bethe–Salpeter equation, first given in the axiomatic approach [9] has been implemented recently [16] in all constructible (small coupling) field models.

(*iii*) On the bordering set δ of the physical sheet, the following relation holds for all values of E such that $\mathcal{D}(E) \neq 0$:

$$F_+ - F_- = F_+ * F_-, \qquad \text{with } F_-(p, p'; E) = -\overline{F_+(\bar{p}', \bar{p}; E)}. \qquad (7)$$

The restriction T_+ of F_+ to the manifold $\mathcal{M} = \{(p, p', E); p \in M_E, p' \in M_E\}$ is the physical "scattering amplitude" (see below). By restricting Eq. (7) to $\mathcal{M}$, one obtains the following relation, which is shown to be valid on the whole set δ:

$$T_+ - T_- = T_+ * T_-, \qquad \text{where } T_-(p, p') = -\overline{T_+(p', p)}. \qquad (8)$$

The possible occurrence of poles for $F_\pm$ in δ (corresponding to $\mathcal{D}(E) = 0$), interpreted as "bound states embedded in the continuum" will be discarded for simplicity in this paper (although our parallel between W.M. and Q.F.T. can be shown to include this case) [17].

The previous results hold in W.M. (where $B = (-2i\pi)^{-1}\, \tilde{V}$ is an input of the theory) and in the "constructive approach" of Q.F.T., for all models in which a Bethe–Salpeter kernel B can be constructed [15, 16]. In the axiomatic approach of Q.F.T., the previous results are valid if and only if the relevant two-field state subspace satisfies the asymptotic completeness property. In the "abnormal case" when the latter is not valid, the function B can always be defined (namely in terms of F via equation (2)), but it results from the Q.F.T. axioms that B is (like F) only analytic [9] in the set $D \times D \times \{\Delta \backslash \delta\}$; it also follows from equations (5) and (6) (written with two different boundary values B_+, B_- of B) that the discontinuities $\Delta F = F_+ - F_-$ and $\Delta B = B_+ - B_-$ on δ are related by the following equation which generalizes equation (7):

$$\Delta F - F_+ * F_- = (1 + F_+) \circ_+ \Delta B \circ_- (1 + F_-) \qquad (9)$$

$$(\text{or } \Delta B = (1 - B_+) \circ_+ (\Delta F - F_+ * F_-) \circ_- (1 - B_-)), \qquad (9')$$

where 1 denotes the identity operator on the space $\mathcal{A}$.

This analytic background, which is the basis of the so-called "time-independent" (i.e., energy-momentum) formalism of scattering theory must be completed by the following "Tauberian property," which (in both W.M. and Q.F.T. frameworks) performs the passage to the "time-dependent" formalism.

Lemma 1. *Let* $\mathbf{t} = (t_1, \ldots, t_m) \in \mathbf{R}^m$, $\boldsymbol{\lambda} = (\lambda_1, \ldots, \lambda_m) \in \mathbf{R}^m$, $\boldsymbol{\lambda} \cdot \mathbf{t} = \sum_{i=1}^m \lambda_i t_i$ *and* $(\boldsymbol{\lambda}^{-1})_+ = \prod_{i=1}^m (\lambda_i^{-1})_+$, *with* $(\lambda^{-1})_+ = \lim_{\varepsilon > 0, \varepsilon \to 0} (\lambda + i\varepsilon)^{-1}$. *Then if* $\hat{\Psi}(\boldsymbol{\lambda})$

belongs to $S(\mathbf{R}^m)$, the corresponding function $f_+(\mathbf{t}) = \int_{-\infty}^{+\infty} e^{-i\lambda \cdot t}(\lambda^{-1})_+ \hat{\Psi}(\lambda)\, d\lambda$ admits the following radial limits: $\lim_{\tau \to +\infty} f_+(\alpha\tau) = (-2i\pi)^m \hat{\Psi}(0)$, if $\alpha = (\alpha_1, \ldots, \alpha_m)$, $\alpha_i > 0$, $1 \le i \le m$, and 0 if $\varepsilon_i \alpha_i > 0$, $\varepsilon_i = \pm 1$, $\sum \varepsilon_i < m$.

3.   Scattering States in Wave-Mechanics

3.1.   The Wave-Function Representation of States

Let

$$\Psi(\mathbf{p}, E) = \frac{1}{(2\pi)^{d+1}} \int e^{i(Et - \mathbf{p} \cdot \mathbf{x})} \psi(\mathbf{x}, t)\, d\mathbf{x}\, dt$$

be the momentum-space wave-function, corresponding to a solution ψ of the Schrödinger equation $i\, \partial\psi/\partial t = [-\Delta_x \to + V(\mathbf{x})]\psi$ and representing a given W.M. state. Ψ is a temperate distribution that satisfies the integral equation (1) together with the condition that for all t, the function

$$f_t(\mathbf{p}) = \int e^{-i(E-\mathbf{p}^2)t} \Psi(\mathbf{p}, E)\, dE \tag{10}$$

belongs to $L^2(\mathbf{R}_{(\mathbf{p})}^d)$.

3.1.1.   *Scattering States*

Equation (1) leads one to consider the states Ψ such that $\hat{\Psi}(\mathbf{p}, E) = (E - \mathbf{p}^2)\Psi(\mathbf{p}, E)$ belongs to the space $\mathcal{A}$ (see Section 2). The distribution Ψ can then be written as follows:

$$\Psi = (s^{-1})_+ \hat{\Psi} + f_{in}\delta(s), \qquad \text{where } s = E - \mathbf{p}^2. \tag{11}$$

In view of equation (10) and of Lemma 1 (for $m = 1$), each state Ψ of the form (11) satisfies the asymptotic properties

(*i*) $\lim_{t \to -\infty} f_t(\mathbf{p}) = f_{in}(\mathbf{p})$ \hfill (12)

(*ii*) $\lim_{t \to +\infty} f_t(\mathbf{p}) = f_{out}(\mathbf{p})$,   with $f_{out} = f_{in} - 2i\pi \hat{\Psi}|_M$. \hfill (13)

In view of (11), equation (1) yields

$$\Psi = (s^{-1})_+ \tilde{V} \cdot \Psi + f_{in}\delta(s), \tag{14}$$

$\tilde{V} \cdot \Psi$ being defined, according to the analysis and notations of Section 2, by $\tilde{V} \cdot \Psi \equiv B \circ_+ \hat{\Psi} - (2i\pi)^{-1}(B * f_{in})$.

By introducing the Green function F and using equation (6) (in the + case), one solves the integral equation (14) and obtains

$$\Psi = (-2i\pi)^{-1}(s^{-1})_+ F_+ * f_{in} + f_{in}\delta(s). \tag{15}$$

The set $\mathscr{A}'$ of "wave-functions" Ψ generated via equation (15) by all functions $\hat{f}_{in}$ in $L^2(\mathbf{R}^d_{(\mathbf{p})})$ characterizes the set of *scattering states*; $\mathscr{A}'$ is the set of all distributions satisfying equation (1) and such that $s\Psi$ belongs to $\mathscr{A}$.

The outgoing limit f_{out} corresponding to any Ψ in $\mathscr{A}'$ is given by equation (13), and can then be reexpressed in view of equation (15) in terms of the scattering function $T_+ = F_+|_{\mathscr{M}}$ (see Section 2) as follows:

$$f_{out} = f_{in} + T_+ * f_{in} = Sf_{in}. \tag{16}$$

Equation (8) is then equivalent to the *unitarity* of the operator S with respect to the norm of $L^2(\mathbf{R}^d_{(\mathbf{p})})$.

3.1.2.   The Wave-Function at Fixed Time

By exploiting equation (14) and the (Fourier) inverse of equation (10), one obtains the following relation (generalized below in the test-function representation—see equation (23)):

$$f_{in} = f_0 + (F_- \circ_- f_0)|_M. \tag{17}$$

Equation (7) then implies that $|f_0| = |f_{in}|$ (in $L^2(\mathbf{R}^d_{(\mathbf{p})})$).

3.1.3.   The Hamiltonian Operator Formalism

Let $(E - H)\Psi = 0$ be the operator form of equation (1), with $H = H_0 + V$ ($H_o = \mathbf{p}^2$ being the free Hamiltonian operator). The operator forms of the kernels $[1/s(p, E)]\delta(\mathbf{p} - \mathbf{p}')$ and $(-2i\pi)F(\mathbf{p}, \mathbf{p}'; E)$ are then respectively $(E - H_o)^{-1}$ and $(E - H_o)(E - H)^{-1}V$; equation (2) corresponds to the "amputated form" of the equation for the resolvent of H (implied by $H = H_o + V$), namely

$$[(E - H_o)(E - H)^{-1}V] = V + [(E - H_o)(E - H)^{-1}V] \times (E - H_o)^{-1}V.$$

Equation (15) then reads $\Psi = (1 + (E_+ - H)^{-1}V)\Psi_{in}$, if $\Psi_{in} = \delta(E - H_o)\hat{f}_{in}$.

Equation (10) is equivalent to $f_t = e^{iH_o t} e^{-iHt}f_0$ and the limits (12), (13) correspond to the existence of the Møller operators [12] $\Omega_\pm = \lim_{t \to \pm\infty} e^{iH_o t} e^{-iHt}$.

3.2. The Test-Function Representation of Scattering States

We define the following mapping α from the space $\mathscr{A}$ onto a subspace $\mathscr{A}'$ of $S'(\mathbf{R}^{d+1}_{(p,E)})$:

$$\alpha(f) = \Psi^{(f)} = \Psi^{(f)}_+ - \Psi^{(f)}_-, \qquad \text{where } \Psi^{(f)}_\pm = -(2i\pi)^{-1}(s^{-1})_\pm(f + F_\pm \circ_\pm f), \tag{18}$$

or, in operator form,

$$\Psi^{(f)}_\pm = -(2i\pi)^{-1}(E_\pm - H)^{-1}f \qquad \text{and} \qquad \Psi^{(f)} = \delta(E - H)f. \tag{19}$$

(Note that one re-obtains the free case: if $H = H_\mathrm{o}$, $\alpha_{\mathrm{fr}}(f) = \delta(E - \mathbf{p}^2)f$.)

Let σ denote here the Schrödinger operator: $\sigma = E - H$. The following properties of α follow from equations (18) or (19):

Proposition 1. (*i*) *The kernel $\mathscr{B}$ of α is the set of all functions f of the form $f = \sigma\chi$, with χ in $\mathscr{A}$.*

(*ii*) *Every Ψ in $\mathscr{A}'$ satisfies the equation $\sigma\Psi = 0$ and is such that $s\Psi$ belongs to $\mathscr{A}$.*

Remark. The mapping α extends to the set of functions f of the form $f_\mathrm{o}(\mathbf{p}) \otimes 1$, with f_o in $S(\mathbf{R}^d)$; f_o is then interpreted as the wave-function at $t = 0$ of the state $\Psi^{(f)}$ (see formula (10)). In fact, the Fourier inversion of equation (10) yields: $\Psi = (2\pi)^{-1} \int e^{i(E-H)t} f_\mathrm{o} \, dt = \delta(E - H)f_\mathrm{o}$, which is an extension of equation (19).

3.2.1. The Hilbertian Structure

From equations (10) and (19), we deduce the following expression (independent of t) for $|f_t|$ and take it as the definition of the norm $\|\Psi^{(f)}\|$ of $\Psi^{(f)} = \alpha(f)$:

$$\|\Psi^{(f)}\| = |f_t| = \left[\int (f, \delta(E - H)f) \, dE \right]^{1/2} \tag{20}$$

(where (,) denotes integration on $\mathbf{R}^d_{(\mathbf{p})}$).

By using equation (18) one obtains the following expression in terms of the Green's function F:

$$\|\Psi^{(f)}\|^2 = |f_{|M}|^2 - \frac{1}{4\pi^2} \int_0^\infty dE (\bar{f} \circ_+ F_+ \circ_+ f - \bar{f} \circ_- F_- \circ_- f). \tag{21}$$

Formula (20) (or (21)) defines a PreHilbertian structure on $\mathscr{A}$; the Hilbert space of states $\mathscr{H}$ can be defined as the closure (for this norm) of $\dot{\mathscr{A}} = \mathscr{A}/\mathscr{B}$, isomorphic to $\mathscr{A}'$ via the mapping $\Psi^{(\dot{f})} = \dot{\alpha}(\dot{f}) = \alpha(f)$.

3.2.2. The Asymptotic Representation of States and the Mappings $\dot{\tau}_{\mathrm{ex}}$ and w_{ex}

Proposition 2. *Every scattering state Ψ generated by an incoming wave-packet $\hat{f}_{\mathrm{in}}$ via equation (15) admits a representation of the form $\Psi = \dot{\alpha}(\dot{f})$, with $\dot{f}$ in $\dot{\mathscr{A}}$; every element f of $\dot{f}$ is given by the following formula:*

$$f = f_{\mathrm{in}} - B \circ_{-} f_{\mathrm{in}}, \tag{22}$$

where $f_{\mathrm{in}}(\mathbf{p}, E)$ is any function in $\mathscr{A}$ such that $f_{\mathrm{in}}|_{M} = \hat{f}_{\mathrm{in}}$.

Conversely, for every $\dot{f}$ in $\dot{\mathscr{A}}$, the corresponding scattering state $\Psi^{(\dot{f})} = \dot{\alpha}(\dot{f})$ admits the following incoming wave-function $\Psi_{\mathrm{in}} = \dot{\tau}_{\mathrm{in}}(\dot{f})$ (independent of the representative f in $\dot{f}$):

$$\Psi_{\mathrm{in}} = f_{\mathrm{in}} \delta(s), \qquad \text{where } f_{\mathrm{in}} = f + F_{-} \circ_{-} f. \tag{23}$$

Analogous formulas hold for the corresponding outgoing wave-function Ψ_{out}, by replacing $\dot{\tau}_{\mathrm{in}}, f_{\mathrm{in}}, F_{-}$ and $\circ_{-}$ respectively by $\dot{\tau}_{\mathrm{out}}, f_{\mathrm{out}}, F_{+}$ and $\circ_{+}$. Moreover one has:

$$\|\Psi^{(\dot{f})}\| = |\hat{f}_{\mathrm{in}}| = |\hat{f}_{\mathrm{out}}|. \tag{24}$$

The proof of formulas (22), (23) is based on equations (6), and equations (24) are shown to be equivalent to equations (7). As a matter of fact, Proposition 2 (and in particular equation (24)) expresses the property of *Asymptotic Completeness* of the space $\mathscr{H}$.

In order to complete the commutative diagram of Fig. 1, it remains to define the mappings w_{ex} from $\mathscr{A}'$ into $\mathscr{A}'_{\mathrm{fr}}$ (via equation (14)):

$$\Psi_{\mathrm{in}} = w_{\mathrm{in}}(\Psi) = \Psi - (s^{-1})_{+} \tilde{V} \cdot \Psi, \tag{25}$$

and similarly,

$$\Psi_{\mathrm{out}} = w_{\mathrm{out}}(\Psi) = \Psi - (s^{-1})_{-} \tilde{V} \cdot \Psi. \tag{25'}$$

In the latter, the notation $\tilde{V} \cdot \Psi$ may be considered either in the sense of distributions (Ψ acting on the test-function $\tilde{V}(\mathbf{p} - \bullet)$), or in the sense of analytic functionals, via a decomposition $\Psi = \Psi_{+} - \Psi_{-}$ of Ψ (as in equation

(18)), since $\tilde{V}(\mathbf{p}^{-\cdot})$ belongs to the space $\mathscr{A}$ (see our remark after formula (5)). The latter viewpoint will prevail in the Q.F.T. case below (see Section 4.3).

3.3.3. *Scattering States as "Haag–Ruelle–Hepp Limits"*

With any f in $\mathscr{A}$, one associates the family of states $\Psi(f, t) = \alpha(f^{(t)})$, where $f_t(\mathbf{p}, E) = f(\mathbf{p}, E)\, e^{i(E - \mathbf{p}^2)t}$. In view of equation (18), one has:

$$\Psi(f, t) = f\delta(s) - (2i\pi)^{-1}(s^{-1})_+ F_+ \circ_+ f^{(t)} + (2i\pi)^{-1}(s^{-1})_- F_- \circ_- f^{(t)}.$$

Then, in view of Lemma 1, the following limits exist in $\mathscr{H}$:

$$|\hat{f}, \text{in}) = \lim_{t \to -\infty} \Psi(f, t) = f\delta(s) - (2i\pi)^{-1}(s^{-1})_+ F_+ * f, \qquad (26)$$

$$|\hat{f}, \text{out}) = \lim_{t \to +\infty} \Psi(f, t) = f\delta(s) - (2i\pi)^{-1}(s^{-1})_- F_- * f, \qquad (26')$$

where $\hat{f} = f|_M$. By comparing equation (26) with equation (15), we conclude that $f\delta(s) = w_{\text{in}}[|\hat{f}, \text{in})] = w_{\text{out}}[|\hat{f}, \text{out})]$. The scattering operator $\mathbf{S}$ in $\mathscr{H}$, defined by $\mathbf{S} = w_{\text{in}}^{-1} \circ w_{\text{out}}$ (i.e. $|\hat{f}, \text{in}) = \mathbf{S}|\hat{f}, \text{out})$ is shown to be represented by the same kernel as the operator S of equation (16), namely (with our normalization conventions) by $\delta(\mathbf{p} - \mathbf{p}') - 4\pi^2 T_+(\mathbf{p}, \mathbf{p}')$.

4. Two-particle Scattering States in Quantum Field Theory

4.1. Scattering Theory for the Standard "Two-Field States"

The axiomatic approach of Q.F.T. allows one to consider in the Hilbert space $\mathscr{H}$ the set of two-field states $\Phi(f)$ of the following form:[3]

$$\Phi(f) = \left[\int T(p_1, p_2) f(p_1, p_2)\, dp_1\, dp_2 \right] \Omega, \qquad f \in S(\mathbf{R}^{2(d+1)}),$$

where $T(p_1, p_2)$ denotes the Fourier transform of the two-field "chronological product," namely (formally)

$$T(p_1, p_2) = \int e^{i(p_1 \cdot x_1 + p_2 \cdot x_2)}$$

$$\times [\theta(x_1^{(0)} - x_2^{(0)}) A(x_1) A(x_2) + \theta(x_2^{(0)} - x_1^{(0)}) A(x_2) A(x_1)]\, dx_1\, dx_2,$$

(with $p \cdot x = p^{(0)} x^{(0)} - \mathbf{p} \cdot \mathbf{x}$, and $\theta(x^{(0)}) = 1$ for $x^{(0)} \geq 0$ and 0 for $x^{(0)} < 0$).

---

[3] Here, one adopts the extension of the Gårding–Wightman axiomatic framework [2], performed in [13, 14].

Let $H_1^+ = \{p = (p^{(0)}, \mathbf{p}); p^{(0)} = \omega(\mathbf{p}) \equiv (\mathbf{p}^2 + 1)^{1/2}\}$ and $M = H_1^+ \times H_1^+$. With every test-function f with support in a (suitable) neighbourhood of M, one associates

(i)　The "Hepp family" of states $\{\Phi(f^{(t)}); t \in \mathbf{R}\}$, where $f^{(t)}(p_1, p_2) = f(p_1, p_2) e^{i[(p_1^0 - \omega_1)t + (p_2^0 - \omega_2)t]}$ (with $\omega_i = \omega(\mathbf{p}_i)$).

(ii)　The function $\hat{f} = (\text{sym} \cdot f)|_M$, where $(\text{sym} \cdot f)(p_1, p_2) = (1/\sqrt{2})[f(p_1, p_2) + f(p_2, p_1)]$.

The sets of two-particle asymptotic states $|\hat{f}, \text{ex}\rangle$ are then introduced (under the additional postulate (Sp)) as the Haag–Ruelle limits

$$|\hat{f}, \text{in}\rangle = \lim_{t \to -\infty} \Phi(f^{(t)}), \qquad |\hat{f}, \text{out}\rangle = \lim_{t \to +\infty} \Phi(f^{(t)}). \qquad (27)$$

For our purpose, we only consider here a closed subspace $\mathcal{H}^{(2)}$ of $\mathcal{H}$ spanned by the two-field states $\Phi(f)$ such that $\text{supp} f \subset \Sigma_\varepsilon^{(2)}$, where $\Sigma_\varepsilon^{(2)} = \{(p_1, p_2); p_1 + p_2 \in \sigma_\varepsilon^{(2)}\}$, $\varepsilon \geq 0$ and $\sigma_\varepsilon^{(2)} = \{p \in \mathbf{R}^{d+1}; 2^2 \leq p^{(0)^2} - \mathbf{p}^2 \langle (3 - \varepsilon)^2, p^{(0)} > 0\}$.

We similarly define the closed subspaces $\mathcal{H}_{\text{ex}}^{(2)}$ (ex = in or out) of $\mathcal{H}^{(2)}$ spanned by the states $|\hat{f}, \text{ex}\rangle$ such that $\text{supp} \cdot \hat{f} \subset \Sigma_\varepsilon^{(2)} \cap M$. Since the definition of $\mathcal{H}^{(2)}$ excludes scattering states of more than two particles (for kinematical reasons), $\mathcal{H}^{(2)}$ is comparable with the space $\mathcal{H}$ of positive energy states of the W.M. framework, considered in Section 3. However, as analyzed below, it may happen that $\mathcal{H}_{\text{ex}}^{(2)} \neq \mathcal{H}^{(2)}$; we shall denote by π_{ex} the orthogonal projection from $\mathcal{H}^{(2)}$ onto $\mathcal{H}_{\text{ex}}^{(2)}$ and say that $\mathcal{H}^{(2)}$ satisfies *asymptotic completeness* if $\mathcal{H}_{\text{ex}}^{(2)} = \mathcal{H}^{(2)}$.

The theory of "reduction formulas" [5, 6] implies the following.

Proposition 3. *For every state $\Phi(f)$ in $\mathcal{H}^{(2)}$, the corresponding scattering state $|\hat{f}_{\text{in}}, \text{in}\rangle = \pi_{\text{in}}[\Phi(f)]$ is defined by the following wave packet:*

$$\hat{f}_{\text{in}}(p_1, p_2) = \hat{f}(p_1, p_2) + \int [(p_1^2 - 1)(p_2^2 - 1)\tau_-(p_1, p_2; p_1', p_2')]|_{(p_1, p_2) \in M}$$

$$\times f(p_1', p_2') \, dp_1' \, dp_2', \qquad (28)$$

where $\tau_-(p_1, p_2; p_1', p_2') = \langle \Omega, \bar{T}(p_1, p_2; p_1', p_2')\Omega \rangle^{tr}$, $\bar{T}$ denoting the Fourier transform of the four-field antichronological product, and the superscript tr denoting the "truncation operation" (whose effect is to exclude from $\langle \Omega, \bar{T}\Omega \rangle$ terms involving "partial energy-momentum conservation δ-functions").

A similar result holds for the state $|\hat{f}_{\text{out}}, \text{out}\rangle = \pi_{\text{out}}[\Phi(f)]$, in which τ_- is replaced by $\tau_+ = \langle \Omega, T\Omega \rangle^{tr}$, T being the Fourier transform of the four-field chronological product.

On the other hand, it follows from the (locality and spectrum) axioms of Q.F.T., that one can introduce the analytic amputated four-point Green function of the field $F(p, p'; P)$ (with the kinematical notations of Section 2) in a certain complex domain D_4, and that the boundary values F_+, F_- of F on the real set $\{(p, p', P); P \in \sigma_0^{(2)}\}$ from the respective subsets $\{\text{Im } P^{(0)} > 0\}$ and $\{\text{Im } P^{(0)} < 0\}$ of D_4 are related to the distributions τ_+, τ_- by the following formulas:

$$\tau_+ = (s^{-1})_+ F_+(s'^{-1})_+ \delta(p_1 + p_2 - p_1' - p_2'),$$
$$\tau_- = -(s^{-1})_- F_-(s'^{-1})_- \delta(p_1 + p_2 - p_1' - p_2'), \qquad (29)$$

where $s = s_+ \cdot s_-$, $s' = s_+' \cdot s_-'$, $s_\pm = p_i^2 - 1$, $s_\pm' = p_i'^2 - 1$, $i = 1, 2$.

By taking equation (29) into account, we can then rewrite equation (28) as follows (with the convention that $f(p; P) = f(p_1, p_2)$, etc.):

$$\hat{f}_{\text{in}}(p; P) = \hat{f}(p, P) - \int F_-(p, p'; P)\Big|_{(p, P) \in M} (s'^{-1})_- f(p', P)\, dp'. \quad (30)$$

Apparently, equation (30) presents some analogy with equation (23), which expresses, in the W.M. framework, the incoming wave-packet of a given state in terms of a test-function representative f of this state. However, a complete similarity with the formalism of Section 3.2 can only be obtained for a class of two-field states $\Phi^{(f)}$ associated in an appropriate way with the space of holomorphic test-functions $\mathscr{A}$ described in Section 2.

4.2. The Representation of "Barycentric Two-Field States" in the Space $\mathscr{A}$

In order to use the simple kinematics of the barycentric Lorentz frame, described in Section 2, we consider the direct integral decomposition $\mathscr{H}^{(2)} = \int_{u \in H_1^+}^{\oplus} \mathscr{H}_u^{(2)}$ associated with the energy-momentum spectral region $\sigma_\varepsilon^{(2)}$, the latter being parametrized by a union of time-like rays of the form $\{P = 2(E+1)^{1/2}u; E \in \delta\}$ with u varying in H_1^+. By using the Lorentz covariance of the theory (implying in particular the smoothness of appropriate field-product matrix elements with respect to the coordinates of $P = p_1 + p_2$, on each Lorentz orbit $P^2 = 4(E+1)$), one can legitimately replace the test-functions $f(p, P)$ of Section 4.1 by test-functions $f(p, E)$ and define corresponding "barycentric states" $\Phi_{u_0}(f)$ in a given space $\mathscr{H}_{u_0}^{(2)}$.

4.2.1. The Mapping β

We shall consider a fixed Lorentz-frame with time-axis along u_0 ($u_0 \in H_1^+$), and shall construct a dense set of barycentric states $\Phi^{(f)} = \beta(f)$, associated with the test-functions $f(p, E)$ that belong to the space $\mathscr{A}$ described in Section 2.

The integration cycles $\Gamma_+(E)$, $\Gamma_-(E)$, respectively involved in the definition of the operations $\circ_+$ and $\circ_-$ (see Section 2), can be chosen as follows (inside D) [9]: $\Gamma_\pm = \Gamma_0^\pm + \Gamma_1^\pm + \Gamma_2^\pm$, where $\Gamma_i^\pm = \mathbf{R}^d \times \gamma_i^\pm$ for $i = 1, 2, \pm\Gamma_0^\pm$ being obtained by "$\pm i\varepsilon$-distortions" from the chain $\Gamma_0 = \mathbf{R}^d \times \gamma_0$ around the manifold $S = S_+ \cup S_-$; $\gamma_0, \gamma_1^\pm, \gamma_2^\pm$ are paths in the p_0-plane represented on Fig. 3 (the sections of S_+, S_- have been indicated there by $\sigma_+ = \sigma_+(\mathbf{p}, E)$, $\sigma_- = \sigma_-(\mathbf{p}, E)$).

By using the locality and spectrum axioms, one first shows that there exists a vector-valued measure, denoted by $\mathscr{R}(p, E)\Omega$, analytic with respect to p in $(D \times \delta)\backslash S$ and taking its values in $\mathscr{H}_{u_0}^{(2)}$, such that

(i) For $\operatorname{Im} p^{(0)} > |\operatorname{Im} \mathbf{p}|$ (resp. $\operatorname{Im} p^{(0)} < -|\operatorname{Im} \mathbf{p}|$), $\mathscr{R}(p, E)\Omega$ coincides with the Fourier–Laplace transform $R_+(p, E)\Omega$ (resp. $R_-(p, E)\Omega$) of the "retarded" (resp. "advanced") vector-distribution $\theta(x_1^{(0)} - x_2^{(0)})[A(x_1), A(x_2)]\Omega$ (resp. $\theta(x_2^{(0)} - x_1^{(0)})[A(x_2), A(x_1)]\Omega$).

(ii) In the complement of the latter set, $\mathscr{R}(p, E)\Omega$ is defined through the edge-of-the-wedge theorem (in view of the relation b.v.$R_+\Omega = $ b.v.$R_-\Omega$, valid on the set $\{p; \operatorname{Im} p = 0; (p, E) \notin S\}$).

One can then prove

Proposition 4. (i) *For every function f in $\mathscr{A}$, there exists a state $\Phi^{(f)}$ in $\mathscr{H}_{u_0}^{(2)}$, defined by either one of the following expressions:*

$$\Phi^{(f)} = \int_\delta dE \int_{\Gamma_+(E)\backslash\Gamma_0^+} f\mu(p, E)\mathscr{R}(p, E)\Omega + \int_\delta dE \int_{\Gamma_0} f\mu(p, E)T(p, E)\Omega$$

$$= \int_\delta dE \int_{\Gamma_-(E)\backslash\Gamma_0^-} f\mu(p, E)\mathscr{R}(p, E)\Omega + \int_\delta dE \int_{\Gamma_0} f\mu(p, E)\bar{T}(p, E)\Omega,$$

$$(31)$$

(where $T(p, E) \equiv T(p_1, p_2)$, $\bar{T}(p, E) \equiv \bar{T}(p_1, p_2)$ in the Lorentz frame considered).

(ii) *The Hilbertian norm $\|\Phi^{(f)}\|$ of $\Phi^{(f)}$ in $\mathscr{H}_{u_0}^{(2)}$ admits the following expression in terms of f:*

$$\|\Phi^{(f)}\|^2 = |\hat{f}|^2 - \frac{1}{4\pi^2} \int_\delta dE(\bar{f} \circ_+ F_+ \circ_+ f - \bar{f} \circ_- F_- \circ_- f),\qquad(32)$$

where $|\hat{f}|$ denotes the L^2-norm of $\hat{f}(\mathbf{p}) = \sqrt{2}^{-1}[f(p, E) + f(-p, E)]|_M$ (with $M = \{(p, E); p^{(0)} = 0, E = \mathbf{p}^2\}$).

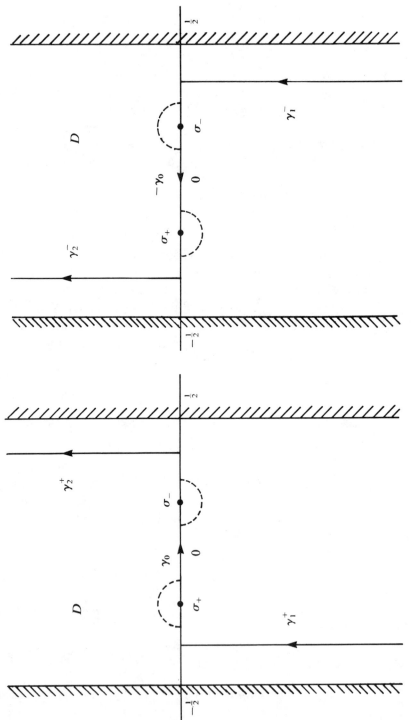

FIGURE 3

The equality of the two expressions of $\Phi^{(f)}$ in equation (31) follows from the algebraic identity (obvious for field products in x-space): $T - \bar{T} - R_+ - R_- = 0$, the latter being used on Γ_0 after taking into account the homological relations: $\Gamma_2^+ - \Gamma_2^- = -\Gamma_0 + i\varepsilon u_0$ and $\Gamma_1^+ - \Gamma_1^- = -\Gamma_0 - i\varepsilon u_0$. The formula (32) is obtained by a computation based on the algebra of four-point field-operators.

Remark. According to formula (32), which is to be compared with formula (21), the states $\Phi^{(f)}$ in Proposition 4 are the exact analogues of the W.M. states $\Phi^{(f)}$ considered in Section 3.2.

4.3. Scattering Theory for the Barycentric States $\Phi^{(f)}$

Let π_{ex} be the orthogonal projection from $\mathcal{H}_{u_0}^{(2)}$ onto the corresponding subspace of (Haag–Ruelle) asymptotic states $\mathcal{H}_{u_0,ex}^{(2)}$ (ex = in, out), and let i_{ex} be the canonical isomorphism from $L^2(\mathbf{R}_{(\mathbf{p})}^d)$ onto $\mathcal{H}_{u_0,ex}^{(2)}$ ($|\hat{f}, ex\rangle = i_{ex}(\hat{f})$). In order to implement the diagram of Fig. 2 (where the subscript u_0 has been omitted for simplicity), it remains to define the mappings τ_{ex} from $\mathcal{A}$ into $\mathcal{A}'_{fr} \approx L^2(\mathbf{R}_{(\mathbf{p})}^d)$.

By performing a straightforward extension of the reduction formulas to the case of states of the form $\int_\delta dE \int_{\Gamma_i^\pm} f(p, E) \mathcal{R}(p, E) \Omega \times \mu(p; E)$ and using equation (31), one obtains the following substitute to Proposition 3 (and to equation (28)).

Proposition 5. *For every state $\Phi^{(f)} = \beta(f)$ in $\mathcal{H}_{u_0}^{(2)}$, the corresponding scattering states $|\hat{f}_{ex}, ex\rangle = \pi_{ex}[\Phi^{(f)}]$ are characterized by the following wavefunctions*:

$$\tau_{ex}(f) = \Psi_{ex}^{(f)} = f_{ex}\delta(s), \qquad \text{where } f_{in} = f + F_- \circ_- f, \quad \text{and} \quad f_{out} = f + F_+ \circ_+ f. \tag{33}$$

We shall now distinguish two cases (as in the W.M. case, we discard here the possible occurrence of two-field bound states embedded in the continuum, corresponding to poles in δ, as it has been assumed in Section 2).

4.3.1. "Normal Analyticity" of B and Asymptotic Completeness

In this paragraph, we consider the case when $B(p, p'; E)$ is analytic in $D \times D \times \Delta$ and show its complete similarity with the W.M. case.

Let $\mathscr{B} = \{f \in \mathscr{A}; f = \sigma \cdot \chi, \chi \in \mathscr{A}\}$, where the operator $\sigma^{\cdot} = (s - B) \cdot$ is defined as follows:

$$[\sigma \cdot \chi](p, E) = s_+(p, E)s_-(p, E)\chi(p, E) - \int_{\dot{\Gamma}} B(p, p'; E)\chi(p', E)\mu(p; E).$$

$\dot{\Gamma}$ denotes the common homology class of $\Gamma_+(E)$ and $\Gamma_-(E)$ in D.

By using formula (32) and taking equations (6) into account, one proves

Proposition 6. *For every f in $\mathscr{B}$, $\beta(f) = 0$.*

One can then define the mapping $\dot{\beta}$ from $\dot{\mathscr{A}} = \mathscr{A}/\mathscr{B}$ in $\mathscr{H}_{u_0}^{(2)}$ such that $\Phi^{(\dot{f})} = \dot{\beta}(\dot{f}) = \beta(f)$ (f being any representative of the class $\dot{f}$ in $\mathscr{A}$). Moreover, one checks by using equation (6) that if $f^{(1)}$ and $f^{(2)}$ belong to the same class $\dot{f}$, the corresponding functions $f_{\mathrm{ex}}^{(1)}, f_{\mathrm{ex}}^{(2)}$ given by equation (33) are such that: $f_{\mathrm{ex}}^{(1)}\delta(s) = f_{\mathrm{ex}}^{(2)}\delta(s)$ and thus define the same element of $\mathscr{A}_{\mathrm{fr}}'$ which we denote by $\Psi_{\mathrm{ex}}^{(\dot{f})} = \dot{\tau}_{\mathrm{ex}}(\dot{f})$ (or $\hat{f}_{\mathrm{ex}}$, in $L^2(\mathbf{R}_{(\mathbf{p})}^d)$). Moreover, by taking equations (5) and (7) into account (the latter being a consequence of the "normal analyticity" of B), one deduces from equation (33) that the norm of $\hat{f}_{\mathrm{ex}}$ (in $L^2(\mathbf{R}_{(\mathbf{p})}^d)$) coincides with the norm of $\Phi^{(\dot{f})}$ in $\mathscr{H}_{u_0}^{(2)}$, given by equation (32). Since (in view of equation (6)), equation (33) can be inverted by formulas similar to equation (22), one can state

Theorem 1. *Under the "normal analyticity" assumption for B, the following diagrams hold (with $\mathrm{ex} = \mathrm{in}$, or out), in which all the mappings represent Hilbertian isomorphisms and express the "asymptotic completeness property" of $\mathscr{H}_{u_0}^{(2)}$.*

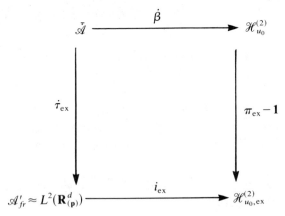

(We note that $\dot{\beta}$ and $\dot{\tau}_{\mathrm{ex}}$ have been extended, via a straightforward completion argument, to the closure $\bar{\dot{\mathscr{A}}}$ of $\dot{\mathscr{A}}$.)

Since $\dot{\tau}_{\mathrm{ex}}$ is the analogue of the mapping of Proposition 2, we are now led to implement the diagram of Fig. 1 exactly as in the W.M. case, by introducing a "field-theoretical wave-function" $\Psi^{(\dot{f})} = \dot{\alpha}(\dot{f})$, for every $\dot{f}$ in $\dot{\mathscr{A}}$. By definition, we put

$$\Psi^{(\dot{f})} = \Psi_+^{(f)} - \Psi_-^{(f)}, \tag{34}$$

where

$$\Psi_\pm^{(f)} = (2i\pi)^{-2}(s^{-1})_\pm (f + F_\pm \circ_\pm f), \qquad \text{for any } f \text{ in } \dot{f}. \tag{35}$$

The notation (34) has to be understood in the sense of analytic functionals on $\mathscr{A}$, namely: for every g in $\mathscr{A}$,

$$g \cdot \Psi^{(\dot{f})} = g \circ_+ [s\Psi_+^{(f)}] - g \circ_- [s\Psi_-^{(f)}]. \tag{35'}$$

(Note that for a function h in the class $\dot{h} = 0$, namely $h = \sigma \cdot \chi$, one has $\Psi_+^{(h)} = \Psi_-^{(h)} = \chi$, and therefore $\Psi^{(\dot{h})} = 0$.)

We then deduce from equations (34), (35) (and, concerning (ii), (33)) the following

Theorem 2. (i) *For every $\dot{f}$ in $\dot{\mathscr{A}}$, the corresponding "wave-function" $\Psi^{(\dot{f})} = \dot{\alpha}(\dot{f})$ satisfies the "Schrödinger- (or Bethe–Salpeter)-type equation"*

$$\sigma \cdot \Psi^{(\dot{f})} \equiv (s - B) \cdot \Psi^{(\dot{f})} = 0. \tag{36}$$

(ii) *the asymptotic wave-functions $\Psi_{\mathrm{ex}}^{(\dot{f})} = \dot{\tau}_{\mathrm{ex}}(\dot{f})$ can alternatively be expressed in terms of $\Psi^{(\dot{f})}$ by the following formulas (in the sense of analytic functionals on $\mathscr{A}$):*

$$\Psi_{\mathrm{in}}^{(\dot{f})} = \Psi^{(\dot{f})} - (s^{-1})_+ B \cdot \Psi^{(\dot{f})}, \qquad \Psi_{\mathrm{out}}^{(\dot{f})} = \Psi^{(\dot{f})} - (s^{-1})_- B \cdot \Psi^{(\dot{f})}, \tag{37}$$

whose inverses are respectively

$$\Psi^{(\dot{f})} = \Psi_{\mathrm{in}}^{(\dot{f})} + (s^{-1})_+ F_+ \cdot \Psi_{\mathrm{in}}^{(\dot{f})}, \qquad \Psi^{(\dot{f})} = \Psi_{\mathrm{out}}^{(\dot{f})} + (s^{-1})_- F_- \cdot \Psi_{\mathrm{out}}^{(\dot{f})}. \tag{38}$$

4.3.2. "Abnormal Analyticity" of B and Violation of Asymptotic Completeness

We now assume that B is only analytic in $D \times D \times (\Delta \backslash \delta)$, and give a Hilbertian interpretation of its discontinuity $\Delta B = B_+ - B_-$ on the cut δ, the equations (6) being now replaced by

$$F_+ = B_+ + F_+ \circ_+ B_+, \qquad F_- = B_- + F_- \circ_- B_-. \tag{39}$$

We can now define two operators $\sigma_+ \cdot = (s - B_+) \cdot$ and $\sigma_- \cdot = (s - B_-) \cdot$ and the corresponding subspaces $\mathscr{B}_+$, $\mathscr{B}_-$ of $\mathscr{A}$, namely: $\mathscr{B}_\pm = \{f \in \mathscr{A}; f = \sigma_\pm \cdot \chi,$ $\chi \in \mathscr{A}\}$. From equations (32) and (39), one then proves

Proposition 7. *For every $f = \sigma_+ \cdot \chi$ (resp. $\sigma_- \cdot \chi$) in $\mathscr{B}_+$ (resp. $\mathscr{B}_-$), the corresponding state $\Phi^{(f)} = \beta(f)$ is orthogonal to $\mathscr{H}^{(2)}_{u_0,\mathrm{out}}$ (resp. $\mathscr{H}^{(2)}_{u_0,\mathrm{in}}$). Moreover, one has $\|\Phi^{(f)}\| = \int_\delta dE(\chi, \Delta B \cdot \chi)$.*

It follows that for each class $\dot{f}$ in $\dot{\mathscr{A}}_+ = \mathscr{A}/\mathscr{B}_+$ (resp. $\dot{\mathscr{A}}_- = \mathscr{A}/\mathscr{B}_-$), all the states $\Phi^{(f)}$ with f in $\dot{f}$ have the same orthogonal projection $\pi_{\mathrm{out}}(\Phi^{(f)})$ (resp. $\pi_{\mathrm{in}}(\Phi^{(f)})$) on $\mathscr{H}^{(2)}_{u_0,\mathrm{out}}$ (resp. $\mathscr{H}^{(2)}_{u_0,\mathrm{in}}$). Similarly, the expressions (33) of the corresponding wave-packets f_{out} (resp. f_{in}) in terms of f are valid for all functions f in $\dot{f}$. Moreover, one can show

Proposition 8. *A state $\Phi^{(f)}$ belongs to $\mathscr{H}^{(2)}_{u_0,\mathrm{out}}$ (resp. $\mathscr{H}^{(2)}_{u_0,\mathrm{in}}$) if and only if the function f satisfies the following integral equation:*

$$\Delta B \circ_+ f + \Delta B \circ_+ F_+ \circ_+ f = 0 \qquad (resp.\ \Delta B \circ_- f + \Delta B \circ_- F_- \circ_- f = 0),$$

or equivalently $\Delta B \circ_+ f_{\mathrm{out}} = 0$ (resp. $\Delta B \circ_- f_{\mathrm{in}} = 0$), with f_{out} (resp. f_{in}) given in terms of f by equation (33), the latter being inverted through formulas similar to equation (22), and involving B_+ (resp. B_-) instead of B.

References

[1] M. Sato, Theory of hyperfunctions, *Journal. Fac. Sci. Univ. Tokyo* **8** (1959–1960) 139–193 and 387–437.

[2] R. F. Streater and A. S. Wightman; *PCT, Spin and Statistics and All That*. Benjamin, New York, 1964.

[3] H. J. Borchers, *Nuovo Cim.* **24** (1962) 214.

[4] R. Haag, *Phys. Rev.* **112** (1958) 669; D. Ruelle, *Helv. Phys. Acta* **35** (1962) 147.

[5] K. Hepp, in *Axiomatic Field Theory* (Chretien and Deser, eds.). Gordon and Breach, New York, 1965.

[6] H. Lehmann, K. Symanzik, and W. Zimmermann, *Nuovo Cimento* **1** (1955) 205; **6** (1957) 319.

[7] F. Pham, *Introduction à l'étude topologique des singularités de Landau, Mémorial des Sciences Mathématiques 164*, Paris, Gauthier-Villars, 1967.

[8] J. Leray, *Bull. Soc. Math. Fr.* **87** (1959) 81–180.

[9] J. Bros, in *Analytic Methods in Mathematical Physics* (Gilbert and Newton, eds.), Gordon and Breach, New York, 1970; J. Bros and M. Lassalle, in *Structural Analysis of Collision Amplitudes* (R. Balian and D. Iagolnitzer, eds.), Amsterdam, North-Holland, 1976.

[10] J. Bros and D. Pesenti, *J. Math. Pures Appl.* **58** (1980) 375 and **62** (1983) 215.

[11] J. Bros and D. Iagolnitzer, *Phys. Rev. D* **27** (1983) 811 and *Comm. Math. Phys.* **88** (1983) 235.

[12] See, e.g., R. G. Newton, *Scattering Theory of Waves and Particles*, McGraw-Hill, New York, 1966; H. M. Nussenzweig, *Causality and Dispersion Relations*, Academic Press, New York, 1972.

[13] O. Steinmann, *Comm. Math. Phys.* **10** (1968) 245.

[14] H. Epstein, V. Glaser, and R. Stora, in *Structural Analysis of Collision Amplitudes* (R. Balian and D. Iagolnitzer, eds.), Amsterdam, North-Holland, 1976.

[15] T. Spencer and F. Zirilli, *Comm. Math. Phys.* **49** (1976) 1.

[16] D. Iagolnitzer and J. Magnen, *Comm. Math. Phys.* **110** (1987) 51 and **111** (1987) 81.

[17] I wish to thank Professor K. Chadan for a useful discussion concerning bound states embedded in the continuum.

Two Remarks on Recent Developments in Solvable Models

Etsuro Date

Department of Mathematics
College of General Education
Kyoto University
Kyoto, Japan

Michio Jimbo, Tetsuji Miwa and Masato Okado

Research Institute for Mathematical Sciences
Kyoto University
Kyoto, Japan

1. Introduction

In the exact study of two-dimensional statistical mechanics and quantum field theory, two approaches are presently available: solvable lattice models (SLMs) [1] and conformal field theory (CFT) [2]. The ideas and methods in these approaches are quite independent of each other. On the one hand, the SLM deals with generically non-critical models on the lattice, and is built upon solutions to a set of algebraic equations called the star–triangle relation. The CFT on the other hand deals with critical and continuous systems. The main tool here is the symmetry under infinite-dimensional Lie algebras, notably the Virasoro algebra.

Algebraic Analysis, Volume I

Notwithstanding their conceptual difference these theories exhibit unexpected similarities. On the basis of the recent works on SLM [3–9], we point out in this article two similar structures that occur in different contexts: One is the representation of the braid group, and the other is the modular covariance. The true nature of these phenomena is still unknown. We believe that once understood properly it will shed new light on the theory of integrable systems.

2. Star-Triangle Relation and the Braid Group

In [4, 8, 9] we have constructed a family of solvable lattice models labeled by three positive integers n, l, N with $l > N$. Leaving aside its physical content, let us briefly explain the setting in order to make contact with the braid group.

Consider partitions $(f_0, \cdots, f_{n-1})$, $f_i \in \mathbf{Z}$, satisfying

$$f_0 \geq \cdots \geq f_{n-1} \geq 0, \qquad f_0 - f_{n-1} \leq l.$$

Denote by $P_+(n; l)$ the set of their equivalence classes under the relation $(f_0, \ldots, f_{n-1}) \sim (f_0 + 1, \ldots, f_{n-1} + 1)$. Let Λ_μ $(0 \leq \mu \leq n - 1)$ denote the fundamental weights of the affine Lie algebra $A_{n-1}^{(1)}$. An element $a = [f_0, \ldots, f_{n-1}]$ of $P_+(n; l)$ is identified with a level l dominant integral weight $\sum_{\mu=0}^{n-1} (f_{\mu-1} - f_\mu)\Lambda_\mu$ $(f_{-1} = l - f_0 + f_{n-1})$. If $a, b \in P_+(n; l)$ are related by $a = [f_0, \ldots, f_{n-1}]$ and $b = [f_0, \ldots, f_\mu + 1, \ldots, f_{n-1}]$ with some $\mu = 0, 1, \ldots, n - 1$, we write $b = a + \hat\mu$. An ordered pair (a, b) is said to be N-admissible if there exist $\xi \in P_+(n; l - N)$ and $\eta \in P_+(n; N)$ such that

$$a = \xi + \eta, \qquad b = \xi + \sigma(\eta), \tag{1}$$

where σ denotes the cyclic diagram automorphism $\sigma(\Lambda_\mu) = \Lambda_{\mu+1}$. For $N = 1$ (1) simply means $b = a + \hat\mu$ with some μ. Now let $\mathscr{P}_m$ be the set of "paths" consisting of N-admissible pairs

$$\mathbf{a} = (a_1, \ldots, a_m),$$

where (a_i, a_{i+1}) is N-admissible for $1 \leq i \leq m - 1$. Let $V_m = \oplus \mathbf{C} v_\mathbf{a}$ be the span of orthonormal vectors $v_\mathbf{a}$ indexed by $\mathbf{a} \in \mathscr{P}_m$. By a local face operator we mean a matrix U_i $(2 \leq i \leq m - 1)$ that acts on V_m in the following manner:

$$U_i v_\mathbf{a} = \sum_{\mathbf{a}'} W\begin{pmatrix} a_{i-1} & a_i \\ a_i' & a_{i+1} \end{pmatrix} v_{\mathbf{a}'}.$$

Here **a'** runs over the paths that differ from **a** only at the ith component a_i. The scalar coefficients $W(\begin{smallmatrix} a & b \\ d & c \end{smallmatrix})$ represent the Boltzmann weights in the language of statistical mechanics. Clearly $U_i U_j = U_j U_i$ if $|i-j| \geq 2$.

The algebraic content of the construction of SLM amounts to the following result. Suppose U_i depends on an extra variable $u \in \mathbf{C}$.

Theorem [4, 8, 9]. *For each n, l, N there exist face operators $U_i(u)$ acting on V_m such that the star-triangle relation (STR) is valid:*

$$U_i(u) U_{i+1}(u+v) U_i(v) = U_{i+1}(v) U_i(u+v) U_{i+1}(u), \qquad u, v \in \mathbf{C}. \quad (2)$$

The STR guarantees the commutativity of the transfer matrix for different values of u [1]. Our solution is parametrized by an elliptic theta function,

$$\theta_1(u, p) = 2|p|^{1/8} \sin u \prod_{k=1}^{\infty} (1 - 2p^k \cos(2u) + p^{2k})(1 - p^k).$$

For instance in the case $N = 1$, the $U_i(u)$ are given explicitly by

$$W\left(\begin{array}{cc} a & a+\hat{\mu} \\ a+\hat{\mu} & a+2\hat{\mu} \end{array} \middle| u\right) = \frac{[1+u]}{[1]},$$

$$W\left(\begin{array}{cc} a & a+\hat{\mu} \\ a+\hat{\mu} & a+\hat{\mu}+\hat{v} \end{array} \middle| u\right) = \frac{[a_{\mu v} - u]}{[a_{\mu v}]} \quad (\mu \neq v), \qquad (3)$$

$$W\left(\begin{array}{cc} a & a+\hat{v} \\ a+\hat{\mu} & a+\hat{\mu}+\hat{v} \end{array} \middle| u\right) = \frac{[u]}{[1]} \frac{[a_{\mu v}+1]}{[a_{\mu v}]} \quad (\mu \neq v),$$

where $[u] = \theta_1(\pi u/L, p)$, $L = l + n$, and $a_{\mu v} = f_{\mu} - f_{v} + v - \mu$ for $a = [f_0, \ldots, f_{n-1}]$. The models with $n = 2$, $N = 1$ have been introduced and studied in detail by Andrews–Baxter–Forrester [10] under the name "eight-vertex SOS models."

The models become critical at $p = 0$. The local face operators U_i then reduce to polynomials in the variable $x = e^{2\pi i u/L}$,

$$\text{const. } U_i(u) = g_i x^N + g_i' x^{N-1} + \cdots + g_i'', \qquad (4)$$

and the STR (2) together with the commutativity $U_i U_j = U_j U_i$ ($|i-j| \geq 2$) imply that their leading coefficients g_i give a representation of the braid group on V_m:

$$g_i g_{i+1} g_i = g_{i+1} g_i g_{i+1}, \qquad g_i g_j = g_j g_i \qquad \text{if } |i-j| \geq 2.$$

In the case $N = 1$ we have an additional relation,

$$(g_i - 1)(g_i + q) = 0, \qquad q = e^{2\pi i/L},$$

which means that the representation factors through the Iwahori-Hecke algebra. In fact, when restricted to the subspace $V_m^\phi =$ the linear span of $\{v_a | a_1 = [0, 0, \ldots, 0]\}$, formulas (3), (4) with $[u]$ replaced by $\sin(\pi u / L)$ reproduce the irreducible representation studied by Wenzl [11]. (For $n = 2$ this has been noted in [12].)

A very similar structure has been encountered by Tsuchiya-Kanie [13] in their study of the CFT with the gauge symmetry with respect to $A_1^{(1)}$. Let V_j (resp. $\mathcal{H}_j$) denote the irreducible A_1 module (resp. $A_1^{(1)}$ module) of spin $j = 0, 1/2, 1, \ldots, l/2$, i.e. the one with highest weight $2j\Lambda_1$ (resp. $(l - 2j)\Lambda_0 + 2j\Lambda_1$). Put $\mathcal{H} = \bigoplus_{0 \le j \le l/2} \mathcal{H}_j$. It is known that on $\mathcal{H}$ there is a natural action of the Virasoro algebra $\mathcal{Vir} = (\bigoplus_{n \in \mathbf{Z}} \mathbf{C} L_n) \oplus \mathbf{C} c$:

$$[L_m, L_n] = (m - n) L_{m+n} + \frac{c}{12} m(m^2 - 1) \delta_{m+n,0}, \qquad [L_m, c] = 0. \qquad (5)$$

Given a triple $\mathbf{v} = \binom{\;\;j\;\;}{j_2\;\;j_1}$ we consider an operator $\Phi_{\mathbf{v}}(u; z)$ on $\mathcal{H}$, linearly parametrized by $u \in V_j$ which satisfies $\pi_{j_2} \Phi_{\mathbf{v}}(u; z) \pi_{j_1} = \Phi_{\mathbf{v}}(u; z)$, where π_j signifies the projection $\mathcal{H} \to \mathcal{H}_j$. It is called a vertex operator if

$$[X \otimes t^m, \Phi_{\mathbf{v}}(u; z)] = z^m \Phi_{\mathbf{v}}(Xu; z), \qquad X \otimes t^m \in A_1^{(1)},$$

$$[L_m, \Phi_{\mathbf{v}}(u; z)] = z^m \left(z \frac{d}{dz} + (m + 1)\Delta_j \right) \Phi_{\mathbf{v}}(u; z), \qquad \Delta_j = \frac{j(j+1)}{l+2}.$$

In the above we identified $A_1^{(1)}$ with $\mathfrak{sl}(2, \mathbf{C}) \otimes \mathbf{C}[t, t^{-1}] \oplus \mathbf{C} l$. A nontrivial vertex operator exists if and only if [13]

$$|j_1 - j_2| \le j \le j_1 + j_2, \qquad j \equiv j_1 + j_2 \bmod \mathbf{Z} \qquad \text{and} \qquad j_1 + j_2 + j \le l. \quad (6)$$

In our terminology this condition states that $b = [2j_1, 0]$ and $c = [2j_2, 0] \in P_+(2; l)$ are $2j$-admissible. Tsuchiya–Kanie derived differential equations and supplementary algebraic relations for the correlation functions $\langle \Phi_{\mathbf{v}_m}(u_m; z_m) \cdots \Phi_{\mathbf{v}_1}(u_1; z_1) \rangle$, and showed that those equations admit a basis of solutions (in fact the correlations themselves) indexed by the set $\{\mathbf{p} = (p_m, \ldots, p_0) | p_i \in \frac{1}{2}\mathbf{Z},\ 0 \le p_i \le l/2,\ p_m = p_0 = 0$ and $\mathbf{v}_i = \binom{\;\;j_i\;\;}{p_i\;\;p_{i-1}}$ satisfies (6)$\}$. When $j_1 = \cdots = j_m = N/2$, the corresponding monodromy representation realizes the braid relation. They verified that in the case $N = 1$ there results Wenzl's representation of the Iwahori-Hecke algebra for $q = e^{2\pi i/(l+2)}$. A similar result has been obtained also by Kohno [14] when q is not a root of unity.

It is noteworthy that the same representation appears in totally different contexts.

3.   Modular Covariant Characters

Our second remark is on different roles of modular covariance in solvable models.

Kac–Peterson [15] established the close connection between the characters of affine Lie algebras, theta functions and modular forms. Of significance here is an object called branching coefficients. Let $(\mathfrak{g}, \mathfrak{h})$ be a pair consisting of an affine Lie algebra and its subalgebra. Consider the irreducible decomposition of a highest weight $\mathfrak{g}$-module $L^{\mathfrak{g}}(\Lambda)$ with respect to $\mathfrak{h}$:

$$L^{\mathfrak{g}}(\Lambda) = \sum \Omega_{\Lambda\Lambda'} \otimes L^{\mathfrak{h}}(\Lambda').$$

Rewriting it as an identity of characters one gets

$$\chi_{\Lambda}^{\mathfrak{g}} = \sum_{\Lambda'} b_{\Lambda\Lambda'}(q)\chi_{\Lambda'}^{\mathfrak{h}},$$

which defines the branching coefficients $b_{\Lambda\Lambda'}(q)$. Goddard–Kent–Olive [16] provided a method to construct a representation of the Virasoro algebra (5) on the space $\Omega_{\Lambda\Lambda'}$. The $b_{\Lambda\Lambda'}(q)$ can be regarded as its characters. Thanks to the transformation formula for the characters $\chi_{\Lambda}^{\mathfrak{g}}$, $\chi_{\Lambda'}^{\mathfrak{h}}$, the branching coefficients also enjoy automorphic properties.

In SLM the branching coefficients appear as the local state probabilities (LSPs). By an LSP we mean the probability that the local fluctuation variable a_1 on a lattice site 1 takes a given state a: $P_a = \text{prob}(a_1 = a)$.

Consider the models in Section 2 in the region $0 < p < 1$, $-n/2 < u < 0$, and put $p = e^{2\pi i \tau}$, $q = e^{-2\pi i n/L\tau}$. For $a \in P_+(n; l)$ we denote by $L(a)$ the irreducible $A_{n-1}^{(1)}$ module with highest weight a. Let $(\mathfrak{g}, \mathfrak{h}) = (A_{n-1}^{(1)} \oplus A_{n-1}^{(1)}, A_{n-1}^{(1)})$, and denote by $b_{\xi\eta a}(q)$ the branching coefficients associated with the decomposition of the tensor product $L(\xi) \otimes L(\eta)$, $\xi \in P_+(n; l - N)$, $\eta \in P_+(n; N)$, with respect to the diagonal subalgebra $A_{n-1}^{(1)}$.

Theorem [5–8].   *Suppose $n = 2$ or $N = 1$. Then the LSP is given by*

$$P_a = b_{\xi\eta a}(q)\chi_a^{\text{sp}}/\chi_{\xi}^{\text{sp}}\chi_{\eta}^{\text{sp}}, \tag{7}$$

where χ_a^{sp} signifies the principally specialized character of $L(a)$, and ξ, η are related to the choice of the boundary condition.

For a more precise statement, see [7, 8].

In the evaluation of the LSP the crucial tool is Baxter's corner transfer matrix (CTM) defined as follows. Divide the lattice into four quadrants A, B, C, D.

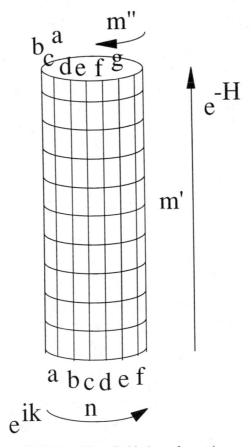

FIGURE 1   The cylindrical transfer matrix.

Choose the states on the two axes $\mathbf{a} = (\lambda_1, \ldots, \lambda_m)$ and $\mathbf{a}' = (\mu_1, \ldots, \mu_m)$. We denote by $\mathscr{A}_{\mathbf{a}\mathbf{a}'}$ the partition function for the quadrant A with this boundary condition, the sum being taken over the internal variables (open circles in Fig. 1). The CTM $\mathscr{A}$ is the matrix with entries $\mathscr{A}_{\mathbf{a}\mathbf{a}'}$. Other CTMs $\mathscr{B}, \mathscr{C}, \mathscr{D}$ are defined similarly. They are block diagonal in the sense that $\mathscr{A}_{\mathbf{a}\mathbf{a}'} = 0$ if $\lambda_1 \neq \mu_1$. Let e^{-E_j} $(j = 0, 1, \ldots)$ be the eigenvalues of $\mathscr{A}$ in a fixed block $\lambda_1 = \mu_1 = a$. Baxter found in the solvable cases the following behavior of E_j as $m \to \infty$:

$$E_j - E_0 \sim \text{const. } N_j u, \qquad N_j: \text{integer}, \qquad (8)$$

where u is the "spectral parameter" entering in the STR (2). The heart of the result (7) is that (up to an overall shift) the N_j constitute the spectrum

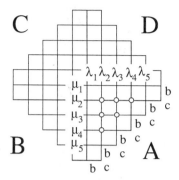

FIGURE 2   The corner transfer matrix.

of L_0 in the space $\Omega_{\Lambda\Lambda'}$. The automorphic properties of the $b_{\xi\eta a}(q)$ and of the specialized characters afford us the complete knowledge about the behavior of the LSP in the critical limit $p \to 0$, $q \to 1$.

Another modular invariance shows up when one considers lattice systems at criticality. Consider a lattice $\mathscr{C}_n$ wound on a cylinder by the periodic boundary condition $(i_1 + n, i_2) = (i_1, i_2)$ (Fig. 2).

Let e^{-H} be the transfer matrix in the vertical direction of a *critical* statistical model on $\mathscr{C}_n$. We denote by V_n the space on which H acts, and denote by e^{ik} the horizontal shift operator on V_n. Let E_j $(j = 0, 1, \dots)$ be eigenvalues of H. The criticality of the model implies the following scaling behaviors of the lowest eigenvalue E_0 and of low-lying excitations (*cf.* (8)) as the width of the cylinder n becomes large:

$$E_0 \sim fn - \frac{\pi c}{6n},$$

$$E_j - E_0 \sim \frac{2\pi x_j}{n}, \qquad x_j: \text{integer}.$$

(9)

On the basis of the conformal invariance of critical systems, Cardy [17] related the constants c, x_j to the representation of the Virasoro algebra. Let V be the space, considered in the limit of $n \to \infty$, spanned by the vectors corresponding to eigenvalues satisfying (9). Cardy's assertion is the following: There exist *two* Virasoro operators $\{L_n\}$ and $\{\bar{L}_n\}$, mutually commutative and acting on V with the common value of the central charge c. The operators L_0 and $\bar{L}_0$ are related to H and k by

$$H - E_0 \sim \frac{2\pi(L_0 + \bar{L}_0)}{n}, \qquad k \sim \frac{2\pi(L_0 - \bar{L}_0)}{n}.$$

(10)

Next consider the lattice $\mathcal{T}_{mn}$ on the torus obtained by imposing further the twisted boundary condition $(i_1, i_2 + m') = (i_1 + m'', i_2)$ in the vertical direction $(m = m' + \sqrt{-1} m'')$. We introduce $\tau = \sqrt{-1} m/n$ as the moduli parameter of the torus. The finite part $Z(\tau)$ of the partition function is defined by the following in the limit $m, n \to \infty$ with τ fixed:

$$Z(\tau) = \lim \frac{Z_{mn}}{e^{-fm'n}} = \lim \text{trace}_{V_n}(e^{-m'(H-fn)} e^{-im''k}),$$

where Z_{mn} is the partition function on $\mathcal{T}_{mn}$. By (10) it is written as a sesqui-linear form of Virasoro characters $\psi_{c,h}(q)$:

$$Z(\tau) = \text{trace}_V(q^{L_0 - c/24} q^{* \bar{L}_0 - c/24})$$

$$= \sum_{(h, \bar{h})} N_{h, \bar{h}} \psi_{c,h}(q) \psi_{c, \bar{h}}(q^*),$$

where $q = e^{2\pi i \tau}$, q^* is the complex conjugate, and $N_{h, \bar{h}}$ are nonnegative integers.

By the definition it is clear that $Z(\tau + 1) = Z(\tau)$. In the non-twisted case $m'' = 0$ (i.e. when τ is purely imaginary), one must have $Z(-1/\tau) = Z(\tau)$ because of the invariance under the rotation of the lattice through 90 degrees. For the Ising model this relation has been verified directly by Ferdinand-Fisher [18] via an exact computation of $Z(\tau)$. If one accepts the invariance of $Z(\tau)$ under the full modular group $SL_2(\mathbf{Z})$, the possible choice of V is strongly restricted.

When V decomposes into a direct sum of a finite number of irreducible representations of $\mathcal{V}i\dot{\iota} \oplus \overline{\mathcal{V}i\dot{\iota}}$, the theory is called minimal. As an example, consider the eight-vertex SOS models (the case $n = 2$ and $N = 1$). In this case the branching coefficients $b_{\xi\eta a}(q)$ give irreducible unitary characters [19] of $\mathcal{V}i\dot{\iota}$ with the central charge

$$c = 1 - \frac{6}{L(L-1)} < 1.$$

Set $\xi = (L-2-s)\Lambda_0 + (s-1)\Lambda_1$, $a = (L-1-t)\Lambda_0 + (t-1)\Lambda_1$, $\eta = \Lambda_0$ or Λ_1 according as $s \equiv t$ or $t + 1$ mod 2. Put further

$$h = h_{st} = \frac{(Ls - (L-1)t)^2 - 1}{4L(L-1)}, \qquad 1 \le t \le s \le L-2.$$

We have then

$$b_{\xi\eta a}(q) = \psi_{c, h_{st}}(q)$$

$$= \eta(\tau)^{-1} \sum_{\varepsilon = \pm 1} \varepsilon \sum_{\gamma \in \mathbf{Z} + s/2(L-1) - \varepsilon t/2L} q^{L(L-1)\gamma^2},$$

where $\eta(\tau) = q^{1/24} \prod_{j=1}^{\infty} (1 - q^j)$, $q = e^{2\pi i \tau}$. It is argued [17] that the torus partition function is given by

$$Z(\tau) = \sum_{1 \le t \le s \le L-2} |\psi_{c,h_{st}}(q)|^2.$$

The branching coefficients used in the LSP result arise here as the constituents of $Z(\tau)$. The complete list of the torus partition functions of the minimal conformal theories is available [20, 21]. In the case of non-minimal theories the finite reducibility with respect to larger symmetry algebras plays a similar crucial role [22-26].

There are other examples where the coincidence is established of the LSP and the branching coefficients for appropriate choice of the pair $(\mathfrak{g}, \mathfrak{h})$ ([3], [6]). In these cases the corresponding torus partition functions are also supposed to be given in terms of the same set of branching coefficients. This coincidence is quite mysterious because the origins of the torus are totally different. In Cardy's theory the torus appears as the continuum limit of the finite lattice, while it appears as the one that describes the commuting family of transfer matrices of SLM. We expect that there is a unified way of understanding conformal field theory, solvable lattice models and the role of infinite-dimensional Lie algebras.

References

[1] R. J. Baxter, *Exactly Solved Models in Statistical Mechanics*, Academic, London, 1982.

[2] A. A. Belavin, A. M. Polyakov, and A. B. Zamolodchikov, Infinite conformal symmetry in two-dimensional quantum field theory, *Nucl. Phys.* **B241** (1984) 333-380.

[3] M. Jimbo, T. Miwa, and M. Okado, Solvable lattice models with broken Z_N symmetry and Hecke's indefinite modular forms, *Nucl. Phys.* **B275** [FS17] (1986) 517-545.

[4] E. Date, M. Jimbo, T. Miwa, and M. Okado, Fusion of the eight-vertex SOS models, *Lett. Math. Phys.* **12** (1986) 209-215.

[5] E. Date, M. Jimbo, T. Miwa, and M. Okado, Automorphic properties of local height probabilities for integrable solid-on-solid models, *Phys. Rev.* **B35** (1987) 2105-2107.

[6] E. Date, M. Jimbo, T. Miwa, and M. Okado, Exactly solvable SOS models: local height probabilities and theta function identities, *Nucl. Phys.* **B290** [FS20] (1987) 231-273.

[7] E. Date, M. Jimbo, A. Kuniba, T. Miwa, and M. Okado, Exactly solvable SOS models, II. Proof of the star-triangle relation and combinatorial identities, *Adv. Stud. Pure Math.* **16** (1988) 17-122.

[8] M. Jimbo, T. Miwa, and M. Okado, Solvable lattice models whose states are dominant integral weights of $A_{n-1}^{(1)}$, *Lett. Math. Phys.* **14** (1987) 123-131.

[9] M. Jimbo, T. Miwa, and M. Okado, An $A_{n-1}^{(1)}$ family of solvable lattice models, *Mod. Phys. Lett.* **B1** (1987) 73-79.

[10] G. E. Andrews, R. J. Baxter, and P. J. Forrester, Eight-vertex SOS model and generalized Rogers-Ramanujan-type identities, *J. Stat. Phys.* **35** (1984) 193-266.

[11] H. Wenzl, Representation of Hecke Algebras and Subfactors, Ph.D. Thesis, University of Pennsylvania (1985).

[12] A. Kuniba, Y. Akutsu, and M. Wadati, Virasoro algebra, von Neumann algebra and critical eight-vertex SOS models, *J. Phys. Soc. Jpn.* **55** (1986) 3285-3288.

[13] A. Tsuchiya and Y. Kanie, Vertex operators in conformal field theory on $\mathbf{P}^1$ and monodromy representations of braid groups, *Adv. Stud. Pure Math.* **16** (1988) 297-372.

[14] T. Kohno, Hecke algebra representations of braid groups and classical Yang-Baxter equations, *Adv. Stud. Pure Math.* **16** (1988) 255-269.

[15] V. G. Kac and D. H. Peterson, Infinite-dimensional Lie algebras, theta functions and modular forms, *Adv. Math.* **53** (1984) 125-264.

[16] P. Goddard, A. Kent, and D. Olive, Virasoro algebras and coset space models, *Phys. Lett.* **B152** (1985) 88-92.

[17] J. L. Cardy, Operator content of two-dimensional conformally invariant theories, *Nucl. Phys.* **B270** [FS16] (1986) 186-204.

[18] A. E. Ferdinand and M. E. Fisher, Bounded and inhomogeneous Ising Models I. Specific heat anomaly of a finite lattice, *Phys. Rev.* **185** (1969) 832-846.

[19] A. Rocha-Caridi, Vacuum vector representations of the Virasoro algebra, in *Vertex Operators in Mathematical Physics* (J. Lepowsky, S. Mandelstam, and I. M. Singer, eds.), MSRI publications **3**, Springer, New York, 1985.

[20] A. Cappelli, C. Itzykson, and J.-B. Zuber, Modular invariant partition functions in two dimensions, *Nucl. Phys.* **B280** [FS18] (1987) 445-465.

[21] D. Gepner and Z. Qiu, Modular invariant partition functions for parafermionic field theories, *Nucl. Phys.* **B285** [FS19] (1987) 423-453.

[22] M. A. Bershadsky, V. G. Knizhnik, and M. G. Teitelman, Superconformal symmetry in two dimensions, *Phys. Lett.* **151B** (1985) 31-36.

[23] D. Friedan, Z. Qiu, and S. Shenker, Superconformal invariance in two dimensions and the tricritical Ising model, *ibid.* 37-43.

[24] A. B. Zamolodchikov and V. A. Fateev, Nonlocal (parafermion) currents in two-dimensional conformal quantum field theory and self-dual critical points in $\mathbf{Z}_N$-symmetric statistical systems, *Sov. Phys. JETP* **62** (1985) 215-225.

[25] T. Hayashi, Sugawara Operators and Kac-Kazhdan Conjecture, Master's Thesis, Nagoya University (1987), to appear in *Inv. Math.*

[26] V. A. Fateev and S. L. Lykyanov, The models of two dimensional conformal quantum field theory with $\mathbf{Z}_n$-symmetry, *Int. J. Mod. Phys.* **A3** (1988) 507-520.

Hypergeometric Functions

Leon Ehrenpreis

Department of Mathematics
Temple University
Philadelphia, Pennsylvania

Preface

Mikio Sato and I grew up as mathematicians together. In the Summer of 1987 we were reunited at the AMS Summer Institute on theta functions. At this occasion I began to appreciate fully his recent work on infinite grassmannians and integrable systems. Sato's interplay of wide-ranging themes must be compared to a Mozart symphony.

It is my great pleasure to wish Mikio a Happy 60th Birthday.

1. Introduction

It has been said that one of the main goals of mathematics is to isolate and study functions which are both interesting in their own rights and whose influence pervades large areas of mathematics and physics. For the Greeks such special functions were the circular functions. Since the times of Bernoulli, Euler, and Legendre, theta functions and hypergeometric functions have been added to the list. These functions arise from the interplay of group

Algebraic Analysis, Volume I

theory and differential equations, and this may account for their ubiquity.

There are, of course, many other functions which have been greatly studied. Their origins are mainly differential or other functional equations. Perhaps the most prominent amongst these are Mathieu and Lamé functions. They are less understood than hypergeometric and theta functions.

In this paper we shall discuss hypergeometric functions from the analytic and group viewpoint. We shall see that there is a natural hierarchy whose first level is the circular functions, whose second level is the hypergeometric functions, whose third level is the Mathieu and Lamé functions, and whose higher levels are, as yet, virgin territory. In another article we shall discuss theta functions. We shall not discuss basic hypergeometric functions (the q-analog of hypergeometric functions).

The classical hypergeometric function $F(a, b; c; x)$ of Gauss may be thought of as having (at least) five origins:

(a) *Hypergeometric series.* We can write

$$F(a, b; c; x) = \sum \frac{(a)_n (b)_n}{n!(c)_n} x^n, \tag{1.1}$$

where $(a)_n$ is Appell's symbol

$$(a)_n = \frac{\Gamma(a+n)}{\Gamma(a)} = a(a+1) \cdots (a+n-1). \tag{1.2}$$

From the point of view of power series what seems important is that the ratio of the coefficients of x^{n+1} and x^n is a fixed rational function $P(n)/Q(n+1)$, where P and Q are monic polynomials and $Q(0) = 0$. This idea leads to the definition of generalized hypergeometric series

$$_pF_q \left[\begin{matrix} a_1, \ldots, a_p \\ b_1, \ldots, b_q \end{matrix} ; x \right] = \sum \frac{(a_1)_n \cdots (a_p)_n}{(b_1)_n \cdots (b_q)_n} \frac{x^n}{n!}. \tag{1.3}$$

Thus Gauss's hypergeometric series is $_2F_1$.

(b) *Ordinary differential equations.* The fact that the ratio of the coefficients of x^{n+1} to x^n is $P(n)/Q(n+1)$ is equivalent to the ordinary differential equations

$$[xP(\delta) - Q(\delta)]F = 0, \tag{1.4}$$

$$\left[P(\delta) - \frac{d}{dx} Q_1(\delta) \right] F = 0, \tag{1.5}$$

where $\delta = xd/dx$ and $Q_1(x) = Q(x)/x$.

The operators $xP - Q$ or $P - (d/dx)Q_1$ are characterized (up to a simple transformation) by having two degrees of homogeneity. This is also clearly seen in the usual form of Gauss's equation

$$x(1-x)\frac{d^2F}{dx^2} + [c - (a+b+1)x]\frac{dF}{dx} - abF = 0. \tag{1.6}$$

We can consider the relation (1.4) or (1.5) from a purely operator theoretic viewpoint. What is really involved is the commutation relations

$$\left[x, x\frac{d}{dx}\right] = -x, \tag{1.7}$$

$$\left[\frac{d}{dx}, x\frac{d}{dx}\right] = \frac{d}{dx}. \tag{1.8}$$

In fact, if we replace x or d/dx and xd/dx by any pair of operators A, B satisfying (1.7) or (1.8) then there is an analogous theory where "power series" become "eigenfunction expansions" according to the eigenfunctions of B. In fact, Barnes's treatment of hypergeometric functions replaces the series expansion by the "Mellin–Barnes integral," which is the continuous spectral analysis of xd/dx. (Power series is a spectral theory which applies only to some functions—namely holomorphic functions; Mellin transform applies to "generic" functions.)

In Lie algebra theory the meaning of (1.7) and (1.8) is that x is a step up transformation for $x\,d/dx$ and d/dx is a step down transformation for $x\,d/dx$. The operators x and $x\,d/dx$ generate an $Ax + B$ group (affine group of the line) which is the simplest non-abelian Lie group. The relevance of this will appear later.

(c) *Representations of Lie groups.* Hypergeometric functions appear in many genres as matrix coefficients of representations of Lie groups. The most complete result in this direction is Bargmann's monumental paper [1] where it is shown that, except for certain integrality conditions, the most general $_2F_1(a, b; c; x)$ appears as a matrix coefficient of some representation of $SL(2, \mathbf{R})$.

(d) *Hypergeometric integrals.* In addition to the Mellin–Barnes integrals mentioned above, there are other types of integral representations bearing names like Euler, Dirichlet, Legendre, etc. Although the integral representations do not play a central role in our theory, we explain in Section V how they may be incorporated.

(e) *Separation of variables.* Separation of the Laplacian or wave operator in various coordinate systems leads naturally to hypergeometric functions. We mention here three coordinate systems and separations which will serve as paradigms for what follows. (We shall explain below the origin of these coordinate systems.)

(*i*) *Cylindrical coordinates.*   In $\mathbf{R}^3$ we use Cartesian coordinates x, y, t. Then cylindrical coordinates are defined by

$$t$$

$$\rho^2 = x^2 + y^2 \tag{1.9}$$

$$\theta = \tan^{-1}(y/x).$$

Let $\square$ denote the wave operator. Then product solutions $f(t, x, y)$ are of the form

$$f(t, \rho, \theta) = e^{i\lambda t}\, e^{in\theta} B_{in}(i\lambda\rho), \tag{1.10}$$

where B_{in} is a Bessel function. (This and the corresponding results in (*ii*) and (*iii*) will be clarified below.)

(*ii*) *Hyperbolic spherical coordinates.*   These are the non-compact version of usual spherical coordinates. They are defined by

$$t = r \cosh \zeta$$

$$x = r \sinh \zeta \cos \varphi \tag{1.11}$$

$$y = r \sinh \zeta \sin \varphi.$$

Separated solutions of the wave equation are of the form

$$f(r, \zeta, \varphi) = r^{is}\, e^{im\varphi} P_{is}^{m}(\cosh \zeta), \tag{1.12}$$

where P_{is}^{m} is the Legendre function. (We could use the associated Legendre function Q_{is}^{m} in place of P_{is}^{m} or any linear combination of them. This follows from (1.22).)

(*iii*) *Upper half-plane coordinates.*   For this we identify the Poincaré upper half-plane with the upper half of the hyperboloid of two sheets: $t^2 - x^2 - y^2 = 1$, $t > 0$. This identification is made group theoretically, that is, we think of the group $G = SL(2, \mathbf{R})$ acting by adjoint representation of $\mathbf{R}^3$. This representation can also be thought of as the symmetric square of the two-dimensional representation of G. In any case the representation realizes G as a two-sheeted covering of $\tilde{G} = SO(1, 2)$.

We use coordinates ξ, η in the Poincaré upper half-plane and also r as in (ii) to form a coordinate system in the interior of the forward light cone. We find the separated solutions

$$f(r, \xi, \eta) = r^{is} e^{i\lambda\xi} \eta^{1/2} K_{is+1/2}(\lambda\eta). \tag{1.13}$$

(Again K can be replaced by I.) This follows from (1.28).

One might start with any of the five approaches and try to understand how the other four aspects fit into this approach.

The combinatorists, led in our times by Askey and Andrews, emphasize approach (a), that of hypergeometric series. This has the advantage of applying to general $_pF_q$ and to the q-analogs of hypergeometric functions. Approach (b) is favored by those interested in ordinary differential equations. Clearly, (c) is the interest of group theorists; we refer to it as "intrinsic group theory." Integrals of the form (d) as well as generalizations appear often in algebraic geometry (when the parameters take rational values), while (e) is the approach of classical analysts.

We shall use an approach which is somewhat intermediate between (d) and (e), namely, we consider a group together with a fixed finite-dimensional representation. For example, when studying the Laplacian Δ on $\mathbf{R}^n$ we might think of associating the orthogonal group. *But this is the wrong group!* The right group is the conformal group of Δ, that is, the group of transformations leaving the kernel of Δ invariant (rather than leaving Δ invariant). We refer to our method as "extrinsic group theory."

There are certain relations amongst the approaches which are obvious and certain which are profound. The relation between (a) and (b) is clear. The relations between (b) and (d) has been the object of a profound study spearheaded by B. Dwork and the French school. As we mentioned, Bargmann's work (which had many predecessors) shows that (c) leads to (b); but Bargmann's result is computational and not conceptual.

We show how our approach of extrinsic group theory leads conceptually to (b), to (c), and to (d). (We do not consider (e), which is essentially built into our method.) The most difficult point is (b), which took me many years to conceptualize (see Problem 3, below).

Perhaps more surprising is the passage from (a) or (b) to extrinsic group theory. Thus we start with one equation, e.g., for the Bessel function J_0, and construct the whole hierarchy of Bessel functions, and an appropriate group, and the Laplacian. This is explained at the end of this article.

This last point bears a relation with some of the recent beautiful work of Sato and his school. They start with the KP equation and build a hierarchy which has a Kac-Moody algebra as symmetry. We start with a Bessel equation and construct the Laplacian which has a large symmetry group.

Let D be a differential operator and call G^c the conformal group of D (defined precisely in Section II). For the present, think of G^c as the group of transformations which preserve the kernel of D.

The maximal abelian subgroups A^c of G^c and their normalizers W^c play a crucial role. These ideas are described in Sections 2 and 3.

Yet A^c and G^c do not seem to explain the structure of the hypergeometric series ((a) above) or Gauss's contiguity relations (described below).

For these we must go beyond the conformal group and study double commutators and other commutation relations instead of single commutators as in group theory. This idea is expounded in Section 4. Actually, we can go beyond this. We set up a hierarchy in which the first term is the exponential function, the second the hypergeometric, the third contains Mathieu and Lamé functions, and we can go on into virgin territory.

In Section 5.4 we also go beyond the conformal group by allowing degeneration. This leads to a clarification of the classical ideas of asymptote and confluence.

We can also reverse the procedure. This means that we start with a group G^c and try to realize it as the conformal group of a (system of) differential equation(s). This is carried out in Section 5.

We have discussed separation of variables in (e) above. Let us clarify the type of coordinate systems with which we shall deal.

We start with n vector fields $\omega_1, \omega_2, \ldots, \omega_n$ on $\mathbf{R}^n$, or locally, or on an n-dimensional manifold, and also with a base point p_0. Suppose that p_0 is not a fixed point of ω_1 (and similar non-degeneracies for the other ω_j). Then we move p_0 a distance x_1 (measured in some normalized way) on the integral curve for ω_1 through p_0. We then move $\omega_1(x_1)p_0$ a normalized distance x_2 on the integral curve for ω_2 through $\omega_1(x_1)p_0$, etc. This gives us coordinates $x = (x_1, \ldots, x_n)$ near p_0.

Of course, we need a suitable connection to be able to define the normalization.

For most (though not all) of the cases of interest we shall replace the ω_j by one-parameter groups which act globally. Then the normalization is clear. In particular the coordinate systems (i), (ii), and (iii) arise from one-parameter groups. Let me write this explicitly and show the origins of the separations (1.10), (1.12), and (1.13). (The first two are classical but I shall give a very brief idea of the calculation for the reader who is unfamiliar with it.)

(*i*) Cylindrical coordinates arise from the one parameter groups

$$H_1 = \text{translation in } t,$$

$$H_2 = \text{scalar multiplication by } \rho \text{ in the } xy \text{ coordinates,} \qquad (1.14)$$

$$H_3 = \text{rotation by } \theta \text{ in the } xy \text{ coordinates.}$$

The base point is chosen as $(t, x, y) = (0, 1, 0)$. Then the (t, ρ, θ) coordinates are $(t, \rho \cos \theta, \rho \sin \theta)$ in accordance with (1.9).

The Jacobian of the transformation from (t, x, y) to (t, ρ, θ) is given by

$$\begin{pmatrix} 1 & 0 & 0 \\ 0 & \cos \theta & -\rho \sin \theta \\ 0 & \sin \theta & \rho \cos \theta \end{pmatrix} \qquad (1.15)$$

with determinant ρ. The matrix of signed minors is

$$\begin{pmatrix} \rho & 0 & 0 \\ 0 & \rho \cos \theta & -\sin \theta \\ 0 & \rho \sin \theta & \cos \theta \end{pmatrix}. \qquad (1.16)$$

From this the expression

$$\square = \frac{\partial^2}{\partial t^2} - \frac{\partial^2}{\partial \rho^2} - \frac{1}{\rho} \frac{\partial}{\partial \rho} - \frac{1}{\rho^2} \frac{\partial^2}{\partial \theta^2} \qquad (1.17)$$

follows. (1.10) is an immediate consequence of this and the definition of the Bessel differential equation. (The reader unfamiliar with Bessel's equation can use this as its definition.)

(*ii*) Here we regard $\mathbf{R}^3$ as the space of symmetric 2×2 matrices $\begin{pmatrix} u & y \\ y & v \end{pmatrix}$. The coordinates are given by the action of the three subgroups

$$H_1 = \text{scalar multiplication by } r,$$

$$H_2 = \text{the diagonal group } \begin{pmatrix} e^\zeta & 0 \\ 0 & e^{-\zeta} \end{pmatrix}, \qquad (1.18)$$

$$H_3 = \text{the rotation group } \begin{pmatrix} \cos \varphi & -\sin \varphi \\ \sin \varphi & \cos \varphi \end{pmatrix}.$$

The groups H_2 and H_3 send the matrix X into gXg'. The base point is the identity matrix. Thus we send the identity into

$$r\begin{pmatrix} \cos^2\varphi\,e^{2\zeta}+\sin^2\varphi\,e^{-2\zeta} & \sin\varphi\,\cos\varphi(e^{2\zeta}-e^{-2\zeta}) \\ \sin\varphi\,\cos\varphi(e^{2\zeta}-e^{-2\zeta}) & \sin^2\varphi\,e^{2\zeta}+\cos^2\varphi\,e^{-2\zeta} \end{pmatrix}. \quad (1.19)$$

Setting $t=\tfrac{1}{2}(u+v)$ and $x=\tfrac{1}{2}(u-v)$ yields the coordinates (1.11) (up to a change of $\zeta\to 2\zeta$ and $\varphi\to 2\varphi$ which we shall ignore).

The Jacobian of the transformation $(t,x,y)\to(r,\zeta,\varphi)$ is given by

$$\begin{pmatrix} \cosh\zeta & r\sinh\zeta & 0 \\ \sinh\zeta\,\cos\varphi & r\cosh\zeta\,\cos\varphi & -r\sinh\zeta\,\sin\varphi \\ \sinh\zeta\,\sin\varphi & r\cosh\zeta\,\sin\varphi & r\sinh\zeta\,\cos\varphi \end{pmatrix} \quad (1.20)$$

with determinant $r^2\sinh\zeta$. The matrix of minors is given by

$$\begin{pmatrix} r^2\cosh\zeta\,\sinh\zeta & -r\sinh^2\zeta & 0 \\ -r^2\sinh^2\zeta\,\cos\varphi & r\cosh\zeta\,\sinh\zeta\,\cos\varphi & -r\sin\varphi \\ -r^2\sinh^2\zeta\,\sin\varphi & r\cosh\zeta\,\sinh\zeta\,\sin\varphi & r\cos\varphi \end{pmatrix}. \quad (1.21)$$

This yields the classical expression

$$\square = \frac{\partial^2}{\partial r^2}+\frac{2}{r}\frac{\partial}{\partial r}-\frac{1}{r^2}\frac{\partial^2}{\partial\zeta^2}-\coth\zeta\frac{\partial}{\partial\zeta}-\frac{1}{r^2\sinh^2\zeta}\frac{\partial^2}{\partial\varphi^2}. \quad (1.22)$$

Again the splitting (1.12) follows from the definition of Legendre's differential equation.

(iii)   We use the same idea as in (ii) except that the group H_3 of (1.18) is replaced by

$$H_3'=\begin{pmatrix} 1 & \xi \\ 0 & 1 \end{pmatrix}, \quad (1.23)$$

and we use the notation

$$\begin{pmatrix} \eta^{1/2} & 0 \\ 0 & \eta^{-1/2} \end{pmatrix} \quad (1.24)$$

for the elements of H_2. Then we obtain

$$r\begin{pmatrix} 1 & \xi \\ 0 & 1 \end{pmatrix}\begin{pmatrix} \eta & 0 \\ 0 & \eta^{-1} \end{pmatrix}\begin{pmatrix} 1 & 0 \\ \xi & 1 \end{pmatrix}=r\begin{pmatrix} \eta+\xi^2\eta^{-1} & \xi\eta^{-1} \\ \xi\eta^{-1} & \eta^{-1} \end{pmatrix}. \quad (1.25)$$

The Jacobian of the transformation from (t, x, y) to (r, ξ, η) is given (up to a constant) by

$$
\begin{pmatrix}
\eta + \eta^{-1} + \xi^2\eta^{-1} & r(1 - \eta^{-2} - \xi^2\eta^{-2}) & 2r\xi\eta^{-1} \\
\eta - \eta^{-1} + \xi^2\eta^{-1} & r(1 + \eta^{-2} - \xi^2\eta^{-2}) & 2r\xi\eta^{-1} \\
2\xi\eta^{-1} & -2r\xi\eta^{-2} & 2r\eta^{-1}
\end{pmatrix}
\tag{1.26}
$$

with determinant $8r^2\eta^{-2}$. The (signed) minors of (1.26) are (up to a constant)

$$
\begin{pmatrix}
r^2(\eta^{-1} + \eta^{-3} + \xi^2\eta^{-3}) & -r(1 - \eta^{-3} - \xi^2\eta^{-2}) & -2r\xi\eta^{-1} \\
-r^2(\eta^{-1} - \eta^{-3} + \xi^2\eta^{-3}) & r(1 + \eta^{-2} - \xi^2\eta^{-2}) & 2r\xi\eta^{-1} \\
-4r^2\xi\eta^{-3} & -4r\xi\eta^{-2} & 4r\eta^{-1}
\end{pmatrix}.
$$

$$\tag{1.27}$$

This leads to the expression

$$
\tfrac{1}{4}\Box = \frac{\partial^2}{\partial r^2} + \frac{2}{r}\frac{\partial}{\partial r} - \frac{\eta^2}{r^2}\left(\frac{\partial^2}{\partial \eta^2} + \frac{\partial^2}{\partial \xi^2}\right),
\tag{1.28}
$$

from which (1.13) follows.

At this point we should observe the structure that is common to the examples (1.10), (1.12), and (1.13). Note that in all cases f is a product of three functions, two of which are characters and one of which is hypergeometric. This property will dominate our work.

We observe that the non-trivial functions in (i) and (iii) are the same, namely Bessel functions.

Problem 1.   Why do Bessel functions appear in examples (i) and (iii)?

This problem is part of the general program of understanding why a function appears in various contexts. Another example we shall consider is the Whittaker or Laguerre function. This appears in the representation theory of the Heisenberg group. When the central character is fixed (so the function can be thought of as a function on the plane) and the rotation character is fixed, then a representation function is a Laguerre function of r.

The Laguerre functions also appear in the representation theory of $G = SL(2, \mathbf{R})$. We identify $\mathbf{R}^4$ with $M_2 = 2 \times 2$ real matrices and then let $G \times G$ act on M_2 by $m \to g_1 m g_2^{-1}$. Then the Laguerre functions are the restrictions to the $G \times G$-invariant set $\det m = 1$ of homogeneous solutions f of the

ultrahyperbolic equation corresponding to the determinant. Here f transforms according to a character of the diagonal subgroup

$$A = \left\{ \begin{pmatrix} a & 0 \\ 0 & a^{-1} \end{pmatrix} \right\} \qquad (1.29a)$$

on the left and a non-trivial character of the nilpotent group

$$N = \left\{ \begin{pmatrix} 1 & n \\ 0 & 1 \end{pmatrix} \right\} \qquad (1.29b)$$

on the right. (A trivial character of N would lead to Bessel functions instead
of Laguerre functions. In fact, for a trivial character of N we would be in
a situation equivalent to (iii).) All this is clarified in Section 3.

Problem 2. Why does the same function (i.e., Laguerre) appear in both
contexts?

As mentioned above we shall concern ourselves with the following

Problem 3. Why do the separated solutions have the structure of a
product of two abelian characters with one hypergeometric function? In
particular, what is the structure of the hypergeometric series?

Note that the second part of Problem 3 involves an interpretation of the
explicit choice of parameters. In solving Problem 3 we must go beyond the
conformal group as mentioned above.

In the classical theory of hypergeometric series it was observed by Kummer and later by Riemann that there are 24 hypergeometric series which
solve the hypergeometric equation.

Problem 4. What is the origin of these 24 solutions?

Note that 24 is the order of the Weyl group of $O(3, 3)$. This Weyl group
parametrizes the set of hypergeometric series.

The existence of many hypergeometric series implies that these series
have many relations. Such relations were found by Kummer. In addition,
Gauss found relations amongst contiguous hypergeometric series.
($F(a, b; c; z)$ is *contiguous* to $F(a \pm 1, b; c; z)$ and to $F(a, b \pm 1; c; z)$ and
to $F(a, b; c \pm 1; z)$.) Gauss showed that amongst $F(a, b; c; z)$ and any two
contiguous series there is a linear relation with coefficients which are linear

in z. An example of such a relation is

$$cF(a, b-1; c; z) + (a-b)zF(a, b; c+1; z) = cF(a-1, b; c; z). \quad (1.30)$$

Problem 5.   Explain the contiguity relations.

My interest in the contiguity relations was stimulated by Walter Feit who showed me their importance in constructing Galois extensions of the rational number field with certain Galois groups.

2.   The Conformal Group

In the classical theory the group associated to the wave operator or to the Laplacian was the orthogonal group of the corresponding quadratic form, meaning, respectively, the Minkowski or the rotation group. However, *this is the wrong group*. The reason is that we are interested in the kernels of $\square$ and Δ. Thus the interesting transformations are those which preserve the kernel of the operator rather than the operator itself. This leads to the

Definition.   Let D be a linear operator acting on functions on a manifold M, a transformation φ of M is called D *conformal* if

$$\varphi \circ D \circ \varphi^{-1} = (\alpha D \beta) \quad (2.1)$$

for functions α and β (depending on D and φ). The functions α and β act by multiplication; they are called *conformal weights*.

Thus conformality means that φ maps the kernel of $D\beta$ into the kernel of D. From our point of view the function β has a basically trivial effect so that, roughly speaking, φ preserves the kernel of D.

One formalism which trivializes the effect of β is a "conformal extension M^c" of M (which is a generalization of conformal compactification) obtained by gluing copies of M using φ. We then study the action of D on sections of line bundles over M^c defined by β. We shall not enter into this matter, which has been well studied in works on the conformal compactification of Minkowski space (see, e.g., [4]).

Instead of considering scalar operators D we could apply (2.1) to cross sections of suitable vector bundles. In particular, if $M = \mathbf{R}^n$ and D is an $n \times n$ matrix of differential operators D_{ij} then α and β are $n \times n$ matrices of functions. We shall come back to this point later.

I want to digress to discuss several interesting aspects of conformality. We write α_φ, β_φ for α, β.

Proposition 2.1. *α and β are 1-cocycles for the conformal group. More precisely,*

$$\alpha_{\varphi\psi} = \alpha_\varphi(\varphi\alpha_\psi)$$

$$\beta_{\varphi\psi} = \beta_\varphi(\varphi\beta_\psi). \tag{2.2}$$

Proof. By the definition (2.1),

$$D(\varphi\psi)^{-1}f = (\varphi\psi)^{-1}\alpha_{\varphi\psi}D\beta_{\varphi\psi}f. \tag{2.3}$$

The left side of (2.3) is

$$= D\psi^{-1}(\varphi^{-1}f)$$

$$= \psi^{-1}\alpha_\psi D\beta_\psi\varphi^{-1}f$$

$$= \psi^{-1}\alpha_\psi D\varphi^{-1}[(\varphi\beta_\psi)f]$$

$$= \psi^{-1}\alpha_\psi\varphi^{-1}\alpha_\varphi D[\beta_\varphi(\varphi\beta_\psi)f]$$

$$= (\varphi\psi)^{-1}(\varphi\alpha_\psi)\alpha_\varphi D[\beta_\varphi(\varphi\beta_\psi)f].$$

Note that we have used the property that φ distributes with multiplication. In fact, this is the dominant property that we use in what follows.

The next properties of conformality are the crucial points used in Eichler cohomology. In fact, they show that we can define an analog of Eichler cohomology relevant to any D.

Proposition 2.2. *Suppose H is an integral of a form h which is of weight α with respect to some subgroup B of the conformal group, that is, for $\gamma \in B$*

$$DH = h \tag{2.4}$$

$$\gamma h = \alpha h. \tag{2.5}$$

Then

$$D(\beta\gamma H - H) = 0. \tag{2.6}$$

That is, the (additive) γ periods of H lie in the kernel of D (when suitable use of β is made).

Proposition 2.3. *For any h as above, the map*

$$\gamma \to \beta\gamma H - H = P_\gamma^H \tag{2.7}$$

is a 1-cocycle on B with values in the kernel of D. (γ acts as $\beta\gamma$.) Moreover the cohomology class of P_γ^h depends on h and not on H.

Proof of Proposition 2.2. Note that

$$\alpha^{-1}\gamma D = D\beta\gamma. \tag{2.8}$$

This means that if h is of weight α then

$$DH = h = \alpha^{-1}\gamma DH$$

$$= D\beta\gamma H. \tag{2.9}$$

Thus

$$D(\beta\gamma H - H) = 0, \tag{2.10}$$

which is the result.

Proof of Proposition 2.3. We verify that P_γ^H is a 1-cocycle. For this we need the fact that β is a 1-cocycle. Thus

$$P_{\gamma\gamma'}^H = \beta_{\gamma\gamma'}\gamma\gamma'H - H$$

$$= \beta_\gamma\gamma(\beta_{\gamma'})\gamma\gamma'H - H$$

$$= [\beta_\gamma\gamma H - H] + \beta_\gamma\gamma[\beta_{\gamma'}\gamma'H - H]. \tag{2.11}$$

Thus P_γ^H is a 1-cocycle.

Next suppose that $DH = DH_1 = h$. Then $Q = H - H_1 \in \text{kernel } D$. Moreover, since P_γ^H depends linearly on H,

$$P_\gamma^H - P_\gamma^{H_1} = P_\gamma^Q, \tag{2.12}$$

which is clearly a 1-coboundary. This proves Proposition 2.3.

For our next task we wish to compute the conformal groups associated to some important D. First let $D = \Delta$ be the Laplacian in $\mathbf{R}^n$. Note that

$$\frac{\partial}{\partial x_j}\beta\varphi f = \frac{\partial\beta}{\partial x_j}\varphi f + \beta\sum_k \frac{\partial\varphi^k}{\partial x_j}\varphi\frac{\partial f}{\partial x_k}. \tag{2.13}$$

Hence

$$\Delta(\beta\varphi f) = (\Delta\beta)\varphi f + 2\sum_{k,j}\frac{\partial\beta}{\partial x_j}\frac{\partial\varphi^k}{\partial x_j}\varphi\frac{\partial f}{\partial x_k} + \beta\sum_k(\Delta\varphi^k)\varphi\frac{\partial f}{\partial x_k}$$

$$+ \beta\sum_{k,l,j}\frac{\partial\varphi^k}{\partial x_j}\frac{\partial\varphi^l}{\partial x_j}\varphi\frac{\partial^2 f}{\partial x_k\,\partial x_l}. \tag{2.14}$$

The conditions for φ to be Δ conformal are easily seen to be

$$\Delta\beta = 0, \tag{2.15a}$$

$$2\sum_{k,j}\frac{\partial\beta}{\partial x_j}\frac{\partial\varphi^k}{\partial x_j} + \beta\sum_k\Delta\varphi^k = 0, \tag{2.15b}$$

$$\sum_j\frac{\partial\varphi^k}{\partial x_j}\frac{\partial\varphi^l}{\partial x_j} = a\delta_{kl} \text{ is diagonal.} \tag{2.15c}$$

The analysis of these conditions is naturally subdivided in two cases.

Case 1, n = 1. Then (2.15a) says that β is linear. (2.15c) is trivial; (2.15b) becomes

$$2\beta'\varphi' + \beta\varphi'' = 0. \tag{2.16}$$

Differentiating and using (2.15a) yields

$$3\beta'\varphi'' + \beta\varphi''' = 0, \tag{2.17}$$

which, by (2.16) is

$$2\varphi'\varphi''' = 3(\varphi'')^2. \tag{2.18}$$

The reader will recognize (2.18) as Schwarz's equation for fractional linear transformations.

Actually we can deduce the fractional linear structure of φ from (2.16). For this states

$$\frac{2\beta'}{\beta} = -\frac{\varphi''}{\varphi'}, \tag{2.19}$$

so that

$$\varphi' = \text{const } \beta^{-2}. \tag{2.20}$$

Since β is linear, we again obtain the fractional linear structure of φ.

Case 2, n > 1. In this case it is (2.15c) which is crucial. For we observe that (2.15c) is exactly the condition that φ be conformal in the usual sense of differential geometry, that is, φ maps the euclidean metric ds^2 into a multiple of itself or, what is the same thing, φ preserves angles.

Conversely we must show that every geometrically conformal map is Δ conformal. The simplest proof of this uses the structure of the geometric conformal group $G^c = O(1, n+1)$. G^c is generated by $O(n)$, translation in $\mathbf{R}^n$, scalar multiplication, and inversion in the unit sphere. (The inversion is, of course, orientation reversing.) All the transformations except inversion are trivially Δ conformal. As for inversion, it is a classical fact that it is Δ conformal; this can be proven by expressing Δ in polar coordinates.

The same computation applies to

$$D = \sum_{i=1}^{p} \frac{\partial^2}{\partial x_i^2} - \sum_{i=p+1}^{n} \frac{\partial^2}{\partial x_i^2}.$$

The conformal group is $O(p+1, n-p+1)$.

There are other ways to verify the conformality of inversion. For the wave operator one can examine the effect of inversion on light rays. Since everything is algebraic the result for the wave operator implies the corresponding result for any ultrahyperbolic operator or the Laplacian. This proof is important for studying the higher-dimensional case.

Another proof capable of (non-computational) generalization comes from Fourier analysis. For simplicity we write it for the Laplacian Δ. Let $f(x) = \exp(ix \cdot \hat{x}_0)$ and call g_0 the inversion in the unit sphere. We assume that the conformal weights are powers of r.

The Fourier transform of $\Delta[r^a \exp(ig_0 x \cdot \hat{x}_0)]$ is

$$\hat{r}^2 \int e^{ix \cdot \hat{x} + ig_0 x \cdot \hat{x}_0} r^{a+n} \frac{dr}{r} \, d\theta.$$

We want this to be equal to the Fourier transform of $r^b g_0 \Delta f$, which is

$$\hat{r}_0^2 \int e^{ix \cdot \hat{x} + ig_0 x \cdot \hat{x}_0} r^{b+n} \frac{dr}{r} \, d\theta.$$

Write $x \cdot \hat{x} = r\hat{r}\gamma(\theta, \hat{\theta})$. Then make the change of variables $t = \hat{r}r$ in the first integral and $t = \hat{r}_0/r$ in the second integral. We find equality if $a = 2 - n$ and $b = -2 - n$.

Remark. One should expect the identity of the Δ conformal and geometrically conformal groups to be a consequence of the geometric characterization of harmonic functions, namely they are characterized by the mean value property over spheres, whereas a geometric conformal map preserves the set of infinitesimal spheres. But the more one thinks of this, the more unclear the relation becomes, for geometric conformality is a *first-order* infinitesimal effect, while Δ can be defined in terms of the mean value property on a second-order infinitesimal level.

Problem. Prove the identity of Δ and geometric conformality by purely geometric means.

It seems that one might give a proof using geometric conformality on many small cubes whose size is like dr^2 where dr is the size of the sphere used in the mean value property. Another possibility is to use Brownian motion.

We have remarked that the conformal group corresponding to $O(p, q)$ is $O(p+1, q+1)$. Let me explain how the generators of the conformal group act on $\mathbf{R}^{n+2}$, where $n = p+q$. We use coordinates $x_0, x_1, \ldots, x_n, x_{n+1}$ in $\mathbf{R}^{n+2}$. The matrix corresponding to the effect of $u \in O(p, q)$ plus translation by $c \in \mathbf{R}^n$ is

$$
\begin{pmatrix}
1 & 0 & \cdots & \cdots & \cdots & 0 \\
c_1 & & & & & \vdots \\
\vdots & & (u_{ij}) & & & \vdots \\
\vdots & & & & & 0 \\
c_n & & & & & 0 \\
\tfrac{1}{2} c \cdot c & c \cdot u_1 & \cdots & \cdots & c \cdot u_n & 1
\end{pmatrix}. \tag{2.21}
$$

Here u_k represents the kth row of u.

Inversion is

$$
\begin{pmatrix}
0 & \cdots & \cdots & \cdots & \cdots & 0 & 1 \\
\vdots & & & I & & & \vdots \\
0 & & & & & & \vdots \\
1 & 0 & \cdots & \cdots & \cdots & \cdots & 0
\end{pmatrix}. \tag{2.22}
$$

Scalar multiplication is

$$
\begin{pmatrix}
a^{-1} & 0 & \cdots & \cdots & 0 \\
0 & & & & 0 \\
\vdots & & I & & \vdots \\
0 & & & & 0 \\
0 & & & & a
\end{pmatrix}. \tag{2.23}
$$

It is easy to compute the conformal factors for these generators. Proposition 2.1 then gives them for all of G^c.

Remark. The group P generated by G, translation, and scalar multiplication is a normal parabolic subgroup of G^c. The conformal factors are very simple on P, namely they are 1 except for scalar multiplication where $\beta = 1$, $\alpha = a^2$.

We have noted above the dichotomy that exists between conformality for $n = 1$ and for $n > 1$. For $n = 1$ we are led to Schwarz's equation, whereas for $n > 1$ we find the same equation as for geometric conformality. We wish to make a further probe of Schwarz's relation for other operators.

Let D be an $m \times m$ matrix of first order linear differential operators with no constant term. (We write the coefficients to the left of the differentiation.) Note that

$$\frac{\partial}{\partial x_j} \varphi h = \frac{\partial \varphi}{\partial x_j} \cdot \varphi \nabla h \tag{2.24}$$

for any function h. This means that

$$D\varphi f = \varphi D^{\varphi^{-1}} (\partial \varphi \cdot \nabla) f. \tag{2.25}$$

Our notation of $D(\partial \varphi \cdot \varphi \nabla)$ means each $\partial / \partial x_j$ is replaced in accordance with (2.24), except that in (2.25) we apply φ last. Our notation $D^{\varphi^{-1}}$ means that φ^{-1} is applied to all matrix coefficients of $D(\partial \varphi \cdot \nabla)$.

D conformality of φ (more precisely, of φ^{-1}) means that there are $n \times n$ matrices α_1, β_1 of functions such that

$$D^{\varphi^{-1}} (\partial \varphi \cdot \nabla) f = \alpha_1 D \beta_1 f. \tag{2.26}$$

A simple computation shows that if D is the Cauchy-Riemann system written for real $f = (f_1, f_2)$ then (2.26) is the condition that φ (considered as a map of the complex plane into itself) satisfies $\partial \varphi / \partial \bar{z} = 0$, provided we assume that β_1 is of the form of a multiplication on $f = f_1 + i f_2$, that is, the matrix for β_1 is of the form

$$\begin{pmatrix} \gamma_1 & \gamma_2 \\ -\gamma_2 & \gamma_1 \end{pmatrix}.$$

Next, let us assume that φ is D conformal and decide if φ is also D^2 conformal. In order to do this we have to assume that we have an algebra of functions f and that φ maps this algebra into itself, and that D acts as a derivation of the algebra. Moreover the α_j, β_j belong to the algebra.

Thus

$$D^2(\beta_2\varphi f) = (D^2\beta_2)\varphi f + 2(D\beta_2)D(\varphi f) + \beta_2 D^2\varphi f$$

$$= (D^2\beta_2)\varphi f + 2(D\beta_2)\varphi(\alpha_1 D\beta_1 f) + \beta_2\varphi[\alpha_1 D\beta_1(\alpha_1 D\beta_1 f)]$$

$$= [D^2\beta_2 + 2D\beta_2\varphi\alpha_1 D\beta_1 + \beta_2\varphi\alpha_1^2\beta_1(D^2\beta_1)$$

$$+ \beta_2\varphi\alpha_1 D(\beta_1\alpha_1)(D\beta_1)]\varphi f$$

$$+ [2(D\beta_2)\varphi\alpha_1\beta_1 + 2\beta_2\varphi\alpha_1^2\beta_1 D\beta_1 + \beta_2\varphi\alpha_1 D(\beta_1\alpha_1)\beta_1]\varphi Df$$

$$+ \beta_2[\varphi\alpha_1^2\beta_1^2]\varphi D^2 f. \tag{2.27}$$

Note that when $n = 1$ we have $\beta_1 = 1$, $\alpha_1 = \varphi^{-1}\varphi'$ and $\beta_2 = \beta$, so that (2.27) agrees with (2.14).

For conformality we need vanishing of the coefficients of φf and φDf. This gives two equations which are the generalization of (2.15a) and (2.15b). We regard α_1 and β_1 as given objects, as they are generally easily related to φ. Thus we regard the equations arising from (2.27) as compatibility conditions for the existence of β_2. Differentiating the equation arising from the vanishing of the coefficient of φDf and using the other equation leads to an equation involving only $D\beta_2$ and β_2 which is the analog of (2.16). Thus we may regard this equation as a general form of Schwarz's equation.

Remark. This procedure shows that Schwarz's equation for fractional linear transformation, like many other interesting nonlinear equations, has its origins in a compatibility condition ("Lax equation"). These nonlinear equations arise from operators D^2 (and also D^j for $j > 2$) when D is a derivation. Thus they do not appear when we study conformality for the Laplacian, which is of the form $d\delta$ where δ is the adjoint of d on forms (or else $\partial^2/\partial z\, \partial\bar{z}$). We might try using the Dirac operator, but it is not a derivation.

Before going on, let me illustrate the ubiquitous position of conformality in mathematics. I have already shown its relation to Eichler cohomology and Schwarz's equation. Let me also note its relation to

(a) *Complex multiplication.* Let T be a complex n-dimensional abelian variety with period matrix P. We seek $n \times n$ complex matrices Q which are nonsingular and which map P invariant meromorphic functions $f(z)$ on $\mathbf{C}^n$ into functions $f(Qz)$ which are also P invariant or, at least, algebraic over the field of meromorphic function on T.

If B denotes the $2n \times 2n$ matrix formed from P and $\bar{P}$, then the classical condition is seen to be (see [5], p. 220)

$$BMB^{-1} = \begin{pmatrix} Q & 0 \\ 0 & \bar{Q} \end{pmatrix}, \tag{2.28}$$

where M is a nonsingular rational matrix.

To relate this to our definition of conformality, we want to compute the effect of $P - I$ on $Q^{-1}f$. We have

$$Q(P - I)Q^{-1}f(z) = Q(P - I)f(Q^{-1}z)$$

$$= Q[f(Q^{-1}z + P) - f(Q^{-1}z)]$$

$$= f(z + QP) - f(z). \tag{2.29}$$

This will vanish when

$$QP = PM \tag{2.30}$$

for some integral matrix M. We may call $PM = \tilde{P}$ as $\tilde{P}$ is in the period lattice defined by P. Then we write (2.29) in the form

$$Q(P - I)Q^{-1} = \tilde{P} - I, \tag{2.31}$$

which is basically the condition (2.1) with $\alpha = \beta = 1$ except that, since the solution of $P - I$ is also a solution of $\tilde{P} - I$, we can use this modification of (2.1) which applies when we are dealing with functions which satisfy many equations.

So much for Q preserving invariance. For Q to send P invariants into functions algebraic over the field of P invariant meromorphic functions, we no longer require that M be integral; rationality suffices. The passage from (2.30) to (2.28) is straightforward.

Remark 1. This example emphasizes the spirit of (2.1). For φ is a spacial transformation and α and β are multipliers. Thus the general spirit of conformality is that conjugation of D by an operator of a special kind (such as spacial transformations in (2.1) or $n \times n$ matrices Q in the above) can be expressed as a transformation of D by other special operators (as in α, β in (2.1)).

Remark 2. I don't know if it is possible to produce "complex multiplications" with nontrivial α, β. However, if they exist, it should be of interest to study the related Eichler cohomology.

(b) *Hecke operators.* These bear a similar role to complex multiplication except that the translation group determined by P is replaced by a discrete group, e.g., the modular group. The Hecke operators are no longer spacial but are finite sums of spacial transformations. Again $\alpha = \beta = 1$ and the above Remarks 1 and 2 apply.

(c) *Scattering theory.* It would take me too far afield to discuss scattering theory in detail. Let me remark that the map which goes from scattering data to the wave function satisfies a conformality condition and the resulting nonlinear equations for the potential (which is a conformal factor β) can be interpreted as a suitable cocycle condition.

3. Cartan and Weyl Group and Their Applications

Let us return to our main theme, which is hypergeometric functions. We begin with three dimensions.

We regard $\mathbf{R}^3$ as the space of 2×2 symmetric matrices. As such $G = SL(2, \mathbf{R})$ acts by

$$a \to gag'. \tag{3.1}$$

(This is easily seen to be the adjoint representation of G.) It is clear that G preserves the quadratic form $\det a$. This form has signature $(1, 2)$, so that G becomes essentially (that is, except for a two-sheeted cover) $SO(1, 2)$. As this two-sheeted cover plays no role in our work, we shall not distinguish between G and $SO(1, 2)$.

We denote by G^c the associated (geometric) conformal group, so that $G^c = SO(2, 3)$. By the results of Section 2 we know that G^c is also the conformal group in the sense of (2.1), where D is the wave operator on $\mathbf{R}^3$.

Let A be a maximal abelian subgroup of a real semisimple Lie group H. We call A a *confluent Cartan* subgroup if it is not a usual Cartan.

In general, a maximal abelian subgroup will be called a *Cartan.* It is a standard theorem that all (usual) Cartans are conjugate in the complexification of H.

Theorem 3.1. *The complexifications of any two Cartans with the same Fitting structure are conjugate in the complexification of H.*

The proof of this theorem will appear elsewhere. We shall make no use of this theorem as, in any case in which it applies, we shall exhibit the conjugation explicitly.

Associated with any Cartan A there is a Weyl group W or W_A which is the normalizer of A in the complexification of H modulo its centralizer. Confluent Cartans A tend to have smaller Weyl groups than ordinary Cartans. In fact

Theorem 3.2. *If A is a confluent Cartan, then the order of W_A is strictly less than the order of the usual Weyl group.*

Again, this theorem will be proven elsewhere.

We can now begin to understand the hypergeometric function.

Let A be a Cartan subgroup of the conformal group G^c. If we ignore the conformal factor β, then A acts on the kernel U_D of the wave operator $\square$. (We shall usually consider only those A on which $\beta = 1$.) Since A is abelian we can diagonalize the action of A on U_D.

For future reference we wish to make explicit the following proposition, whose proof is simple.

Proposition 3.3. *If $g \in G^c$, then $x \to gx$ sends solutions of $\square$ which transform under A as a character into solutions (using the β of g) which transform like characters of the conjugate gAg^{-1} of A.*

We call *basic solutions* those solutions which transform like characters of a Cartan A.

There are essentially two possibilities:

(a) A is three-dimensional
(b) A is two-dimensional.

Case (a) corresponds to A being the group $\mathbf{R}^3$ of translations in $\mathbf{R}^3$. The reduction of $U_\square$ under A yields the exponential functions

$$\exp(ix \cdot \hat{x}) \qquad \hat{x} \in \text{light cone}. \qquad (3.2)$$

Every $f \in U_\square$ can be written as an integral of such exponential solutions (see [3]).

Let us pass to two-dimensional A.

Example 1.   As an example of a confluent A we can choose

$$A_1 = \{\text{translates in } t, \text{rotations in } \theta\}. \tag{3.3}$$

We are using cylindrical coordinates as in (1.9).

We now decompose any solution f of the wave equation under the characters of A_1. Thus, formally, f is a linear combination of solutions of the form

$$\exp(it\hat{t} + in\theta)f_{\hat{t},n}(\rho). \tag{3.4}$$

We are left with one interesting function, namely $f_{\hat{t},n}$ of the single variable ρ. By (1.10) we can identify $f_{\hat{t},n}$ with the usual Bessel function

$$f_{\hat{t},n}(\rho) = J_{in}(i\hat{t}\rho), \tag{3.5}$$

as in Section 1 above. (We could use K_n instead of J_n.) This will be understood on a deeper level in Section 4 below.

Example 2.   For a second confluent Cartan A_2 we pass to upper half-plane coordinates as in Section 1.

$$A_2 = \{\text{scalar multiplication, translates in } \xi\}. \tag{3.6}$$

(For scalar multiplication we can choose either α or β to be nontrivial. In any case, scalar multiplication preserves $U_\square$ which is the kernel of $\square$.)

We decompose $U_\square$ under A_2. The basic functions are given by (1.13). We can now understand why the Bessel function appears in both examples A_1 and A_2. For A_1 and A_2 both clearly have the same Fitting structure (one semisimple and one nilpotent generator). Thus they are conjugate by Theorem 3.1. (Do not be concerned by the fact that the semisimple part of A_1 is compact while the corresponding semisimple part of A_2 is not compact; the conjugation is in the complex domain.) The explicit conjugation is

$$t \rightarrow \xi$$

$$\rho \rightarrow i\eta \tag{3.7}$$

$$e^{i\theta} \rightarrow r.$$

To verify that (3.7) is conformal, we verify that it is geometrically conformal, that is, it takes the Minkowski metric ds^2 into a multiple of itself. The Minkowski metric in cylindrical coordinates is clearly $dt^2 - d\rho^2 - \rho^2 d\theta^2$, and in upper half-plane coordinates is $dr^2 - r^2\eta^{-2}(d\xi^2 + d\eta^2)$, from which the conformality of (3.7) follows. This conjugation interchanges characters of A_1 and A_2 by Proposition 3.3, hence it must interchange the non-character factor of (3.4) or (1.13). There is a two-dimensional space of such factors for given characters, hence the two-dimensional spaces are interchanged. It is easy to write the explicit 2×2 matrix for the conjugation (3.7) using the basis J, K for the Bessel function.

Example 3. Let A_3 be the usual Cartan

$$A_3 = \{\text{scalar multiplication, rotation about } t \text{ axis}\} \qquad (3.8)$$

in the notation of (1.11). The basic solutions are given by (1.12).

Thus far we have discussed only the Cartan group. Let us examine the effect of the Weyl groups W_A. If $w \in W_A$, then by Proposition 3.3 w preserves the set of basic solutions corresponding to A.

Let us illustrate this for the following conformal map w:

$$r' = e^{i\varphi}$$

$$e^{i\varphi'} = r \qquad (3.8')$$

$$\cosh \zeta' = \coth \zeta.$$

Using the expression $ds^2 = dr^2 - r^2\, d\zeta^2 - r^2 \sinh^2 \zeta\, d\varphi^2$ it is easy to verify that this map preserves ds^2 up to a factor and so is conformal. Note that this w interchanges φ and r. Hence it interchanges the dual variables n and s. If we compute things explicitly we are led to the basic formula for the quadratic transform of hypergeometric functions (sometimes called Whipple's formula),

$$Q_s^m(\cosh \zeta) = e^{im\pi}(\pi/2)^{1/2}\Gamma(m+s+1)(\sinh \zeta)(\sinh \zeta)^{-1/2}P_{-m-1/2}^{-s-1/2}(\coth \zeta)$$
$$(3.9)$$

(see [2], vol. I, p. 141).

Let us remark that the factor $(\sinh \zeta)^{-1/2}$ comes from conformal factors.

So much for the 3-dimensional theory; let us pass to $\mathbf{R}^4$, which we think of as the space of 2×2 matrices. This is a representation space for $\mathcal{G} = G \times G$ where $G = SL(2, \mathbf{R})$ as above, namely

$$(g_1, g_2): a \to g_1^{-1} a g_2. \qquad (3.10)$$

Again, the quadratic form $\det a$ is invariant. The associated differential operator D is now ultrahyperbolic, i.e., signature $(2, 2)$. Thus $\mathcal{G}$ is embedded in $O(2, 2)$ and, in fact, $\mathcal{G}$ is essentially $O(2, 2)$.

The conformal group $\mathscr{G}^c = O(3, 3)$ has rank 3 and is essentially $SL(4)$. The rank being 3 means that the situation we face here is similar to the one we faced in $\mathbf{R}^3$, namely, diagonalizing the action of a Cartan A of rank 3 leaves us with one interesting function of one variable. We shall see this in detail below.

For later purposes we note that the order of the ordinary Weyl group of $\mathscr{G}^c$ is 24.

As in the case of $\mathbf{R}^3$ we continue our study by choosing various Cartans. The simplest Cartan $A = \mathbf{R}^4$ is the translation group of $\mathbf{R}^4$. This leads to exponential solutions of D as in the case of $\mathbf{R}^3$.

Example 4. Let A_4 denote the ordinary Cartan

$$A_4 = \{\text{left rotation, right rotation, scalar multiplication}\}$$

$$= K_L \times K_R \times \mathbf{R}. \tag{3.11}$$

We need a "natural" coordinate system related to this A_4. If we apply A_4 to the identity we need one more one parameter group, which we choose to be the diagonal group A^0 acting on the left. This means that we use coordinates

$$\begin{pmatrix} \cos \varphi_L & \sin \varphi_L \\ -\sin \varphi_L & \cos \varphi_L \end{pmatrix} \begin{pmatrix} e^\zeta & 0 \\ 0 & e^{-\zeta} \end{pmatrix} \begin{pmatrix} r & 0 \\ 0 & r \end{pmatrix} \begin{pmatrix} \cos \varphi_R & \sin \varphi_R \\ -\sin \varphi_R & \cos \varphi_R \end{pmatrix}. \tag{3.12}$$

The reason for choosing A^0 will become apparent in the next section.

Each hyperboloid $r = \text{const}$ can be identified with $G = G_L$ (left G), and (3.12) means that we are using the polar decomposition $G = KAK$ to set up coordinates.

In terms of this coordinate system the operator D takes the form (see [1])

$$D = \frac{\partial^2}{\partial r^2} + \frac{3}{r} \frac{\partial}{\partial r} - \frac{1}{r^2} \Delta_\theta, \tag{3.13}$$

where Δ_θ is the Laplacian on the hyperboloid. The basic solutions are of the form

$$f(r, \varphi_L, \varphi_R, \zeta) = e^{in\varphi_L + im\varphi_R} r^{is} f_{n,m,s}(\zeta). \tag{3.14}$$

We find from (3.13) that

$$f_{m,n,s}(\zeta) = y^{|m-n|/2}(1+y)^{-|m+n|/2} {}_2F_1(c_1, c_2; c_3; -y), \tag{3.15}$$

where $_2F_1$ is the hypergeometric function, $y = \sinh^2 \zeta$, $c_1 = s + \frac{1}{2}(|m - n| - |m + n|)$, $c_2 = 1 - s + \frac{1}{2}(|m - n| - |m + n|)$, $c_3 = 1 + |m - n|$. The fact that we get a hypergeometric series will be made conceptual in the next section.

Note that, except for integrality conditions (which are not important to us) we obtain the most general hypergeometric function in this fashion.

We can now apply the Weyl group W_{A_4}. This acts on the basic solutions (3.14) and produces 24 basic solutions from any one. In this way we obtain 24 hypergeometric series which satisfy the same hypergeometric equation (when we adjust the new parameters properly). This coincides with Kummer's list (see e.g., [2] Vol. I, p. 105). The meaning of Kummer's list is that we have picked out 24 hypergeometric series in the two-dimensional space of solutions of the hypergeometric equation for which the parameters depend in a simple way on the original parameters. (The sum of hypergeometric series is not necessarily a hypergeometric series.)

It is somewhat complicated to compare our list with that of Kummer so I shall omit it from this article. It will appear elsewhere.

Example 5. We choose for A_5 the confluent Cartan

$$A_5 = K_R \times N_L \times \mathbf{R}. \tag{3.16}$$

Here N is the nilpotent group $\left(\begin{smallmatrix} 1 & x \\ 0 & 1 \end{smallmatrix}\right)$. We again use A_L^0 for the fourth one-parameter group, so we obtain a coordinate system in $\mathbf{R}^4$. On each hyperbolid $r = \text{const}$ we have KAN coordinates.

Since the calculations do not seem to be standard, I shall give them. The coordinates t, x, y, z are determined by

$$\begin{pmatrix} r & 0 \\ 0 & r \end{pmatrix}\begin{pmatrix} 1 & n \\ 0 & 1 \end{pmatrix}\begin{pmatrix} a^{1/2} & 0 \\ 0 & a^{-1/2} \end{pmatrix}\begin{pmatrix} c & s \\ -s & c \end{pmatrix} = \begin{pmatrix} u & q \\ w & v \end{pmatrix},$$

with $t = u + v$, $x = u - v$, $y = q + w$, $z = q - w$. Thus,

$$r^{-1}t = c(a^{1/2} + a^{-1/2}) - a^{-1/2}sn$$

$$r^{-1}x = c(a^{1/2} - a^{-1/2}) - a^{-1/2}sn$$

$$r^{-1}y = s(a^{1/2} - a^{-1/2}) + a^{-1/2}cn \tag{3.17}$$

$$r^{-1}z = s(a^{1/2} + a^{-1/2}) + a^{-1/2}cn.$$

Here $c = \cos \theta$ and $s = \sin \theta$.

The Jacobian of the transformation is given by

$$J(t, x, y, z; r, a, \theta, n) =$$

$$\begin{pmatrix} \overset{+}{c\alpha} - a^{-1/2}sn & \frac{1}{2}a^{-1}(\overset{-}{c\alpha} + a^{-1/2}sn) & -\overset{+}{s\alpha} - a^{-1/2}cn & -a^{-1/2}s \\ \overset{-}{c\alpha} - a^{-1/2}sn & \frac{1}{2}a^{-1}(\overset{+}{c\alpha} + a^{-1/2}sn) & -\overset{-}{s\alpha} - a^{-1/2}cn & -a^{-1/2}s \\ \overset{-}{s\alpha} + a^{-1/2}cn & \frac{1}{2}a^{-1}(\overset{-}{s\alpha} - a^{-1/2}cn) & \overset{+}{c\alpha} - a^{-1/2}sn & a^{-1/2}c \\ \overset{+}{s\alpha} + a^{-1/2}cn & \frac{1}{2}a^{-1}(\overset{+}{s\alpha} - a^{-1/2}cn) & \overset{-}{c\alpha} - a^{-1/2}sn & a^{-1/2}c \end{pmatrix}. \quad (3.18)$$

Here we have written $\overset{+}{\alpha} = a^{1/2} + a^{-1/2}$ and $\overset{-}{\alpha} = a^{1/2} - a^{-1/2}$ and we have omitted a factor r from the second, third, and fourth columns. The determinant of the Jacobian is $4r^3a^{-2}$.

The matrix of signed minors of (3.18) divided by r^3a^{-2} is

$$\begin{pmatrix} -(\overset{+}{c\alpha} - a^{-1/2}sn) & 2ar^{-1}(\overset{-}{c\alpha} + a^{-1/2}sn) & 2ar^{-1}a^{-1/2}s & -2ar^{-1}(\overset{-}{s\alpha} - a^{-1/2}cn) \\ \overset{-}{c\alpha} - a^{-1/2}sn & -2ar^{-1}(\overset{+}{c\alpha} + a^{-1/2}sn) & -2ar^{-1}a^{-1/2}s & 2ar^{-1}(\overset{+}{s\alpha} - a^{-1/2}cn) \\ \overset{-}{s\alpha} + a^{-1/2}cn & -2ar^{-1}(\overset{-}{s\alpha} - a^{-1/2}cn) & 2ar^{-1}a^{-1/2}c & -2ar^{-1}(\overset{+}{c\alpha} + a^{-1/2}sn) \\ -(\overset{+}{s\alpha} + a^{-1/2}cn) & 2ar^{-1}(\overset{-}{s\alpha} - a^{-1/2}cn) & -2ar^{-1}a^{-1/2}c & 2ar^{-1}(\overset{+}{c\alpha} + a^{-1/2}sn) \end{pmatrix}.$$

$$(3.19)$$

This yields the ultrahyperbolic operator (up to a constant)

$$\Box = \frac{\partial^2}{\partial t^2} - \frac{\partial^2}{\partial x^2} - \frac{\partial^2}{\partial y^2} + \frac{\partial^2}{\partial z^2}$$

$$= \frac{\partial^2}{\partial r^2} + \frac{3}{r}\frac{\partial}{\partial r} - \frac{4a^2}{r^2}\left(\frac{\partial^2}{\partial a^2} + \frac{\partial^2}{\partial n^2}\right) + \frac{2a}{r^2}\frac{\partial^2}{\partial \theta \, \partial n}. \quad (3.20)$$

Applying this to $r^s e^{in\theta} e^{i\xi n} f(a)$ gives

$$\left[s(s-1) + 3s - 4a^2\frac{d^2}{da^2} + 4a^2\xi^2 - 2a\xi n \right] f. \quad (3.21)$$

We divide by $4a^2$ and set the result equal to zero. Thus

$$\left[\frac{d^2}{da^2} + \left(-\xi^2 + \frac{\xi n}{2a} - \frac{s(s+2)}{4a^2} \right) \right] f = 0. \quad (3.22)$$

Comparing this with [2], Vol. I, p. 248(4) shows that f is Whittaker's function $M_{\kappa,\mu}(a)$ if we set $\xi = 1/2$, $\kappa = n/4$, and $\mu = (s+1)/2$.

As it stands, $M_{\kappa,\mu}$ is not a hypergeometric series. We must multiply $M_{\kappa,\mu}$ by $e^{a/2}a^{-1-s/2}$. This point will be clarified in the next section.

We can also use the Weyl group W_{A_5} to obtain the Kummer transformation ([2], Vol. I, p. 253) for Whittaker's function.

I hope that these examples illustrate the power of the conformal group.

4. Beyond the Conformal Group

I have just noted the power of the conformal group; yet there is a beyond. This beyond seems necessary to explain Gauss contiguity and also to explain why the non-character factor of the basic solution can be expressed as a hypergeometric series in the form (1.1) or (1.3). Another beyond, relating to asymptotics and confluence, is presented in Section 5.4.

We approach these questions by passing to the infinitesimal analog of (2.1), which is

$$[D, \psi] = \alpha D + D\beta. \tag{4.1}$$

Here ψ is the infinitesimal generator corresponding to a one-parameter group of φ in (2.1), and we have used the letters α, β both for the conformal factors in (2.1) and for their infinitesimal elements in (4.1).

(4.1) can be looked at from two points of view. In the first place it states that the vanishing of Df is the same as the vanishing of $D(-\beta + \psi)f$. This is the point of view presented first and it is related to Gauss contiguity.

In the second place, for $\beta = 0$ and for α a constant, we can regard (4.1) as stating that D is a root for ψ, that is, D is an eigenfunction of ψ (under bracket) with eigenvalue α. This point of view is discussed below; it is related to hypergeometric series.

A consequence of (2.1) is the equivalence of the vanishing of Df and $D\beta\psi f$. Ignoring β (as we usually do in heuristics) we may say that ψ preserves the kernel of D.

Now, suppose that ψ does not preserve the kernel of D but rather that ψ maps the kernel of D into the kernel of D^2. In the cases of interest, D is a Laplacian, so ψ sends harmonics into biharmonics. The natural extension of (4.1) which accomplishes this is the double commutator condition

$$[D, [D, \psi]] = \alpha D + D\beta. \tag{4.2}$$

Otherwise stated, $[D, \psi]$ is conformal. It is clear that if (4.2) is satisfied then $Df = 0$ implies $D^2\psi f - D\beta f = 0$. Hence if $\beta = 0$ we have a map from harmonics to biharmonics. An operator ψ satisfying (4.2) is called *biconformal*. (We could define triconformal, etc. We shall make some remarks about them later.)

Remark.   Unlike the situation we encountered in Section 2, we do not assume any special structure for ψ.

Most of the calculations in Section 3 depend on the following uniqueness property: A solution of D which is a character of some Cartan A is uniquely determined (up to a constant multiple) by some regularity or smallness condition. This is how we pick J or K amongst the Bessel functions and P or Q amongst the Legendre functions, etc. For D^2 we have a two-dimensional space of regular solutions which are characters under A.

What does this mean? Suppose that I can find three operators ψ_1, ψ_2, ψ_3 which are biconformal and each takes some regular basic solution f_j (for an A which is the same for ψ_1, ψ_2, ψ_3) into a regular basic solution. Suppose that the A characters of $\psi_j f_j$ are all the same. Then we obtain a linear relation among $\psi_j f_j$ hence among the f_j since the regular biharmonics with a given A character form a two-dimensional space.

The best choice of such operators ψ_j are those which boost the eigenvalues of A. These lead to the contiguity relations for $_2F_1$.

Let me illustrate these ideas.

For the case of the wave operator $\square$ on $\mathbf{R}^3$ discussed in Section 3 above, we note that the operators

$$
\begin{array}{ll}
\text{(a)} & \text{multiplication by } t, \\
& \hspace{6cm} (4.3) \\
\text{(b)} & \text{multiplication by } x \pm iy,
\end{array}
$$

are biconformal for $\square$. Moreover, both these boost the eigenvalue for $r\partial/\partial r$ and $x \pm iy$ boost (up or down) the eigenvalue for $\partial/\partial\varphi = x\,\partial/\partial y - y\,\partial/\partial x$ while t leaves the eigenvalue of $\partial/\partial\varphi$ unchanged. (See formulae (1.11) and (1.12) above.)

Thus, if we start with the harmonic, i.e., a solution $f_{s-1,n-1}$ of $\square$ which has eigenvalue $s-1$ for $r\,\partial/\partial r$ and $n-1$ for $i\,\partial/\partial\varphi$, then $(x+iy)f_{s-1,n-1}$ is biharmonic with respective eigenvalues s, n. The biharmonic functions $(x-iy)f_{s-1,n+1}$ and $f_{s,n}$ (which is harmonic, hence biharmonic) have the same eigenvalues for $r\,\partial/\partial r$ and $\partial/\partial\varphi$. This leads to a linear relation amongst the corresponding Legendre functions of (1.12). It is not difficult to write this relation explicitly:

$$
P_s^{n+1}(\cosh\zeta) + 2n\coth\zeta P_s^n(\cosh\zeta) = (s+1-n)(s+n)P_s^{n-1}(\cosh\zeta).
$$
$$(4.3')$$

The reader should have no difficulty in using the same idea to derive the other Gauss contiguity relations using $\mathbf{R}^3$ or $\mathbf{R}^4$.

The Gauss contiguity relations arise from biconformal operators which are related to biharmonic functions. Naturally one could obtain other relations from triconformal, etc., operators.

Conjecture. All contiguity relations can be deduced from those of Gauss.

Remark. Grunbaum and Duistermatt have studied the question of finding differential equations in the spectral parameter which are satisfied by eigenfunctions of suitable differential operators. (Contiguity can be thought of as difference relations in the spectral parameters.) They are also led to higher commutators. Similar ideas appear in the PhD thesis of Grunbaum's student, Michael Reach.

The second topic that seems beyond the conformal group is the nature of the power series of the hypergeometric function. More precisely, can we understand how to choose the parameter (such as $\cosh \zeta$ for the Legendre function) so that the non-character factor of the basic solution has a hypergeometric series? (On a personal note let me say that I searched for such an understanding for 17 years.)

To clarify this point, let us first understand the character factor. This comes from studying the eigenfunctions of the generators ψ_j of the Cartan group. Since ψ_j are conformal, relation (4.1) holds. If the annoying term β were not present (we can get around this point without much difficulty), then D behaves like a boost operator, except that α is not necessarily a constant, which is the usual situation with boost operators.

If, in fact, α is a constant (or if we ignore this problem) then D maps fixed eigenspaces of A into other fixed eigenspaces of A. Thus the kernel of D can be studied on each eigenspace of A.

This procedure corresponds, more or less, to decomposing the kernel of D under the action of A.

For α a constant (and $\beta = 0$) we can think of (4.1) as an "eigenfunction" equation. Thus D is an "eigenfunction" or "eigenoperator" of ψ (acting by commutator) with eigenvalue α and this is what accounts for the ability to decompose the kernel of D under A.

The next step in this hierarchy is to consider operators ψ for which D is a sum of two eigenfunctions or eigenoperators. For example, the operator $t \, \partial/\partial t$ has such a property with respect to the wave operator $\square = \partial^2/\partial t^2 - \partial^2/\partial x^2 - \partial^2/\partial y^2$, namely, $\partial^2/\partial t^2$ and $\partial^2/\partial x^2 + \partial^2/\partial y^2$ are eigenoperators with different eigenvalues. Because of (1.4), (1.5) and the remarks following these equations, it is natural to call such operators *hypergeometric*. More generally, we make the

Definition. An operator ψ is called *m geometric* with respect to an operator D if D is the sum of m eigenoperators of ψ acting by commutator.

One geometric operators are called *conformal* or *geometric* and 2 geometric operators are called *hypergeometric*.

We can now understand completely the structure of the recursion relations for the power series coefficients of hypergeometric functions. For suppose we have a Cartan A whose orbits have codimension one, as in Section 3. Let $\psi_1, \psi_2, \ldots, \psi_{n-1}$ be infinitesimal generators for A; let us assume the ψ_j are conformal with $\beta_j = 0$. Suppose we can find a hypergeometric ψ_n which commutes with all the ψ_j. We perform a simultaneous diagonalization of all the ψ_j. Writing any function f in terms of these eigenfunctions, the equation $Df = 0$ becomes a two-step relation for the expansion coefficients in terms of the eigenfunctions of ψ_n. This is exactly the hypergeometric relation discussed in (a) of Section 1 (except that one has to be careful to fulfill the requirement that the relation be rational).

In the case of the wave operator $\square$ on $\mathbf{R}^3$, we can choose

$$\psi_1 = y \frac{\partial}{\partial x} - x \frac{\partial}{\partial y}$$

$$\psi_2 = r \frac{\partial}{\partial r} \tag{4.4}$$

$$\psi_3 = t \frac{\partial}{\partial t}.$$

The operators ψ_1 and ψ_2 are conformal while ψ_3 is hypergeometric. Thus the non-character factor in the basic solutions is a hypergeometric series in $\{t^n\}$ which are the eigenfunctions of ψ_3. Since $t = r \cosh \zeta$, this explains the choice of parameter $\cosh \zeta$.

It should be noted that the Legendre function, as it stands, is not a hypergeometric series, as the Legendre equation involves three degrees of homogeneity. It is only after multiplication by $(\sinh \zeta)^n$ that the Legendre equation becomes one with two degrees of homogeneity. The point is that we should use the vector fields of (4.4) rather than those obtained from (1.19).

One could also use a modified form of (4.4) in which ψ_1 is replaced by

$$z \frac{\partial}{\partial z} = x \frac{\partial}{\partial x} + y \frac{\partial}{\partial y} + i \left(y \frac{\partial}{\partial x} - x \frac{\partial}{\partial y} \right).$$

A similar idea is used in understanding how the Whittaker function is related to hypergeometric series.

Actually, a simpler example occurs in the case of the Laplacian for $n = 2$. We choose

$$\psi_1 = r \frac{\partial}{\partial r}$$

$$\psi_2 = x \frac{\partial}{\partial x}.$$

(4.5)

We are led quickly to the Tchebycheff polynomials.

In fact, the simplest example is for $n = 1$, $D = d/dx - \lambda$ and $\psi = x\, d/dx$. The above explains the recursion for the power series coefficients of $\exp(\lambda x)$.

Added in proof. I have recently discovered a new method of establishing the hypergeometric nature of various power series. The results depend on the concept of *nonlinear Fourier transform.* The usual Fourier transform is based on the exponential of linear functions; the nonlinear Fourier transform is based on the exponential of polynomial functions. The simplest case is the quadratic Fourier transform on $\mathbf{R}^2$ for which the kernel is

$$\exp(ix_1\hat{x}_1 + ix_2\hat{x}_2 + ix_1^2 u + ix_2^2 v + ix_1 x_2 w).$$

To compute the power series coefficients of, e.g. $J_0(r)$, we use the integral representation of $J_0(r)$ namely as the usual Fourier transform of the measure $d\theta$ on the unit circle in the x_1, x_2 plane. Note that the quadratic Fourier transform of this measure is just $\exp(iu)$ on the set $\hat{x}_1 = \hat{x}_2 = w = 0$, $u = v$.

Now, unlike the linear Fourier transform, the quadratic Fourier transform satisfies a system of linear partial differential equations. In the present case of functions of r, this amounts to a heat-like equation relating $J_0(r)$ and $\exp(iu)$. If we write $J_0(r) = \sum c_k r^k$ and apply the explicit formula for the fundamental solution for the heat equation we arrive at the usual solution

$$c_{2k} = (-1)^k 2^{-2k}/(k!)^2$$

(with $c_l = 0$ for l odd).

By a simple modification we can compute the power series coefficients of $J_u(\hat{r})$ and also of the general hypergeometric functions which arise from group representations.

(The first exposition of nonlinear Fourier analysis appears in my article "Fourier Analysis, Partial Differential Equations, and Automorphic Functions" in *Proceedings of the American Math. Society Summer Institute,* Bowdoin, 1987.)

We have explained that 2 geometric (hypergeometric) operators are related to hypergeometric functions. What about 3 geometric operators for the wave operator or Laplacian? The answer is that they are related to Mathieu and Lamé functions (see [2] Vol. III; we shall quote freely from that source).

If we examine the standard form of Mathieu's operator

$$\frac{d^2}{dz^2} + h - 2\theta \cos 2z, \tag{4.6}$$

we note that it is the sum of three eigenoperators of d/dz, namely

$$\frac{d^2}{dz^2} + h \text{ has eigenvalue } 0, \qquad -2\theta \cos 2z = -\theta(e^{2iz} + e^{-2iz}).$$

This is also true for the algebraic form

$$4x(1-x)\frac{d^2}{dx^2} + 2(1-2x)\frac{d}{dx} + (h - 2\theta + 4\theta x). \tag{4.7}$$

Similarly, an algebraic form of Lamé's operator is

$$\frac{d^2}{dx^2} + \frac{1}{2}\left(\frac{1}{x} + \frac{1}{x-1} + \frac{1}{x-k^{-2}}\right)\frac{d}{dx} + \frac{hk^{-2} - n(n+1)x}{4x(x-1)(x-k^{-2})}, \tag{4.8}$$

which again has three degrees of homogeneity (as is seen by multiplying by $4x(x-1)(x-k^{-2})$).

Let us see how this arises from 3 geometric operators. For simplicity we deal with the Mathieu equation; the Lamé equation can be treated along similar lines.

To use the notation of [2], Vol. III, p. 91, we introduce coordinates in $\mathbf{R}^3$: t and u, v where

$$x = c \cosh u \cos v$$
$$y = c \sinh u \sin v. \tag{4.9}$$

Then we can write

$$[\cosh 2u - \cos 2v](\Delta + k^2) = \frac{\partial^2}{\partial u^2} + \frac{\partial^2}{\partial v^2} + [\cosh 2u - \cos 2v]\left(\frac{\partial^2}{\partial t^2} + k^2\right). \tag{4.10}$$

For the operator $[\cosh 2u - \cos 2v](\Delta + k^2) = D$ we can use one 1 geometric operator ψ_1 and two 3 geometric operators ψ_2, ψ_3, where

$$\psi_1 = \frac{\partial}{\partial t}$$

$$\psi_2 = \frac{\partial}{\partial u} \tag{4.11}$$

$$\psi_3 = \frac{\partial}{\partial v}.$$

For the Lamé equation, all three operators are 3 geometric.

Problem. Find special properties of m geometric operators.

The idea of 2 geometric operators has various connections with classical hypergeometric functions. Consider the following relation for Laguerre polynomials or functions (see [2], Vol. II, p. 189):

$$x \frac{d}{dx} L_n^\alpha(x) = n L_n^\alpha(x) - (n + \alpha) L_{n-1}^\alpha(x). \tag{4.12}$$

Here L_n^α is the eigenfunction of the Laguerre operator

$$\mathscr{L}^\alpha = x \frac{d^2}{dx^2} + (\alpha + 1 - x) \frac{d}{dx} \tag{4.13}$$

with eigenvalue n.

We can interpret (4.12) as saying that $x d/dx$ sends an eigenfunction of $\mathscr{L}^\alpha$ into a sum of two eigenfunctions, one with the same value n and one with a boosted (down) eigenvalue $n - 1$. This should mean that the operator $x d/dx$, acting by commutator, is the sum of two eigenoperators of $\mathscr{L}$. Indeed we have the commutator relation

$$\left[\mathscr{L}^\alpha, x \frac{d}{dx} \right] = x \frac{d^2}{dx^2} + (\alpha + 1) \frac{d}{dx}$$

$$= \mathscr{L}^\alpha + x \frac{d}{dx}. \tag{4.14}$$

Clearly $\mathscr{L}^\alpha$ is an eigenoperator of $\mathscr{L}^\alpha$ with eigenvalue 0 while, by (4.14), $\mathscr{L}^\alpha + x \, d/dx$ has eigenvalue 1.

Before concluding this section let me give an example of a boost operator for which the eigenvalue is not a constant (and is even an operator).

In fact,

$$\left[\mathcal{L}^{\alpha}, \frac{d}{dx} \right] = -\frac{d^2}{dx^2} + \frac{d}{dx}$$

$$= \left(-\frac{d}{dx} + 1 \right) \frac{d}{dx}. \tag{4.15}$$

In accordance with general principles (or by a simple direct computation) this means that the operator d/dx transforms solutions of $\mathcal{L}^{\alpha} + n$ into solutions of

$$\mathcal{L}^{\alpha} + n - \left(-\frac{d}{dx} + 1 \right) = \mathcal{L}^{\alpha+1} + (n - 1). \tag{4.16}$$

This is the origin of the formula

$$\frac{d}{dx} L_n^{\alpha} = -L_{n-1}^{\alpha+1}. \tag{4.17}$$

We note that there is an interesting Lie algebra associated to the Laguerre operator. This has the generators

$$x\frac{d}{dx}, \qquad \mathcal{L}^{\alpha} + x\frac{d}{dx}, \qquad x, \tag{4.18}$$

which, by (4.14) and simple calculations, are the standard generators for the Lie algebra of $SL(2)$.

There is a well-known Lie algebra associated to the Hermite operator.

Problem. Which other classical functions or polynomials are associated to Lie algebras?

5. Miscellanea

5.1. Integral Representation

There are classical integral representations for hypergeometric functions. We have mentioned the Mellin–Barnes integrals which are an integral form of the hypergeometric series. Such integrals can be understood using the ideas of 2 geometric operators described in Section 4. Much work has been done on other types of integrals which have their origin in the integrals of Euler and Dirichlet.

These latter integrals can be understood from the viewpoint of the fundamental principle of [3]. For simplicity, let us consider the case of $\mathbf{R}^3$. Then any solution f of the wave equation can be expressed in the form

$$f(t, x, y) = \int e^{it\hat{t} - ix\hat{x} - iy\hat{y}} \, d\mu(\hat{t}, \hat{x}, \hat{y}), \tag{5.1}$$

where μ is a suitable measure on the (complex) light cone $\hat{t}^2 = \hat{x}^2 + \hat{y}^2$. To construct a basic solution we can choose μ so that it transforms according to a character χ of the Cartan group A. Then f will, of necessity, transform according to a simple modification $\tilde{\chi}$ of χ, provided that A is contained in the parabolic part of the conformal group, i.e., A is contained in the group generated by $G = O(1, 2)$ and $\mathbf{R}^3$ acting by translations and $\mathbf{R}^\times$ acting by scalar multiplication.

Consider the Example 3 of Section 3, for which the Cartan is described in (3.8). The simplest μ to use in (5.1) is then

$$d\mu = \hat{t}^{is} \, e^{in\hat{\varphi}} \, d\hat{\varphi} \, d\hat{t}/\hat{t} \tag{5.2}$$

on the positive light cone. There are some analytic difficulties in defining things precisely. The following formal argument can be justified:

$$f(t, x, y) = \int\!\!\int e^{i\hat{t}r \cosh \zeta + i\hat{t}r \sinh \zeta \cos (\varphi + \hat{\varphi})} \hat{t}^{is} \, e^{in\hat{\varphi}} \, d\hat{\varphi} \, d\hat{t}/\hat{t}$$

$$= \Gamma r^{-is} \int [\cosh \zeta + \sinh \zeta \cos (\varphi + \hat{\varphi})]^{-is} \, e^{in\hat{\varphi}} \, d\hat{\varphi}$$

$$= \Gamma r^{-is} e^{-in\varphi} P_{is}^n(\cosh \zeta), \tag{5.3}$$

according to [2], Vol. I, p. 157 formula (15). Here P_{is}^n is a Legendre function and Γ represents some simple factor involving the Γ function and exponential functions. We have thus obtained an integral representation for $P_s^n(\cosh \zeta)$.

The same method works for all the hypergeometric functions.

It would be of interest to solve the following

Problem. Use 2 geometric operators to compare the Mellin–Barnes and the above integral representations.

Integral representations provide one of the most powerful tools for the study of functions. However I don't know how to apply the above method to Lamé, or Mathieu functions.

Problem. Can the above method be used to obtain integral representations of Lamé and Mathieu functions? If not, why?

5.2.   Group Representations

The appearance of hypergeometric functions as matrix coefficients of representations of some classical groups provided one of the early stimuli for the study of these functions. Let me explain how representation theory fits into our work.

Let us begin, as usual, with the action of $G = SL(2, \mathbf{R}) \sim SO(1, 2)$ on $\mathbf{R}^3$. G preserves the kernel of $\square$. Moreover, G commutes with the scalar multiplication group $\mathbf{R}^\times$ which is the maximal abelian subgroup of G^c which commutes with G. Thus we can decompose the kernel of $\square$ under $\mathbf{R}^\times$. The set of f with $\square f = 0$ which transform according to a character χ of $\mathbf{R}^\times$ forms a representation space for G. The representation is the (not necessarily unitary) principal series for G.

A basis for the principal series consists of all basic solutions $f_{s,n}$ for s fixed, n varying. This space can be identified with the non-character components of the $f_{s,n}$, that is, the linear span of P_s^n.

By the theory expounded in the above Section 5.1, we can identify the principal series with the representation of G on functions on the light cone which are homogeneous. As the light cone is G/MN, this corresponds to a usual description of the principal series.

We can perform a similar construction for the group $\mathscr{G} = G \times G$ acting on $\mathbb{R}^4$ as before. We obtain representations of $\mathscr{G}$ in this way. If we want representations of G (left) itself, we note that G_L commutes with $\mathbf{R}^\times \times K_R$ where K_R is a circle in G_R. Moreover, $\mathbf{R}^\times \times K_R$ is a maximal abelian subgroup of $\mathscr{G}$ which commutes with G_L. Considering solutions of D which transform under $\mathbf{R}^\times \times K_R$ like a character $\chi = \{s, n\}$ gives a representation of G. The representation space can be identified, using (3.12) with functions on G which are eigenfunctions of the group Laplacian and which transform on the right under K_R like $\exp(in\varphi_R)$.

We can, naturally, describe this space in terms of functions on the corresponding light cone.

If we consider the compact form G_0 of G and the Laplacian Δ replacing the wave or ultrahyperbolic operators, then the above is the usual theory of spherical harmonics and their relation to group representations.

5.3.   Higher Rank Groups

Naturally, one would like to extend all the above to general non-compact semisimple Lie groups. This is a tremendous task so I shall content myself with a few words.

One of the main ingredients of our theory is the pair G, G^c. In fact, we started with G and passed to G^c. In the general theory we reverse things and pass from G^c to G. Let G^c be a semisimple Lie group and let P^c be a parabolic subgroup. We write $P^c = M^c A^c N^c$ in the usual notation. (In our example of the orthogonal group $G^c = O(p, q)$, P^c is a maximal parabolic, $N^c = \mathbf{R}^n$ under addition, $M^c = O(p-1, q-1)$ and $A^c = \mathbf{R}^\times$.)

In analogy to our main work, we should have a differential operator D on N^c which is $G = M^c$ invariant. G^c should act conformally on D, which means that the kernel of D is G^c invariant.

By what was discussed in Section 5.1 above, the solutions of the equations like $\square$ can be identified with functions on the light cone. For the orthogonal group the light cone is G/MN. Thus, in general, we want to identify the solutions of D with functions on G/MN.

To accomplish this we search for representations of G on some $\mathbf{R}^k$ such that there is a point $p_0 \in \mathbf{R}^k$ whose isotropy group in G is exactly MN. The G^c orbit of p_0 is called a light cone Γ.

It is not difficult to find light cones which, in addition, are "large" pieces of algebraic varieties. Such varieties are generally described as the common zeros of a family of quadratic polynomials $\hat{\mathbf{D}} = (\hat{D}_1, \ldots, \hat{D}_r)$. The Fourier transform (using a G invariant inner product) $\mathbf{D}$ is called a wave operator.

In accordance with what we said above, let us change our viewpoint and study a light cone Γ^c for G^c. Denote by $\mathbf{D}^c$ the corresponding system of differential equations.

Now, G^c acts on Γ^c (say on the left), and A^c acts on Γ^c (on the right) since A^c normalizes $M^c N^c$. Thus G^c acts on functions on Γ^c which are A^c homogeneous, that is, transform according to a representation of A^c. Such functions (perhaps vector valued) have their natural existence on G^c/P^c which can be identified with $K^c/K^c \cap P^c$.

The theory of Shubert cells shows how to decompose G^c/P^c into Weyl transforms of N^c. Thus we can regard G^c as acting "conformally" on N^c. (In our example of the orthogonal group this action is, in fact, the conformal action on $N^c = \mathbf{R}^n$.) The parabolic P^c acts "affinely" on N^c, in particular, P^c preserves N^c.

We can now state the main problem:

Problem.   Find light cones Γ for $G = M^c$ such that G^c acts conformally on the corresponding system $\mathbf{D}$.

We shall return to this problem elsewhere.

Let me mention one interesting case. Let $G_1 = SU(1, 2)$. It is easily seen that the Heisenberg group H is the nilradical of a suitable parabolic P_1. A computation shows that the Laplacian Δ_H of H is G_1 conformal.

We have mentioned in Section 1 that Laguerre (Whittaker) functions arise from the kernel of Δ_H by fixing the central character and the rotational character, that is, the character of a suitable Cartan in G_1. We noted in Section 4 how they arise from $\mathscr{G} = SL(2) \times SL(2) \sim O(2, 2)$.

The conformal group of $\mathscr{G}$ is $O(3, 3)$. This differs in an insignificant way, for us, from $O(2, 4)$, which contains $G_1 = SU(2, 4)$. Thus there is hope of finding a suitable element of $O(2, 4)$ or $O(3, 3)$ which explains the appearance of the Laguerre functions in both contexts.

However, I do not know how to solve this problem.

5.4.   Asymptotics and Confluence

Much of the classical work on hypergeometric functions is concerned with their asymptotic behavior at the singular points and with confluence. The classical concept of confluence consists of setting $x = z/b$ in (1.1) and letting $|b| \to \infty$.

From our point of view there are two approaches to these phenomena:

(a) Geometric.
(b) Analytic.

5.4.1.   *Geometric Approach*

Let us first clarify the geometry related to asymptotics and confluence. For simplicity, let us discuss the three-dimensional representation of $G = SO(1, 2)$ as in Section 3. The natural geometry associated with this representation is that of the orbits of the representation. These orbits are of three types:

(1)   *Hyperboloid of two sheets.*   The isotropy group of a point is conjugate to K. In particular it is compact.

(2)   *Light cone.*   The isotropy group is conjugate to MN, where $N = \{\begin{pmatrix} 1 & n \\ 0 & 1 \end{pmatrix}\}$ and $M = \pm I$. (M is basically unimportant and will be ignored in general.)

(3)   *Hyperboloid of one sheet.*   The isotropy group is conjugate to

$$\tilde{K} = \left\{ \begin{pmatrix} \cosh \zeta & \sinh \zeta \\ \sinh \zeta & \cosh \zeta \end{pmatrix} \right\}.$$

Although $\tilde{K}$ is conjugate to the diagonal group A, this is only true for $SO(1, 2)$ and it is better to think of $\tilde{K}$ as another real form of K.

The analysis of Section 3 was mostly concerned with orbits of type (1). By the geometry of asymptotics we mean the study of what happens, for example, as the hyperboloids approach the light cone. The fact that all orbits appear in a single entity, namely $\mathbf{R}^3$, allows us to make a simple calculation of the asymptotics of the Legendre functions at infinity. For note that by (1.12) the function $f(r, \zeta, \varphi)$ is homogeneous in r. Thus the study of $f(1, x, 0)$ as $x \to 1^-$ is equivalent to the study of $P_{is}^m(\cosh \zeta)$ as $\zeta \to \infty$ because the (r, ζ, φ) corresponding to $(t, x, y) = (1, x, 0)$ is $(\sqrt{1-x^2}$, $\text{ch}^{-1}(1-x^2)^{-1/2}, 0)$.

We can compute the limit of $f(1, x, 0)$ as $x \to 1^-$. It is just $f(1, 1, 0)$. This can be evaluated most simply by using the Fourier representation of f as in Section 5.1 above. Thus, formally, f is the Fourier transform of the function $\hat{t}^{-is} e^{-im\hat{\varphi}}$ on the light cone (up to Γ factors). We can formally compute the value of this Fourier transform at the point $(t, t \cos \varphi, t \sin \varphi)$ on the light cone. As in (5.3) we obtain Γ factors times

$$t^{is} e^{im\varphi} \int (1 + \cos \hat{\varphi})^{is} e^{-im\hat{\varphi}} \, d\hat{\varphi}$$

$$= t^{is} e^{im\varphi} 2^{-is} \frac{\Gamma(1+2is)}{\Gamma(1+2is+2im)\Gamma(1+2is-2im)}, \qquad (5.4)$$

by [2], Vol. I, p. 12 (30).

The integral in (5.4) is an example of Harish-Chandra's c function. It controls the asymptotic value of P_{is}^m. For by the homogeneity of f, calling $\sqrt{1-x^2} = \varepsilon$ we have

$$P_{is}^m(\varepsilon)(\varepsilon^{-1})\varepsilon^{is} \to (5.4) \text{ times the } \Gamma \text{ factor}, \qquad (5.5)$$

as $\varepsilon \to 0$. By a more careful analysis of the above we can obtain the whole asymptotic series for P_{is}^m.

The geometry of confluence takes a somewhat different turn. We are now interested in the passage from a genuine Cartan to a confluent Cartan or, more generally, from a Cartan to one which is more confluent. We can think of this, in the above picture, as the study of what happens when K ($=$ isotropy of $(t, x, y) = (1, 0, 0)$) passes to $K_x =$ isotropy $(1, x, 0)$ with $x \to 1$. Certainly $K_x \to MN =$ isotropy $(1, 1, 0)$ so we have to explain what phenomena we are examining.

The calculations are made more simple when we think of the representation space as 2×2 symmetric matrices. Then we search for the isotropy group K_ε of $\left(\begin{smallmatrix} 1 & 0 \\ 0 & \varepsilon^2 \end{smallmatrix}\right)$. We find easily that

$$K_\varepsilon = \left\{ \begin{pmatrix} \cos \varphi & \varepsilon \sin \varphi \\ -\varepsilon^{-1} \sin \varphi & \cos \varphi \end{pmatrix} \right\}. \tag{5.6}$$

This suggests that we should set $\varphi = \varepsilon n$ so that the limit of K_ε can be identified with N if we drop ε^2 terms.

Technically, it is sometimes more convenient to replaced K by K_{n_0}, which is the conjugate of K by $\left(\begin{smallmatrix} 1 & n_0 \\ 0 & 1 \end{smallmatrix}\right)$. Then

$$K_{n_0} = \left\{ \begin{pmatrix} c + n_0 s & n_0^2 s + s \\ -s & -n_0 s + c \end{pmatrix} \right\}, \tag{5.7}$$

where, as usual, we write s for $\sin \varphi$ and c for $\cos \varphi$. Then $K_{n_0} \to N$ if we set $\varphi \to n / n_0^2$ and we drop n_0^{-1} terms.

What is the effect of this limit process on the Legendre functions? This is best seen from the integral representation [compare (5.2) and (5.3)]. For $K_{n_0} \cdot \left(\begin{smallmatrix} 1 & 0 \\ 0 & 0 \end{smallmatrix}\right) \to N \cdot \left(\begin{smallmatrix} 1 & 0 \\ 0 & 0 \end{smallmatrix}\right)$, so the Fourier transform of the K_{n_0} invariant measure on $K_{n_0} \cdot \left(\begin{smallmatrix} 1 & 0 \\ 0 & 0 \end{smallmatrix}\right)$ approaches the Fourier transform of the N invariant measure on $N \cdot \left(\begin{smallmatrix} 1 & 0 \\ 0 & 0 \end{smallmatrix}\right)$.

However, this is not very interesting because the Fourier transform of this measure is just η^{is}, which is too simple. To get something interesting, as seen from (1.13) we need to use a nontrivial character on N. Now, had we started from $e^{im\theta}$ on K we would have e^{imn/n_0^2}. Thus we must also allow $m = m_0 n_0^2$, in which case we are led to the Bessel function.

This is the

Confluence Principle. When conjugates of a Cartan approach a confluent Cartan, we must also adjust the dual of the conjugates in order to obtain a meaningful result.

5.4.1. Analytic Approach

What appears from the geometric approach, especially from the confluence principle, is that we have to let the dual element of the Cartan approach infinity. Now, as the dual approaches infinity there is higher and higher oscillation, which means that all integrals should approach zero. How are we to obtain a meaningful result?

The point is that the oscillations are least violent near the identity of the Cartan. Thus this process should favor the identity of K. Indeed, this is exactly what happens when we put $\theta = n/n_0^2$.

For a direct example of this phenomenon, let us replace the confluence letting $K \to N$ by letting $r \to t$. (Recall that we have a conformal map interchanging $(K, \{t\}) \leftrightarrow (\{r\}, N)$.) By our present analytic idea this means that we want to let $s \to \infty$.

To understand how this works in detail, let us recall that all functions we deal with are Fourier transforms of measures on the light cone. Thus we must study integrals of the form, e.g.,

$$\int e^{ir\hat{r}[\cosh \zeta + \sinh \zeta \cos \hat{\varphi}]} e^{-im\hat{\varphi}} \hat{r}^{-is} \, d\hat{r} \, d\hat{\varphi}/\hat{r}$$

$$= \Gamma(ir)^{is} \int e^{-im\hat{\varphi}} [\cosh \zeta + \sinh \zeta \cos \hat{\varphi}]^{is} \, d\hat{\varphi}. \qquad (5.8)$$

According to our principle, we should favor $[\cosh \zeta + \sinh \zeta \cos \hat{\varphi}] \sim 1$ since, after normalizing by setting $r = 1$, this makes the exponential have its value close to that for $\hat{r} = 1$, which is the multiplicative identity.

This suggests that we set $\zeta = 1 + \varepsilon$. The correct form of ε to obtain a meaningful result is $\varepsilon = \hat{t}/is$. The last integral in (5.8) then becomes a Bessel function as we expected. In fact, this is identical with the classical confluence procedure.

The same idea applies to the study of integrals of the forms $\int \exp[irh(\theta, \hat{\varphi})]\hat{r}^{-is} \, d\hat{r} \, d\hat{\varphi}/\hat{r}$. We study the variation of h near a point θ_0 where for some values of $\hat{\varphi}$ we have $h(\theta_0, \hat{\varphi}) = 0$.

Remark. We can regard this as a differentiation process and thus as a passage from the hyperboloid to its tangent space at the identity. This replaces $SO(1, 2)$ by the affine group; the significance of Bessel functions for the latter is well known.

Problem. Can one interpret this process as a degeneration of $\{r\}$ to $\{t\}$ using conjugation in the conformal group?

5.5. From Hypergeometric Equations to the Laplacian

All our work up to now started with the Laplacian or wave operator and then passed, via some form of separation of variables, to the hypergeometric functions. We want to attack the inverse question: Can one pass from hypergeometric functions to the Laplacian?

Let us illustrate our method by means of the Bessel functions and Bessel equation. We write the equation for J_0 in the form

$$\left(\frac{d^2}{dr^2}+\frac{1}{r}\frac{d}{dr}\right)J_0=-J_0. \tag{5.9}$$

The left side is a differential operator of order 2. Following Dirac's philosophy it would be better to factor this operator into two first-order operators and thus obtain a first-order system of equations which imply (5.9).

There is an obvious factorization of (5.9), namely

$$\frac{d^2}{dr^2}+\frac{1}{r}\frac{d}{dr}=\left(\frac{d}{dr}+\frac{1}{r}\right)\frac{d}{dr}. \tag{5.10}$$

This suggests defining

$$\frac{d}{dr}J_0=J_{-1}. \tag{5.11}$$

Then J_{-1} satisfies the equation

$$\frac{d}{dr}\left(\frac{d}{dr}+\frac{1}{r}\right)J_{-1}=-\frac{d}{dr}J_0=-J_{-1}. \tag{5.12}$$

Our construction of the "Bessel hierarchy" would stop here if not for the amazing identity

$$\left(\frac{d}{dr}-\frac{(n-1)}{r}\right)\left(\frac{d}{dr}+\frac{n}{r}\right)=\left(\frac{d}{dr}+\frac{(n+1)}{r}\right)\left(\frac{d}{dr}-\frac{n}{r}\right). \tag{5.13}$$

This identity suggests that we can think of $(d/dr-(n-1)/r)$ as the step down operator from $J_{-(n-1)}$ to J_{-n} while $(d/dr+n/r)$ is the step up operator from J_{-n} to $J_{-(n-1)}$.

Thus we "create" J_{-n} from $J_{-(n-1)}$ by applying $d/dr-(n-1)/r$. Identity (5.13) then implies that J_{-n} satisfies

$$\left(\frac{d}{dr}-\frac{(n-1)}{r}\right)\left(\frac{d}{dr}+\frac{n}{r}\right)J_{-n}=-J_{-n}. \tag{5.14}$$

This is, of course, the standard Bessel equation.

To verify (5.14) we note that the left side of (5.14) is

$$\left(\frac{d}{dr}-\frac{(n-1)}{r}\right)\left(\frac{d}{dr}+\frac{n}{r}\right)\left(\frac{d}{dr}-\frac{(n-1)}{r}\right)J_{-(n-1)}$$

$$=\left(\frac{d}{dr}-\frac{(n-1)}{r}\right)\left(\frac{d}{dr}-\frac{(n-2)}{r}\right)\left(\frac{d}{dr}+\frac{n-1}{r}\right)J_{-(n-1)}$$

$$=-\left(\frac{d}{dr}-\frac{(n-1)}{r}\right)J_{-(n-1)}$$

$$=-J_{-n}.$$

Here we have used the definition of J_{-n} in terms of $J_{-(n-1)}$ and our identity (5.13) and induction on (5.14).

Unfortunately, this natural procedure applies to construct J_{-n} for $n \geq 0$. There does not seem to be a similar idea for J_n. However, we observe that the operators $d/dr + n/r$ are "separated," meaning that it seems reasonable to replace n by $i\,d/d\theta$ and to multiply J_{-n} by $\exp(-in\theta)$. Then

$$e^{i\theta}\left(\frac{\partial}{\partial r}-\frac{i}{r}\frac{\partial}{\partial\theta}\right) \tag{5.15}$$

can be regarded as a shift down and

$$e^{i\theta}\left(\frac{\partial}{\partial r}+\frac{i}{r}\frac{\partial}{\partial\theta}\right) \tag{5.16}$$

as a shift up.

With the usual definition

$$J_{-n}=(-1)^n J_n, \tag{5.17}$$

it follows that the double infinite sum

$$\sum J_n(r)\,e^{in\theta}=u(r,\theta) \tag{5.18}$$

is an eigenfunction of both shifts (5.15) and (5.16) which are, of course, $\partial/\partial z$ and $\partial/\partial \bar{z}$. The startling relation (5.13) can now be interpreted as the commutativity of $\partial/\partial z$ and $\partial/\partial \bar{z}$.

Of course, the Laplacian is $\partial^2/\partial z\,\partial \bar{z}$. We have thus constructed from the Bessel operator the Laplacian in the plane, the operators $\partial/\partial z$, $\partial/\partial \bar{z}$, hence the translation group, and the operator $\partial/\partial\theta$, hence the rotation group of the plane.

Since u is an eigenfunction of $\partial^2/\partial z\,\partial\bar{z}$ there is a natural passage to the wave operator. We have thus constructed the wave operator and the affine group of the plane which, as we have already noted, is the natural group associated to the Bessel functions.

An analogue of (5.13) holds for the Legendre operator and enables us to construct $SL(2)$ or its conformal group from a single Legendre equation.

I have not studied the scope of this idea, but, as I mentioned, it seems related in spirit, to Sato's *KP* hierarchy.

Let me conclude by noting that a similar procedure can be used in the study of analogues of hypergeometric functions. Elsewhere I shall show how this leads to a deeper understanding of the Rogers–Ramanujan and allied identities.

Added in proof. Doran Zeilberger pointed out to me that the identity (5.13) and most of the above formalism were discovered by Infeld and Hull and is known as the *Infeld–Hull factorization*. However, the significance of the generating function (5.18) whose q analog is crucial for the study of the Rogers–Ramanujan identities, does not seem to have been noted before.

References

[1] V. Bargmann, Irreducible representations of the Lorentz group, *Ann. Math.* **48** (1947), 568–640.

[2] Bateman Manuscript Project, *Higher Transcendental Functions*, Vol. I, II, III, McGraw-Hill, 1953.

[3] L. Ehrenpreis, *Fourier Analysis in Several Complex Variables*, Wiley-Interscience, 1970.

[4] L. Ehrenpreis, Conformal Geometry, in *Differential Geometry*, R. Brooks, A. Gray, and B. Reinhart, eds), Birkhäuser, 1983.

[5] C. L. Siegel, *Lectures on Advanced Analytic Number Theory*, Tata, 1961.

Quantization of Lie Groups and Lie Algebras

L. D. Faddeev, N. Yu. Reshetikhin,
and L. A. Takhtajan

Steklov Mathematical Institute
Leningrad Branch
Leningrad
USSR

The Algebraic Bethe Ansatz—the quantum inverse scattering method—emerges as a natural development of the following directions in mathematical physics: the inverse scattering method for solving nonlinear equations of evolution [1], the quantum theory of magnets [2], the method of commuting transfer-matrices in classical statistical mechanics [3] and factorizable scattering theory [4, 26]. It was formulated in our papers [5–7]. Two simple algebraic formulae lie at the foundation of the method:

$$RT_1 T_2 = T_2 T_1 R \qquad (*)$$

and

$$R_{12} R_{13} R_{23} = R_{23} R_{13} R_{12}. \qquad (**)$$

Their exact meaning will be explained in the next section. In the original context of the Algebraic Bethe Ansatz, T plays the role of quantum

129

monodromy matrix of the auxiliary linear problem and is a matrix with operator-valued entries, whereas R is the usual "c-number" matrix. The second formula can be considered as a compatibility condition for the first one.

Realizations of the formulae (∗) and (∗∗) for the particular models naturally led to new algebraic objects which can be viewed as deformations of Lie-algebraic structures [8–11]. V. Drinfeld [12, 13] has shown that in the description of these constructions the language of Hopf algebras [14] is quite useful. In this way he obtained a deep generalization of the results of [9–11]. Part of these results were also obtained by M. Jimbo [15, 16].

However, from our point of view, these authors did not use formula (∗) to the full strength. We decided, using the experience gained in the analysis of concrete models, to look again at the basic constructions of deformations. Our aim is to show that one can naturally define the quantization (q-deformation) of simple Lie groups and Lie algebras using exclusively the main formulae (∗) and (∗∗). Following the spirit of the non-commutative geometry [17] we will quantize instead of the Lie group G the algebra of functions $\mathrm{Fun}(G)$ on it. The quantization of the universal enveloping algebra $U(\mathfrak{G})$ of the Lie algebra $\mathfrak{G}$ will be based on a generalization of the relation

$$U(\mathfrak{G}) = C_e^{-\infty}(G),$$

where $C_e^{-\infty}(G)$ is a subalgebra in $C^{-\infty}(G)$ of distributions with support in the unit element e of G.

We begin with some general definitions. After that we treat two important examples and finally we discuss our constructions from the point of view of deformation theory. In this paper we use a formal algebraic language and do not consider the problems connected with topology and analysis. A detailed presentation of our results will be given elsewhere.

1. Quantum Formal Groups

Let V be an n-dimensional complex vector space (the reader can replace the field $\mathbf{C}$ by any field of characteristic zero). Consider a non-degenerate matrix $R \in \mathrm{Mat}(V^{\otimes 2}, \mathbf{C})$, satisfying the equation

$$R_{12}R_{13}R_{23} = R_{23}R_{13}R_{12}, \tag{1}$$

where the lower indices describe the imbedding of the matrix R into $\mathrm{Mat}(V^{\otimes 3}, \mathbf{C})$.

Definition 1. Let $A = A(R)$ be the associative algebra over $\mathbf{C}$ with the generators 1, t_{ij}, $i, j = 1, \ldots, n$, satisfying the relations

$$RT_1 T_2 = T_2 T_1 R, \tag{2}$$

where $T_1 = T \otimes I$, $T_2 = I \otimes T \in \mathrm{Mat}(V^{\otimes 2}, A)$, $T = (t_{ij})_{i,j=1}^n \in \mathrm{Mat}(V, A)$ and I is a unit matrix in $\mathrm{Mat}(V, \mathbf{C})$. The algebra $A(R)$ is called the *algebra of functions on the quantum formal group corresponding to the matrix R.*

In the case $R = I^{\otimes 2}$ the algebra $A(R)$ is generated by the matrix elements of the group $GL(n, \mathbf{C})$ and is commutative.

Theorem 1. *The algebra A is a bialgebra (a Hopf algebra) with comultiplication* $\Delta : A \to A \otimes A$

$$\Delta(1) = 1 \otimes 1$$

$$\Delta(t_{ij}) = \sum_{k=1}^n t_{ik} \otimes t_{kj}, \qquad i, j = 1, \ldots, n.$$

Let $A' = \mathrm{Hom}(A, \mathbf{C})$ be the dual space to the algebra A. Comultiplication in A induces multiplication in A':

$$(l_1 l_2, a) \equiv (l_1 l_2)(a) = (l_1 \otimes l_2)(\Delta(a))$$

where $l_1, l_2 \in A'$ and $a \in A$. Thus A' has a structure of an associative algebra with unit $1'$, where $1'(t_{ij}) = \delta_{ij}$, $i, j = 1, \ldots, n$.

Definition 2. Let $U(R)$ be the subalgebra in $A(R)'$, generated by elements $1'$, and $l_{ij}^{(\pm)}$, $i, j = 1, \ldots, n$, where

$$(1', T_1 \cdots T_k) = I^{\otimes k},$$

$$(L^{(+)}, T_1 \cdots T_k) = R_1^{(+)} \cdots R_k^{(+)}, \tag{3}$$

$$(L^{(-)}, T_1 \cdots T_k) = R_1^{(-)} \cdots R_k^{(-)}.$$

Here $L^{(\pm)} = (l_{ij}^{(\pm)})_{i,j=1}^n \in \mathrm{Mat}(V, U)$,

$$T_i = I \otimes \cdots \otimes \underbrace{T \otimes}_{i} \cdots \otimes I \in \mathrm{Mat}(V^{\otimes k}, A), \qquad i = 1, \ldots, k,$$

the matrices $R_i^{(\pm)} \in \mathrm{Mat}(V^{\otimes(k+1)}, \mathbf{C})$ act nontrivially on factors number 0 and i in the tensor product $V^{\otimes(k+1)}$ and coincide there with the matrices $R^{(\pm)}$, where

$$R^{(+)} = PRP, \qquad R^{(-)} = R^{-1}; \tag{4}$$

here P is the permutation matrix in $V^{\otimes 2}$: $P(v \otimes w) = w \otimes v$ for $v, w \in V$. The left hand side of the formula (3) denotes the values of $1'$ and of the matrices-functionals $L^{(\pm)}$ on the homogeneous elements of the algebra A of degree k. When $k = 0$ the right hand side of formula (3) is equal to I. The algebra $U(R)$ is called the *algebra of regular functionals on $A(R)$*.

Due to the equation (1) and

$$R_{12} R_{23}^{(-)} R_{13}^{(-)} = R_{13}^{(-)} R_{23}^{(-)} R_{12},$$

Definition 2 is consistent with the relations (2) in the algebra A.

Remark 1. The apparent doubling of the number of generators of the algebra $U(R)$ in comparison with the algebra $A(R)$ is explained as follows: due to the formula (3) some of the matrix elements of the matrices $L^{(\pm)}$ are identical or equal to zero. In the interesting examples (see below) matrices $L^{(\pm)}$ are of Borel type.

Theorem 2. (1) *In the algebra $U(R)$ the following relations take place*:

$$R^{(+)} L_1^{(\pm)} L_2^{(\pm)} = L_2^{(\pm)} L_1^{(\pm)} R^{(+)}, \qquad R^{(+)} L_1^{(+)} L_2^{(-)} = L_2^{(-)} L_1^{(+)} R^{(+)}, \qquad (5)$$

where $R^{(+)} = PRP$ and $L_1^{(\pm)} = L^{(\pm)} \otimes I$, $L_2^{(\pm)} = I \otimes L^{(\pm)} \in \mathrm{Mat}(V^{\otimes 2}, A')$.

(2) *Multiplication in the algebra $A(R)$ induces comultiplication δ in $U(R)$*,

$$\delta(1') = 1' \otimes 1'$$

$$\delta(l_{ij}^{(\pm)}) = \sum_{k=1}^{n} l_{ik}^{(\pm)} \otimes l_{kj}^{(\pm)}, \qquad i, j = 1, \ldots, n,$$

so $U(R)$ acquires a structure of a bialgebra.

The algebra $U(R)$ can be considered as a quantization of the universal enveloping algebra, which is defined by the matrix R.

Let us also remark that in the framework of the scheme presented one can easily formulate the notion of quantum homogeneous spaces.

Definition 3. A subalgebra $B \subset A = A(R)$ which is a left coideal, $\Delta(B) \subset A \otimes B$, is called the *algebra of functions on a quantum homogeneous space corresponding to the matrix R*.

Now we shall discuss concrete examples of the general construction presented above.

2. Finite-Dimensional Example

Let $V = \mathbf{C}^n$; a matrix R of the form

$$R = \sum_{\substack{i \neq j \\ i,j=1}}^{n} e_{ii} \otimes e_{jj} + q \sum_{i=1}^{n} e_{ii} \otimes e_{ii} + (q - q^{-1}) \sum_{1 \leq j < i \leq n} e_{ij} \otimes e_{ji}, \tag{6}$$

where $e_{ij} \in \mathrm{Mat}(\mathbf{C}^n)$ are matrix units and $q \in \mathbf{C}$, satisfies equation (1). It is natural to call the corresponding algebra $A(R)$ the algebra of functions on the q-deformation of the group $GL(n, \mathbf{C})$ and denote it by $\mathrm{Fun}_q(GL(n, \mathbf{C}))$.

Theorem 3. *The element*

$$\det{}_q T = \sum_{s \in S_n} (-q)^{l(s)} t_{1 s_1} \cdots t_{n s_n},$$

where summation goes over all elements s of the symmetric group S_n and $l(s)$ is the length of the element s, generates the center of the algebra $\mathrm{Fun}_q(GL(n, \mathbf{C}))$.

Definition 4. The quotient algebra of the algebra $\mathrm{Fun}_q(GL(n, \mathbf{C}))$ defined by an additional relation $\det_q T = 1$ is called the *algebra of functions on the q-deformation of the group $SL(n, \mathbf{C})$* and is denoted by $\mathrm{Fun}_q(SL(n, \mathbf{C}))$.

Theorem 4. *The algebra $\mathrm{Fun}_q(SL(n, \mathbf{C}))$ has an antipode γ, which on the generators t_{ij} is given by:*

$$\gamma(t_{ij}) = (-q)^{i-j} \tilde{t}_{ji}, \qquad i, j = 1, \ldots, n,$$

where

$$\tilde{t}_{ij} = \sum_{s \in S_{n-1}} (-q)^{l(s)} t_{1 s_1} \cdots t_{i-1 s_{i-1}} t_{i+1 s_{i+1}} \cdots t_{n s_n}$$

and $s = (s_1, \ldots, s_{i-1}, s_{i+1}, \ldots, s_n) = s(1, \ldots, j-1, j+1, \ldots, n)$.   *The antipode γ has the properties $T\gamma(T) = \gamma(T)T = I$ and $\gamma^2(T) = DTD^{-1}$, where $D = \mathrm{diag}(1, q^2, \ldots, q^{2(n-1)}) \in \mathrm{Mat}(\mathbf{C}^n)$.*

In the case $n = 2$, the matrix R in explicit form is given by

$$R = \begin{pmatrix} q & 0 & 0 & 0 \\ 0 & 1 & 0 & 0 \\ 0 & q - q^{-1} & 1 & 0 \\ 0 & 0 & 0 & q \end{pmatrix}, \tag{7}$$

and the relations (2) reduce to the following simple formulae:

$$t_{11}t_{12} = qt_{12}t_{11}, \qquad\qquad t_{11}t_{21} = qt_{21}t_{11},$$

$$t_{12}t_{21} = t_{21}t_{12}, \qquad\qquad t_{12}t_{22} = qt_{22}t_{12},$$

$$t_{21}t_{22} = qt_{22}t_{21}, \qquad t_{11}t_{22} - t_{22}t_{11} = (q - q^{-1})t_{12}t_{21},$$

and

$$\det{}_q T = t_{11}t_{22} - qt_{12}t_{21}.$$

In this case

$$\gamma(T) = \begin{pmatrix} t_{22} & -q^{-1}t_{12} \\ -qt_{21} & t_{11} \end{pmatrix}.$$

Remark 2. When $|q| = 1$ relations (2) admit the following $*$-anti-involution: $t_{ij}^* = t_{ij}$, $i, j = 1, \ldots, n$. The algebra $A(R)$ with this anti-involution is nothing but the algebra $\mathrm{Fun}_q(SL(n, \mathbf{R}))$. In the case $n = 2$ this algebra and the matrix R of the form (7) appeared for the first time in the paper [18]. The subalgebra $B \subset \mathrm{Fun}_q(SL(n, \mathbf{R}))$, generated by the elements 1 and $\sum_{k=1}^{n} t_{ik}t_{jk}$, $i, j = 1, \ldots, n$ is the left coideal and may be called the algebra of functions on the q-deformation of the symmetric homogeneous space of rank $n - 1$ for the group $SL(n, \mathbf{R})$. In the case $n = 2$ we obtain the q-deformation of the Lobachevsky plane.

Remark 3. When $q \in \mathbf{R}$ the algebra $\mathrm{Fun}_q(SL(n, \mathbf{C}))$ admits the following $*$-anti-involution: $\gamma(t_{ij}) = t_{ji}^*$, $i, j = 1, \ldots, n$. The algebra $\mathrm{Fun}_q(SL(n, \mathbf{C}))$ with this anti-involution is nothing but the algebra $\mathrm{Fun}_q(SU(n))$. In the case $n = 2$ this algebra was introduced in the papers [19–20].

Remark 4. The algebras $\mathrm{Fun}_q(G)$, where G is a simple Lie group, can be defined in the following way. For any simple group G there exists a corresponding matrix R_G satisfying equation (1) which generalize the matrix R of the form (6) for the case $G = SL(n, \mathbf{C})$. This matrix R_G depends on the parameter q and as $q \to 1$

$$R_G = I + (q - 1)r_G + O((q - 1)^2),$$

where

$$r_G = \sum_i \frac{\rho(H^i) \otimes \rho(H^i)}{2} + \sum_{\alpha \in \Delta_+} \rho(X_\alpha) \otimes \rho(X_{-\alpha}).$$

Here ρ is the vector representation of Lie algebra $\mathfrak{G}$, H^i, X_α its Cartan–Weyl basis, and Δ_+ the set of positive roots. The explicit form of the matrices R_G can be extracted from [16], [21]. The corresponding algebra $A(R)$ is defined by the relations (2) and an appropriate anti-involution compatible with them. It may be called the algebra of functions on the q-deformation of the Lie group G.

Let us discuss now the properties of the algebra $U(R)$. It follows from the explicit form (6) of the matrix R and Definition 2 that the matrices-functionals $L^{(+)}$ and $L^{(-)}$ are, correspondingly, the upper- and lower-triangular matrices. Their diagonal parts satisfy the simple relation:

$$\operatorname{diag}(L^{(+)}) \operatorname{diag}(L^{(-)}) = I.$$

Theorem 5. *The following equality holds:*

$$U(R) = U_q(\mathfrak{sl}(n, \mathbf{C})),$$

where $U_q(\mathfrak{sl}(n, \mathbf{C}))$ is the q-deformation of the universal enveloping algebra $U(\mathfrak{sl}(n, \mathbf{C}))$ of the Lie algebra $\mathfrak{sl}(n, \mathbf{C})$, introduced in the papers [12] and [15]. The center of $U(R)$ is generated by the elements

$$C_k = \sum_{s,s' \in S_n} (-q)^{l(s)+l(s')} l^{(+)}_{s_1 s_1'} \cdots l^{(+)}_{s_k s_k'} l^{(-)}_{s_{k+1} s_{k+1}'} \cdots l^{(-)}_{s_n s_n'}, \qquad k = 0, \ldots, n-1.$$

Remark 5. It is instructive to compare the relations for the elements $l^{(\pm)}_{ij}$, $i, j = 1, \ldots, n$, which follow from (5), with those given in the papers [12] and [15]. The elements $l^{(\pm)}_{ij}$ can be considered as the q-deformation of the Cartan–Weyl basis, whereas the elements $l^{(+)}_{ii+1}$, $l^{(-)}_{i-1i}$, $l^{(+)}_{ii}$ are the q-deformation of the Chevalley basis. It was this basis that was used in [12] and [15]; the complicated relations between the elements of the q-deformation of the Chevalley basis, presented in these papers, follow from the simple formulae (3), (5) and (6).

Remark 6. It follows from the definition of the algebra $U_q(\mathfrak{sl}(n, \mathbf{C}))$ that it can be considered as the algebra $\operatorname{Fun}_q((G_+ \times G_-)/H)$, where $G_\pm$ and H are, correspondingly, Borel and Cartan subgroups of Lie group $SL(n, \mathbf{C})$. Moreover in the general case the q-deformation $U_q(\mathfrak{G})$ of the universal enveloping algebra of the simple Lie algebra $\mathfrak{G}$ can be considered as the quantization of the group $(G_+ \times G_-)/H$. For the infinite-dimensional case (see the next section) this observation provides a key to the formulation of the quantum Riemann problem.

In the case $n = 2$ we have the following explicit formulae:

$$L^{(+)} = \sqrt{q} \begin{pmatrix} e^{-hH/2} & hX \\ 0 & e^{hH/2} \end{pmatrix},$$

$$L^{(-)} = \frac{1}{\sqrt{q}} \begin{pmatrix} e^{hH/2} & 0 \\ hY & e^{-hH/2} \end{pmatrix}, \qquad q = e^{-h},$$

where the generators 1 and $e^{\pm hH/2}$, X, Y of the algebra $U_q(\mathfrak{sl}(2, \mathbf{C}))$ satisfy the relations

$$e^{\pm hH/2} X = q^{\pm 1} X e^{\pm hH/2}, \qquad e^{\pm hH/2} Y = q^{\mp 1} Y e^{\pm hH/2},$$

$$XY - YX = -\frac{(q - q^{-1})}{h^2} (e^{hH} - e^{-hH}),$$

which appear for the first time in the paper [8].

3.   Infinite-Dimensional Example

Replace in the general construction of Section 1 the finite-dimensional vector space V by the infinite-dimensional $\mathbf{Z}$-graded vector space $\tilde{V} = \bigoplus_{n \in \mathbf{Z}} \lambda^n V = \bigoplus_{n \in \mathbf{Z}} V_n$, where λ is a formal variable (spectral parameter). Denote by $\mathbf{S}$ the shift operator (multiplication by λ) and as matrix R consider an element $\tilde{R} \in \mathrm{Mat}(\tilde{V}^{\otimes 2}, \mathbf{C})$, satisfying equation (1) and commuting with the operator $\mathbf{S} \otimes \mathbf{S}$. Infinite-dimensional analogs of the algebras $A(R)$ and $U(R)$—the algebras $A(\tilde{R})$ and $U(\tilde{R})$—are introduced as before by Definitions 1 and 2; in addition elements $\tilde{T} \in \mathrm{Mat}(\tilde{V}, A(\tilde{R}))$ and $\tilde{L}^{(\pm)} \in \mathrm{Mat}(\tilde{V}, U(\tilde{R}))$ commute with $\mathbf{S}$. Theorems 1 and 2 are valid for this case.

Let us discuss a meaningful example of this construction. Choose the matrix $\tilde{R}$ to be a matrix-valued function $R(\lambda, \mu)$, defined by the formula

$$R(\lambda, \mu) = \frac{\lambda q^{-1} R^{(+)} - \mu q R^{(-)}}{\lambda q^{-1} - \mu q},$$

where the matrices $R^{(\pm)}$ are given by (4). The role of the elements $\tilde{T}$ now played by the infinite formal Laurent series

$$T(\lambda) = \sum_{m \in \mathbf{Z}} T_m \lambda^m,$$

with the relations

$$R(\lambda, \mu)T_1(\lambda)T_2(\mu) = T_2(\mu)T_1(\lambda)R(\lambda, \mu).$$

The comultiplication in the algebra $A(\tilde{R})$ is given by the formula

$$\Delta(t_{ij}(\lambda)) = \sum_{k=1}^{n} t_{ik}(\lambda) \otimes t_{kj}(\mu), \qquad i, j = 1, \ldots, n.$$

The following result is the analog of Theorem 3.

Theorem 6. *The element*

$$\det{}_q T(\lambda) = \sum_{s \in S_n} (-q)^{l(s)} t_{1s_1}(\lambda) \cdots t_{ns_n}(\lambda q^{n-1})$$

generates the center of the algebra $A(\tilde{R})$.

Let us now briefly discuss the properties of the algebra $U(\tilde{R})$. It is generated by the formal Taylor series

$$L^{(\pm)}(\lambda) = \sum_{m \in \mathbf{Z}_+} L_m^{(\pm)} \lambda^{\pm m},$$

which act on the elements of the algebra $A(\tilde{R})$ by the formulae (3). The relations in $U(\tilde{R})$ have the form

$$R^{(+)}(\lambda, \mu)L_1^{(\pm)}(\lambda)L_2^{(\pm)}(\mu) = L_2^{(\pm)}(\mu)L_1^{(\pm)}(\lambda)R^{(+)}(\lambda, \mu),$$

$$R^{(+)}(\lambda, \mu)L_1^{(+)}(\lambda)L_2^{(-)}(\mu) = L_2^{(-)}(\mu)L_1^{(+)}(\lambda)R^{(+)}(\lambda, \mu).$$

Due to the shortage of space we shall not discuss here an interesting question about the connection of the algebra $U(\tilde{R})$ with the q-deformation of the Kac–Moody algebras, introduced in the paper [13]. We shall only point out that the algebra $U(\tilde{R})$ has a natural limit when $q \to 1$. In this case the subalgebra, generated by elements $L^{(+)}(\lambda)$, coincides with the Yangian $Y(\mathfrak{sl}(n, \mathbf{C}))$, introduced in the papers [12, 22].

4. Deformation Theory and Quantum Groups

Consider the contraction of the algebras $A(R)$ and $U(R)$ when $q \to 1$. For definiteness we have in mind a finite-dimensional example. The algebra $A(R) = \mathrm{Fun}_q(G)$ when $q \to 1$ goes into the commutative algebra $\mathrm{Fun}(G)$ with the Poisson structure given by the formula

$$\{g \underset{,}{\otimes} g\} = [r_G, g \otimes g]. \tag{8}$$

Here g_{ij}, $i, j = 1, \ldots, n$ are the coordinate functions on the Lie group G. Passing from the Lie group G to its Lie algebra $\mathfrak{G}$, from (8) we obtain the Poisson structure on the Lie algebra

$$\{h \underset{\text{>}}{\otimes} h\} = [r_G, h \otimes I + I \otimes h]. \tag{9}$$

Here $h = \sum_{i=1}^{\dim \mathfrak{G}} h_i X^i$, where X^i, $i = 1, \ldots, \dim \mathfrak{G}$, form a basis of $\mathfrak{G}$. If we define $h^{(\pm)} = h_\pm + h_{\mathfrak{H}}/2$, where $h_\pm$ and $h_{\mathfrak{H}}$ are respectively the nilpotent and Cartan components of h, we can rewrite the formula (9) in the form

$$\{h^{(\pm)} \underset{\text{>}}{\otimes} h^{(\pm)}\} = [r_G, h^{(\pm)} \otimes I + I \otimes h^{(\pm)}],$$

$$\{h^{(\pm)} \underset{\text{>}}{\otimes} h^{(\mp)}\} = 0. \tag{10}$$

This Poisson structure and its infinite-dimensional analogs were studied in [23] (see also [24–25]). Thus the Lie algebra $\mathfrak{G}$ has a structure of a Lie bialgebra, where the cobracket $\mathfrak{G} \to \mathfrak{G} \wedge \mathfrak{G}$ is defined by the Poisson structure (10).

Analogously the contraction $q \to 1$ of the algebra $U(R)$ leads to the Lie bialgebra structure on the dual vector space $\mathfrak{G}^*$ to the Lie algebra $\mathfrak{G}$. The Lie bracket on $\mathfrak{G}^*$ is dual to the Poisson structure (10) and the cobracket $\mathfrak{G}^* \to \mathfrak{G}^* \wedge \mathfrak{G}^*$ is defined by the canonical Lie–Poisson structure on $\mathfrak{G}^*$.

This consideration explains in more detail in what senses the algebras $A(R)$ and $U(R)$ define deformations of the corresponding Lie group and Lie algebra. Moreover it shows how an additional structure is defined on these "classical" objects. Returning to the relations (∗) and (∗∗) we can say now that (∗) constitutes a deformation of the Lie-algebraic relations with t_{ij} playing the role of generators, R being the array of "quantum" structure constants, and (∗∗) generalizes the Jacobi identity.

References

[1] C. Gardner, J. Green, M. Kruskal, and R. Miura, *Phys. Rev. Lett.* **19** (1967) 19, 1095–1097.

[2] H. Bethe, *Z. Phys.* **71** (1931) 205–226.

[3] R. Baxter, *Exactly Solved Models in Statistical Mechanics.* Academic Press, London, 1982.

[4] C. N. Yang, *Phys. Rev. Lett.* **19** (1967) 23, 1312–1314.

[5] E. Sklyanin, L. Takhtajan, and L. Faddeev, *TMF* **40** (1979) 2, 194–220 (in Russian).

[6] L. Takhtajan and L. Faddeev, *Usp. Math. Nauk* **34** (1979) 5, 13–63 (in Russian).

[7] L. Faddeev, Integrable Models in 1 + 1-Dimensional Quantum Field Theory, in *Les Houches Lectures 1982*, Elsevier, Amsterdam, 1984.

[8] P. Kulish and N. Reshetikhin, *Zap. Nauch. Semin. LOMI* **101** (1981) 101–110 (in Russian).

[9] E. Sklyanin, *Func. Anal. and Appl.* **16** (1982) 4, 27–34 (in Russian).

[10] E. Sklyanin, *ibid,* **17**, 4, 34–48 (in Russian).

[11] E. Sklyanin, *Usp. Mat. Nauk* **40** (1985) 2, 214 (in Russian).

[12] V. Drinfeld, *DAN SSSR* **283** (1985) 5, 1060-1064 (in Russian).

[13] V. Drinfeld, Talk at IMC-86, Berkeley, 1986.

[14] E. Abe, *Hopf Algebras*, Cambridge Tracts in Math., 74, Cambridge University Press, Cambridge-New York, 1980.

[15] M. Jimbo, *Lett. Math. Phys.* **11** (1986) 247-252.

[16] M. Jimbo, *Commun. Math. Phys.* **102** (1986) 4, 537-548.

[17] A. Connes, *Noncommutative Differential Geometry*. Extract Publ. Math., I.H.E.S., **62**, 1986.

[18] L. Faddeev and L. Takhtajan, Liouville model on the lattice. Preprint, Université Paris VI, June 1985; *Lect. Notes in Physics* **246** (1986) 166-179.

[19] L. Varksman and J. Soibelman, *Funct. Anal. and Appl.* (to appear).

[20] S. Woronowicz, *Publ. RIMS, Kyoto Univ.* **23** (1987) 117-181.

[21] V. Bazhanov, *Phys. Lett.* **159B** (1985) 4-5-6, 321-324.

[22] A. Kirillov and N. Reshetikhin, *Lett. Math. Phys.* **12** (1986) 199-208.

[23] N. Reshetikhin and L. Faddeev, *TMF* **56** (1983) 323-343 (in Russian).

[24] M. Semenov-Tian-Shansky, *Publ. RIMS Kyoto Univ.* **21** (1985) 6, 1237-1260.

[25] L. Faddeev and L. Takhtajan, *Hamiltonian Methods in the Theory of Solitons*. Springer-Verlag, 1987.

[26] A. Zamolodchikov and A. Zamolodchikov, *Annals of Physics* **120** (1979) 2, 253-291.

The Toda Lattice in the Complex Domain

H. Flaschka

Department of Mathematics
The University of Arizona
Tucson, Arizona

1.   Introduction

The *Toda lattice* is an integrable Hamiltonian system. Its many special properties can be explained by various analytical, algebraic, and geometric constructions. Because of its rich and rigid structure, it is a paradigm for integrability: most features of integrable systems are likely to be revealed in the clearest possible way by the Toda lattice.

This note deals with the non-periodic Toda lattice in the complex domain: time, position, and momentum are allowed to assume complex values. The goal is to explain, in a simple setting, some implications of the so-called Kovalevskaya–Painlevé method. This is a test for integrability, first used by S. Kovalevskaya in her pioneering work on the motion of a rigid body. The idea is simple, the consequences are profound and not yet understood.

Consider a system of differential equations

$$\frac{dz_i}{dt} = f_i(z_1, \ldots, z_l), \qquad i = 1, \ldots, 2l, \tag{1}$$

Algebraic Analysis, Volume I

in which the f_i are polynomials. One says that (1) has the *Kovalevskaya property* if for every i, there exists a vector solution $z^{(i)}: \mathbf{C} \to \mathbf{C}^{2l}$ of (1), meromorphic near $t = 0$, such that

(*i*) $z_i^{(i)}(t)$ has a pole at $t = 0$, and
(*ii*) $z^{(i)}(t)$ depends on $2l - 1$ arbitrary constants.

All known systems with this property are "integrable" in a strong sense. An example is afforded by the equations

$$\dot{a}_i = a_i \sum_{j=1}^{l} N_{ij} b_j, \qquad \dot{b}_i = \sum_{j=1}^{l} N_{ji} a_j, \qquad N_{ij} \in \mathbf{R}, \qquad (2)$$

where $a_1, \ldots, a_l, b_1, \ldots, b_l \in \mathbf{C}$, and (N_{ij}) is an $l \times l$ invertible matrix. It was shown by M. Adler and P. van Moerbeke and by H. Yoshida in 1981, see (Yoshida, 1983), that (2) has the Kovalevskaya property if and only if the rows of the matrix N are the simple roots, viewed as vectors in $\mathbf{R}^l$, of a semisimple complex Lie algebra; there are then l constants of motion in involution with respect to a certain natural Poisson bracket. This note is concerned with a special case of (2): we take

$$N = \begin{pmatrix} \sqrt{2} & 0 & \cdots & & 0 \\ -\sqrt{\tfrac{1}{2}} & \sqrt{\tfrac{3}{2}} & \cdots & & 0 \\ \vdots & \ddots & & \ddots & \vdots \\ 0 & \cdots & & -\sqrt{(l-1)/l} & \sqrt{(l+1)/l} \end{pmatrix}; \qquad (3)$$

the rows are the simple roots of the Lie algebra $A_l = \mathfrak{sl}(l+1, \mathbf{C})$ of $(l+1) \times (l+1)$ matrices of trace zero (see Humphreys, 1972).

Examples. $l = 1$.

$$\dot{a} = \sqrt{2} ab, \qquad \dot{b} = \sqrt{2} a.$$

$l = 2$.

$$\dot{a}_1 = \sqrt{2} a_1 b_1, \qquad \dot{a}_2 = a_2 \left(-\frac{\sqrt{2}}{2} b_1 + \frac{\sqrt{6}}{2} b_2 \right)$$

$$\dot{b}_1 = \sqrt{2} a_1 - \frac{\sqrt{2}}{2} a_2, \qquad \dot{b}_2 = \frac{\sqrt{6}}{2} a_2.$$

The equations (2), with N as in (3), will be called the *Toda equations*. They are the perhaps more familiar "finite" Toda equations (Moser, 1975) in center-of-mass coordinates. The physical interpretation of the equations will play no role in this paper, and therefore will not be discussed (but see (Toda, 1981)).

Before describing some of the consequences of a Kovalevskaya–Painlevé analysis, we introduce some definitions. A *balance* is a class of Laurent series solutions of (1), in which the most singular behavior is fixed and the arbitrary parameters enter at fixed powers of t; those powers are called *resonances*.

Example.   For the Toda lattice (2) (with N as in (3)), *some* of the possible balances and resonances are as follows:

(i)   Laurent series in which $a_i(t) = a_i^{(0)} \times t^{-2} + \cdots$, with $a_i^{(0)} = 1$ for $i = i_0$, and $=0$ for $i \neq i_0$. Those series contain $2l - 1$ arbitrary parameters, which enter at certain (computable) powers of t.

(ii)   Laurent series in which $a_i(t) = a_i^{(0)} \times t^{-2} + \cdots$, with all $a_i^{(0)}$ nonzero. These series have l arbitrary parameters, which enter at the resonances $t^0, t^1, \ldots, t^{l-1}$.

Balances will be denoted by the letter Θ. A balance Θ with m arbitrary parameters may degenerate into a balance Θ' with $m - 1$ parameters, when one of the m parameters in Θ tends to ∞. In that case, we say that $\Theta > \Theta'$. The balances are partially ordered by $>$. (In the example above, any balance of type (i) is $>$ the balance of type (ii).) A balance with the maximal number of free parameters (here, $2l - 1$) is called *principal*, and a balance with the least number of free parameters (here, the balance (ii) with l free parameters) is called a *lowest balance*. In the Toda example it turns out to be unique.

Analysis of balances has provided geometric and algebraic information about the systems being studied. If (1) is an integrable Hamiltonian system, there are l constants of motion in involution. A level set of these constants of motion is an affine variety. It is compactified by the addition of ideal points at infinity: one adds varieties of dimension $l - 1$, parametrized by $l - 1$ of the $2l$ free parameters of the principal balances. Roughly speaking, one considers l of those $2l$ parameters to be used up to specify the values of the l constants of motion; this is only a heuristic statement, and we hope to make it precise in this paper. At the boundaries of these "principal" varieties, one finds the next-lowest balances, and the lower ones at their boundaries, etc. (Ercolani and Siggia, 1987) show how this construction goes in general. The beautiful geometric consequences, particularly for Abelian surfaces defined by two Poisson-commuting functions, are explored in a series of papers by Adler, van Moerbeke, and their collaborators, see (Adler and van Moerbeke, 1985; Haine, 1983).

Rather less understood, except at a computational level, are the algebraic implications of the Kovalesvskaya analysis. (Newell, Tabor and Zeng, 1987), following earlier work by (Weiss, Tabor and Carnevale, 1983) and many papers by Weiss, explore Bäcklund transformations, find Lax pairs from generalized Laurent series, and show that Schur functions play an important rôle in the analysis of the poles of solutions of the Koreteweg–de Vries equation.

All the integrable systems that have, to date, been subjected to a Kovalevskaya analysis, define flows on Lie algebras (possibly infinite-dimensional). Their solutions are provided by triangular factorization in the associated Lie group (see Section 2 below). In collaboration with Yun-bo Zeng of the University of Science and Technology of China, I have undertaken a study of the Lie-algebraic meaning of the Laurent series solutions of the Toda lattices associated to semisimple complex Lie algebras. Most of our results will be published elsewhere. In this note, I will use a part of our analysis to *clarify the geometric significance of the parameters entering the Laurent series solutions for the various balances of the Toda equations* (2). The application of Schubert cells, described below, is an addition to our joint work; it draws on the ideas developed by Professor Sato in his analysis of soliton equations.

Throughout this paper, I employ the language of Lie algebras, but keep all computations within the setting of $\mathfrak{sl}(l+1, \mathbf{C})$. The final section will describe further results and open problems, which suggest that our geometric methods should help to explain the Kovalevskaya properties of other popular integrable systems.

2. Toda Lattice Background

We will study equations (2), with N as given in (3); this is assumed henceforth. Many results carry over to general N (associated to semisimple complex Lie algebras); some do not (or not yet). We will not comment on the nature of these generalizations.

The Toda equations $(2, 3)$ admit a *Lax representation*: define

$$
X = \begin{pmatrix}
\beta_1 & a_1 & 0 & \cdots & 0 & 0 \\
-1 & \beta_2 & a_2 & \cdots & 0 & 0 \\
\vdots & \vdots & \vdots & \cdots & \vdots & 0 \\
0 & 0 & 0 & \cdots & 0 & a_l \\
0 & 0 & 0 & \cdots & -1 & \beta_{l+1}
\end{pmatrix}, \tag{4}
$$

and

$$
B = \begin{pmatrix}
0 & -a_1 & 0 & \cdots & 0 \\
0 & 0 & -a_2 & \cdots & 0 \\
\vdots & \vdots & \vdots & \cdots & \vdots \\
0 & 0 & 0 & \cdots & -a_l \\
0 & 0 & 0 & \cdots & 0
\end{pmatrix}.
$$

The diagonal entries β_j are linear functions of the Toda variables b_k; their explicit form is not needed in this paper. One can then show that the matrix equation $\dot{X} = [X, B]$ is equivalent to (2, 3). The matrices X, B, and the commutator equation (or *Lax equation*) for X have intrinsic Lie-algebraic and Poisson-geometric meaning, see (Kostant, 1979; Symes, 1980). We call a matrix of the form (4) a *Jacobi matrix*, and we denote the set of Jacobi matrices by $\mathscr{J}$. It is a subvariety, of complex dimension $2l$, of A_l.

The Lax pair X, B is associated with a certain decomposition of the Lie algebra A_l, and this decomposition gives rise to a reasonably explicit solution of the Toda equations. From given initial values for a_i, b_i, build X. Compute e^{-tX}, which belongs to the Lie group $G = SL(l+1, \mathbf{C})$ of $(l+1) \times (l+1)$ matrices of determinant 1. Factor $e^{-tX} = n_-(t)b_+(t)$, where n_- is lower triangular, with 1's on the diagonal, and b_+ is upper triangular. Then

$$
X(t) = n_-(t)^{-1} X n_-(t) \tag{5}
$$

solves $\dot{X}(t) = [X(t), B(t)]$. It is easy to show this, and the reader who is not familar with the idea might want to carry out the calculation. (One must note that (5) is automatically a Jacobi matrix. This is seen as follows: $e^{-tX} X e^{tX} = X$, and so $n_-^{-1} X n_- = b_+ X b_+^{-1}$; the matrix on the left side of this equation has zeros above the first superdiagonal, while the right side has -1's on the first subdiagonal, and zeros below it.) The solution of the Toda lattice by factorization has been used by many authors: the most complete formulas are to be found in (Kostant, 1979); see also (Goodman and Wallach, 1982, 1984; Ol'shanetsky and Perelomov, 1979; Symes, 1980).

When e^{-tX} is factored as described, the diagonal entries τ_k of the upper-triangular factor $b_+(t)$ are just the determinants of the principal $k \times k$ minors of e^{-tX}. After computing formula (5), one finds that

$$
a_i(t) = \frac{\tau_{i-1}(t)\tau_{i+1}(t)}{\tau_i(t)^2} a_i.
$$

Not every matrix e^{-tX} admits a lower–upper factorization. When such a factorization is not possible, $n_-(t)$ will have a pole, and so the *solution*

$a_i(t)$, $b_i(t)$ of (2) will have a pole. Zeng and I have found the possible balances for the Toda equations (2), for an arbitrary semisimple complex Lie algebra; our computation is algebraic and does not use the factorization at all.

Proposition 1. *There are $2^l - 1$ balances; given a subset $\Theta = \{i_1, \ldots, i_r\}$ of $\{1, \ldots, l\}$ there is a Laurent series solution of (2) in which a_i has a pole, $\text{const} \times t^{-2}$, when $i \in \Theta$, and a_i is holomorphic (near $t = 0$) when $i \notin \Theta$. Moreover, the series depends on $2l - r$ arbitrary parameters.*

We also compute the resonances (see Section 1) in terms of the roots and weights of the given Lie algebra; only one part of those results will be needed:

Proposition 2. *In a balance $\Theta = \{i_1, \ldots, i_r\}$, the functions τ_{i_j} vanish to order n_{i_j}, where the integers n_{i_j} are characterized by either of the following conditions[1]:*

(i)   *Let $\mathscr{G}^\Theta$ be the Lie algebra built from the subset $\alpha_{i_1}, \ldots, \alpha_{i_r}$ of the simple roots of A_l. For $\mathscr{G}^\Theta$, one can write the sum of all positive roots as the sum $\sum_{j=1}^{r} n_{i_j} \alpha_{i_j}$ of simple roots.*

(ii) *For the lowest balance, the integers $n_1, \ldots, n_l$ are the dimensions of the coset spaces $G/P_{\hat{i}}$, where $P_{\hat{i}}$ is the subgroup of G of matrices of the form*

$$\begin{pmatrix} GL_i & * \\ 0 & GL_{l+1-i} \end{pmatrix}.$$

For the general balance Θ, let $\{j_1, \ldots, j_s\}$ be the complement of Θ, and let P_Θ be the subgroup of matrices of the form

$$\begin{pmatrix} GL_{j_1} & * & * & * & * \\ 0 & GL_{j_2 - j_1} & * & * & * \\ \vdots & \vdots & \cdots & * & * \\ 0 & 0 & \cdots & 0 & GL_{l+1-j_s} \end{pmatrix}; \qquad (6)$$

(of course, the matrix (6) is required to have determinant 1). The integers n_{i_j} are the dimensions of certain coset spaces of P_Θ.

One can already tell from this proposition that the leading terms of the Laurent series solutions to (2) have a geometric significance. This will

---

[1] These abstract statements will be made more explicit later.

become clearer in a moment. For the time being, I note that the *resonances* for the lowest balance, which are known to play a special rôle (Ercolani and Siggia, 1987; Newell, Tabor and Zeng, 1987), are the degrees of the invariant polynomials of the Lie algebra $\mathcal{G}$ (in the present example, A_l). The proof uses Kostant's principal $\mathfrak{sl}(2)$ subalgebra, and the eigenvalues of its Casimir operator. Evidently, the lowest balance is intimately related to the constants of motion of the Toda lattice; the same point is made in most of the papers devoted to the Kovalevskaya analysis of other systems. We will understand this in a different and simpler way in Section 4.

3.   Schubert Cells for A_l

We now state some facts about the Lie group $SL(l+1, \mathbf{C})$, denoted by G in the sequel. The group of upper-triangular matrices is denoted by B_+, and the group of lower-triangular matrices by B_-. A particularly important subgroup of G is the so-called *Weyl group* W, which is the subgroup of *permutation matrices*. (A permutation matrix w has precisely one $+1$ or -1 in each row and column, and its determinant must be 1).

Proposition 3.   *G is the disjoint union*

$$\bigcup_{w \in W} B_- w B_+.$$

This is called the *Birkhoff decomposition* of G. It expresses the familiar fact that not all matrices can be factored into lower–upper triangular form; some of them may need to have their rows permuted in order to bring nonzero pivots to the diagonal. A set of the form $B_- w B_+$, with some $w \in W$, is called a *Schubert cell*. For more information about this theory, see (Hiller, 1982) or (Seshadri, 1985).

The purpose of this paper is to explain the structure of the balances of the Toda equations by relating them to the Schubert cells of G. In other words, a given balance arises from a specific obstruction to the factorization of e^{-tX}, and is characterized by the complexity of the permutation w defining the Schubert cell $B_- w B_+$ to which e^{-tX} belongs at the value of t for which factorization becomes impossible. A permutation w has a *length*: the number of transpositions of adjacent numbers necessary to produce w. The Schubert cell $B_- w B_+$ has dimension $(l^2 - 1) - length\ of\ w$. Matrices in a dense open set (of dimension $l^2 - 1$) can be factored into lower times upper (or, equivalently, belong to $B_- B_+$, with $w = identity$); this set is called the *big cell*. At the other extreme, when

$$w = \begin{pmatrix} 0 & 0 & \cdots & 0 & 1 \\ 0 & 0 & \cdots & 1 & 0 \\ \vdots & \vdots & \cdots & 0 & 0 \\ 1 & 0 & \cdots & 0 & 0 \end{pmatrix} \tag{7}$$

(a 1 may need to be changed to -1 to make the determinant equal to 1), the cell $B_- w B_+$ has dimension $\frac{1}{2}(l+1)(l+2) - 1$. We will see in Section 4 that this, the *smallest cell*, is related to the lowest balance. It will be convenient for us to use the cells of cosets, $B_- w B_+ / B_+$; the smallest cell is then a single point with coset representative (7).

Two more ingredients enter the geometric description of balances. First, we need the *standard parabolic subgroups* of G. These are the groups P_Θ defined in (6). In particular, since $B_+ \subseteq P_\Theta$, we have a projection of coset spaces, $G/B_+ \to G/P_\Theta$, with P_Θ/B_+ as the pre-image of a point in G/P_Θ.

Second, it will be useful to know something about the representations of the group G.

A *representation* of G is a group isomorphism ρ from G to the invertible linear transformations on a vector space V. The elements of G are thereby represented as matrices. The most obvious representation exhibits them as $(l+1) \times (l+1)$ matrices, but there are others. Let $v_1, \ldots, v_{l+1}$ be the natural basis vectors in $\mathbf{C}^{l+1}$. The so-called *fundamental representations* of G are defined on the k-fold exterior powers $V^k := \bigwedge^k (V)$:

$$\rho_k(g)(w_1 \wedge w_2 \wedge \cdots \wedge w_k) = g w_1 \wedge g w_2 \wedge \cdots \wedge g w_k.$$

Let v^k be the vector $v_1 \wedge \cdots \wedge v_k$ in V^k. There is a natural inner product on V^k, and it is not hard to see that the tau-functions of the Toda lattice are given by

$$\tau_k(t) = (e^{-tX} v^k, v^k). \tag{8}$$

Example. Take $k = 1$. Then v^1 is the vector $(1, 0, \ldots, 0)^t$, and $(e^{-tX} v^1, v^1)$ is the $(1, 1)$ entry of e^{-tX}, i.e., $\tau_1(t)$.

Take $k = 2$. Then $v^2 = (1, 0, \ldots, 0)^t \wedge (0, 1, \ldots, 0)^t = v_1 \wedge v_2$, and $e^{-tX} v^2 = e^{-tX} v_1 \wedge e^{-tX} v_2$. The $v_1 \wedge v_2$ component of the last expression is just the determinant of the principal 2×2 minor of e^{-tX}, i.e., $\tau_2(t)$.

More generally, to each subset Θ of $\{1, \ldots, l\}$ there is associated a representation ρ_Θ of G on a vector space V^Θ. There is a distinguished vector

v^{Θ}, analogous to the v^k introduced above, and one has the following (known) result:

Proposition 4. *There is a one-to-one correspondence between cosets in G/P_{Θ} and lines from the origin of V^{Θ} to points on the highest weight orbit $\{\rho_{\Theta}(g)(v^{\Theta})\,|\,g \in G\}$.*

The reference to "lines" means that the orbit should be considered as a subvariety of projective space. For discussion, see for example (Bernstein, Gel'fand, and Gel'fand, 1972).

4.   The Lowest Balance

It is now time to put all this technology to use. First, for practice, we describe its relevance to the lowest balance of the Toda equations $(2, 3)$. The basic information about the lowest balance can be obtained by calculation, with no appeal to factorization or Schubert cells. The result, as found by Zeng and me, is a special case of Proposition 2:

Proposition 5. *The lowest balance of the Toda equations $(2, 3)$ consists of Laurent series of the form*

$$a_i(t) = t^{-2}(n_i + a_i^{(2)}t^2 + \cdots);$$

the integers n_i are the dimensions of the coset spaces $G/P_{\hat{i}}$,[2] where $\hat{i} = \{1, \ldots, \hat{i}, \ldots, l\}$. Moreover, the Laurent series have l arbitrary parameters that enter at the powers $t^0, \ldots, t^{l-1}$. These parameters are determined by the l constants of motion of the Toda equations, which are the traces of powers of the Jacobi matrix X in (4): $f_i(a_1, \ldots, b_l) = \text{trace } X^{i+1}$.

As was mentioned earlier, the integers n_i, and the resonances $0, 1, \ldots, l-1$ have a Lie-algebraic meaning. This feature may be peculiar to the Toda equations, because in most of the other systems that have been subjected to a Kovalevskaya analysis, the Lie-theoretical aspects are not so near the surface. It has been noted by several authors that the arbitrary constants in the lowest balance of all popular equations are determined by the constants of motion; this is obvious in one sense: if one has only l parameters

---

[2] These varieties are the *Grassmannians*, the sets of i-dimensional subspaces of $\mathbf{C}^{l+1}$. G/P_1, for example, is the l-dimensional projective space $\mathbf{P}_{\mathbf{C}}^l$.

to play with, and if there are l constants of motion, there had better be a connection. The Schubert-cell picture will show this yet again, but in a way that generalizes to other balances.

Given Proposition 5, one can deduce this next result.

Proposition 6. *If the Toda equations* $(2, 3)$, *with initial conditions encoded in the Jacobi matrix* X, *pass through the lowest balance at* $t = t_0$, *then* $e^{-t_0 X}$ *lies in the smallest cell.*

This proposition has an interesting heuristic interpretation. Whatever the initial condition, X, the factorization $e^{-tX} = n_-(t)b_+(t)$ breaks down *in exactly one way* when t passes through the lowest balance at some value $t_0{}^3$: the only way to distinguish the different Toda solutions (5), $n_-(t)^{-1}Xn_-(t)$, at $t = t_0$, is by the invariants of the initial Jacobi matrix X, i.e., by its eigenvalues or by the traces of its powers.

Example. It may help to see how this works in practice. Consider the tau-function $\tau_1(t) = (e^{-tX})_{1,1}$. If this solution passes through the lowest balance, say at $t = t_0$, then τ_1 and the derivatives of τ_1 up through the $(l-1)^{\text{st}}$ will vanish at t_0. Now, by (8),

$$\tau_1(t) = (e^{-tX}v_1, v_1),$$

and

$$\frac{d^k}{dt^k}\tau_1(t) = (e^{-tX}v_1, (-X^t)^k v_1).$$

Because X is tridiagonal, one checks easily that at $t = t_0$,

$$e^{-t_0 X} = \begin{pmatrix} 0 & * & \cdots \\ 0 & * & \cdots \\ \vdots & \vdots & \cdots \\ 0 & * & \cdots \\ * & * & \cdots \end{pmatrix}.$$

One can now show that the second column of $e^{-t_0 X}$ must be zero down to the next-to-last entry, and so forth. Therefore $e^{-t_0 X}$ has the form

$$\begin{pmatrix} 0 & 0 & \cdots & 0 & * \\ 0 & 0 & \cdots & * & * \\ \vdots & \vdots & \cdots & \vdots & \vdots \\ * & * & \cdots & * & * \end{pmatrix},$$

[3] For most X, this will never happen; the typical initial condition will develop a pole in one of the principal balances.

and it lies in the smallest Schubert cell in G/B_+. In the next section, we give the analog of this result for the general balance.

5.   Other Balances

We saw in Proposition 2 that for each index set $\Theta = \{i_1, \ldots, i_r\}$, there is a Laurent series solution that has $2l - r$ arbitrary parameters. The function a_i has a pole, or equivalently τ_i vanishes, if and only if $i \in \Theta$. The lowest balance has $r = l$, and the l remaining arbitrary parameters are directly related to the constants of motion of the Toda lattice. There have been attempts to generalize this, by viewing l of the $2l - r$ parameters in the general balance as being determined by the constants of motion, with the other $l - r$ playing the rôle of angle-type variables (Newell, Tabor and Zeng, 1987). The attempts were not really successful: one knows no useful computational way of grouping the arbitrary parameters that arise in the Laurent series. Our geometric picture will show that there is, in fact, such a grouping, but it seems hard to make it explicit.

To understand a balance, one must know the degree of obstruction to the $n_- b_+$ factorization: in which coset of G/B_+ does e^{-tX} lie? Using the projection $G/B_+ \to G/P_\Theta$ whose fiber is P_Θ/B_+, we split the problem into two easier sub-problems.

Proposition 7. *Suppose that the Toda solution passes through the balance* Θ *at* $t = t_0$. *Then* $e^{-t_0 X}$ *lies in the smallest cell of* P_Θ/B_+ *and in the largest cell of* G/P_Θ. *Moreover, the collection of cosets* $\{e^{tY}P_\Theta \mid t \in \mathbf{C}, Y \in \mathcal{J}\}$ *depends on* $l - |\Theta|$ *parameters.*

Examples.   (*i*) Let $\Theta = \{1, \ldots, l\}$ be the lowest balance. Then $P_\Theta = G$, and $G/P_\Theta = G/G$ is a single point, while $P_\Theta/B_+ = G/B_+$. No parameters come from the singleton G/G, no parameters come from the smallest cell in G/B_+, which is again a single point, and so all the free parameters in the lowest balance come from the initial condition X, as explained in Section 4.

(*ii*)   Let $\Theta = \hat{i} = \{1, \ldots, \hat{i}, \ldots, l\}$ be a next-lowest balance. No parameters come from the smallest cell in $P_{\hat{i}}/B_+$, which is a point. We show that one parameter comes from $G/P_{\hat{i}}$; together with the l constants of motion encoded in the initial X, this will account for the $l + 1$ parameters in this balance.

First, take $i = 1$. We think of G/P_i as a projectivized highest weight orbit, $G \cdot v_1$, as in Proposition 4. Define a map $j : \mathbf{C} \times \mathcal{J} \to G/P_i$, $j(t, Y) = e^{-tY}v_1$. We need the generic dimension of the image, and to this end we must compute the rank of dj at $(t, Y) = (0, X)$. The easy calculation, which strongly uses the fact that X is tri-diagonal, shows that the image has (projective) dimension *one*. For $i \geq 2$, the dimension is still one (the vector v_1 is now to be replaced by $v_1 \wedge \cdots \wedge v_i$). For general Θ, one proceeds by taking tensor products of the fundamental representations.

6. Conclusion

This paper has made two points. First, *the singularities of solutions of many familiar integrable systems occur when a certain factorization cannot be performed.* This is probably evident to people who think of the vanishing of "tau functions" as an obstruction to factorization, but that point of view has played no role in Kovalevskaya–Painlevé analysis so far. Second, the arbitrary parameters arising in the Laurent series solutions are of two different types: $l - r$ of them serve as coordinates in the Schubert cell in which factorization breaks down, and l of them determine the conjugacy class of the Jacobi matrix X (the constants of motion).

Both points bear equally well on integrable systems associated to loop groups, such as the periodic Toda lattice, the stationary Korteweg–deVries equation, and so on. The richer structures found in the infinite-dimensional Lie groups make the problems more interesting.

This paper has not addressed the most obvious question. Once the common level set of the constants of motion is compactified by the addition of ideal points at infinity, what can one say about the resulting complex manifold? The study of the mutual disposition of Schubert cells in G/P_Θ is difficult (Hiller, 1982; Seshadri, 1985), and I do not know whether intersecting the cells with exponentials of Jacobi matrices will make matters easier or harder. Using methods unrelated to factorization, Zeng and I have shown that the compactified level sets are obtained from complex projective space $\mathbf{P}_\mathbf{C}^l$ by blowing up certain linear subspaces. These objects are, as one might expect from the algebraic curve approach to the Toda lattice, compactified generalized Picard varieties of rational curves with double points. In the periodic Toda lattice with two degrees of freedom, the compactified level surfaces are Abelian surfaces (Adler and van Moerbeke, 1985). One can see, in that construction, a hint of the Schubert calculus; some aspects of the structure of the divisor at infinity can be read off the Dynkin diagram of the associated affine Lie algebra. Recently, Luc Haine (private communication) has checked that for the $A_3^{(1)}$ periodic Toda lattice, the intersections of the irreducible components of the divisor at infinity are encoded in the Dynkin diagram of $A_3^{(1)}$. The unification of Lie theory and algebraic geometry in integrable systems remains an attractive and largely unexplored problem.

Acknowledgements

I am grateful to Nick Ercolani, Alan Newell, and Yunbo Zeng for many discussions about the Kovalevskaya–Painlevé method. I also thank Doug Pickrell for providing private tutoring on Lie groups. This research was supported by the AFOSR through the Arizona Center for Mathematical Sciences, and by a grant from the NSF.

References

M. Adler and P. van Moerbeke, *Algebraic Completely Integrable Systems: A Systematic Approach.* University of Louvain preprint, to appear as a book, 1985.

I. N. Bernstein, I. M. Gel'fand, and S. I. Gel'fand, Schubert cells and cohomology of the spaces G/B, *Russian Math. Surveys* **28** (1972) 1–26.

N. M. Ercolani and E. D. Siggia, Painlevé property and geometry, 1987 (to appear in *Physica D*).

R. Goodman and N. R. Wallach, Classical and quantum-mechanical systems of Toda lattice type. I. *Comm. Math. Phys.* **83** (1972) 355–386.

R. Goodman and N. R. Wallach, Classical and quantum mechanical systems of Toda lattice type. II. Solutions of the classical flows, *Comm. Math. Phys.* **94** (1984) 177–217.

L. Haine, Geodesic flow on $SO(4)$ and Abelian surfaces, *Math. Ann.* **263** (1983) 435–472.

H. Hiller, *Geometry of Coxeter Groups.* Pitman Research Notes in Mathematics **54** 1982.

J. E. Humphreys, *Introduction to Lie Algebras and Representation Theory.* Springer Graduate Texts in Mathematics, 1972.

B. Kostant, The solution to a generalized Toda lattice and representation theory, *Adv. Math.* **34** (1979) 195–338.

J. Moser, Three integrable Hamiltonian systems connected with isospectral deformation, *Adv. Math.* **16** (1975) 197–220.

A. C. Newell, M. Tabor, and Y.-B. Zeng, A unified approach to Painlevé expansions, *Physica* **29D** (1987) 1–68.

M. A. Ol'shanetsky and A. M. Perelomov, Explicit solutions of classical generalized Toda models, *Invent. Math.* **54** (1979) 261–269.

C. S. Seshadri, *Introduction to the Theory of Standard Monomials.* Brandeis Lecture Notes **4**, 1985.

W. W. Symes, Systems of Toda type, inverse spectral problems, and representation theory, *Invent. Math.* **59** (1980) 13–51.

M. Toda, *Theory of Nonlinear Lattices.* Springer-Verlag, Berlin, 1981.

J. Weiss, M. Tabor, and G. Carnevale, The Painlevé property for partial differential equations, *J. Math. Phys.* **28** (1983) 522–526.

H. Yoshida, Integrability of generalized Toda lattice systems and singularities in the complex t-plane, in *Non-linear Integrable Systems—Classical Theory and Quantum Theory,* Proceedings of a RIMS Colloquium in 1981, World Scientific, Singapore, 1983.

Zoll Phenomena in (2+1) Dimensions

Victor Guillemin

Department of Mathematics
Massachusetts Institute of Technology
Cambridge, Massachusetts

1. Introduction

The results which I will describe below are biproducts of a (largely unsuccessful) effort to extend the twistor program of Penrose to non-symmetric models of space-time. Therefore, I will begin with a brief review of twistor theory.

I will denote the conformal compactification of ordinary Minkowski space by $M_{n-1,1}$. Geometrically $M_{n-1,1}$ is $\mathbf{R}^n$ with a light-cone attached at infinity. (See the figure on p. 156.) Analytically it is the (unique) compactification of $\mathbf{R}^n$ on which the causal group, $SO(n, 2)$, acts in a global fashion. (Recall that certain causal transformations like "time-involution" are not globally defined on $\mathbf{R}^n$ itself.)

It is not hard to see from the figure (where $(-x, t) = (x, t + \pi)$) that light rays, which look like straight lines in $\mathbf{R}^n$, look like circles in $M_{n-1,1}$. The manifold whose points are these light-like circles is called *twistor space.*

Copyright © 1988 by Academic Press, Inc.

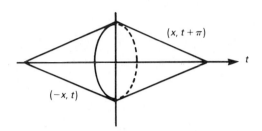

This space is a $(2n-3)$-dimensional contact manifold, and $SO(n, 2)$ acts transitively on it. One model for it is the co-sphere bundle of S^{n-1} with its standard contact structure. However, for $n = 4$, there is another model for it: The double cover of $SO(4, 2)$ is the group of **C**-linear transformations of $\mathbf{C}^4$ which leave fixed the Hermitian form

$$H(z) = |z_1|^2 + |z_2|^2 - |z_3|^2 - |z_4|^2. \tag{1.1}$$

The set of $z \in \mathbf{C}^4 - 0$ for which $H(z) > 0$ projects onto a three-dimensional complex domain, T_+, in $\mathbf{C}P^3$, and twistor space is just its boundary. $M_{3,1}$ itself has a simple description in terms of (1.1). It is the Grassmannian of complex two-dimensional subspaces of $\mathbf{C}^n$ on which $H = 0$. The twistor program of Penrose consists, *grosso modo*, of describing causally invariant objects on $M_{3,1}$ in terms of the complex geometry of T_+. (For example, solutions of the massless field equations of spin $k/2$ on $M_{3,1}$ have a very simple description in terms of holomorphic sheaves on T_+.) Unfortunately I will not have time to go into this theory in detail; however, I would like to mention some efforts to extend the ideas of twistor theory to other four-manifolds besides $M_{3,1}$.

To begin with, $M_{4,0} = S^4$ can be thought of as the quaternionic complex plane; so points of S^4 correspond to one-dimensional quaternionic subspaces of Q^2. However, a one-dimensional quaternionic subspace of Q^2 is a two-dimensional complex subspace of $\mathbf{C}^4$ and hence corresponds to a complex line in $\mathbf{C}P^3$. Moreover, no two of these lines can intersect; so $\mathbf{C}P^3$ is fibered by these lines; i.e., there is a fibration

$$\mathbf{C}P^3 \to S^4, \tag{1.2}$$

whose fibers are these $\mathbf{C}P^1$'s. This fibration is the $(4+0)$ version of the twistorial picture described above.

More generally, let M be a compact four-manifold with a conformal structure of type $(4+0)$. M is said to be conformally self-dual if the anti-self-dual part of its Weyl curvature tensor vanishes. Atiyah, Hitchin, and Singer have shown that for such a manifold there exists a fibration analogous to (1.2),

$$T \to M,$$

T being a three-dimensional complex manifold and the fibers being CP^1's. For details see [1].

Lorentzian examples of twistor space seem to be harder to come by. Hitchin has found a few examples of objects which can be legitimately described as Lorentzian twistor spaces by making use of the "nonlinear graviton" construction of Penrose. (See [8].) However, there seems to be no systematic way of obtaining such examples or knowing when a given Lorentzian manifold comes from such a construction.

Another approach in $(3+1)$ dimensions is to focus on the "Zoll" property of $M_{3,1}$: the fact that all null-geodesics are S^1's. Given any Lorentzian four-manifold, M, with this property, one can define its twistor space, T, to be the set of these null-geodesics. It is easy to show that T has the structure of a compact 5-dimensional orbifold. In fact, if one excludes the situation depicted in the picture below, this orbifold is a manifold, and is equipped with a canonical contact structure.

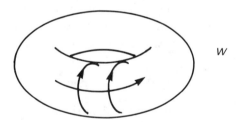

W is a solid torus in the cotangent bundle of M, Seifert-fibered by circular null-geodesics.

As we will see in Section 4, T has some features of Penrose's twistor space. (For instance the massless field equations of spin $k/2$ on M give rise to geometric data on T.) However, what it seems to lack is a natural complex structure. In other words, the most interesting feature of the Penrose correspondence, "causality on $M_{3,1}$ = holomorphicity on T_+," seems to have no analogue for such M's. A more down-to-earth problem is that there seem to be very few examples of such M's. The only examples I know of, other than $M_{3,1}$ itself, are product metrics of the form $g \times (-dt^2)$, on $S^3 \times S^1$,

g being a Zoll metric on S^3 and t the periodic time variable on S^1. In particular I have tried unsuccessfully to construct examples by deforming the standard Einstein metric on $M_{3,1}$ and am beginning to suspect that this is impossible. As a biproduct of these efforts, however, I noticed that the problem one dimension down, on $M_{2,1}$, is much more tractable. I will report on what I know about this problem in Section 3.

I will define a Zoll–Einstein manifold to be a compact Lorentzian manifold all of whose null-geodesics are periodic. Notice that this definition is conformally invariant: if g is a Zoll–Einstein metric, all conformal multiples of it are Zoll–Einstein. In Section 2, I will discuss Zoll–Einstein metrics and their deformations. In Section 3, I will apply the results of Section 2 to $M_{2,1}$; and in Section 4, I will show that there is a kind of "Penrose correspondence" for Zoll–Einstein manifolds which describes scattering-theoretical properties of the massless field equations of spin $k/2$ on M in terms of geometric data on T.

2. Zoll–Einstein Deformations

In this section I will explain why Zoll–Einstein manifolds are relatively deformable in dimension three and not so in higher dimensions. First, however, I will review a few elementary facts about deformations of metrics and conformal classes of metrics. Let M be an n-manifold and g a pseudometric on M. The bundle of symmetric, contravariant two-tensors splits as a direct sum

$$S^2(T^*) = \Theta + S^2(T^*)_0$$

the first summand being multiples of g and the second summand the traceless two-tensors. Moreover, Θ is isomorphic to the bundle of $(2/n)$ densities, $\Omega^{2/n}$. Now let

$$g_t, \qquad -\varepsilon < t < \varepsilon \tag{2.1}$$

be a family of pseudometrics deforming g, and let $\dot{g}$ be the traceless part of dg/dt at $t = 0$. Notice that if one replaces (2.1) by a conformally equivalent deformation, $\rho_t g_t$, $\dot{g}$ gets multiplied by ρ_0. Therefore, the deformation, (2.1), gives rise to a section of

$$\mathrm{Hom}(\Theta, S^2(T^*)_0),$$

or, alternatively, a section of

$$S^2(T^*)_0 \otimes \Omega^{-2/n}. \tag{2.2}$$

Now, let v be a vector field on M, and consider the traceless part of $D_v g$. If one replaces g by ρg the traceless part of $D_v g$ gets multiplied by ρ; so $D_v g$ has a conformally invariant interpretation as a section of (2.2). We will denote this section by δv, and we will denote by δ the operator

$$\delta : C^\infty(T) \to C^\infty(S^2(T^*)_0 \otimes \Omega^{-2/n}), \tag{2.3}$$

mapping v to δv.

Definition. The cokernel of (2.3) is the space of infinitesimal conformal deformations of (M, g).

Remarks: 1. The density bundle, $\Omega^{-2/n}$, is often omitted from this definition. If one does this, however, the definition will not be intrinsic: It will depend on the choice of g.

2. There seems to be no standard notation for this space. For the duration of this article we will denote it by $D(M, g)$.

Now suppose that g is a Zoll–Einstein metric. Let T be its twistor space, and let $\Omega^{-1/(n-1)}T$ be the bundle of $-1/(n-1)$ densities on T. I will show that there is a canonical transformation,

$$R : D(M, g) \to C^\infty(\Omega^{-1/(n-1)}T), \tag{2.4}$$

the data on the right being the "periods" of the data on the left over null-geodesics of g. Indeed, given an element, σ, of $D(M, g)$, let $\dot{g}$ be a section of (2.2) representing it. If one chooses a representative in the set of metrics conformally equivalent to g, e.g., g itself, this gives one a trivialization of $\Omega^{2/n}$; so one can think of $\dot{g}$ as a section of $S^2(T^*)_0$. Now let $\gamma = \gamma(s)$ be a closed null-geodesic and q the element of twistor space represented by γ. Let

$$(R\sigma)(q) = \int_\gamma \dot{g}\left(\frac{d\gamma}{ds}, \frac{d\gamma}{ds}\right) ds. \tag{2.5}$$

The number on the left changes if one reparametrizes γ by replacing s by a constant multiple of s; and it is not hard to see that, in order for the left-hand side to have an "intrinsic status," one has to interpret it as an element of $\Omega^{-1/(n-1)}$. It is also not hard to show that (2.5) does not depend on the choice of the representative, $\dot{g}$, of σ.

I can now state my first main result.

Theorem 2.1. *Suppose that (2.1) is a Zoll–Einstein deformation of g. Let σ be the element of $D(M, g)$ represented by $\dot{g}$. Then $R\sigma = 0$.*

The converse of this theorem is not true in general. However, if R is surjective, which is quite likely to be the case when $n = 3$, one has the following formal result.

Theorem 2.2. *Let σ be an element of $D(M, g)$ satisfying $R\sigma = 0$. Then there exists a family, g_i, $i = 1, 2, \ldots$, of traceless two-tensors such that g_1 is a representative of σ and such that the formal power series*

$$g_t = g + tg_1 + t^2 g_2 + \cdots \tag{2.6}$$

is formally a Zoll–Einstein deformation of g.

If R is not surjective, this theorem is false; in fact, it is quite hard even to find a g_2 such that $g + tg_1 + t^2 g_2$ is a Zoll–Einstein deformation "up to third order" in the parameter t. The problem of finding g_2 involves showing that a certain object,

$$Q(\sigma) \in C^\infty(\Omega^{-1/(n-1)} T),$$

is in the range of R. This object is a quadratic function of σ and is called the "quadratic obstruction" to the deformation of g by σ. Results of Kiyohara [9] indicate that it rarely (if ever) vanishes when $n > 3$.

Even if $n = 3$, Theorem 2.2 merely asserts that there is a formal deformation of g corresponding to σ. Built into the proof of Theorem 2.2 is a fairly simple algorithm for determining the ith term in the series (2.6) from the preceding terms. Unfortunately the series constructed using this algorithm seems almost always to diverge. For certain choices of (M, g), e.g., for $M = M_{2,1}$, this algorithm can be "souped up" by adding a *KAM* component to it; and I have been able to show, at least for $M_{2,1}$, that the souped-up algorithm converges. (In the souped-up algorithm one prescribes at stage i the terms up to order 2^i in the series (2.6), and then, at the next stage, one not only prescribes the next 2^i terms, but makes new choices of the first 2^i terms as well. In other words this algorithm is a "fast" algorithm and is equipped with an internal feed-back mechanism.)

Setting aside these (very difficult) convergence questions, it is clear from Theorem 2.1 that this theory is interesting only if the kernel of R is non-zero. There are compelling physical reasons for one to expect this to be the case (in fact not only in three dimensions but in higher dimensions as well). Namely, imagine space-time as being filled with a very thin gruel of astral debris. The matter tensor of such a space-time is equal to the Minkowski metric plus a small perturbation, $\dot{g}$. The only way to determine $\dot{g}$ physically is by seeing how it deflects light rays coming from sources at infinity, and one can show [6] that this problem is basically a problem in tomography: What the scattering of light rays determines is precisely the integrals (2.5). On the other hand one does not expect to be able to determine $\dot{g}$ completely from this experiment, i.e., relativity, unlike classical physics, has built-in limits on what is epistemologically accessible to an isolated observer. In particular an isolated observer cannot have access to data located in that part of the universe which is time-like in relation to himself.

There are a number of ways of making these crude heuristic remarks precise. For instance the operator (2.4) is a Fourier integral operator and operates in a natural way on microfunctions. One can show:

Theorem 2.3. *If a microfunction, f, is supported on the time-like part of relativistic phase space, then $Rf = 0$.*[1]

3. Deformations of $M_{2,1}$

I pointed out in Section 1 that $M_{3,1}$ is (in one of its many guises) the Grassmannian of complex two-dimensional subspaces of $\mathbf{C}^4$ on which the Hermitian form (1.1) vanishes. $M_{2,1}$ turns out to have a similar "twistorial" description: Let ω be the standard symplectic form on $\mathbf{R}^4$, and let $Sp(2, \mathbf{R})$ be the group of linear transformations of $\mathbf{R}^4$ which preserve ω. It turns out that the group of causal transformations of $M_{2,1}$ has $Sp(2, \mathbf{R})$ as double cover. Moreover, regarded as $Sp(2, \mathbf{R})$ entities, $M_{2,1}$ is just the Grassmannian of Lagrangian subspaces of $\mathbf{R}^4$, and its twistor space is the contact manifold, $\mathbf{R}P^3$, obtained by projectivizing $(\mathbf{R}^4, \omega)$. The mechanism by which points of $\mathbf{R}P^3$ get identified with null-geodesics in $M_{2,1}$ is the following: Given $q \in \mathbf{R}P^3$ let W be the one-dimensional subspace of $\mathbf{R}^4$ corresponding to it. Then the null-geodesic represented by q is the set of all Lagrangian subspaces of $\mathbf{R}^4$ containing W.

In this "twistor" picture of $M_{2,1}$, symplectic geometry seems to have co-opted the role played by Kaehlerian geometry in the twistor picture above. However, this picture turns out to have an important Kaehlerian component as well. Let $\omega_{\mathbf{C}}$ be the $\mathbf{C}$-linear extension of ω to $\mathbf{C}^4$, and let $CM_{2,1}$ be the Grassmannian of complex Lagrangian subspaces of $\mathbf{C}^4$. A Lagrangian subspace, V, of $\mathbf{C}^4$ is said to be positive-definite if $i\omega_{\mathbf{C}}(v, \bar{v}) > 0$

---

[1] A point, (q, p), in relativistic phase space is time-like if the length-squared of p is less than zero.

for all $v \in V - 0$. The set of positive-definite complex Lagrangian subspaces of $\mathbf{C}^4$ is a domain, $M_{2,1}^+$, in $\mathbf{C}M_{2,1}$ on which $Sp(2, \mathbf{R})$ acts transitively and in a holomorphic fashion. (In fact it is isomorphic to the generalized Siegel upper half-space: the set of symmetric complex 2×2 matrices, C, with $\operatorname{Im} C > 0$.)

Null-geodesics, in this picture, are related to complex-geometric objects in the following way: Let γ be a null-geodesic on $M_{2,1}$ and let q be the point in twistor space (i.e. in $\mathbf{R}P^3$) corresponding to γ. Let W be the one-dimensional subspace of $\mathbf{R}^4$ sitting above q, and let $\mathbf{C}\gamma$ (resp. γ^+) be the set of all complex (resp. positive-definite) Lagrangian subspaces of $\mathbf{C}^4$ containing W. The relation between γ and γ^+ is depicted in the figure below.

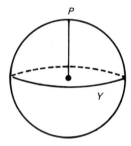

In this figure, $\mathbf{C}\gamma$ is S^2, γ^+ is the upper hemisphere, and γ is the equatorial circle.

We will show that the existence of Zoll–Einstein deformations of $M_{2,1}$ has to do with the existence of a certain type of holomorphic data on $M_{2,1}^+$. Namely, let σ be an infinitesimal conformal deformation of $M_{2,1}$ and let $\dot{g}$ be a traceless symmetric two-tensor representing σ. Suppose there exists a holomorphic traceless symmetric two-tensor, $\dot{h}$, defined on an open neighborhood, U, of $M_{2,1}^+$ in $\mathbf{C}M_{2,1}$, such that $\dot{g}$ is equal to the real part of $\dot{h}$ on $M_{2,1}$. Then, in computing the integral (2.5), one can deform the contour, γ, in the above figure over the region, γ^+, to the point p; so the right-hand side of (2.5) has to be zero. In other words, if there exists an $\dot{h}$ of the type above, $R\sigma = 0$. We will see shortly that the entire kernel of R is accounted for by such $\dot{h}$'s.

Now let g be any metric, globally defined on $M_{2,1}$, which is compatible with the standard Minkowski metric on $\mathbf{R}^3$, and consider the space $D(M, g)$ as an abstract $Sp(2, \mathbf{R})$ module. The representation of $Sp(2, \mathbf{R})$ on $D(M, g)$ is a subquotient of a principal series representation. Therefore, $D(M, g)$ possesses at most a finite number of irreducible subquotients (counting multiplicity). Thanks to some recent work of Vogan on the Langlands program for $Sp(2, \mathbf{R})$, one knows exactly what these irreducibles are. (See [13], pages 251–253.) In fact, using Vogan's results, Luis Casian has compiled a definitive list of lattice diagrams for these $Sp(2, \mathbf{R})$ modules. Vogan was able to locate $D(M_{2,1})$ on this list for me and to show that it has the lattice diagram

In this diagram, a slash indicates that the module represented by the bottom dot is of finite co-dimension in the module represented by the top dot.

This diagram says that there are just two irreducible submodules of $D(M_{2,1}, g)$ and that the only other submodule, besides these two irreducibles, is their direct sum. It also says that one of these irreducibles has to be finite dimensional. It is easy to see, incidentally, that such a representation has to occur on deformation-theoretic grounds. Namely, consider the space of conformally flat infinitesimal deformations of $M_{2,1}$. The representation of $Sp(2, \mathbf{R})$ on this space can be easily computed using the machinery of the Gasqui–Goldschmidt paper, [3], and turns out just to be the adjoint representation.

This leaves very few options for the other irreducible. One can compute R on the finite-dimensional irreducible directly and show that it is non-zero; so the other irreducible has to be identical with the kernel of R and hence also has to be the image in $D(M_{2,1}, g)$ of the space of holomorphic traceless symmetric two-tensors on $M_{2,1}^{+}$ (confirming the assertion which we made earlier).

Summarizing, we have proved

Theorem. *The two irreducible subspaces of $D(M_{2,1}, g)$ are the space of infinitesimal Zoll–Einstein deformations of $M_{2,1}$ and the space of infinitesimal conformally flat deformations of $M_{2,1}$.*

Notice that, since these two spaces are disjoint, this theorem implies:

Corollary. *There are no conformally flat Zoll–Einstein deformations of $M_{2,1}$.*

This corollary is a little misleading: It turns out that there are lots of conformally flat Zoll–Einstein metrics on $M_{2,1}$ arbitrarily close to g. However, these metrics have null-geodesics with much longer periods than the null-geodesics of g (i.e. null-geodesics which "wind around" many times before closing up). Therefore, they cannot be deformations of g.

Thanks to Vogan's results, the irreducible subrepresentation of $D(M_{2,1}, g)$ corresponding to the space of Zoll–Einstein deformations of $M_{2,1}$ can be easily identified: It is the restriction to $Sp(2, \mathbf{R})$ of an irreducible representation of $SU(2, 2)$: the unitary representation describing massless spin 2 fields (or "gravitons") in $(3+1)$ dimensions. I have not been able to decide whether this is an accident or a manifestation of some hidden connection between GR in $(3+1)$ dimensions and Zoll–Einstein deformations of $M_{2,1}$. That such a hidden connection exists is not entirely out of the question. For instance $M_{2,1}$ can be canonically imbedded in $M_{3,1}$, and its complement is the domain of definition of the anti-DeSitter metric. Moreover, if one multiplies the anti-DeSitter metric by a function which vanishes exactly to second order on $M_{2,1}$, one gets a metric which is defined on all of $M_{3,1}$ and is conformally equivalent to g on $M_{2,1}$. Is it true that Zoll–Einstein deformations of $M_{2,1}$ correspond to deformations of the anti-DeSitter metric? I am indebted to Mike Eastwood for calling my attention to the fact that something of this nature is true locally in the real-analytic category. (See C. LeBrun, [10].)

4.   Floquet Theory

Let (M, g) be a compact Lorentzian manifold and let D be a hyperbolic differential operator, such as the wave operator or Dirac operator, intrinsically associated with g. Because of the compactness of M it rarely happens that there are nontrivial solutions of the equation, $Du = 0$. However, it is not unlikely that there exist lots of solutions of this equation on the universal cover of M. For instance, if M is compactified Minkowski four-space, its universal cover is $S^3 \times \mathbf{R}$, which is globally hyperbolic and spatially compact; so all of the standard hyperbolic operators defined on M have large numbers of solutions on $S^3 \times \mathbf{R}$. One recaptures from these solutions data relevant to M itself by looking at the action of the fundamental group, $\pi_1(M)$ on the space

$$\{u \in C^{\infty}(\tilde{M}), Du = 0\}.$$

I will call this representation the *Floquet representation* of $\pi_1(M)$ associated with D. Paneitz and Segal have made an extensive study of this representation for $M = M_{n-1,1}$ taking for D a conformally invariant hyperbolic operator of the type above. In the $M_{n-1,1}$ case $\pi_1(M) = Z$; so the Floquet representation is completely determined by what it does to the generator of Z. Paneitz and Segal show that the Floquet representative of this generator is identical with what the physicists would call the scattering matrix of the equation, $Du = 0$. They do this by observing that there exist metrics on $M_{n-1,1}$ which are conformally compatible with the standard Minkowski metric on $\mathbf{R}^n$ (the simplest of these being the Einstein metric, which is the only metric of this kind whose group of isometries is the maximal compact subgroup of $SO(n, 2)$). The operator, D, appears to have complicated singularities at infinity when one expresses it in terms of the Minkowski metric; however, one can show that these singularities are not "intrinsic" singularities by expressing D in terms of a metric of the type described above. (For further details, see [11].)

I have given this brief sketch of Floquet theory with rather different applications in mind than those I have just described: Namely, I will show below that if (M, g) is a Zoll–Einstein manifold, the massless spin $k/2$ field equations on M give rise to conformal Floquet invariants which can, to a certain extent, be interpreted as twistorial invariants of M. For simplicity I will confine myself to (M, g)'s which are deformations of $M_{2,1}$ and I will take D be to the Dirac operator. However, much of what I say is true for the other massless spin $k/2$ equations as well and is true for spaces which are not deformations of $M_{2,1}$ (and for other dimensions besides $(2+1)$).

I will begin by recalling how the Dirac operator is defined in $(2+1)$ dimensions. The co-frame bundle of M is a principal $SL(2, \mathbf{R})$ bundle. Let ρ be the standard representation of $SL(2, \mathbf{R})$ on $\mathbf{R}^2$, and let V be the vector bundle over M associated with ρ. Then $S^2(V)$ is the cotangent bundle of M. The Dirac operator is usually defined to be a first-order partial differential operator mapping sections of V to sections of V^*; i.e. an operator of the form

$$\partial : C^{\infty}(V) \to C^{\infty}(V^*). \tag{4.1}$$

It is defined by observing that the anti-symmetrization operation

$$\mathbf{R}^2 \otimes S^2(\mathbf{R}^2) \to \Lambda^2(\mathbf{R}^2) \otimes \mathbf{R}^2 = [\Lambda^2(\mathbf{R}^2)]^2 \otimes (\mathbf{R}^2)^* \tag{4.2}$$

gives rise to a morphism of induced vector bundles

$$V \otimes T^* \to V^* \otimes [\Lambda^2(V)]^2. \tag{4.3}$$

The line bundle on the far right is isomorphic to the density bundle, $\Omega^{2/3}$; so fixing a metric in the conformal class of g has the effect of trivializing this bundle, in which case (4.3) becomes a bundle morphism

$$V \otimes T^* \to V^*. \tag{4.4}$$

The metric also gives rise to a Levi–Civita connection on the co-frame bundle and on its spin double cover, and hence gives rise to a connection on V, which is a first-order differential operator

$$D: C^\infty(V) \to C^\infty(V \otimes T^*).$$

Composing this with (4.4) one ends up with an operator going from $C^\infty(V)$ to $C^\infty(V^*)$, and this is precisely the Dirac operator (4.1). As it stands it is not conformally invariant. However, if one does not eliminate the density bundle, $\Omega^{2/3}$ from the right-hand side of (4.3) and tensors both sides of (4.3) by $\Omega^{1/6}$, one gets a bundle morphism

$$V \otimes \Omega^{1/6} \otimes T^* \to V^* \otimes \Omega^{5/6}$$

and, by the same route as before, a first-order differential operator

$$\partial: C^\infty(V \otimes \Omega^{1/6}) \to C^\infty(V^* \otimes \Omega^{5/6}) \tag{4.5}$$

which is conformally invariant (see [7]). Notice, by the way, that if s_1 and s_2 are sections of $V \otimes \Omega^{1/6}$, then by pairing ∂s_1 with s_2 at $p \in M$, one gets a density, $(\partial s_1, s_2) \in \Omega_p$ and, by integrating this density over M, a scalar quantity,

$$\int (\partial s_1, s_2).$$

Thus the Dirac operator, in its conformally invariant form, defines a conformally invariant inner product on the space $C^\infty(V \otimes \Omega^{1/6})$.

Coming back to the Floquet theory, let $\tilde{M}$ be the universal cover of M. Since M has the diffeotype of $M_{2,1}$,

$$\tilde{M} = S^2 \times \mathbf{R}, \tag{4.6}$$

and $\pi_1(M) = Z$. Let σ be the generator of Z. The deck transformation

$$\tilde{\sigma}: \tilde{M} \to \tilde{M} \tag{4.7}$$

associated with σ is just the map

$$\sigma(x, t) = (-x, t + \pi)$$

(see [11]). For metrics, g, which are small deformations of the Einstein metric on $M_{2,1}$, (4.6) is a space-time splitting of $\tilde{M}$; so ∂ is globally hyperbolic on $\tilde{M}$.

I will now give an unintrinsic but fairly concrete description of the Floquet operator, $S(g, \partial)$ representing σ: Let $M_0 = S^2 \times \{0\}$ and let i be the inclusion map of M_0 into $\tilde{M}$. The space of solutions of the equation, $\partial u = 0$, can be identified with $C^\infty(i^*(V \otimes \Omega^{1/6}))$ by the restriction map, $u \to i^*u$. Given $\varphi \in C^\infty(i^* V \otimes \Omega^{1/6}))$, let u be the solution of $\partial u = 0$ associated with φ. Then $\tilde{\sigma}^* u$ is also a solution, and, by definition

$$S(g, \partial)\varphi = i^* \tilde{\sigma}^* u. \tag{4.8}$$

In this representation, $S(g, \partial)$ is an operator on sections of vector bundle:

$$S(g, \partial): C^\infty(i^*(V \otimes \Omega^{1/6})) \to C^\infty(i^*(V \otimes \Omega^{1/6})). \tag{4.9}$$

The main result which I have been able to prove about this operator is the following

Theorem. *The metric, g, is Zoll–Einstein if and only if $S(g, \partial) = \pm I + K$, K being a compact operator. Moreover, if g is Zoll–Einstein, K is a pseudodifferential operator of order (-1) in the representation* (4.9).

One can use this result to define an intriguing conformal invariant of M (which I suspect should have a simple physical interpretation, though I have not been able to find one). Let $\sigma(K)(x, r\xi)$ be the symbol of K at (x, ξ) and let $\det \sigma(K)(x, X)$ be its determinant. Since K is of order (-1), this expression is a homogeneous function of degree (-2) in ξ; so the integral

$$\int \det \sigma(K)(x, \xi) \, dx \, d\xi \tag{4.10}$$

over the cotangent bundle of M_0 is well-defined providing one interprets the integrals

$$\int \det \sigma(K)(x, \xi) \, d\xi$$

over the cotangent fibers as residues in the sense of Gelfand–Shilov (see [4], page 305). It is not hard to show that (4.10) is a spectral invariant of

$S(g, \partial)$; and one can, in fact, interpret it as a kind of regularized determinant of $S(g, \partial)$. I suspect that it is a non-local invariant of the metric, g, however, I have been able to show that its variation with respect to g can be written in a semi-local manner: as periods over null-geodesics of the curvature tensor of g and its first covariant derivatives.

There are other interesting conformal invariants of g which are spectral invariants of $S(g, \partial)$. Namely, for all $N > 2$, K^N is of trace class, so

$$\text{trace } K^N \tag{4.11}$$

is such an invariant. These invariants seem to be much more subtle and harder to get one's hands on than (4.10).

The representation, (4.9), of the Floquet operator is aesthetically somewhat unappealing. Since this operator is itself an intrinsic object, one would like to have a more intrinsic description of it than that afforded by (4.9). I will try, in the last few paragraphs of this article, to give a "twistorial" description of this Floquet operator. First of all note that, in (4.9), M_0 can be replaced by essentially any space-like hypersurface in $\tilde{M}$, or, what amounts to the same thing, by any space-like hypersurface in M. I will now show that there is a one-to-one correspondence between space-like hypersurfaces in M and certain polarizations of twistor space. Indeed, consider the double fibration

the elements of Z being pairs (γ, p), p being a point of M and γ a null-geodesic containing p, and π and ρ being the projection mappings sending (γ, p) to γ and p respectively. It is easy to show that if M_0 is a space-like hypersurface in M and Z_0 its preimage in T, then π maps Z_0 diffeomorphically onto T; so we can compose π^{-1} with ρ to get a map

$$\rho \pi^{-1} : T \to M_0. \tag{4.12}$$

It turns out that (4.12) is a fibration and that the fibers are "Legendrian" submanifolds of T, (i.e., the canonical contact form vanishes on these fibers). In other words, (4.12) is a "polarization" of twistor space. One can interpret $S(g, \partial)$ as a quantum object on twistor space, (4.9) being the description of this object in terms of the polarization (4.12) (see [12] and [14]).

Concluding remark. A more detailed account of the results described in this article can be found in [5].

References

[1] M. Atiyah, N. Hitchin, and I. M. Singer, Self-duality in four-dimensional Riemannian geometry, *Proc. R. Soc. Lond. A* **362** (1978) 425-461.

[2] M. Eastwood, R. Penrose, and R. O. Wells Jr., Cohomology and massless fields, *Comm. Math. Phys.* **78** (1980) 305-351.

[3] J. Gasqui and H. Goldschmidt, *Deformations infinitesmales des Structures Conformes Plates.* Birkhauser, Boston, 1984.

[4] I. M. Gelfand and G. E. Shilov, *Generalized Functions*, Vol. 1. Academic Press, New York (1964).

[5] V. Guillemin, Cosmology in $(2+1)$ dimensions, cyclic models and deformations of $M_{2,1}$, to appear.

[6] V. Guillemin and S. Sternberg, Souriau scattering and Yang-Mills dust, *Annals of Physics* **165** (1985) 259-279.

[7] N. Hitchin, "Harmonic spinors," *Advances in Math.* **14** (1974) 1-55.

[8] N. Hitchin, *Complex Manifolds and Einstein's Equations*, Lecture Notes in Math. 970, Springer Verlag, 1982, pp. 73-100.

[9] K. Kiyohara, C_1-metrics on spheres, *Proc. Jap. Acad.* **58**, ser. A (1982).

[10] C. LeBrun, H-space with a cosmological constant. *Proc. R. Soc. Lond., A* **380** (1982) 171-185.

[11] I. Segal, Induced bundles and non-linear wave equations, Conference in honor of George Mackey, MSRI, Berkeley, 1984.

[12] D. J. Simms and N. Woodhouse, *Lectures on Geometric Quantization.* Springer Lecture Notes in Physics, 53 Springer-Verlag, New York, 1976.

[13] D. Vogan, The Kazhdan-Lusztig conjecture for real reductive Lie groups, in *Representation Theory of Reductive Groups*, Progress in Math. **40**, Birkhauser, Boston, 1983, pp. 223-264.

[14] A. Weinstein, Symplectic geometry, *Bull. Amer. Math. Soc.* **5** (1981) 1-13.

A Proof of the Bott Inequalities

B. Helffer

Department of Mathematics
Université de Nantes
Nantes, France

and

Centre de Mathématiques
Ecole Polytechnique
Palaiseau, France

J. Sjöstrand

Department of Mathematics
Université Paris-Sud
Orsay, France

0. Introduction

The purpose of this paper is to give a short proof of the Bott inequalities using the ideas of E. Witten. Let M be a compact Riemannian manifold and let $f \in C^\infty(M, \mathbf{R})$ be a function such that

$$(df)^{-1}(0) = \bigcup_1^N \Gamma_j, \tag{0.1}$$

where Γ_j are smooth connected submanifolds such that the transversal

Algebraic Analysis, Volume I

Hessian of f at every point of Γ_j is nondegenerate, or equivalently, that

$$\|\nabla f(x)\| \approx \operatorname{dist}\left(x, \bigcup_1^N \Gamma_j\right). \tag{0.2}$$

The Bott inequalities then relate the Betti numbers of M with the Betti numbers of certain twisted cohomologies of the manifolds Γ_j (see Section 3 for the precise statement). In the case when the Γ_j are points they reduce to the Morse inequalities. In [11], Witten proposed a semi-classical proof of the Morse inequalities that was carried out by B. Simon [10] (see also Henniart [9]). In [6] we showed that the ideas of Witten could be used to obtain the full cohomology with real coefficients.

In the more general situation when Γ_j are manifolds, Witten proposed an approach leading to the spectral analysis of certain Laplacians on Γ_j and a potential well problem where Γ_j are the potential wells.

In this order of ideas Bismut gave a probabilistic proof of the Bott inequalities (cf. [2]). In [8] we also developed the necessary machinery to prove the Bott inequalities in this spirit (see [5] for details).

However this approach ([8], [5]) turns out to be quite technical (more than one could expect from Witten's paper) and we shall develop here a slightly different approach. Following a classical idea, we shall replace f by a more generic function in the following way: Let g_j be a generic Morse function on Γ_j and let g_j also denote some suitable extension to a neighborhood of Γ_j. If χ_j is a smooth cutoff function equal to 1 on G_j, then for $\varepsilon > 0$ small enough the function $f_\varepsilon = f + \varepsilon \sum \chi_j g_j$ is a Morse function on M with critical set

$$\bigcup_1^N (dg_j)^{-1}(0) \subset \bigcup_1^N \Gamma_j.$$

The Witten complex $d_{f_\varepsilon} = e^{-f_\varepsilon/h} hd\, e^{f_\varepsilon/h}$ restricted to the space associated to the exponentially small eigenvalues of the corresponding Hodge Laplacians converges, when h tends to 0, after suitable conjugations to a complex which can be decomposed into a direct sum of complexes associated to each Γ_j. These subcomplexes turn out to be precisely the ones that describe the twisted cohomology of the Γ_j and the Bott inequalities follow. Most of the arguments here are purely geometric, and by some additional work one could certainly obtain a purely geometric proof, using only the fact that the cohomology is given by the orientation complex obtained from a generic Morse function. The interesting fact from an analyst's point of view is, however, that the basic idea of Witten and some analysis lead to a proof that can be developed without any previous insight in topology.

1. Some Complements to [6]

Let M be compact (not necessarily orientable) Riemannian manifold and let $\mathcal{O}$ be a real line bundle with transition functions ± 1. We can then define the de Rham complex d:

$$\to C^\infty(M;(\Lambda^{l-1}T^*M)\otimes\mathcal{O}) \to C^\infty(M;(\Lambda^l T^*M)\otimes\mathcal{O}) \to \cdots. \quad (1.1)$$

Let $f \in C^\infty(M;\mathbf{R})$ be a Morse function, i.e., a function such that all critical points are nondegenerate and hence isolated. If dx^2 denotes the metric on M, we consider the Agmon metric $\|\nabla f(x)\|^2\, dx^2$ and we let d_f denote the corresponding distance. (It will always be clear from the context whether d_f denotes the Witten complex or the Agmon distance.) Then (cf. [6]) $|f(x)-f(y)| \le d_f(x,y)$.

We assume in this section

(H1) If U_j, U_k are two critical points of f and $d_f(U_j,U_k) = f(U_j)-f(U_k)$, then $\operatorname{ind}_f U_j \ge \operatorname{ind}_f U_k + 1$.

Here $\operatorname{ind}_f x_0$ is by definition the number of negative eigenvalues of $f''(x_0)$, when x_0 is a critical point of f. We also assume

(H2) If U_j, U_k are critical points of index l, $l+1$ and if $d_f(U_j,U_k) = f(U_j)-f(U_k)$, then there are only finitely many minimal d_f-geodesics from U_k to U_j. Moreover, V_k^+ and V_j^- intersect transversally along each such minimal geodesic.

Here $V_k^\pm$ denotes the stable outgoing/incoming ∇f-invariant manifold passing through U_k. Then $T_{U_k}V_k^\pm$ is the space associated to the positive/negative eigenvalues of $f''(U_k)$, so $\dim V_k^- = l$, $\dim V_k^+ = n-l$, $\dim V_j^- = l+1$. It is known that (H1) and (H2) are generically satisfied and we shall call f a *generic* Morse function.

As in the case of the ordinary de Rham complex we can follow the ideas of Witten and conjugate the complex (1.1) into $d_f = e^{f/h}hd\,e^{f/h}$. If $F^{(l)} \subset C^\infty(M;(\Lambda^l T^*M)\otimes\mathcal{O})$ is the space associated to the exponentially small eigenvalues of the corresponding Hodge Laplacian in degree l: $d_f^* d_f + d_f d_f^*$, then for $h>0$ small enough the dimension of $F^{(l)}$ is equal to the number of critical points of index l. If U_k, $k \in J$ are the critical points, we put $C^{(l)} = \{k;\ \operatorname{ind}_f U_k = l\}$. As in [6], $F^{(l)}$ has an orthonormal basis e_k, $k \in C^{(l)}$ which satisfies

$$e_k = e^{-\phi(x)/h}\omega(x,h), \qquad \omega(x,h) = \omega_0(x,h) + O(h^{1-n/4}), \quad (1.2)$$

where $\phi(x) = d_f(x, U_k)$ and

$$\omega_0 = (|\lambda_1| \ldots |\lambda_n|)^{1/4} (\pi h)^{-n/4} \, dy_1 \wedge \cdots \wedge dy_l \otimes u_{U_k} \qquad \text{on } V_k^+, \quad (1.3)$$

$$\omega_0 = (-1)^{l(n-l)} (|\lambda_1| \cdots |\lambda_n|)^{1/4} (\pi h)^{-n/4} J_{n-l} \qquad (1.4)$$

$$(dz_{l+1} \wedge \cdots \wedge dz_n) \otimes u_{U_k} \qquad \text{on } V_k^+.$$

Here $\lambda_1, \ldots, \lambda_n$ are the eigenvalues of $f''(U_k)$ and u_{U_k} is one of the two possible local sections of $\mathcal{O}$ of absolute value 1. $y_1, \ldots, y_l$ are functions with linearly independent differentials vanishing on V_k^+ and with the property that their differentials restricted to $T_{U_k} V_k^-$ form an orthonormal system.

For small $h > 0$, the cohomology of the complex (1.1) is the same as that of

$$d_f : \cdots \rightarrow F^{(l)} \rightarrow F^{(l+1)} \rightarrow \cdots. \quad (1.5)$$

After suitable conjugations with invertible h-dependent matrices, this complex converges to the orientation complex

$$\beta^{(l)} : \cdots \rightarrow \mathbf{R}^{m(l)} \rightarrow \mathbf{R}^{m(l+1)} \rightarrow \cdots. \quad (1.6)$$

where $m(v) = \#(C^{(v)})$, and where for each critical point, we have chosen an orientation on the corresponding manifold V^- and a local section of modulus 1 of $\mathcal{O}$. Here the matrix $\beta^{(l)}$ is of the form

$$\beta_{j,k}^{(l)} = \Sigma_\gamma \varepsilon_\gamma \alpha_\gamma, \quad \text{if } d_f(U_j, U_k) = f(U_j) - f(U_k), \quad \text{and} = 0 \text{ otherwise.} \quad (1.7)$$

Here the summation is over the finite set of the minimal d_f-geodesics from U_k to U_j. By definition, $\alpha_\gamma = \pm 1$ if u_{U_k} and u_{U_j} continuously extended to γ are equal/unequal. To define $\varepsilon_\gamma = \pm 1$ we choose $y_1, \ldots, y_l$ associated to U_k and $z_{l+2}, \ldots, z_n$ associated to U_j. The definition of these coordinates extend to a neighborhood of the interior of γ so that $y_1, \ldots, y_l = 0$ on V_k^+ and $z_{l+2}, \ldots, z_n = 0$ on V_j^-. Then $\varepsilon_\gamma = 1$ if

$$d\chi_{j,k} \wedge dy_1 \wedge \cdots \wedge dz_{l+2} \wedge \cdots \wedge dz_n \quad (1.8)$$

has the same sign at a point of γ where $\nabla f \cdot \nabla \chi_{j,k} > 0$ as the orientation of $T_{U_j} M$ transported along γ. Otherwise $\varepsilon_\gamma = -1$. Notice that ε_γ does not depend on the choice of the orientation on $T_{U_j} M$. If we choose different orientations on the various V^- or if we choose different local sections of $\mathcal{O}$, then the complex (1.6) will only change by conjugation with invertible matrices. Apart from these minor modifications all arguments of [6] go through and we conclude that the complexes (1.1) and (1.6) have the same cohomology.

We now give a more geometric description of ε_γ. Fix a point x_0 in the interior of γ where $\nabla f \cdot \nabla \chi_{j,k} > 0$. We have already noticed that $dy_1 \wedge \cdots \wedge dy_l$ gives the orientation on $T_{x_0} M / T_{x_0} V_k^+$ obtained by transportation along γ from U_k. Choose also $y_1', \ldots, y_l'$ vanishing on V_k^+ such that $d\chi_{j,k} \wedge dy_1' \wedge \cdots \wedge dy_l'$ induces the orientation on $T_{x_0} V_j^-$ obtained by transportation from U_j. By the choice of $z_{l+2}, \ldots, z_n$ we then know that

$$d\chi_{j,k} \wedge dy_1' \wedge \cdots \wedge dz_{l+2} \wedge \cdots \wedge dz_n \qquad (1.9)$$

has the same orientation on $T_{x_0} M$ as the one obtained from $T_{U_j} M$. Hence $\varepsilon_\gamma = 1$ iff $dy_1 \wedge \cdots \wedge dy_l$ and $dy_1' \wedge \cdots \wedge dy_l'$ have the same sign. We can interpret $dy_1' \wedge \cdots \wedge dy_l'$ as the orientation of $T_{x_0} V_j^- / (\gamma') \approx T_{x_0} M / T_{x_0} V_k^+$. Hence $\varepsilon_\gamma = 1$ iff we get the same orientation on $T_{x_0} M / T_{x_0} V_k^+$ from $T_{U_k} V_k^-$ and from $T_{x_0} V_j^- / (\gamma')$.

2. Geometric Preparations

Let M, f, Γ_j be as in Section 1. We fix one component Γ_j that we shall denote simply by Γ and we shall work in a suitable small tubular neighborhood Ω of Γ. We also assume that $f|_\Gamma = 0$. Let (d_+, d_-) be the signature of the transversal Hessian of f on Γ and let n', n be the dimensions of Γ and M respectively. By the stable manifold theorem (see [1]) we have smooth ∇f-invariant closed submanifolds of Ω: $F_+, F_- \subset \Omega$ such that for every $x \in \Gamma$

$$T_x(F_\pm) = T_x\Gamma \oplus \{\text{the sum of eigenspaces corresponding}$$
$$\text{to positive/negative eigenvalues of } f''(x)\}.$$

Moreover $F_\pm = \bigcup_{x\in\Gamma} F_{\pm,x}$, where $F_{\pm,x} = \{y \in \Omega; \exp(\pm t\nabla f)(y)$ tends to x exponentially fast as t tends to $-\infty\}$.

$F_{\pm,x}$ are smooth submanifolds of dimension $d_\pm$ which vary smoothly with x, and $T_x(F_{\pm,x})$ is the sum of eigenspaces of $f''(x)$ corresponding to positive/negative eigenvalues.

Let $\phi(x) = d_f(x, \Gamma)$ be the distance from x to Γ for the Agmon metric $\|df\|^2 dx^2$. Then, as we have seen in [7], [8], ϕ is a C^∞ function vanishing precisely to the second order on Γ, and $-\phi(x) \leq f(x) \leq \phi(x)$ with $\pm\phi = f$ precisely on $F^\pm$.

If $F_x = \{y \in \Omega; \exp(-t\nabla\phi)(x) \to x, t \to \infty\}$, then F_x is a smooth submanifold of Ω of dimension $d_+ + d_-$ depending smoothly on x and we have $F_{\pm,x} \subset F_x$.

In particular $T_xF_x = T_x\Gamma^\perp$ (since $T_x\Gamma$ is the kernel of $f''(x)$). The F_x form a fibration of Ω so we can define a smooth map $\pi: \Omega \to \Gamma$ by $\pi(x) = y$ if $x \in F_y$.

Let $g \in C^\infty(\Gamma; \mathbf{R})$ be a Morse function and put $\tilde{g} = g \circ \pi \in C^\infty(\Omega; \mathbf{R})$. Let $\chi \in C^\infty(\Omega; [0, 1])$ be equal to 1 in a neighborhood of Γ. For $\varepsilon > 0$ small, we put

$$f_\varepsilon = f(x) + \varepsilon\chi(x)\tilde{g}(x). \tag{2.1}$$

If $x_0 \in \Gamma$ is a critical point of g, then it is also a critical point of f_ε, and we have the direct sum decomposition

$$f_\varepsilon(x_0)'' = f''(x_0)|_{T_xF_{x_0}} \oplus \varepsilon g''(x_0)|_{T_x\Gamma}. \tag{2.2}$$

In particular,

$$\mathrm{ind}_{f_\varepsilon}(x_0) = \mathrm{ind}_g(x_0) + d_-, \tag{2.3}$$

where $\mathrm{ind}_g(x_0)$ denotes the number of negative eigenvalues of $g''(x_0)$ and similarly for $\mathrm{ind}_{f_\varepsilon}(x_0)$. It also follows from (2.2), that if $G_{x_0}^\pm$ are the stable outgoing/incoming ∇g-invariant manifolds passing through x_0 and $V_{\varepsilon,x_0}^\pm$ are the corresponding objects for f_ε, then

$$T_{x_0}(V_{\varepsilon,x_0}^\pm) = T_{x_0}G^\pm \oplus T_{x_0}(F_{x_0}^\pm). \tag{2.4}$$

Lemma 2.1. *If $\varepsilon > 0$ is sufficiently small, the critical points of f_ε are precisely those of g.*

Proof. Since $\|df(x)\| \sim \mathrm{dist}(x, \Gamma)$ it is clear that the critical points of f_ε have to be in a region $\mathrm{dist}(x, \Gamma) \leq \mathrm{Const.}\ \varepsilon$, and here $f_\varepsilon(x) = f(x) + \varepsilon\tilde{g}(x)$. Considering the gradients in the F_x direction we see that the critical points must be in Γ. Since $f_\varepsilon|_\Gamma = \varepsilon g$ they then have to coincide with those of g.

In the following we replace Ω by a smaller tubular neighborhood of Γ so that $f_\varepsilon(x) = f(x) + \varepsilon\tilde{g}(x)$. We shall next compare Agmon-distances in Γ and in Ω. Let d_{f_ε} be the distance in Ω with respect to $\|\nabla f_\varepsilon\|^2 dx^2$ and let d_g be the distance in Γ with respect to $\|\nabla g\|^2 dx^2$. Put $V_\varepsilon = \|\nabla f_\varepsilon\|^2 = \|\nabla f\|^2 + 2\varepsilon\nabla f \cdot \nabla\tilde{g} + \varepsilon^2\|\nabla\tilde{g}\|^2$. Here $\nabla\tilde{g}$ is orthogonal to the fibers F_x while ∇f is tangential to F_x to the second order at x. Hence $\nabla f \cdot \nabla\tilde{g} = O(D_\Gamma^2\|\nabla\tilde{g}\|^2)$, where $D_\Gamma(x)$ is the Riemannian distance from x to Γ. Put $W = \|\nabla g\|^2 \circ \pi$, $V_0 = \|\nabla f\|^2 \sim D_\Gamma^2$. Then

$$V_\varepsilon = V_0 + \varepsilon^2 W + O(\varepsilon D_\Gamma^2 + \varepsilon^2 D_\Gamma W).$$

Let $\alpha:[0,1]\ni t\to x(t)\in\Omega$ be a C^1-curve and let $\beta:[0,1]\ni t\to y(t)=\pi(x(t))\in\Gamma$ be the corresponding projected curve. Then $\|y'(t)\|\le(1+CD_\Gamma(x(t)))\|x'(t)\|$ (with Riemannian norms evaluated at the appropriate places).

The energy of α is then

$$E(\alpha)=\int_0^1 V_\varepsilon(x(t))\|x'(t)\|^2\,dt$$

$$\ge\int_0^1[\varepsilon^2 W(y(t))+C^{-1}D_\Gamma(x(t))^2-C\varepsilon^2 W(y(t))D_\Gamma(x(t))]$$

$$\times(1-C\cdot D_\Gamma(x(t)))\|y'(t)\|^2\,dt$$

$$\ge\int_0^1\left[\varepsilon^2 W(y(t))+\left(\frac{1}{C_1}\right)D_\Gamma(x(t))^2\right.$$

$$\left.-C_1\varepsilon^2 W(y(t))D_\Gamma(x(t))\right]\|y'(t)\|^2\,dt.$$

Here $(1/C_1)D^2-C_1\varepsilon^2 WD\ge-C_2\varepsilon^4 W^2$, so with a new constant C, we obtain

$$E(\alpha)\ge(1-C\varepsilon^2)\varepsilon^2 E(\beta),$$

where $E(\beta)=\int_0^1 W(y(t))\|y'(t)\|^2\,dt$ is the energy of y with respect to g. If $|\alpha|_{f_\varepsilon}$ and $|\beta|_g$ denote the corresponding Agmon distances, then $|\alpha|\le E(\alpha)^{1/2}$ and assuming that y avoids the critical points of g, we reparametrize α so that $|\alpha|=E(\alpha)^{1/2}$. Hence

$$|\alpha|_{f_\varepsilon}\ge(1-C\varepsilon^2)^{1/2}\varepsilon|\beta|_g\ge(1-C_1\varepsilon^2)|\beta|_{\varepsilon g}. \tag{2.5}$$

We have then proved:

Proposition 2.2. *There is a constant $C>0$ such that for $x,y\in\Gamma$ and $\varepsilon>0$ small enough*

$$d_{f_\varepsilon}(x,y)\ge(1-C\varepsilon^2)\varepsilon d_g(x,y). \tag{2.6}$$

Notice here that d_{f_ε} is obtained by taking the infimum over C^1 curves that stay in Ω. For a C^1-curve with end points in Γ that does not stay in Ω, the corresponding d_{f_ε}-length will be $\ge C_0\gg\varepsilon$ where C_0 is independent of ε so (2.6) will remain valid later also when f_ε is globally defined on M.

Let $x_0\in\Gamma$ be a critical point of g. Let $G_+\subset\Gamma$ be the stable outgoing manifold w.r.t. ∇g. We clearly have $G_+\subset V_\varepsilon^+$. We shall localize V_ε^+ better in a neighborhood of Γ. Let us therefore consider a ∇f_ε-trajectory $]-\infty, 0] \ni t \to x(t) \in \Omega$ with $x(t) \to x_0$, $t \to -\infty$. As in [6] we introduce $f_+ = \phi(x) - f(x)$, $f_- = \phi(x) + f(x)$, where $\phi(x) = d_f(x, \Gamma)$. Then $f_\pm \sim \text{dist}(\cdot, F_\pm)^2$ and

$$\phi(x) = \tfrac{1}{2}(f_+(x) + f_-(x)), \qquad f(x) = \tfrac{1}{2}(f_-(x) - f_+(x)).$$

From $\|\nabla \phi\|^2 = \|\nabla f\|^2$ we get: $\nabla f_+ \cdot \nabla f_- = 0$ and hence $\nabla_f(f_+) = -\tfrac{1}{2}(\nabla f_+)^2$, $\nabla_f(f_-) = \tfrac{1}{2}(\nabla f_-)^2$. With $f'_\pm = (d/dt)(f_\pm(x(t)) = \nabla f_\varepsilon(f_\pm)$, we get

$$f'_+ = -\tfrac{1}{2}(\nabla f_+)^2 + \varepsilon \nabla \tilde{g} \cdot \nabla f_+ = -\tfrac{1}{2}(\nabla f_+)^2 + O(\varepsilon(f_+ + f_-)),$$

so

$$f'_+ \leq -f_+ C + O(\varepsilon(f_+ + f_-)), \qquad f'_- \geq \frac{f_-}{C} + O(\varepsilon(f_+ + f_-)).$$

From these two inequalities, we see that if $C_1 \gg 0$, then $t \to f_+$ is exponentially decreasing (when t increases) when $f_+ \geq C_1 \varepsilon f_-$, and $t \to f_-$ is exponentially increasing when $f_- \geq C_1 \varepsilon f_+$. From this we see that $f_+(x(t)) \leq C_1 \varepsilon f_-(x(t))$ for all $t \in]-\infty, 0]$. (In fact, if $f_+(x(t_0)) > C_1 \varepsilon f_-(x(t_0))$, then $[0, \infty[\ni s \to f_+(x(t_0 - s))$ is exponentially increasing and we stay in the region $f_+ \geq C_1 \varepsilon f_-$ because $[0, \infty[\ni s \to f_-(x(t_0 - s))$ is decreasing when $f_+ = C_1 \varepsilon f_-$.) This proves the first part of the following proposition:

Proposition 2.3. *Let V^+_{ε, x_0} be the stable outgoing manifold in Ω through $x_0 \in \Gamma$, for the flow of ∇f_ε. Then there is a constant $C > 0$ such that*

$$V^+_{\varepsilon, x_0} \subset \{x \in \Omega; f_+(x) \leq C \varepsilon f_-(x)\}. \tag{2.7}$$

We have analogous results for V^-_{ε, x_0}. (Replace f_ε by $-f_\varepsilon$ in the proposition.) We now assume that g is a generic Morse function (satisfying (H1) and (H2) of Section 1). Then we have

Proposition 2.4. *Let U_k and $U_j \in \Gamma$ be critical points of g of index l and $l+1$ respectively.*

(a) *If $d_g(U_k, U_j) > g(U_j) - g(U_k)$, then for $\varepsilon > 0$ sufficiently small, $d_{f_\varepsilon}(U_k, U_j) \geq f_\varepsilon(U_j) - f_\varepsilon(U_k) + \varepsilon/C_0$.*

(b) *If $d_g(U_k, U_j) = g(U_j) - g(U_k)$, then $d_{f_\varepsilon}(U_k, U_j) = f_\varepsilon(U_j) - f_\varepsilon(U_k)$, and the minimal d_g-geodesics from U_k to U_j coincide with the minimal d_{f_ε}-geodesics. Moreover $V^+_{\varepsilon,k}$ and $V^-_{\varepsilon,j}$ intersect transversally along each such geodesic.*

Proof. The part (a) follows immediately from Proposition 2.2. For the proof of (b), we have the general inequality:

$$d_{f_\varepsilon}(U_k, U_j) \geq f_\varepsilon(U_j) - f_\varepsilon(U_k). \tag{2.8}$$

Let γ be a minimal d_g-geodesic from U_k to U_j. Since the d_{f_ε}-length of γ is equal to $\varepsilon d_g(U_k, U_j) = f_\varepsilon(U_j) - f_\varepsilon(U_k)$, we also have the opposite inequality in (2.12), hence

$$d_{f_\varepsilon}(U_k, U_j) = f_\varepsilon(U_j) - f_\varepsilon(U_k), \tag{2.9}$$

and all minimal d_g-geodesics are also minimal d_{f_ε}-geodesics. Let γ' be a minimal d_{f_ε}-geodesic from U_k to U_j. Then from the proof of Proposition 2.2 (see (2.5)) we know that γ' stays close to a corresponding minimal d_g-geodesic γ. Moreover, by (2.9), $\gamma' \subset V^+_{\varepsilon,k} \cap V^-_{\varepsilon,j}$, so we can apply (2.7) to $V^+_{\varepsilon,k}$ and its analogue to $V^-_{\varepsilon,j}$. Then at a point x of the interior of γ' we have both $f_+(x) \leq C\varepsilon f_-(x)$ and $f_-(x) \leq C\varepsilon f_+(x)$, hence $f_+(x) = f_-(x) = 0$ and $x \in \Gamma$ so: $\gamma' \subset \Gamma$. Then necessarily $\gamma' = \gamma$.

It follows from (2.4) that $V^+_{\varepsilon,k}$ intersects Γ cleanly near γ with $V^+_{\varepsilon,k} \cap \Gamma = G^+_k$ (the ∇_g-invariant stable outgoing manifold through U_k), and we have the analogous result for $V^-_{\varepsilon,j}$. The transversal intersection of $V^+_{\varepsilon,k}$ and $V^-_{\varepsilon,j}$ along γ then follows from the transversal intersection of G^+_k and G^-_j along γ in Γ and from (2.7) and its analogue for $V^-_{\varepsilon,j}$.

3. Proof of the Bott Inequalities

Let f and Γ_j be as in Section 1, let (d^+_j, d^-_j) be the signature of the transversal Hessian of f on Γ_j. Then d^-_j is the fiber dimension of the tangent bundle $T^-_j = (TF^-_j)|_{\Gamma_j}$, where F^-_j is the incoming stable ∇f-invariant manifold. The orientation bundle $\mathcal{O}^-_j$ of T^-_j is then the line bundle over Γ_j of d^-_j-differential forms in the fiber variables of T^-_j, constant on each fiber. We then have a "twisted" de Rham complex on each Γ_j (as in (1.1)) and we denote by $b_k(\Gamma_j, f)$ the corresponding Betti numbers. Let $b_k(M)$ be the Betti numbers of M for the ordinary de Rham complex. We introduce the Poincaré polynomials,

$$P^M(t) = \sum b_l(M) t^l, \tag{3.1}$$

$$P^\Gamma(t) = \sum b_k(\Gamma_j, f) t^k. \tag{3.2}$$

Then the Bott inequalities ([3], [4]) or the degenerate Morse inequalities can then be stated in the following form:

Theorem 3.1. *There exists a polynomial with positive integer coefficients,* $Q(t)$, *such that*

$$\sum t^{d_j} P^{\Gamma_j}(t) = P^M(t) + (1+t)Q(t). \tag{3.3}$$

On each Γ_j we choose a generic Morse function g_j and we extend it to a function $\tilde{g}_j$ in a neighborhood of Γ_j as in Section 2 (where the index j was suppressed for convenience). If $\chi_j \in C_0^\infty(M)$ has its support close to Γ_j and is equal to 1 near Γ_j, we put

$$f_\varepsilon = f + \varepsilon \sum \chi_j \tilde{g}_j. \tag{3.4}$$

As we saw in Section 2, for $\varepsilon > 0$ small enough, the critical points of f_ε are precisely the critical points of the various functions g_j. If $x \in \Gamma_j$ is such a critical point, then

$$\mathrm{ind}_{f_\varepsilon} = \mathrm{ind}_{g_j} + d_j^-. \tag{3.5}$$

We choose $\varepsilon > 0$ sufficiently small so that all the results of Section 2 apply to all the Γ_j and so that $d_f(\Gamma_j, \Gamma_k) \gg \varepsilon$ for $j \neq k$. Put $d_{f_\varepsilon} = \exp(-f_\varepsilon/h)(hd)\exp(f_\varepsilon/h)$ (with ε small but independent of h) and let $P = d_{f_\varepsilon}^* d_{f_\varepsilon} + d_{f_\varepsilon} d_{f_\varepsilon}^*$ be the corresponding Hodge Laplacian. Let $F(l) \subset C^\infty(M; \Lambda^l T^*M)$ be the space associated to the exponentially small eigenvalues of P in degree l. "Exponentially small" means here $O_\varepsilon(\exp(-1/C(\varepsilon)h)$ for some $C(\varepsilon) > 0$ independent of h.

As we recalled in Section 1, the de Rahm cohomology of M is equal to that of the complex

$$d_{f_\varepsilon} : \cdots \to F^{(l)} \to F^{(l+1)} \to \cdots, \tag{3.6}$$

and we have bases in $F^{(1)}$ so that (3.6) reduces to

$$A : \cdots \to \mathbf{R}^{m(l)} \to \mathbf{R}^{m(l+1)} \to \cdots, \tag{3.7}$$

where in degree l,

$$A_{\nu,\mu}^{(l)} = (\beta_{\nu,\mu}^{(l)} + O(h^{1/2})) \exp[-(f_\varepsilon(U_\nu) - f_\varepsilon(U_\mu))/h],$$

$$\text{respectively } O(\exp(-C_0/h)), \quad (3.8)$$

depending on whether the critical points of f_ε; U_ν and U_μ belong to the same component Γ_j or to different components of $(df)^{-1}(0)$. Here (as in Section 1), we have

$$\beta_{\nu,\mu}^{(l)} = \sum_\gamma \varepsilon_\gamma(f_\varepsilon), \qquad \text{if } d_{g_j}(U_\nu, U_\mu) = g_j(U_\nu) - g_j(U_\mu) \qquad \text{and}$$

$$= 0 \qquad \text{otherwise,} \tag{3.9}$$

where $\varepsilon_\gamma(f_\varepsilon)$ is defined as in Section 1. If we conjugate the complex (3.7) by $\exp(\varepsilon \hat{g}/h)$, where $\hat{g}(\nu) = g_j(U_\nu)$, $U_\nu \in \Gamma_j$ in all degrees, we get the new equivalent complex,

$$B : \cdots \to \mathbf{R}^{m(l)} \to \mathbf{R}^{m(l+1)} \to \cdots, \qquad (3.10)$$

where in degree l: $B_{\nu,\mu}^{(l)} = b_{\nu,\mu}^{(l)} + O(h^{1/2})$, and $b_{\nu,\mu}^{(l)} = \beta_{\nu,\mu}^{(l)}$ if U_ν and U_μ belong to the same component Γ_j while $b_{\nu,\mu}^{(l)} = 0$ otherwise. Put $B_0^{(l)} = (b_{\nu,\mu}^{(l)}) = \lim_{h\to 0} B^{(l)}$. Then B_0 is a complex which is decomposed into a direct sum of complexes $C_1, \ldots, C_N$, one for each Γ_j. For B and B_0 we write down the corresponding Hodge Laplacians: $\square_B$ and $\square_{B_0}$ (using the standard inner products on $\mathbf{R}^{m(l)}$). If $E_0^{(l)}$ is the kernel of $\square_{B_0}$ in degree l, then $E_0^{(l)} = \lim_{h\to 0} E_h^{(l)}$, where $E_h^{(l)} \subset \mathbf{R}^{m(l)}$ is the space associated to the eigenvalues of $\square_B$ (in degree l) that converge to 0 when h tends to 0. In particular,

$$\dim E_h^{(l)} = \dim E_0^{(l)} \qquad (3.11)$$

for small h. The cohomology of (3.10) (which is also the cohomology of M) is given by that of the restriction of (3.10):

$$\cdots \to E_h^{(l)} \to E_h^{(l+1)} \to \cdots, \qquad (3.12)$$

so by an easy argument, we see that there exists a polynomial $Q(t)$ with non-negative integer coefficients, such that

$$\sum (\dim E_0^{(l)}) t^l = P^M(t) + (1+t)Q(t). \qquad (3.13)$$

Theorem 3.1 then follows from (3.13) and

Proposition 3.2. *We have*

$$\dim E_0^{(l)} = \sum_{l = d_j + k} b_k(\Gamma_j, f).$$

Proof of Proposition 3.2. As noted above, the complex B_0 decomposes into the direct sum of $C_1, \ldots, C_N$ which are associated to the various components of Γ. It is then enough to show that the $(l + d_j^-)^{\text{th}}$ cohomology space of the complex C_j has the same dimension as the lth twisted cohomology space of Γ_j. To do this it is enough to identify the complex C_j with the orientation complex of Γ_j associated to g_j, translated in degree by d_j^-. From now on we fix a Γ_j and suppress the corresponding subscript j.

Let U_j, $U_k \in \Gamma$ be critical points of g of index $l+1$ and l respectively. Let $\alpha_{j,k}$ be the corresponding coefficient in degree l of the orientation complex for the twisted cohomology on Γ and let $\beta_{j,k}$ be the corresponding coefficient in degree $l + d^-$ for the complex C. We shall see that with suitable choices of orientations and sections in the orientation bundle O_-, these two coefficients are equal up to a factor ± 1, which only depends on the component Γ where we work. A first trivial case is when $g(U_j) - g(U_k) < d_g(U_j, U_k)$. Then both coefficients vanish. Assume then that $g(U_j) - g(U_k) = d_g(U_j, U_k)$. Let γ be a minimal d_g-geodesic from U_k to U_j. On $T_{U_k}(G_k^-)$ we fix an orientation given by an l-form ω_k and on $F_{U_k^-}$ we fix an orientation form f_k. We choose ω_j, f_j analogously at U_j. Then $\omega_k \wedge f_k$ and $\omega_j \wedge f_j$ give orientations on $V_{\varepsilon,k}^-$ and $V_{\varepsilon,j}^-$. We now recall from Section 1 that the contributions from γ to $\alpha_{j,k}$ and $\beta_{j,k}$ are respectively $\varepsilon_\gamma(g)\alpha_\gamma$ and $\varepsilon_\gamma(f_\varepsilon)$, where $\varepsilon_\gamma(g) = \pm 1$, $\alpha_\gamma = \pm 1$ are such that $\varepsilon_\gamma(g) = +1$ iff the orientations on TM/TG_k^+ induced from G_k^- and $G_j^-(\gamma')$ respectively agree, and where $\alpha_\gamma = +1$ iff ω_k and ω_j extended to continuous non-vanishing d^--forms over γ, have the same sign.

In computing $\varepsilon_\gamma(f_\varepsilon)$, we notice that this index will not change if we continuously deform $V_{\varepsilon,k}^+$ and $V_{\varepsilon,j}^-$ in such a way that they continue to intersect transversally along γ. It is easy to do such a deformation to the case when $V_{\varepsilon,k}^+ = V_k = \{x \in F_+; \ \pi(x) \in G_k^+\}$, $V_{\varepsilon,j}^- = \{x \in F_-; \ \pi(x) \in G_j^-\}$ (using the notation of Section 2). Then $\varepsilon_\gamma(f_\varepsilon) = 1$ iff the orientation of $T_{U_k}G_k^- \times T_{U_k}F_{U_k}$ transported along γ and the orientation on $T_{x_0}G_j^- \times TF_{x_0}^-/(\gamma'), x_0 \in \gamma$, obtained by transportation from U_j, induce the same orientation on $T_{x_0}M/(T_{x_0}G_k^+ \oplus T_{x_0}F_{x_0}^+)$. Committing an error which only depends on the component Γ of $(df)^{-1}(0)$, we may identify the last orientation by that induced from $T_{x_0}G_j^-/(\gamma') \times TF_{x_0}^-$. In other words, $\varepsilon_\gamma(f_\varepsilon) = 1$ iff we get the same orientation on $(T_{x_0}\Gamma/T_{x_0}G_k^+) \oplus T_{x_0}F_{x_0}^-$ from those of $T_{U_k}G_k^- \times T_{U_k}F_{U_k}^-$ and $T_{x_0}G_j^-/(\gamma') \times TF_{x_0}^-$. This is the case iff $\varepsilon_\gamma(g)\alpha_\gamma = 1$. This completes the proof of the proposition and of the theorem.

References

[1] R. Abraham and J. Robbin, *Transversal Mappings and Flows*, W. A. Benjamin Inc. (1967). Appendix C by A. Kelley.

[2] J. M. Bismut, The Witten complex and the degenerate Morse inequalities, *J. Diff. Geom.* **23** (3) (1986) 207–241.

[3] R. Bott, Non degenerate critical manifolds, *Ann. of Math.* **60** (2) (1954) 248–261.

[4] R. Bott, Lectures on Morse theory, old and new, *B.A.M.S.* **7** (2) (1982).

[5] B. Helffer, In *Séminaire de EDP à Nantes*, (1985–86).

[6] B. Helffer and J. Sjöstrand, Puits multiples en limite semiclassique IV, Etude du complexe de Witten, *Comm. P.D.E.*, **10** (3) (1985) 245–340.

[7] B. Helffer and J. Sjöstrand, Puits multiples . . . V, Etude des mini-puits, *Current topics in PDE* (Festschrift for Prof. Mizohata), Kinokuniya Co Ltd., Tokyo (1986) 133–186.

[8] B. Helffer and J. Sjöstrand, Puits multiples . . . VI. (Cas des puits sous-variétés), *Ann. Inst. H. Poincaré,* **46** (4) (1987) 353–373.

[9] G. Henniart, Les inegalités de Morse (d'après E. Witten), *Sém. Bourbaki, 36ème année,* 1983–84, n° 617.

[10] B. Simon, Semiclassical analysis of low lying eigenvalues I, *Ann. I.H.P.* **38**, n⁰ 3 (1983) 225–307.

[11] E. Witten, Super symmetry and Morse theory, *J. Diff. Geom,* **17** (1982) 661–692.

What is the Notion of a Complex Manifold with a Smooth Boundary?

C. Denson Hill

Department of Mathematics
State University of New York at Stony Brook
Stony Brook, New York

The point of this article is to show how there are two distinct notions of what constitutes a complex manifold with a smooth (C^∞) boundary. The first, which we shall call a *concrete boundary*, consists of prescribing an atlas of holomorphic coordinate charts. The second, which we shall call an *abstract boundary*, consists of prescribing an integrable almost complex structure that is C^∞ up to the boundary. For a complex manifold without boundary, these two prescriptions are equivalent, according to the well known Newlander–Nirenberg theorem [7]. But when a boundary is present, we show here by examples that the two concepts do not coincide; the abstract boundary is the more general concept.

The distinction we are making here has a picturesque feature: One could imagine that in dealing with a complex manifold with a smooth boundary, the classical complex analyst has at his disposal two maximal atlases: the atlas of all C^∞ charts, and the atlas of all holomorphic charts. If there is a purely abstract boundary point p, it will not appear on any chart in his

185

atlas of all holomorphic charts, but it will appear on some chart in his C^∞ atlas. He may thus navigate in a C^∞ manner to p along any smooth path, using eventually only one C^∞ coordinate patch. But in order to navigate to p along the same path in a complex analytic manner, it is conceivable that he might need to use an infinite number of holomorphic patches. So he might be inclined to think of p as being an "abstract boundary point" when there is really nothing "abstract" about p.

These considerations are actually of interest for a complex manifold with a pseudoconcave boundary.

The abstract boundary is really the intrinsic notion; the concrete boundary has some extrinsic flavor.

We do insist that the structures under discussion are compatible with some fixed and given underlying differentiable structure on a manifold M with C^∞ boundary ∂M.

Let M be such a complex manifold with a smooth abstract boundary ∂M. Then a particular point $p \in \partial M$ will be called a *concrete boundary point of M* if some neighborhood of p in ∂M forms the concrete boundary of a corresponding neighborhood of p in M. This is an open condition in ∂M. Set

$$\partial_c M = \{\text{the open set of concrete boundary points in } \partial M\},$$

and let

$$\partial_{ab} M = \{\text{the closed set of purely abstract points in } \partial M\},$$

be its complement. We have the disjoint union:

$$\partial M = \partial_c M \cup \partial_{ab} M.$$

In this paper we first give an example of such an M which is diffeomorphic to a closed half-space of $\mathbf{R}^6$, with $\partial M = \mathbf{R}^5$, which has the following property: Initially $\partial_{ab} M = \{\varnothing\}$, so $\partial M = \partial_c M$; but when the complex structure on M is subjected to certain arbitrarily small random perturbations, one obtains $\partial_c M = \{\varnothing\}$ and $\partial M = \partial_{ab} M$. Second we show how this example can be modified to obtain a compact manifold M, whose boundary ∂M is diffeomorphic to $S^3 \times S^2$, which has the same properties.

The author would like to acknowledge some very fruitful discussions with R. Penrose [9], and especially the incisive suggestions of G. Sparling.

Later in the paper we discuss the influence of real analyticity and the influence of pseudoconvexity. We also put forth a conjecture about the example of L. Nirenberg [8].

Finally, for a complex manifold M with an abstract smooth pseudo-concave boundary ∂M, we mention the following *open problem*: What is the geometric structure of the closed set $\partial_{ab} M$ of purely abstract boundary points within ∂M?

1. Complex Manifolds with a Concrete Boundary $(\dim_c M = n)$

Since this concept does not seem to be standard in the literature, we must make it explicit: It means that it is possible to specify the differentiable structure on M by a special C^∞ atlas of holomorphic charts $\{(U_\alpha, \varphi_\alpha)\}$: The $\{U_\alpha\}$ give an open cover of M, and each $\varphi_\alpha : U_\alpha \to \varphi_\alpha(U_\alpha)$ is a homeo-morphism of U_α onto a coordinate patch $\varphi_\alpha(U_\alpha)$ which is either an *interior coordinate patch*:

(i) $\varphi_\alpha(U_\alpha) =$ an open neighborhood of the origin in $\mathbf{C}^n$, or else is a *boundary coordinate patch*:

(ii) $\varphi_\alpha(U_\alpha) =$ an open neighborhood of the origin in Ω_α. Here Ω_α is a set of the form $\{z \in \mathbf{C}^n$ with $\rho_\alpha(z) \le 0\}$, for some C^∞ real-valued function ρ_α defined on $\mathbf{C}^n$, such that $\rho_\alpha(0) = 0$, and $d\rho_\alpha \ne 0$ when $\rho_\alpha = 0$. We also require that for every α and β, the coordinate transformations

(iii) $\varphi_\alpha \circ \varphi_\beta^{-1} : \varphi_\beta(U_\alpha \cap U_\beta) \to \varphi_\alpha U_\alpha \cap U_\beta)$ be C^∞ diffeomorphisms which are biholomorphic on the interrior of each $\varphi_\beta(U_\alpha \cap U_\beta)$. Note that this means that each $\varphi_\alpha \circ \varphi_\beta^{-1}$ is required to be C^∞ up to the boundary, but it is *not* required that the $\varphi_\alpha \circ \varphi_\beta^{-1}$ have holomorphic extensions into the region where $\rho_\beta > 0$.

In this way M is equipped with the complex structure that one naturally sees on each $\varphi_\alpha(U_\alpha)$; local holomorphic coordinates have been prescribed "up to the boundary."

2. Complex Manifolds with an Abstract Boundary $(\dim_R M = 2n)$

This is just to require that a C^∞ integrable almost complex structure has been prescribed on the interior $\overset{\circ}{M} = M - \partial M$, which is C^∞ up to the boundary ∂M. In other words, one prescribes a subbundle $\mathcal{H}$, of complex fiber dimension n, of the complexified tangent bundle $CT(M)$ such that:

    1°. $\mathcal{H}_x \cap \bar{\mathcal{H}}_x = 0$ for each $x \in M$,

and

    2°.   If P and Q are arbitrary smooth sections of $\mathcal{H}$, then the Lie bracket $[P, Q]$ is also a section of $\mathcal{H}$.

This is equivalent to the consistent local prescription, in each C^∞ coordinate chart, of n complex vector fields

$$P_k(x) = \sum_{j=1}^{2n} a_k^j(x) \frac{\partial}{\partial x^j}, \qquad (k = 1, \ldots, n)$$

$$a_k^j(x) \in C^\infty \text{ and complex-valued,}$$

which on M satisfy: (1) $P_1, \ldots, P_n, \bar{P}_1, \ldots, \bar{P}_n$ are linearly independent over $\mathbf{C}$ at each point, and (2) For all i and j,

$$[P_i, P_j](x) = \sum_{k=1}^{n} c_{ij}^k(x) P_k(x),$$

with C^∞ complex-valued coefficients c_{ij}^k.

    If M is a complex manifold with a concrete boundary, then one can take

$$P_k = \frac{1}{2}\left(\frac{\partial}{\partial x_k} + \sqrt{-1}\,\frac{\partial}{\partial y_k}\right), \qquad (k = 1, \ldots, n)$$

in each holomorphic coordinate chart; hence M is also a complex manifold with an abstract boundary.

3. Induced *CR* Structure on the Boundary ($\dim_{\mathbf{R}} \partial M = 2n - 1$)

This refers to the *CR* structure of hypersurface type on ∂M determined by $\mathcal{H}|_{\partial M}$. In a local C^∞ boundary coordinate patch one can take a suitable linear combination of the $P_1, \ldots, P_n$ (with C^∞ complex-valued coefficients) so that the integrable almost complex structure on M is equivalently given by $\tilde{P}_1, \ldots, \tilde{P}_{n-1}, \tilde{P}_n$ satisfying (1) and (2), with $\tilde{P}_1, \ldots, \tilde{P}_{n-1}$ tangent to ∂M at points of ∂M. The *CR* structure induced on ∂M is then locally consistently defined by

$$Q_1 = \tilde{P}_1|_{\partial M}, \ldots, Q_{n-1} = \tilde{P}_{n-1}|_{\partial M};$$

these Q_k's satisfy (1) and (2).

A *CR* structure on ∂M is called *embeddable* at a point p if some neighborhood of p in ∂M can be embedded in $\mathbf{C}^N$, for some N, in such a way that the given structure agrees with the one induced by the ambient complex structure in $\mathbf{C}^N$, see [1].

4.   Noncompact Examples

We shall describe a family M_ω of complex manifolds with abstract C^∞ boundaries ∂M_ω that depend on a parameter ω. Each M_ω is diffeomorphically a closed half-space in $\mathbf{R}^6$, whose boundary ∂M_ω in $\mathbf{R}^5$. We first describe the family of induced *CR* structures on $\mathbf{R}^5$:

Let $(t_1, t_2, \ldots, t_5)$ denote the real coordinates in $\mathbf{R}^5$. Set $t = (t_1, t_2, t_3)$, $z = t_4 + it_5$, and $z = t_4 - it_5$; we identify a point in $\mathbf{R}^5$ with the point $(t, z) \in \mathbf{R}^3 \times \mathbf{C}$. $C^\infty(\mathbf{R}^3)$ denotes the space of complex-valued smooth functions on $\mathbf{R}^3$. The points $\omega \in C^\infty(\mathbf{R}^3)$ are our parameters; i.e., $\omega = \omega(t) \equiv \omega(t_1, t_2, t_3)$. On $\mathbf{R}^3$ we employ the operator of H. Lewy,

$$L = \frac{1}{2}\left(\frac{\partial}{\partial t_1} + i\frac{\partial}{\partial t_2}\right) - i(t_1 + it_2)\frac{\partial}{\partial t_3}.$$

On $\mathbf{C}$ we have

$$\frac{\partial}{\partial z} = \frac{1}{2}\left(\frac{\partial}{\partial t_4} - i\frac{\partial}{\partial t_5}\right) \quad \text{and} \quad \frac{\partial}{\partial \bar{z}} = \frac{1}{2}\left(\frac{\partial}{\partial t_4} + i\frac{\partial}{\partial t_5}\right).$$

The inhomogenous equation

$$LX = \omega \tag{1_ω}$$

will be called *solvable* at a point $t_0 \in \mathbf{R}^3$ if, in some neighborhood of t_0, there exists a complex-valued C^∞ solution $\chi = \chi(t)$.

For each ω consider the complex vector fields

$$P = \frac{\partial}{\partial \bar{z}}$$

$$Q = L + \omega(t)\frac{\partial}{\partial z}. \tag{2_ω}$$

It is clear that P, Q, $\bar{P}$, and $\bar{Q}$ are linearly independent and $[P, Q] = 0$. Hence they define a CR structure on $\mathbf{R}^5$, which will be our boundary structure on ∂M_ω.

Lemma 1. *The CR structure (2_ω) is embeddable at a point $(t_0, z_0) \in \mathbf{R}^5$ if and only if (1_ω) is solvable at t_0.*

Let $\mathbf{R}_+^6 = \{(t, z, \tau) \in \mathbf{R}^3 \times \mathbf{C} \times \mathbf{R}$ with $\tau \geq 0\}$.

Theorem 1. *For each $\omega \in C^\infty(\mathbf{R}^3)$, $\mathbf{R}_+^6$ can be given the structure of a complex manifold M_ω with an abstract boundary. The CR structure on the boundary ∂M_ω is defined by (2_ω).*

Recall that a subset of the complete metric space $C^\infty(\mathbf{R}^3)$ is said to be of the first Baire category if it can be written as the countable union of nowhere dense sets; otherwise it is called of the second category.

Theorem 2. *For a set of $\omega \in C^\infty(\mathbf{R}^3)$ of the second Baire category, $\partial M_\omega = \partial_{ab} M_\omega$; i.e., there are no concrete boundary points on M_ω.*

Note that when $\omega \equiv 0$ we get the product structure on $\mathbf{C} \times \mathbf{R}^3$, which is globally embeddable.

Corollary 1. *Almost every deformation (2_ω) of the product structure is not embeddable at any point.*

Theorem 3. *There is a set of $\omega \in C^\infty(\mathbf{R}^3)$ of the second category such that any smooth local solution u of $Pu = 0$ and $Qu = 0$, near any point $\mathbf{R}^5$, is a constant function of t_1 and t_5.*

Remark. Theorem 3 bears some analogy to the example [8] of L. Nirenberg of a C^∞ complex vector field P in $\mathbf{R}^3$ which has the property that any smooth local solution of $Pu = 0$ near the origin is constant. In his example P, $\bar{P}$, and $[P, \bar{P}]$ are linearly independent at every point.

5.   Proof of Lemma 1

The CR structure on $\mathbf{R}^3$ given by L alone has a global embedding in $\mathbf{C}^2$ given by the map $\mathbf{R}^3 \ni t \mapsto v(t) \equiv (v_1(t), v_2(t)) \in \mathbf{C}^2$, where $v_1(t) = t_1 + it_2$ and

$v_2(t) = t_3 + i(t_1^2 + t_2^2)$. After embedding, the CR structure on $\mathbf{R}^3$ becomes the induced CR structure on the Heisenberg group; i.e., the boundary $\partial\Omega$ of the region $\Omega = (\zeta_1, \zeta_2) \in \mathbf{C}^2$ with Im $\zeta_2 \leq |\zeta_1|^2$. These *characteristic coordinates* [1] satisfy $Lv_j = 0$ ($j = 1, 2$) and $dv_1 \wedge dv_2 \neq 0$. The existence, locally, of such a maximal set of functionally independent characteristic coordinates is equivalent to the local embeddability of a CR structure.

Suppose first that (1_ω) has a C^∞ solution $\chi(t)$ for t near t_0. Set $v_3(t, z) = z - \chi(t)$. Then $Pv_j = 0$, $Qv_j = 0$ ($j = 1, 2, 3$) and $dv_1 \wedge dv_2 \wedge dv_3 \neq 0$, so v_1, v_2, v_3 give a maximal set of functionally independent characteristic coordinates for (2_ω). The map $\mathbf{R}^3 \times \mathbf{C} \ni (t, z) \mapsto (v(t), v_3(t, z)) \in \mathbf{C}^2 \times \mathbf{C}$ provides the embedding near any point (t_0, z_0).

For the converse suppose that (2_ω) is embeddable at some point (t_0, z_0). This means that some neighborhood of (t_0, z_0) is embeddable in $\mathbf{C}^3$ (for if the local embedding can be done in $\mathbf{C}^N$, for some N, then it can be accomplished in $\mathbf{C}^3$, see [1]); we can assume that (t_0, z_0) corresponds to the origin in $\mathbf{C}^3$. So there exist complex-valued C^∞ functions $u_1(t, z)$, $u_2(t, z)$, $u_3(t, z)$ which satisfy $Pu_j = 0$ and $Qu_j = 0$ ($j = 1, 2, 3$), $du_1 \wedge du_2 \wedge du_3 \neq 0$, and give the embedding. In particular each u_j is holomorphic in z. So the Jacobian matrix of the embedding map $(t, z) \mapsto u \equiv (u_1, u_2, u_3)$ has a block decomposition of the form

$$\begin{bmatrix} \dfrac{\partial u}{\partial t} & \dfrac{\partial u}{\partial z} & 0 \\[2ex] \dfrac{\partial \bar{u}}{\partial t} & 0 & \dfrac{\partial \bar{u}}{\partial \bar{z}} \end{bmatrix}$$

with six rows and five linearly independent columns. Hence $\partial u_j / \partial z(t_0, z_0) \neq 0$ for some j; we will assume that it is $\partial u_3 / \partial z \neq 0$. Since $Pv_1 = Pv_2 = Pu_3 = 0$, $Qv_1 = Qv_2 = Qu_3 = 0$, with $dv_1 \wedge dv_2 \wedge du_3 \neq 0$, we arrive at the fact that the functions $v_1(t)$, $v_2(t)$, $u_3(t, z)$ provide a new maximal set of functionally independent characteristic coordinates.

Thus we have another embedding $\varphi : (t, z) \mapsto (v(t), u_3(t, z))$ of some neighborhood V of (t_0, z_0) into $\mathbf{C}^3$; $\varphi(V)$ is a portion of a smooth 5-dimensional hypersurface in $\mathbf{C}^3$. Denote the coordinates in this $\mathbf{C}^3$ by $(\zeta_1, \zeta_2, \zeta_3)$, and consider the function

$$F(\zeta_1, \zeta_2, \zeta_3) = \varphi_*\left(-\left[\frac{\partial u_3}{\partial z}(t, z)\right]^{-1}\right),$$

defined for $(\zeta_1, \zeta_2, \zeta_3) \in \varphi(V)$. Here φ_* denotes the push-forward by the diffeomorphism φ of V onto $\varphi(V)$. Since $u_3(t, z)$ is a CR function on V, differentiation shows that $\partial u_3/\partial z$, and hence $-[\partial u_3/\partial z]^{-1}$ is a CR function on V. This means exactly that F is a CR function on $\varphi(V)$. In particular F is holomorphic in the variable ζ_3 because, for each fixed t, the holomorphic map $z \mapsto u_3(t, z)$ has a holomorphic inverse, and $[\partial u_3/\partial z]^{-1}$ is holomorphic in z.

Consider next the function

$$G(\zeta_1, \zeta_2, \zeta_3) = \int_0^{\zeta_3} F(\zeta_1, \zeta_2, \eta)\, d\eta,$$

defined for $(\zeta_1, \zeta_2, \zeta_3) \in \varphi(V)$, where the integration is a contour integral in the ζ_3-plane, taken along some path from the origin to ζ_3. For fixed t, as z varies, an open neighborhood (in the ζ_3-plane) of points of the form $\zeta_1 = \text{const.}$, $\zeta_2 = \text{const.}$, $\zeta_3 = u_3(t, z)$, is swept out, and that neighborhood lies on $\varphi(V)$, where F is known. Hence G is well-defined. Moreover G is a CR function on $\varphi(V)$ and its pull-back

$$g(t, z) = \varphi^* G,$$

to V is a CR function there: Clearly $Pg = 0$ since g is holomorphic in z. An easy way to see that $Qg = 0$ is to write

$$g(t, z) = \int_0^{u_3(t,z)} \tilde{F}(v_1(t), v_2(t), \eta)\, d\eta,$$

where $\tilde{F}$ is a smooth extension of F off of $\varphi(V)$ such that $\bar{\partial}\tilde{F}|_{\varphi(V)} = 0$, and then differentiate.

Now

$$\frac{\partial g}{\partial z} = G_{\zeta_3}(v_1, v_2, u_3)\frac{\partial u_3}{\partial z}(t, z)$$

$$= F(v_1(t), v_2(t), u_3(t, z))\frac{\partial u_3}{\partial z}(t, z) \equiv -1,$$

so $g(t, z) = \chi(t) - z$, where the C^∞ "constant of integration" $\chi(t)$ could be taken to be $g(t, z_0)$.

The equation $Qg = 0$ therefore says that $LX - \omega = 0$, so (1_ω) is solvable at t_0, and the proof of Lemma 1 is complete.

6. Proof of Theorem 1

We set $M = \mathbf{R}_+^6 = \{\tau \geq 0\}$, $\partial M = \mathbf{R}^5 = \{\tau = 0\}$, and introduce the diffeomorphism $\Phi: (t, z, \tau) \mapsto (z_1, z_2, z_3)$ of M onto $\Phi(M) \subset \mathbf{C}^3$ defined by $\Phi: \{z_1 = t_1 +$

it_2, $z_2 = t_3 + i(t_1^2 + t_2^2 - \tau)$, $z_3 = z = t_4 + it_5$}. Note that $\Phi(M) = \Omega \times \mathbf{C}$ and $\Phi(\partial M) = \partial \Omega \times \mathbf{C}$, where $\Omega = \{(z_1, z_2) \in \mathbf{C}^2 \text{ with Im } z_2 \leq |z_1|^2\}$. Thus $\varphi = \Phi|_{\tau=0}$ is the map which embeds the product CR structure.

Set $\rho = |z_1|^2 - \text{Im } z_2$. Then $\partial \Omega$ is given by $\{\rho = 0\}$, with Ω corresponding to $\rho \geq 0$. A basis for forms of type $(0, 1)$ in $\mathbf{C}^2$ is given by $d\bar{z}_1$ and $\bar{\partial}\rho = z_1 d\bar{z}_1 + 1/2i d\bar{z}_2$, so any $(0, 1)$ form $\psi = \psi_1 d\bar{z}_1 + \psi_2 d\bar{z}_2$ can be written as

$$\psi = [\psi_1 - 2iz_1\psi_2] d\bar{z}_1 + 2i\psi_2 \bar{\partial}\rho.$$

We shall consider "tangential" $(0, 1)$ forms $\psi_b dz_1$, where ψ_b is a C^∞ function defined on $\partial \Omega$.

We need the following basic fact: Any tangential $(0, 1)$ form is the "boundary value" on $\partial \Omega$ of a C^∞, $\bar{\partial}$-closed $(0, 1)$ form ψ on Ω; i.e.,

$$\psi_b = [\psi_1 - 2iz_1\psi_2]|_{\partial\Omega}, \tag{6.1}$$

and

$$\frac{\psi_1}{\partial \bar{z}_2} = \frac{\psi_2}{\partial \bar{z}_1} \quad \text{on } \Omega. \tag{6.2}$$

To see this we adopt temporarily the notation of [2]: $S = \partial \Omega$, $\Omega^+ = \Omega$, $\Omega^- = \mathbf{C}^2 - \overset{\circ}{\Omega}$. Since $H^1(\mathbf{C}^2, \mathcal{O}) = H^2(\mathbf{C}^2, \mathcal{O}) = 0$ the Mayer-Vietoris sequence of [2] yields the isomorphism $H^1(\Omega^+) \oplus H^1(\Omega^-) \xrightarrow{\approx} H^1(S)$ given by the jump in the boundary values. But $H^1(\Omega^-) = 0$; i.e., every C^∞ $\bar{\partial}$-closed $(0, 1)$ form on Ω^- is the $\bar{\partial}$ of a C^∞ function on Ω^-; see [2]. Thus we arrive at the isomorphism $H^1(\Omega^+) \xrightarrow{\approx} H^1(S)$, which asserts that our Cauchy problem is solvable in terms of cohomology classes; whence it is solvable in terms of forms (see [2], p. 345).

Now fix an $\omega(t) \in C^\infty(\mathbf{R}^3)$ and consider its push forward $\psi_b = \varphi_*\omega$. Set $N = \partial/\partial \tau + i \partial/\partial t_3$, and let ψ be as above. Introduce the functions $\omega_1(t, \tau) = \Phi^*(\psi_1 - 2iz_1\psi_2)$, $\omega_2(t, \tau) = \Phi^*(2i\psi_1)$, defined on $\mathbf{R}^3 \times \mathbf{R}^+$. Obviously $\omega_1(t, 0) = \omega(t)$. Consider the system of complex vector fields

$$P_1 = L + \omega_1(t, \tau) \frac{\partial}{\partial z}$$

$$P_2 = N + \omega_2(t, \tau) \frac{\partial}{\partial z} \tag{6.3}$$

$$P_3 = \frac{\partial}{\partial \bar{z}},$$

defined on M. We claim that (6.3) provides the desired integrable almost complex structure M_ω on M, and when $\tau = 0$, the abstract boundary structure ∂M_ω on ∂M is the CR structure (2_ω).

Clearly (6.3) induces (2_ω) on ∂M, since P_1 and P_3 are tangent to ∂M, while P_2 is not. It is clear that $P_1, P_2, P_3, \bar{P}_1, \bar{P}_2, \bar{P}_3$ are linearly independent at each point of M. It remains to verify the integrability condition: All the Lie brackets of the P_k's are zero, except possibly for

$$[P_1, P_2] = [L, N] + \left[L, \omega_2 \frac{\partial}{\partial z} \right] + \left[\omega_1 \frac{\partial}{\partial z}, N \right] + \left[\omega_1 \frac{\partial}{\partial z}, \omega_2 \frac{\partial}{\partial z} \right]$$

$$= \{ L(\omega_2) - N(\omega_1) \} \frac{\partial}{\partial z}.$$

Straightforward calculation yields

$$L(\omega_2) = \Phi^* \left(2i \frac{\partial \psi_2}{\partial \bar{z}_1} + 4z_1 \frac{\partial \psi_2}{\partial \bar{z}_2} \right),$$

$$N(\omega_1) = \Phi^* \left(2i \frac{\partial \psi_1}{\partial \bar{z}_2} + 4z_1 \frac{\partial \psi_2}{\partial \bar{z}_2} \right).$$

The desired result now follows because of (6.2); this completes the proof Theorem 1.

7. Proof of Theorem 2

The H. Lewy equation (1_ω) has the following well known property: For each point $t_0 \in \mathbf{R}^3$, the set of all $\omega(t)$ in the metric space $C^\infty(\mathbf{R}^3)$ for which (1_ω) is solvable at t_0 is a set of the first Baire category. More general results can be found in [2]; the proof is most clearly outlined in [6].

Choose a countable dense sequence of points $\{t^{(k)}\}$ in $\mathbf{R}^3$, and set $\Lambda^{(k)} = \{\omega \in C^\infty(\mathbf{R}^3)$ such that (1_ω) is solvable at $t^{(k)}\}$. Then $\Lambda = \bigcup_{k=1}^\infty \Lambda^{(k)}$ is also of the first category. Now fix an $\omega(t) \in C^\infty(\mathbf{R}^3)$ and suppose that the CR structure were embeddable at some point (t_0, z_0). Then, for some k, it is also embeddable at a point of the form $(t^{(k)}, z_0)$, because being embeddable at a point is an open condition. By Lemma 1, (1_ω) must be solvable at $t^{(k)}$. Hence $\omega(t) \in \Lambda^{(k)} \subset \Lambda$. Thus for any $\omega(t)$ in $C^\infty(\mathbf{R}^3)$ minus the exceptional set Λ, the CR structure (2_ω) is not embeddable at any point. This means that no boundary point can be a concrete boundary point, completing the proof of Theorem 2.

8.   Proof of Theorem 3

Consider an $\omega(t) \notin \Lambda$, and let $u = u(t, z)$ be any smooth solution of $Pu = Qu = 0$ near some point of $\mathbf{R}^5$. Then $\partial u/\partial \bar{z} \equiv 0$ and $dv_1 \wedge dv_2 \wedge du \equiv 0$, because otherwise the CR structure (2ω) would be embeddable at some point. But v_1 and v_2 are functions of t only, and $dv_1 \wedge dv_2 \neq 0$. Therefore $\partial u/\partial z \equiv 0$ and hence u is a constant function of $z = t_4 + it_5$.

9.   Compact Examples

The examples constructed above have the advantage that the entire argument can be made explicit in a single C^∞ coordinate patch. We now indicate how to modify our construction to obtain a family of *compact* M_ω which have the same basic properties.

These compact M_ω can all be taken to be diffeomorphic to $M \approx \Omega^4 \times S^2$, with abstract boundaries $\partial M \approx S^3 \times S^2$, where $\Omega^4 = \{$the total space of a 2-disc bundle over S^2, twisted by the Hopf fibration of S^3 over $S^2\}$. So $\dim_{\mathbf{C}} M_\omega = 3$, and the 5-dimensional manifold ∂M_ω forms the boundary of M_ω as follows: Each 2-disc has as its boundary a fiber S^1 in the Hopf fibration of S^3 over S^2.

The second factor S^2 above is just $\mathbf{CP}^1$: On the unit sphere S^2 we use the standard holomorphic charts $z_+ \in \mathbf{C} \cong V_+ \equiv S^2 - \{\infty\}$, $z_- \in \mathbf{C} \cong V_- \equiv S^2 - \{0\}$, given by stereographic projection from $\{\infty\}$ and $\{0\}$, with $z_+ z_- = 1$ on $V_+ \cap V_-$. The associated integrable almost complex structure on the Riemann sphere is determined by $P_\beta = \partial/\partial \bar{z}_\beta$ in V_β ($\beta = +, -$). Our parameters $\omega = \omega(t)$ are now complex-valued functions $\omega \in C^\infty(S^3)$. We will identify S^3: $|\zeta_1|^2 + |\zeta_2|^2 = 1$, $t \leftrightarrow \zeta = (\zeta_1, \zeta_2) \in \mathbf{C}^2$, with the boundary of the unit ball in $\mathbf{C}^2$, in order not to overburden the exposition by making explicit choices of coordinates patches for S^3. Thus on $\partial M = S^3 \times S^2$ we may think of (t, z_β), $\beta = +, -$, as local coordinates. Let $L(t) = \zeta_2 \partial/\partial \bar{\zeta}_1 - \zeta_1 \partial/\partial \bar{\zeta}_2$ denote the tangential Cauchy-Riemann operator to S^3, and let $(1'_\omega)$ be the analogue of (1_ω) on S^3.

In place of (2_ω) we have

$$P_\beta = \frac{\partial}{\partial \bar{z}_\beta}$$

$$Q_\beta = L(t) + \beta \omega(t) z_\beta \bar{P}_\beta \qquad (\beta = +, -), \qquad (2'_\omega)$$

in $S^3 \times V_\beta$. It is easy to see that this defines a smooth CR structure ∂M_ω on ∂M. Let $\rho = \zeta_1 \bar{\zeta}_1 + \zeta_2 \bar{\zeta}_2 - 1$.

Finally we regard $B^4 \subset C^2 \to CP^2$, by ignoring a hyperplane at infinity, and set anew $M = \Omega \times S^2$, where $\Omega = \Omega^4 \equiv CP^2 - \mathring{B}^4$, and B^4 is the unit ball.

Theorem 4

(a)   For each $\omega \in C^\infty(S^3)$, M can be given the structure of a compact complex manifold M_ω with abstract boundary ∂M_ω.

(b)   For a set of ω's of the second Baire category, $\partial M_\omega = \partial_{ab} M_\omega$; i.e., there are no concrete boundary points on M_ω.

The analogues of Corollary 1 and Theorem 3 are also valid in this compact case.

Proof of Theorem 4.   Consider first part (b). The operator $L(t)$ has the same first category local solvability property as the H. Lewy operator L. Since the proof of Theorem 2 employed Lemma 1 only on a countable dense set of points, it will suffice to verify that the proof of Lemma 1 goes through at points (t_0, z_0) on ∂M with $z_0 \in V_+ \cap V_-$ (choose $z = z_+$). Then $z_0 \neq 0$, so for z in a neighborhood of z_0, we may make the change of variables $w = \log z$. In place of $v_1(t)$ and $v_2(t)$ we may use the natural embedding functions ζ_1 and ζ_2. In the new local coordinates (t, w) we obtain

$$P_+ = e^{-\bar{w}} \frac{\partial}{\partial \bar{w}} \qquad \text{and} \qquad z_+ \bar{P}_+ = \frac{\partial}{\partial w}.$$

Since $e^{-\bar{w}} \neq 0$, the proof of Lemma 1 goes through if one reads w in place of z.

Now consider part (a). We cover CP^2 with the standard coordinate patches $U_0 = C^2$ with coordinates $\zeta = (\zeta_1, \zeta_2)$, $U_1 = C^2$ with $z = (z_1, z_2)$, $U_2 = C^2$ with $w = (w_1, w_2)$, and take U_0 as the C^2 that contains the S^3. In $U_0 - \mathring{B}^4$ we shall use the linearly independent $(0, 1)$ forms

$$l = \bar{\zeta}_2 \, d\bar{\zeta}_1 - \bar{\zeta}_1 \, d\bar{\zeta}_2 \qquad \text{and} \qquad \bar{\partial}\rho = \zeta_1 \, d\bar{\zeta}_1 + \zeta_2 \, d\bar{\zeta}_2,$$

and the antiholomorphic vector field

$$N = \bar{\zeta}_1 \frac{\partial}{\partial \bar{\zeta}_1} + \bar{\zeta}_2 \frac{\partial}{\partial \bar{\zeta}_2},$$

which is linearly independent of $L(t)$. Given an $\omega \in C^\infty(S^3)$, we associate to it the tangential $(0, 1)$ form $\omega(t)l(t)$ on S^3, which is tangentially $\bar{\partial}$-closed by reason of degree. There is a C^∞ $\bar\partial$-closed $(0, 1)$ form ψ on Ω that has this form as its boundary values on $\partial\Omega = S^3$. To see this we again adopt temporarily the notation of [2]: $S = S^3$, $\Omega^+ = \Omega$, $\Omega^- = B^4$. By Hodge theory we have $H^1(\mathbf{CP}^2, \mathcal{O}) = H^2(\mathbf{CP}^2, \mathcal{O}) = 0$; hence the Mayer-Vietoris sequence of [2] again gives the isomorphism $H^1(\Omega^+) \oplus H^1(\Omega^-) \approx H^1(S)$. But $H^1(\Omega^-) = 0$, using boundary regularity in the $\bar\partial$-Neuman problem, since B^4 is strictly pseudoconvex. Thus $H^1(\Omega^+) \approx H^1(S)$ and, as before, the Cauchy problem is solvable not only for cohomology classes, but also at the level of forms. This gives the desired ψ.

We represent ψ in each coordinate patch: $\psi = \psi_1\, d\bar\zeta_1 + \psi_2\, d\bar\zeta_2$ in $U_0 \cap \Omega$, $\psi = \varphi_1\, d\bar z_1 + \varphi_2\, d\bar z_2$ in $U_1 \cap \Omega$, $\psi = \theta_1\, d\bar w_1 + \theta_2\, d\bar w_2$ in $U_2 \cap \Omega$. In $U_0 \cap \Omega = U_0 - \mathring{B}^4$ we can also write $\psi = \psi_l + \psi_n\,\bar\partial\rho$, with $\psi_l(t) = \omega(t)$ on S^3.

For $\gamma = 1, 2, 3$ ($\alpha = 0, 1, 2$ and $\beta = +, -$) we introduce the complex vector fields $P_{\alpha\beta}^{(\gamma)}$ on $(U_\alpha \cap \Omega) \times V_\beta$, as follows: $P_{\alpha\beta}^{(3)} \equiv P_\beta = \partial/\partial\bar z_\beta$ for all α and β, on $(U_0 - \mathring{B}^4) \times V_\beta$ we set

$$P_{0\beta}^{(1)} = \frac{\partial}{\partial\bar\zeta_1} + \beta\psi_1 z_\beta \bar P_\beta \quad \text{and} \quad P_{0\beta}^{(2)} = \frac{\partial}{\partial\bar\zeta_2} + \beta\psi_2 z_\beta \bar P_\beta, \tag{0}$$

on $(U_1 \cap \Omega) \times V_\beta$ we use

$$P_{1\beta}^{(1)} = \frac{\partial}{\partial\bar z_1} + \beta\varphi_1 z_\beta \bar P_\beta \quad \text{and} \quad P_{1\beta}^{(2)} = \frac{\partial}{\partial\bar z_2} + \beta\varphi_2 z_\beta \bar P_\beta, \tag{1}$$

and on $(U_2 \cap \Omega) \times V_\beta$

$$P_{2\beta}^{(1)} = \frac{\partial}{\partial\bar w_1} + \beta\theta_1 z_\beta \bar P_\beta \quad \text{and} \quad P_{2\beta}^{(2)} = \frac{\partial}{\partial\bar w_2} + \beta\theta_2 z_\beta \bar P_\beta. \tag{2}$$

We claim that this prescription provides an integrable almost complex structure M_ω on M, whose induced CR structure on ∂M is the CR structure ∂M_ω defined by $(2'_\omega)$.

To see that (0) gives the correct boundary structure on $S^3 \times V_\beta$, multiply the system (0) on $(U_0 - \mathring{B}^4) \times V_\beta$ by the nonsingular matrix

$$\begin{bmatrix} \zeta_2 & -\zeta_1 \\ \bar\zeta_1 & \bar\zeta_2 \end{bmatrix},$$

whose determinant is $\|\zeta\|^2$, to obtain

$$L + \beta\|\zeta\|^2\psi_l z_\beta \bar P_\beta \quad \text{and} \quad N + \beta\|\zeta\|^2\psi_n z_\beta \bar P_\beta.$$

On S^3 we have $L\rho = 0$ and $N\rho = 1$; hence the first complex vector field above is tangential, and the second one is not. When $\|\zeta\|^2 = 1$ the first one becomes $L(t) + \beta\omega(t)z_\beta\bar{P}_\beta = Q_\beta$.

Next we verify that we have indeed defined an almost complex structure on M. On each $(U_\alpha \cap \Omega) \times V_\beta$ it is clear that $P^{(1)}_{\alpha\beta}, P^{(2)}_{\alpha\beta}, P^{(3)}_{\alpha\beta}, \bar{P}^{(1)}_{\alpha\beta}, \bar{P}^{(2)}_{\alpha\beta}, \bar{P}^{(3)}_{\alpha\beta}$ are linearly independent. To check that our definitions are consistent on overlapping coordinate patches, it will suffice to consider a typical overlap of the form, say, $(U_0 \cap U_1 \cap \Omega) \times V_\beta$. On it one has that $z_1 = 1/\zeta_1$, $z_2 = \zeta_2/\zeta_1$ and $z_1 \neq 0$. Straightforward calculation reveals that (0) is obtained from (1) via multiplication by the matrix

$$\begin{bmatrix} -\bar{z}_1^2 & -\bar{z}_1\bar{z}_2 \\ 0 & \bar{z}_1 \end{bmatrix},$$

whose determinant $= -\bar{z}_1^3 \neq 0$.

Finally we verify that the almost complex structure is integrable: Consider a typical coordinate patch, say, $(U_0 \cap \Omega) \times V_\beta$. Again all the Lie brackets are zero except, possibly, for

$$[P^{(1)}_{0\beta}, P^{(2)}_{0\beta}] = 0 + \left[\frac{\partial}{\partial\bar{\zeta}_1}, \beta\psi_2 z_\beta\bar{P}_\beta\right] + \left[\beta\psi_1 z_\beta\bar{P}_\beta, \frac{\partial}{\partial\bar{\zeta}_2}\right] + [\beta\psi_1 z_\beta\bar{P}_\beta, \beta\psi_2 z_\beta\bar{P}_\beta]$$

$$= \beta\left\{\frac{\partial\psi_2}{\partial\bar{\zeta}_1} - \frac{\partial\psi_1}{\partial\bar{\zeta}_2}\right\}z_\beta\frac{\partial}{\partial z_\beta}.$$

But the last term is zero because ψ is $\bar{\partial}$-closed. The proof of Theorem 4 is complete.

Remark. The unit ball B^4 is not an essential ingredient for the construction of these compact examples. Instead of Ω we could have employed $\Omega' = \mathbf{CP}^2 - \mathring{G}$, where $G \subset \mathbf{C}^2 \to \mathbf{CP}^2$ is any bounded strictly pseudoconvex domain with smooth boundary ∂G. *There is an analogous family of CR structures on $\partial G \times S^2$ with the same properties as in Theorem 4.*

10. Influence of Real Analyticity

The manifolds M in our examples all have an underlying differentiable structure that is real analytic up to the boundary; it is the integrable almost complex structure on M that is only C^∞ up to the boundary. For such an M the following result is straightforward:

Theorem 5. *If M is a complex manifold with an abstract boundary defined by an integrable almost complex structure that is real analytic up to the* *boundary, then $\partial M = \partial_c M$; i.e., every boundary point of M is a concrete boundary point.*

Note that this result has nothing to do with the shape of the boundary locally in the sense of Levi convexity.

11.   Influence of Pseudoconvexity

The examples we have constructed are all weakly pseudoconcave in the sense that at each boundary point the Levi form has one zero and one negative eigenvalue.

Suppose now that M is a complex manifold with an abstract boundary ∂M, such that both the underlying differentiable structure and the integrable almost complex structure are only C^∞ up to the boundary. Some years ago the author had conjectured that in the presence of enough pseudoconvexity, an up-to-the-boundary version of the Newlander–Nirenberg theorem should be true, and had suggested a possible approach patterned after Theorem 1, p 775, of [2]. That theorem involved a hybridization of Hormander's interior results using three weight functions, and Kohn's boundary regularity results. However recently, using such combined techniques, D. Catlin has proved the following beautiful and much needed theorem, which assumes only weak pseudoconvexity:

Theorem 6 (Catlin [3]). *Let M be a complex manifold with abstract boundary ∂M, and let $p \in \partial M$ be a point such that ∂M is pseudoconvex in a neighborhood of p. Then p is a concrete boundary point.*

In a simultaneous development N. Hanges and H. Jacobowitz [4] have given a particularly simple and elegant proof of the weaker result which assumes that p is a point of strict pseudoconvexity. Their proof employs the trick of L. Boutet de Monvel [10], and uses only the classical estimates for $\bar{\partial}$.

12.   Remark on Local Embeddability

In [10] L. Boutet de Monvel proved that a compact CR manifold of real dimension ≥ 5 is locally embeddable at any point if the structure is strictly pseudoconvex, and raised the question of relaxing that assumption. Our compact nowhere embeddable examples of § 9 have one zero and one nonzero eigenvalue for the Levi form at every point. Hence the result of Boutet de Monvel cannot be significantly improved.

13. A Conjecture

In view of our examples of Sections 4 and 9, one might ask when do *CR* manifolds of hypersurface type form the abstract boundary of some complex manifold, even when they may not be locally embeddable? Manifolds which have been called nonrealizable in the literature may be perfectly realizable as "one-sided boundaries," and that could possibly be their natural role in mathematics. In fact Theorem 6 implies that our examples cannot form the abstract boundary of any complex manifold from the "wrong side."

Consider, for example, the lowest-dimensional case of one complex vector field in three variables:

$$P = \sum_{j=1}^{3} a^j(x) \frac{\partial}{\partial x_j}.$$

Assume that the $a^j(x) \in C^\infty(\mathbf{R}^3)$ and that $P, \bar{P}, [P, \bar{P}]$ are linearly independent at each point. This *CR* structure on $\mathbf{R}^3$ would correspond locally to a strictly pseudoconvex hypersurface in $\mathbf{C}^2$, if the structure were locally embeddable. The example [8] found by L. Nirenberg gives a nonembeddable structure; it has the property that any smooth local solution of $Pu = 0$ in a neighborhood of the origin must be a constant. According to the up-to-the-boundary version of the Newlander–Nirenberg theorem mentioned above, such a 3-dimensional structure cannot be the abstract boundary of some complex manifold from the pseudoconvex side. But what about the other side? We conjecture that, at least locally, *such a structure forms the abstract boundary of a complex manifold from the pseudoconcave side.*

References

[1] A. Andreotti and C. D. Hill, Complex characteristic coordinates and tangential Cauchy–Riemann equations, *Ann. Scuola Norm. Sup. Pisa* **26** (1972), 299–324.
[2] A. Andreotti and C. D. Hill, E. E. Levi convexity and the Hans Lewy problem, Part I: Reduction to vanishing theorems, *Ann. Scuola Norm. Sup. Pisa* **26** (1972), 325–363; Part II: Vanishing theorems, 747–806.

[3] D. Catlin, A Newlander, Nirenberg theorem for manifolds with boundary, to appear.

[4] N. Hanges and H. Jacobowitz, A remark on almost complex structures with boundary, to appear.

[5] C. D. Hill, The Cauchy problem for $\bar{\partial}$, *Symposia in Pure Math.*, **23** AMS, Providence, R.I. (1973), 135-143.

[6] C. D. Hill, A hierarchy of non-solvability examples, *Symposia in Pure Math.*, **27** AMS, Providence, R.I. (1975), 301-305.

[7] A. Newlander and L. Nirenberg, Complex coordinates in almost complex manifolds, *Ann. Math.* **65** (1957), 391-404.

[8] L. Nirenberg, On a question of Hans Lewy, *Russian Math. Surveys* **29** (1974), 251-262.

[9] R. Penrose, Physical space-time and nonrealizable CR-structures, *Bull. AMS*, **8** (1983), 427-448.

[10] L. Boutet de Monvel, Integration des equations de Cauchy-Riemann induites formelles, *Sem. Goulaouic-Lions-Schwartz*, **9** (1975).

Toda Molecule Equations

Ryogo Hirota

Department of Applied Mathematics
Faculty of Engineering
Hiroshima University
Higashi-Hiroshima, Japan

1. Introduction

The studies of the nonlinear wave (soliton) equations have been exciting and challenging subjects in recent years. Among them, the KP (Kadomtsev-Petviashvili) equation and the Toda equation play the fundamental role in the soliton theory.

It is known [1, 2, 3] that the hierarchy of the KP equation reduces by "reductions" to a variety of soliton equations, such as the KdV equation, the modified KdV (or Gardner) equation, the Boussinesq equation, the coupled KdV equation, the classical Boussinesq (Kaup's higher order water wave) equation, the nonlinear Schrödinger equation exhibiting dark-soliton solution, and the Drinfel'd–Sokolov–Wilson equation, etc.

The hierarchy of the KP equation is transformed into the following bilinear equations through the dependent variable transformation [1]:

$$(D_1^4 - 4D_1 D_3 + 3D_2^2)\tau \cdot \tau = 0, \qquad (1.1)$$

203

$$[(D_1^3 + 2D_3)D_2 - 3D_1D_4]\tau \cdot \tau = 0, \tag{1.2}$$

where

$$D_i^m \tau \cdot \tau = \left(\frac{\partial^m}{\partial y_i^m}\right)\tau(x_i + y_i)\tau(x_i - y_i)\big|_{y_i = 0}. \tag{1.3}$$

Sato discovered that the above bilinear equations are nothing but the Plücker relations [4, 5].

The Plücker relations are proved by using the Laplace expansion of the determinant. We note that the Laplace expansion was discovered by Yosihiro Kurusima (?–1757) [6] before P. S. Laplace.

In 1967 Toda [7] introduced a nonlinear equation describing a one-dimensional nonlinear lattice, for which he found an exact two-soliton solution. Since then its importance was quickly and widely recognized. Soliton phenomena have been observed using nonlinear electrical networks simulating the Toda lattice [8, 9, 10]. We found an N-soliton solution to it for the first time [11]. The Toda lattice equation with two time variables called two-dimensional Toda equation has been introduced [12]

$$\frac{\partial^2 \hat{Q}_n}{\partial x \, \partial y} = \hat{V}_{n+1} - 2\hat{V}_n + \hat{V}_{n-1}, \tag{1.4}$$

where

$$\hat{V}_n = \exp(\hat{Q}_n) - 1, \tag{1.5}$$

for $n = -\infty, \ldots, -1, 0, 1, \ldots, \infty$.

Kametaka [13] has noticed that the same equation can be seen in the famous book of G. Darboux.

Equation (1.4) is transformed into the bilinear differential equation

$$\frac{\partial^2 \hat{\tau}_n}{\partial x \, \partial y}\hat{\tau}_n - \frac{\partial \hat{\tau}_n}{\partial x}\frac{\partial \hat{\tau}_n}{\partial y} = \hat{\tau}_{n+1}\hat{\tau}_{n-1} - \hat{\tau}_n^2, \tag{1.6}$$

through the dependent variable transformation

$$\hat{V}_n = \frac{\partial^2}{\partial x \, \partial y}\log(\hat{\tau}_n). \tag{1.7}$$

A discrete analogue [15] of the two-dimensional Toda equation in the bilinear form was obtained:

$$[Z_1 \exp(D_1) + Z_2 \exp(D_2) + Z_3 \exp(D_3)]f \cdot f = 0, \tag{1.8}$$

where Z_i and D_i for $i = 1, 2, 3$, are arbitrary parameters and linear combinations of the bilinear operators D_t, D_x, D_y and D_n, respectively. The equation generates various types of soliton equations including the KdV equation, the KP equation, the modified KdV equation, the sine-Gordon equation, a nonlinear Klein–Gordon equation by Fordy and Gibbons, and Mikhailov's equation, the Benjamin–Ono equation, etc.

Several authors [15, 16, 17] have considered the Toda lattice with free ends:

$$\frac{\partial^2 Q_n}{\partial x \, \partial y} = V_{n+1} - 2V_n + V_{n-1}, \tag{1.9}$$

where

$$V_n = \exp(Q_n), \tag{1.10}$$

for $n = 1, 2, \ldots, N$, with the boundary conditions

$$V_0 = V_{N+1} = 0, \tag{1.11}$$

which is called the *two-dimensional Toda molecule equation*. Note that V_n of equation (1.10) is different from $\hat{V}_n$ of equation (1.5).

In Section 2 we transform equation (1.9) into the bilinear differential equation and show that the bilinear differential equation is nothing but the Jacobi formula for the determinant. We describe exact solutions to the two- and one-dimensional Toda molecule equations. In Section 3 we start with the Jacobi formula for the Casorati determinant and construct a discrete analogue [18] of the two-dimensional Toda molecule equation.

In Section 4 we consider a cylindrical Toda molecule equation [19],

$$\left[\frac{\partial^2}{\partial r^2} + \frac{1}{r}\frac{\partial}{\partial r}\right] \log(V_n) = V_{n+1} - 2V_n + V_{n-1}, \tag{1.12}$$

with the boundary conditions

$$V_0 = V_{N+1} = 0. \tag{1.13}$$

The equation is transformed into the one-dimensional Toda molecule equation through the dependent and independent variable transformations.

In Section 5 we consider a $(3+1)$-dimensional (spherical) Toda molecule equation introduced by Nakamura [20],

$$\Delta \log V_n = V_{n+1} - 2V_n + V_{n-1}, \tag{1.14}$$

where

$$\Delta = \frac{\partial^2}{\partial r^2} + \left(\frac{2}{r}\right)\left(\frac{\partial}{\partial r}\right) + r^{-2}\left[\frac{\partial^2}{\partial \theta^2} + \cot\theta\frac{\partial}{\partial \theta} + \frac{1}{\sin^2\theta}\frac{\partial^2}{\partial \varphi^2}\right], \tag{1.15}$$

with the boundary conditions

$$V_0 = V_{N+1} = 0. \tag{1.16}$$

The equation is transformed into the two-dimensional Toda molecule equation under the assumption that

$$V_n(r, \theta, \varphi) = r^{-2}V_n(\theta, \varphi). \tag{1.17}$$

2. Transformation of Toda Molecule Equations into Bilinear Forms

We transform the two- and one-dimensional Toda molecule equations into the bilinear forms and describe their exact solutions.

2.1. Two-Dimensional Toda Molecule Equation

We have the two-dimensional Toda molecule equation

$$\frac{\partial^2}{\partial x\, \partial y}\log(V_n) = V_{n+1} - 2V_n + V_{n-1}. \tag{2.1}$$

Let

$$V_n = \frac{\partial^2}{\partial x\, \partial y}\log(\tau_n). \tag{2.2}$$

Then equation (2.1) is integrated with respect to x and y,

$$\frac{\partial^2}{\partial x\, \partial y}\log(\tau_n) = \frac{\tau_{n+1}\tau_{n-1}}{\tau_n^2}, \tag{2.3}$$

where we have chosen integration constants to be zero. Equation (2.3) is equivalent to the bilinear differential equation

$$\frac{\partial^2\tau_n}{\partial x\, \partial y}\tau_n - \frac{\partial\tau_n}{\partial x}\frac{\partial\tau_n}{\partial y} = \tau_{n+1}\tau_{n-1}. \tag{2.4}$$

The boundary conditions (1.11) are satisfied with

$$\tau_0 = 1, \tag{2.5a}$$

$$\tau_{N+1} = \Phi(x)\chi(y), \tag{2.5b}$$

where $\Phi(x)$ and $\chi(y)$ are arbitrary functions of x and y respectively.

Leznov and Saveliev [15] have obtained an exact solution to equation (1.9). Their solution is obtained as follows.

Suppose that τ_n is expressed with a Wronskian of order n,

$$\tau_n = \det \left| \left(\frac{\partial}{\partial x}\right)^{i-1} \left(\frac{\partial}{\partial y}\right)^{j-1} \psi(x, y) \right|_{1 \le i, j \le n}, \tag{2.6}$$

where $\psi(x, y)$ is an arbitrary function of x and y.

We introduce symbols D, $D\binom{i}{j}$ and $D\binom{i,k}{j,l}$ which are the determinants of the $(n+1) \times (n+1)$, $n \times n$ and $(n-1) \times (n-1)$ matrices respectively, with the respective definitions

$$D = \det \left| \left(\frac{\partial}{\partial x}\right)^{i-1} \left(\frac{\partial}{\partial y}\right)^{k-1} \psi(x, y) \right|_{1 \le i, k \le n+1} \tag{2.7a}$$

$$= \tau_{n+1},$$

$$\begin{aligned} D\binom{i}{j} = \text{ same as } D \text{ except that the } i\text{-th and} \\ j\text{-th column are removed from it,} \end{aligned} \tag{2.7b}$$

$$\begin{aligned} D\binom{i,k}{j,l} = \text{ same as } D \text{ except that the } i\text{-th and} \\ k\text{-th rows, and } j\text{-th and } l\text{-th columns} \\ \text{are removed from it.} \end{aligned} \tag{2.7c}$$

With these symbols we find that equation (2.4) is expressed as

$$D\binom{n}{n}D\binom{n+1}{n+1} - D\binom{n}{n+1}D\binom{n+1}{n} = D \cdot D\binom{n, n+1}{n, n+1}, \tag{2.8}$$

which is nothing but the Jacobi formula for the determinant:

$$D\binom{i}{i}D\binom{k}{k} - D\binom{i}{k}D\binom{k}{i} = D \cdot D\binom{i, k}{i, k}, \tag{2.9}$$

for integers $1 \le i, k \le n+1$. Hence τ_n of equation (2.6) is proved to be a solution to equation (2.4).

The functional form of ψ is determined by the boundary condition (2.5). It is known that the following form of ψ satisfies the boundary condition:

$$\psi(x, y) = \sum_{j=1}^{N+1} u_j(x) v_j(y), \tag{2.10}$$

where $u_j(x)$ and $v_j(y)$ for $j = 1, 2, 3, \ldots, N+1$ are arbitrary functions of x and y respectively. Substituting equation (2.10) into equation (2.6) we find that τ_n is expressed with a determinant of an $n \times n$ matrix which is a product of $n \times (N+1)$ matrix A_n and $(N+1) \times n$ matrix B_n:

$$\tau_n = \det|A_n \times B_n|, \tag{2.11}$$

where the matrix elements of A_n and B_n are defined by

$$(A_n)_{ij} = \left(\frac{\partial}{\partial x}\right)^{i-1} u_j(x), \tag{2.12a}$$

for $i = 1, 2, \ldots, n$, and $j = 1, 2, \ldots, N+1$, and

$$(B_n)_{jk} = \left(\frac{\partial}{\partial y}\right)^{k-1} v_j(y), \tag{2.12b}$$

for $k = 1, 2, \ldots, n$, and $j = 1, 2, \ldots, N+1$, respectively.

At $n = N+1$ the matrices A_n and B_n become square matrices. Consequently τ_{N+1} is expressed with a product of determinants

$$\tau_{N+1} = \det|A_{N+1}| \times \det|B_{N+1}|, \tag{2.13}$$

where $\det|A_{N+1}|$ is a function of x only and $\det|B_{N+1}|$ is a function of y only. Hence the boundary condition (2.5) is satisfied with the $\psi(x, y)$ of equation (2.10).

2.2. One-Dimensional Toda Molecule Equation

We have the one-dimensional Toda molecule equation

$$\frac{\partial^2}{\partial t^2} \log(V_n) = V_{n+1} - 2V_n + V_{n-1}, \tag{2.14}$$

with the boundary conditions

$$V_0 = V_{N+1} = 0. \tag{2.15}$$

Let

$$V_n = \frac{\partial^2}{\partial t^2} \log(\tau_n). \tag{2.16}$$

Then equation (2.14) is integrated with respect to t,

$$\frac{\partial^2}{\partial t^2} \log(\tau_n) = \frac{\tau_{n+1}\tau_{n-1}}{\tau_n^2}, \tag{2.17}$$

where we have chosen integration constants to be zero. Equation (2.17) is equivalent to the bilinear differential equation

$$\frac{\partial^2 \tau_n}{\partial t^2} \tau_n - \left(\frac{\partial \tau_n}{\partial t}\right)^2 = \tau_{n+1}\tau_{n-1}. \tag{2.18}$$

The boundary conditions (2.15) are satisfied with

$$\tau_0 = 1, \tag{2.19a}$$

$$\tau_{N+1} = C \times \exp(Pt), \tag{2.19b}$$

where C and P are constants.

Farwell and Minami [16] have found an exact solution to equation (2.14). Their solution is obtained by a similar procedure as that used in the two-dimensional Toda molecule equation.

Suppose that τ_n is expressed with a Wronksian of order n,

$$\tau_n = \det \left| \left(\frac{\partial}{\partial t}\right)^{i+j-2} \Psi(t) \right|_{1 \le i,j \le n}, \tag{2.20}$$

where $\Psi(t)$ is an arbitrary function of t.

We use the same symbols as those of equation (2.7) except that D, the determinant of the $(n+1) \times (n+1)$ matrix, is defined by

$$D = \det \left| \left(\frac{\partial}{\partial t}\right)^{i+j-2} \Psi(t) \right|_{1 \le i,j \le n+1}, \tag{2.21}$$

$$= \tau_{n+1}. \tag{2.22}$$

Equation (2.18) is expressed as

$$D\binom{n}{n} D\binom{n+1}{n+1} - D\binom{n}{n+1} D\binom{n+1}{n} = D \cdot D\binom{n, n+1}{n, n+1}, \tag{2.23}$$

which clearly holds because of the Jacobi formula.

The functional form of $\psi(t)$ is determined by the boundary condition $V_{N+1} = (\partial^2/\partial t^2) \log \tau_{N+1} = 0$. It is known that the following form of $\psi(t)$,

$$\psi(t) = \sum_{i=1}^{N+1} \exp(\eta_i), \tag{2.24}$$

where $\eta_i = p_i t + c_i$, and p_i and c_i for $i = 1, 2, \ldots, N+1$ are arbitrary constants, satisfies the boundary condition.

Substituting equation (2.24) into (2.20) we find that τ_n is expressed with a determinant of an $n \times n$ matrix which is a product of an $n \times (N+1)$ matrix A_n and an $(N+1) \times n$ matrix B_n:

$$\tau_n = \det |A_n \times B_n|, \tag{2.25}$$

where the matrix elements of A_n and B_n are defined by

$$(A_n)_{ij} = p_j^{i-1}, \tag{2.26a}$$

for $i = 1, 2, \ldots, n$, and $j = 1, 2, \ldots, N+1$, and

$$(B_n)_{jk} = p_j^{k-1} \exp(\eta_j), \tag{2.26b}$$

for $k = 1, 2, \ldots, n$, and $j = 1, 2, \ldots, N+1$, respectively.

At $n = N+1$ the matrices A_n and B_n become square matrices. Consequently τ_{N+1} is expressed with a product of determinants,

$$\tau_{N+1} = \det |A_{N+1}| \times \det |B_{N+1}|, \tag{2.27}$$

$$= V^2(p_1, p_2, \ldots, p_{N+1}) \exp\left(\sum_{i=1}^{N+1} \eta_i\right), \tag{2.28}$$

where $V(p_1, p_2, \ldots, p_{N+1})$ is the Vandermonde determinant. Clearly the boundary condition $V_{N+1} = 0$ is satisfied with the $\Psi(t)$ of equation (2.24).

3. Discrete Two-Dimensional Toda Molecule Equation

We have the bilinear differential equation of the Toda molecule equation

$$\frac{\partial \tau_n}{\partial x \, \partial y} \tau_n - \frac{\partial \tau_n}{\partial x} \frac{\partial \tau_n}{\partial y} = \tau_{n+1} \tau_{n-1}. \tag{3.1}$$

In this section we investigate a difference analogue [18] of equation (3.1). We know that τ_n is expressed with the Wronksian of order n. It is known that a difference analogue of a Wronksian is a Casorati determinant, which is used in the theory of linear difference equations. Correspondingly we assume that a solution of the bilinear difference equation is expressed with a Casorati determinant of the form

$$\hat{\tau}_n(l, m) = \det |\psi(l+i-1, m+j-1)|_{1 \le i,j \le n}, \tag{3.2}$$

where $\psi(l, m)$ is a function of integers l and m, and is determined by the boundary condition.

Let

$$D \equiv \hat{\tau}_{n+1}(l, m). \tag{3.3a}$$

Then we find the following relations:

$$D\binom{1}{1} = \hat{\tau}_n(l+1, m+1), \tag{3.3b}$$

$$D\binom{n+1}{n+1} = \hat{\tau}_n(l, m), \tag{3.3c}$$

$$D\binom{1}{n+1} = \hat{\tau}_n(l+1, m), \tag{3.3d}$$

$$D\binom{n+1}{1} = \hat{\tau}_n(l, m+1), \tag{3.3e}$$

$$D\binom{1, n+1}{1, n+1} = \hat{\tau}_{n-1}(l+1, m+1). \tag{3.3f}$$

Consequently the Jacobi formula

$$D\binom{1}{1}D\binom{n+1}{n+1} - D\binom{1}{n+1}D\binom{n+1}{1} = D \cdot D\binom{1, n+1}{1, n+1} \tag{3.4a}$$

becomes the bilinear difference equation of $\hat{\tau}_n(l, m)$,

$$\hat{\tau}_n(l+1, m+1)\hat{\tau}_n(l, m) - \hat{\tau}_n(l+1, m)\hat{\tau}_n(l, m+1)$$
$$= \hat{\tau}_{n+1}(l, m)\hat{\tau}_{n-1}(l+1, m+1). \tag{3.4b}$$

Let

$$x = l\delta, \tag{3.5a}$$

$$y = m\varepsilon, \tag{3.5b}$$

$$\hat{\tau}_n(l, m) = (\delta\varepsilon)^{n(n-1)/2}\tau_n(x, y), \tag{3.5c}$$

where δ and ε are parameters specifying the intervals. Then equation (3.4b) is transformed into

$$\tau_n(x+\delta, y+\varepsilon)\tau_n(x, y) - \tau_n(x+\delta, y)\tau_n(x, y+\varepsilon)$$
$$= \delta\varepsilon\tau_{n+1}(x, y)\tau_{n-1}(x+\delta, y+\varepsilon). \tag{3.6}$$

Expanding equation (3.6) in a power series of δ and ε we find that equation (3.6) becomes the bilinear differential equation (3.1) in the second order of δ and ε.

The boundary condition of $\hat{\tau}_n(l, m)$ is satisfied by choosing the following ψ. Let

$$\psi(l, m) = \sum_{j=1}^{N+1} \hat{u}_j(l)\hat{v}_j(m). \tag{3.7}$$

Then following the same procedure as that used in proving equation (2.13) we find that $\hat{\tau}_n(l, m)$ becomes a product of a function of l and a function of m at $n = N + 1$.

The bilinear difference equation (3.6) can be transformed back into the nonlinear difference equation in the ordinary form [18].

Let us introduce the forward difference operators Δ_x and Δ_y operating on a function $f(x, y)$ by

$$\Delta_x f(x, y) = \delta^{-1}[f(x + \delta, y) - f(x, y)], \tag{3.8a}$$

$$\Delta_y f(x, y) = \varepsilon^{-1}[f(x, y + \varepsilon) - f(x, y)], \tag{3.8b}$$

and the quantities $V_n(x, y)$ and $Q_n(x, y)$ by

$$V_n(x, y) = (\delta\varepsilon)^{-1} \log\left[\frac{\tau_n(x + \delta, y + \varepsilon)\tau_n(x, y)}{\tau_n(x + \delta, y)\tau_n(x, y + \varepsilon)}\right], \tag{3.9}$$

$$Q_n(x, y) = \log\left[\frac{\tau_{n+1}(x, y)\tau_{n-1}(x + \delta, y + \varepsilon)}{\tau_n(x + \delta, y)\tau_n(x, y + \varepsilon)}\right]. \tag{3.10}$$

Then using equations (3.6), (3.9) and (3.10) we obtain an equation

$$\Delta_x \Delta_y Q_n(x, y) = V_{n+1}(x, y) - V_n(x + \delta, y) - V_n(x, y + \varepsilon) + V_{n-1}(x + \delta, y + \varepsilon), \tag{3.11}$$

where

$$V_n(x, y) = (\delta\varepsilon)^{-1} \log[1 + \delta\varepsilon \exp[Q_n(x, y)]], \tag{3.12}$$

with the boundary conditions

$$V_0(x, y) = 0, \tag{3.13a}$$

$$V_{N+1}(x, y) = 0, \tag{3.13b}$$

which is a discrete analogue of the two-dimensional Toda molecule equation.

4. Cylindrical Toda Molecule Equation

We show that the cylindrical Toda molecule equation introduced by the author and Nakamura [19],

$$\left[\frac{\partial^2}{\partial r^2} + \frac{1}{r}\frac{\partial}{\partial r}\right] \log(V_n) = V_{n+1} - 2V_n + V_{n-1}, \tag{4.1}$$

with the boundary conditions

$$V_0 = 0, \tag{4.2a}$$

$$V_{N+1} = 0, \tag{4.2b}$$

is transformed into the one-dimensional Toda molecule equation through the dependent and independent variable transformations.

Let

$$V_n = \frac{1}{r}\frac{\partial}{\partial r}\frac{r\partial}{\partial r}\log[r^{-n^2}\tau_n]. \tag{4.3}$$

Then equation (4.1) is integrated with respect to r,

$$\frac{1}{r}\frac{\partial}{\partial r}\frac{r\partial}{\partial r}\log(\tau_n) = r^{-2}\frac{\tau_{n+1}\tau_{n-1}}{\tau_n^2}, \tag{4.4}$$

where an integration constant is chosen to be zero. We introduce a new independent variable y by

$$y = \log(r). \tag{4.5}$$

Then equation (4.4) becomes

$$\frac{\partial^2 \tau_n}{\partial y^2}\tau_n - \left(\frac{\partial \tau_n}{\partial y}\right)^2 = \tau_{n+1}\tau_{n-1}, \tag{4.6}$$

which is the bilinear form of the one-dimensional Toda molecule equation. Accordingly we have the exact solution to equation (4.1)

$$V_n = \frac{1}{r}\frac{\partial}{\partial r}\frac{r\partial}{\partial r}\log[r^{-n^2}\tau_n], \tag{4.7a}$$

$$\tau_n = \det\left|\left(\frac{r\partial}{\partial r}\right)^{i+j-2}\psi(r)\right|_{1\le i,j\le n}, \tag{4.7b}$$

$$\psi(r) = \sum_{i=1}^{N+1} c_i y^{p_i}, \tag{4.7c}$$

where c_i and p_i for $i = 1, 2, \ldots, N+1$ are arbitrary constants.

5. (3+1)-Dimensional Toda Molecule Equation

We consider a $(3+1)$-dimensional (spherical) Toda molecule equation introduced by Nakamura [20]:

$$\Delta \log V_n = V_{n+1} - 2V_n + V_{n-1}, \tag{5.1}$$

where

$$\Delta = \frac{\partial^2}{\partial r^2} + \frac{2}{r} \frac{\partial}{\partial r} + r^{-2} \left[\frac{\partial^2}{\partial \theta^2} + \cot \theta \frac{\partial}{\partial \theta} + \frac{1}{\sin^2 \theta} \frac{\partial^2}{\partial \varphi^2} \right] \tag{5.2}$$

with the boundary conditions

$$V_0 = 0, \tag{5.3a}$$

$$V_{N+1} = 0. \tag{5.3b}$$

The equation is shown to be transformed into the two-dimensional Toda molecule equation under the assumption that

$$V_n(r, \theta, \varphi) = r^{-2} V_n(\theta, \varphi). \tag{5.4}$$

Substituting equation (5.4) into equation (5.1) we obtain

$$\Delta(\theta, \varphi) \log V_n(\theta, \varphi) - 2 = V_{n+1}(\theta, \varphi) - 2V_n(\theta, \varphi) + V_{n-1}(\theta, \varphi), \tag{5.5}$$

where

$$\Delta(\theta, \varphi) = \frac{\partial^2}{\partial \theta^2} + \cot \theta \frac{\partial}{\partial \theta} + \frac{1}{\sin^2 \theta} \frac{\partial^2}{\partial \varphi^2}. \tag{5.6}$$

Now we consider the dependent variable transformation

$$V_n(\theta, \varphi) = c_1(n) + \Delta(\theta, \varphi) \log f_n(\theta, \varphi), \tag{5.7a}$$

$$c_1(n) = -n^2 + a_1 n + b_1, \tag{5.7b}$$

where a_1 and b_1 are constants to be determined. Substituting equation (5.7) into equation (5.5) we obtain

$$\Delta(\theta, \varphi) \log[c_1(n) + \Delta(\theta, \varphi) \log f_n] = \Delta(\theta, \varphi) \log \left[\frac{f_{n+1} f_{n+1}}{f_n^2} \right], \tag{5.8}$$

where we have used f_n instead of $f_n(\theta, \varphi)$ for simplicity. Integrating equation

(5.8) we obtain

$$c_1(n) + \Delta(\theta, \varphi) \log f_n = c_2(n) \frac{f_{n+1} f_{n-1}}{f_n^2}, \tag{5.9}$$

where $c_2(n)$ is an integration constant with respect to θ, φ.

Equation (5.9) is equal to the bilinear form

$$\frac{\partial^2 f_n}{\partial \theta^2} f_n - \left(\frac{\partial f_n}{\partial \theta}\right)^2 + \cot \theta f_n \frac{\partial f_n}{\partial \theta}$$

$$+ \sin^{-2} \theta \left[\frac{\partial^2 f_n}{\partial \varphi^2} f_n - \left(\frac{\partial f_n}{\partial \varphi}\right)^2\right] - c_2(n) f_{n+1} f_{n-1} + c_1(n) f_n^2 = 0. \tag{5.10}$$

We introduce a dependent variable $\tau_n(\theta, \varphi)$ by the relation

$$f_n(\theta, \varphi) = \sin^{c_1(n)} \theta \, \tau_n(\theta, \varphi). \tag{5.11}$$

Substituting equation (5.11) into equation (5.10) we find that equation (5.10) is transformed into

$$\left[\left(\sin \theta \frac{\partial}{\partial \theta}\right)^2 \tau_n\right] \tau_n - \left[\sin \theta \frac{\partial \tau_n}{\partial \theta}\right]^2 + \frac{\partial^2 \tau_n}{\partial \varphi^2} \tau_n - \left[\frac{\partial \tau_n}{\partial \varphi}\right]^2 = c_2(n) \tau_{n+1} \tau_{n-1}. \tag{5.12}$$

Now we introduce an independent variable z. Let

$$z = \log(\tan(\theta/2)). \tag{5.13}$$

Then equation (5.12) becomes

$$\frac{\partial^2 \tau_n}{\partial z^2} \tau_n - \left(\frac{\partial \tau_n}{\partial z}\right)^2 + \frac{\partial^2 \tau_n}{\partial \varphi^2} \tau_n - \left(\frac{\partial \tau_n}{\partial \varphi}\right)^2 = c_2(n) \tau_{n+1} \tau_{n-1}. \tag{5.14}$$

The factor $c_2(n)$ in equation (5.14) can be eliminated by replacing τ_n by $\hat{\tau}_n$. Let

$$\tau_n(\theta, \varphi) = c(n) \hat{\tau}_n(\theta, \varphi), \tag{5.15}$$

where $c(n)$ satisfies the relation

$$\frac{c(n+1)c(n-1)}{c(n)^2} = \frac{1}{c_2(n)}. \tag{5.16}$$

Then equation (5.14) reduces to

$$\frac{\partial^2 \hat{\tau}_n}{\partial z^2} \hat{\tau}_n - \left(\frac{\partial \hat{\tau}_n}{\partial z}\right)^2 + \frac{\partial^2 \hat{\tau}_n}{\partial \varphi^2} \hat{\tau}_n - \left(\frac{\partial \hat{\tau}_n}{\partial \varphi}\right)^2 = \hat{\tau}_{n+1} \hat{\tau}_{n-1}, \tag{5.17}$$

which is the bilinear form of the two-dimensional Toda molecule equation. Hence we have succeeded in transforming the $3+1$-(spherical) Toda molecule equation introduced by Nakamura into the two-dimensional Toda molecule equation via the dependent and independent variable transformations [21].

References

[1] M. Jimbo and T. Miwa, *Publ. RIMS, Kyoto Univ.* **439** (1983).

[2] R. Hirota, *Physica* **18D** (1986) 161.

[3] R. Hirota, B. Grammaticos, and A. Ramani, *J. Math. Phys.* **27** (1986) 1499.

[4] M. Sato and M. Noumi, *Soliton Equations and the Universal Grassmann Manifold*, Sugaku Kokyuroku, Sophia Univ. no. 18, 1984 [in Japanese].

[5] R. Hirota, *J. Phys. Soc. Jpn.* **55** (1986) 2137.

[6] *Iwanami Sugaku Ziten*, p. 450, ed. Math. Soc. Jpn. (Iwanami, 1985).

[7] M. Toda, *J. Phys. Soc. Jpn.* **22** (1967) 431.

[8] R. Hirota and K. Suzuki, *J. Phys. Soc. Jpn.* **28** (1970) 1336.

[9] H. Nagashima and Y. Amagishi, *J. Phys. Soc. Jpn.* **45** (1978) 680.

[10] S. Watanabe, M. Miyakawa, and M. Toda, *J. Phys. Soc. Jpn.* **45** (1978) 2030.

[11] R. Hirota, *J. Phys. Soc. Jpn.* **35** (1973) 286.

[12] A. Mikhailov. *Pis'ma Zh. Eks. Theor. Fiz.* **30** (1979) 443.

[13] Y. Kametaka, *RIMS Kokyuroku* **554**, *Development of Soliton Theory* (Kyoto, March 1985) p. 26.

[14] R. Hirota, *J. Phys. Soc. Jpn.* **50** (1981) 3785.

[15] A. N. Leznov and M. V. Saveliev, *Physica* **3D** 1&2 (1981) 62.

[16] R. Farwell and M. Minami, *Prog. Theor. Phys.* **69** (1983) 1091.

[17] Z. Popowicz, *J. Math. Phys.* **25** (1984) 2212.

[18] R. Hirota, Discrete Two-Dimensional Toda Molecule Equation, submitted to *J. Phys. Soc. Jpn.*

[19] R. Hirota and A. Nakamura, to appear in *J. Phys. Soc. Jpn.*

[20] A. Nakamura, The $3+1$ Dimensional Toda equation and Its Exact Solution, submitted to *J. Phys. Soc. Jpn.*

[21] R. Hirota, Exact Solutions of the Spherical Toda Molecule Equation, submitted to *J. Phys. Soc. Jpn.*

Microlocal Analysis and Scattering in Quantum Field Theories

D. Iagolnitzer

Service de Physique Théorique
CEN-Saclay
Gif-sur-Yvette, France

1. Introduction

This text aims to provide a unified presentation of results on the momentum space analytic structure of the multiparticle S matrix in relativistic quantum theories describing systems of massive particles with short-range interactions, a domain which has been closely linked with purely mathematical developments, e.g., in microlocal analysis. The scattering operator S, or S-matrix, relates for each physical system the two free-particle states with which it coincides asymptotically before and after interactions. For, e.g., a theory with only one type of stable particle of mass $\mu > 0$ (and without spin), the Hilbert space $\mathscr{F}$ of free-particle states is the direct sum of spaces of square integrable "wave functions" φ_n depending on n real energy-momentum variables $p_1, \ldots, p_n$, $p_i = (p_{i,0}, \mathbf{p}_i)$, restricted to the mass-shell $p_i^2 \equiv p_{i,0}^2 - \mathbf{p}_i^2 = \mu^2$, $p_{i,0} > 0$, $i = 1, \ldots, n$ ($p_{i,0}$ and $\mathbf{p}_i$ are the energy and momentum components of p_i and units such that $c = 1$ are used). S is a unitary operator from $\mathscr{F}$ to $\mathscr{F}$ (conservation of probabilities) whose kernels

$S_{m,n}(-p_1, \ldots, -p_m, p_{m+1}, \ldots, p_{m+n})$ can be written (almost everywhere), in view of energy–momentum conservation (e.m.c.), in the form $s_{m,n}\delta(p_1 + \cdots + p_{m+n})$, where $s_{m,n}$ is defined in the physical region of the $m \to n$ process (mass shell and e.m.c. constraints). It is on the other hand useful to introduce connected components $S^c_{m,n}$. By inductive definition, $S_{m,n}$ is equal to $S^c_{m,n}$ plus a sum, over nontrivial partitions of the sets of initial and final particles, of tensorial products of $S^c_{m_\alpha, n_\alpha}$. Connected kernels still contain a global e.m.c. δ-function but in contrast to $s_{m,n}$ will contain no partial e.m.c. δ-function. The "scattering function" $T_{m,n}$ is the distribution $S^c_{m,n}/\delta(\sum p_k)$.

In field theories, scattering functions are derived quantities obtained by restriction to the mass shell of N-point distributions $F_N(p_1, \ldots, p_N)$, ($\sum p_k = 0$), Fourier transforms of distributions related to the fields and depending on N space-time variables $x_1, \ldots, x_N$ (up to global space-time translation). Energy and momentum are dual to time and space respectively. For each N, scattering functions of various "crossed" processes with $m + n = N$ are obtained by specifying initial and final particles. In the perturbative approach to field theory models (see Section 4), which is only formal but is an important heuristic guide, F_N appears as an infinite divergent sum of (possibly "renormalized") Feynman distributions $I_G(p_1, \ldots, p_N)$ associated with connected multiple scattering graphs with N external lines: there is one coupling constant λ and one e.m.c. δ-function at each vertex of G, one Feynman propagator $[k_l^2 - m^2 + i\varepsilon]^{-1}$ for each internal line and integration is made over the internal energy–momenta k_l, with possibly a "renormalization" procedure to avoid divergences at infinity. Results on the analytic structure of these integrals have been obtained in successive steps, with, in particular, general results on characteristic varieties (modified Landau singularities) and regular holonomicity in the second part of the seventies, following ideas of M. Sato. However, a more complete understanding of the multisheeted structure of these integrals has still to be obtained: e.g., conjectures of physical interest on some of their discontinuities remain to be established.

Non-perturbative theories include on the one hand a pure S-matrix approach outlined in Section 2 and on the other hand axiomatic and constructive theory approaches outlined in Sections 3 and 4 respectively. The purpose of Section 2 is also to present some general properties of the S matrix that one will aim to establish in field theory approaches in a more precise way. Sections 2, 3, 4 are related but can be read to a large extent independently. The starting point is more and more detailed as one moves from Section 2 to 4. For conciseness, references have been omitted: see the short bibliography given at the end.

2. S-Matrix Theory

2.1 Unitarity Equations and Macrocausality

Unitarity ($SS^{-1} = SS^\dagger = 1$) yields an infinite system of nonlinear equations between scattering functions such as:

$$(1)$$

where $+$ and $-$ refer to connected kernels of S and of $-S^{-1} = -S^\dagger$ respectively, and terms in the right-hand side are (on-mass-shell) convolution integrals, with $r' < r$ internal lines if the squared center-of-mass energy s ($= (p_1 + p_2)^2 = (p_3 + p_4)^2$ in the channel $1, 2 \to 3, 4$) is $< (r\mu)^2$. Another example at s ($= (p_1 + p_2 + p_3)^2) < 9\mu^2$ is:

$$(2)$$

It has been believed since the beginning of the sixties that, starting from some "maximal analyticity" principle and some causality idea, it should be possible to extract from these equations detailed information on the structure of scattering functions. A precise causality condition has been stated at the end of the sixties in macroscopic space-time in terms of particles (macrocausality). It entails that the essential support of scattering functions corresponds, at each real point, to (relative) configurations of external trajectories of connected classical multiple scattering diagrams in space-time. As well known, the essential support ($=$ at each real point, the set of "singular directions" along which a generalized Fourier transform does not fall off exponentially in a well-specified sense) and the singular spectrum of hyperfunction theory (as also the analytic wave front) coincide and define the (first) *microsupport*, which characterizes by duality local analyticity properties. In view of macrocausality, $T_{m,n}$ is in particular analytic outside (codimension 1 or more) $+\alpha$-Landau surfaces $L^+(G)$ of connected graphs and is at almost all $+\alpha$-Landau points P the boundary value of an analytic function from the "plus $i\varepsilon$" directions dual to the "causal" direction(s) at P. (Exceptions that do exist, e.g., for the $3 \to 3$ process discussed below, and cannot be neglected are, in particular, points that lie on several surfaces with conflicting plus $i\varepsilon$ directions.)

2.2. Pole Factorization Theorem, Local Discontinuity Formulae and Related Results

Therefrom, a first program is to derive results on the structure of scattering function at $+\alpha$-Landau points, which will now be related to properties of macrocausal factorization (and measurement theory). The simplest example is the "pole-factorization theorem." It asserts that along the $+\alpha$-Landau surface $k^2 = \mu^2$, $k_0 > 0$, $k = p_1 + p_2 + p_4$ of the graph

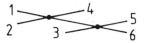

$T_{3,3}$ has a pole in $[k^2 - \mu^2 + i\varepsilon]^{-1}$ with the factorized residue $T_{2,2}(-p_1, -p_2; p_4, k)\,T_{2,2}(-k, -p_3; p_5, p_6)$, i.e., the local discontinuity is

$$T_{2,2}\,T_{2,2}\,\delta(k^2 - \mu^2) \equiv \quad /\delta(p_1 + \cdots + p_6).$$

The proof relies on the study of the microsupports of the various integrals involved in (2) and on a related algebraic analysis. (Note that the term

in (2) has itself a singularity in $\delta(k^2 - \mu^2)$.) However, previous information on the microsupport of scattering functions yields no information on terms such as   ($u = 0$ situation). A second problem is to

"separate singularities" of different contributions which are expected to have a different structure but whose microsupports do interfere. These two problems were overlooked at the Nice 1973 meeting. As a consequence, the proof given there is both incomplete and not correct. To get an actual proof, one approach is recourse to an ad hoc assumption of (microlocal) separation of singularities in unitarity equations with respect to classes of topological diagrams. Via a nontrivial algebraic work, it yields general (micro-)local discontinuity formulae of scattering functions. Other approaches are based on more detailed assumptions on scattering functions that correspond to a refined form of macrocausality or to part of "maximal analyticity" and are probably linked with information on the "second" microsupport. Besides the pole factorization theorem, a structure equation has then been obtained for $T_{3,3}$ in the low energy region: it expresses it as a sum of contributions associated with singularities encountered and having corresponding discontinuities. However, it is not clear that the method can be extended. On the other hand, solutions of the $u = 0$ problem are not so far fully satisfactory in the present context. One approach, based on refined macrocausality, deserves more investigation. An alternative one, based on $u = 0$ results on products of regular holonomic microfunctions, provides best results but relies on an assumption that may seem too strong and detailed and whose aspects (convergence) are also questionable in general. (The S matrix is assumed from the outset to be near any physical point an infinite, convergent sum of regular holonomic contributions with some of the properties of Feynman integrals.) The possible application of further "$u = 0$"-type results obtained more recently by various methematicians has not been examined so far.

2.3. Global Analyticity Properties

More global properties have been conjectured in S-matrix theory from heuristic arguments based on "maximal analyticity":

(a) "physical-sheet" analyticity in a (one-sheeted) domain of the complex mass-shell ($p_k \in \mathbf{C}^d$, $p_k^2 \equiv p_{k,0}^2 - \mathbf{p}_k^2 = \mu^2$, $\forall k$, $\sum p_k = 0$). Namely there should exist for each N a unique analytic function T_N from which scattering functions $T_{m,n}$ and $T_{m,n}^-$ of crossed processes are (almost everywhere) specified boundary values in their respective (disconnected) physical regions. It is useful to work here with Lorentz invariant variables. In the simplest case $N = 4$, there is for each $2 \to 2$ channel (obtained by specifying the two initial and two final indices) a cut along $s = 4\mu^2 + \rho$, $\rho \geq 0$; $T_{2,2}$ and $T_{2,2}^-$ are plus $i\varepsilon$ and minus $i\varepsilon$ boundary values along the cut (from Im $s > 0$ and Im $s < 0$ respectively), $T_{2,2}^-$ being obtained from $T_{2,2}$ by analytic continuation around $s = 4\mu^2$. Equation (1) is thus a discontinuity formula, useful in "dispersion relations" ($=$ Cauchy theorem applied to T_4).

(b) "maximal" analytic continuations into "unphysical" sheets, e.g., at $N = 4$ across the real intervals separated by the $+\alpha$-Landau singularities $s \equiv 4\mu^2, 9\mu^2, 16\mu^2, \ldots$ (in each given $2 \to 2$ channel). Poles corresponding to "unstable" particles and other related singularities can occur in these unphysical sheets, generated, e.g., by the Fredholm-type equation $T_+ - T_- = T_+ T_-$ (equation (1) at $s < 9\mu^2$).

(c) Multiparticle dispersion relations have been proposed also at $N > 4$, e.g., $N = 6$. They involve a much larger number of discontinuity formulae, now differences between boundary values of analytic continuations "above" or "below" various sets of "normal threshold" singularities. The discontinuities are again expressed only in terms of physical scattering functions, as sums of on-mass-shell convolution integrals.

2.4. Nature of Singularities and Holonomicity Properties

In view of previous results, singularities associated with a particular class of graphs have been shown (under some conditions) to be regular holonomic. In particular, $T_{2,2}$ has a (holonomic) two-sheeted square-root type singularity at the 2-particle threshold $s = 4\mu^2$ if the dimension d of space-time is even, as follows from the equation $T_+ - T_- = T_+ T_-$. However, this cannot be expected in general. E.g., in the case just mentioned, $T_{2,2}$ is non-holonomic at $s = 4\mu^2$ if d is odd. It is then shown to be locally an infinite (convergent) sum of regular holonomic terms of the form $a_n(\ln[s - 4\mu^2])^n$, $n = 1, 2, \ldots$, with a_n locally analytic. The simplest example of a non-holonomic singularity at d even is the 3-particle threshold $s = 9\mu^2$ in a $2 \to 2$ or $3 \to 3$ process. Sato's conjecture on holonomicity properties of the S matrix has thus to be understood in the sense of expansions in terms of specified holonomic contributions. This is in agreement with the perturbative viewpoint of Feynman integrals and appears in a more precise way in Sections 3.6 and 4.2. In the previous case, the Landau singularity $s = 4\mu^2$ is generated by the graphs ⋈, ⋙, ⋙⋙, ... and different structures, with higher and higher powers of ln, will correspond to each of them at d odd. Non-holonomicity will more generally result from an accumulation of terms with different structures, corresponding either to a common singularity or to an accumulation of singularities, possibly in successive sheets.

3. Axiomatic Field Theory

3.1. Axioms and Haag–Ruelle Theory

A local interacting field A is an operator-valued distribution which to each test function f of the space-time variable $x = (x_0, \mathbf{x})$ where x_0 is time, associates an operator $A(f)$ acting in an Hilbert space $\mathcal{H}$ of states. The *Wightman axioms* include microcausality, Lorentz invariance and positivity of the energy. Microcausality asserts that $[A(x), A(y)] = 0$ if $u = x - y$ is space-like ($u^2 \equiv u_0^2 - \mathbf{u}^2 < 0$). It is assumed that there is a continuous unitary representation of the inhomogeneous Lorentz group $g \to U(g)$ acting on $\mathcal{H}$, with a related property of the fields, and a unique "vacuum" vector Ω (up to scalar multiples) invariant under $U(g)$, in the domain of any polynomial in the field operators and such that the subspace generated by vectors of the form $A(f_1) \cdots A(f_n)\Omega$ is dense in $\mathcal{H}$. Positivity of the energy is the assertion that the energy-momentum generators $(H = P_0, \mathbf{P})$ of the subgroup of space-time translations have spectrum in the forward cone $p^2 (\equiv p_0^2 - \mathbf{p}^2) \geq 0$, $p_0 > 0$.

For later purposes, a more detailed spectral condition, including the existence of a "mass gap" and isolated hyperboloids, is needed: spectrum will be, e.g., assumed to be composed of the origin, a positive energy hyperboloid $p^2 = \mu^2$, $p_0 > 0$ for some $\mu > 0$ and a continuum $p^2 > M$, $p_0 > 0$, $M > \mu$, with M, e.g., equal to $4\mu^2$ (see below: this will correspond to a theory with one type of particle, of mass $\mu > 0$ and no other type of stable particle that might be generated by products of fields). Haag–Ruelle theory then allows one to define in $\mathcal{H}$ subspaces $\mathcal{F}_{\text{in}}$, $\mathcal{F}_{\text{out}}$ corresponding to particles of mass μ and a corresponding S matrix. The axiom of asymptotic completeness will assert that $\mathcal{H} = \mathcal{F}_{\text{in}} = \mathcal{F}_{\text{out}}$.

The Wightman N-point functions $W_N(x_1, \ldots, x_N)$ are the distributions $\langle \Omega, A(x_1) \ldots A(x_N)\Omega \rangle$. Related functions involving various combinations of products of fields are useful below.

3.2. The Linear Program

The linear program is based on the interplay of support properties in x-space and in p-space provided by microcausality and the spectral condition respectively. It yields for each N the existence of an N-point function $\underline{F}_N(p_1, \ldots, p_N)$ analytic in a "primitive domain" which is a union of tubes in $\mathbf{C}^{(N-1)d}$ ($p_k \in \mathbf{C}^d$, $\Sigma p_k = 0$), specifying where the $\text{Im } p_k$ are ($\text{Re } p_k$ arbitrary). Real points do not belong in general to these domains but to their closures. Boundary values obtained from various tubes coincide in certain real regions. This leads to a first trivial improvement of the analyticity domain, which will contain in particular *euclidean space* (purely imaginary energy components and real momentum components). One boundary value of $\underline{F}_N$, from directions that may vary with the real region considered, is the distribution F_N mentioned in Section 1. Others will correspond to part of the boundary values needed in multiparticle dispersion relations (see Section 2.3).

The derivation of analyticity properties *on the complex mass-shell*, which are those of physical interest, is nontrivial because the intersection of the primitive domain with the complex mass-shell is empty. At $N = 4$, techniques of holomorphy envelopes allow one to enlarge this domain and to establish physical sheet analyticity properties on the complex mass-shell (see Section 2.3) that include the crossing property (analytic continuation between crossed physical regions) and dispersion relations. This is not so at $N > 4$ and in fact, as mentioned in Section 2.1, $T_{m,n}$ cannot be expected to be the boundary value of an analytic function (from the complex mass-shell) at all physical points. A more subtle microlocal analysis has provided results on the microsupport of F_N that confirm that mass-shell restrictions exist, with related information on their microsupports. Results obtained at that stage remain, however, very weak compared to expected ones, and, e.g., do not include the macrocausality property of the S matrix (Section 2) in terms of particles or a related property that can be conjectured on F_N.

3.3. Nonlinear Program: From Microcausality to Macrocausality

To get better results, including suitable macrocausality and macrocausal factorization properties, recourse to AC (asymptotic completeness) and to regularity assumptions is needed. AC provides relations which, as the unitary relations of Section 2, involve on-mass-shell convolution integrals. For $2 \to 2$ processes in the low energy region, AC relations coincide with equation (1) except that *external* energy-momenta are allowed to vary off-shell. Starting, e.g., from the assumption that $T_{2,2}$ is continuous, results analogous to those given in Section 2 are then obtained for T_4 and in turn for F_4. For $3 \to 3$ processes in the low-energy region, AC relations have then been shown, via microlocal analysis, to yield structure equations of the form:

$$ F = \sum \;\raisebox{-0.5ex}{\scriptsize\boxed{}}\; + \sum \;\raisebox{-0.5ex}{\scriptsize\boxed{}}\; + R \qquad (3) $$

where individual terms are Feynman-type convolution integrals with "complete propagators" ($= 2$-point functions with a pole at the physical mass μ) on internal lines, and R has specified analyticity properties (outside in particular the 3-particle threshold $s = 9\mu^2$). Each explicit term is singular along a corresponding $+\alpha$-Landau surface and is along that surface a plus $i\varepsilon$ boundary value in the sense of Section 2 on shell or in a similar sense off-shell. However, problems remain in this approach and it is not clear how it can be extended to more general cases.

3.4. Nonlinear Program: Irreducible Kernels

We now outline another method, based on "irreducible kernels," which has produced analogous results with some more detailed information and seems potentially better adapted to a more general analysis. (It involves *a priori* somewhat stronger regularity assumptions). Irreducible kernels in perturbative theory are formal sums of Feynman integrals of graphs with specified properties: e.g., a graph is *r*-particle irreducible in a given channel if it cannot be separated into two parts with respect to that channel by cutting *r* internal lines or less. At $N = 4$, the 2-*p.i.* Bethe–Salpeter kernel G in a given $2 \to 2$ channel satisfies perturbatively, in a theory without renormalization, the equation

$$F = G + \;\;\equiv\!\!\bigcirc\!\!\!\!-\!\!\!\!\bigcirc\!\!\equiv \tag{4}$$

In the axiomatic framework, a class of irreducible kernels has been defined by integral Fredholm-type equations, e.g., the B.S. kernel G is defined by equation (4), with possibly some analytic cut-off factor on internal lines in order to cover theories with renormalization (in which case G does not have a simple perturbative content). An algebraic equivalence modulo possible poles is then established between AC in the low-energy region and the 2-particle irreducibility of G in the analytic sense, namely analyticity in complex momentum space in a "strip" around euclidean space that goes up to the 3-particle threshold in real (Minkowski) space. This analyticity of G when reinjected in equation (4), together with properties of the 2-point function occurring on internal lines (isolated pole at the physical mass, i.e. at $s = \mu^2$), then yields (via a suitable extension of Fredholm theory in complex space) results on F_4 analogous to those found in Section 3.3 (meromorphy in a two-sheeted or multisheeted domain around the 2-particle threshold, . . .).

Structure equations analogous to (3), with more detailed results, have then been obtained by an extension of this method. However, technical and conceptual problems have prevented so far the analysis of more general situations (apart from preliminary results). A further heuristic analysis will be outlined in Section 3.6.

3.5. Global Analyticity for Multiparticle Processes

By combining methods of the linear and nonlinear programs, further results on global analyticity properties of the S matrix on the complex mass-shell, such as the existence of asymptotic crossing domains, have been obtained in particular at $N = 5$.

On the other hand, a subset of the discontinuity formulae needed in multiparticle dispersion relations follows from algebraic formulae of the linear program together with AC, provided one can ultimately show that the boundary values involved (= those mentioned in Section 3.1) can be realized as suitable limits from within the complex mass-shell.

3.6. Heuristic Investigations

The following heuristic analysis has been proposed on semi-perturbative bases. Formal expansions in terms of Feynman-type integrals with suitable irreducible kernels at each vertex and 2-point functions on internal lines have been conjectured. They are a generalization (with new features) of the Neumann series $F = G + \;\text{—}\!\bigcirc\!\text{—}\!\bigcirc\!\text{—}\; + \;\text{—}\!\bigcirc\!\text{—}\!\bigcirc\!\text{—}\!\bigcirc\!\text{—}\; + \;....$ of F in equation (4) and should be well suited to the analysis of higher and higher energy regions. In view of the analyticity properties assumed for irreducible kernels, each individual integral (which is analytic in euclidean energy–momentum space) should, after analytic continuation away from euclidean space, have a specified analytic and monodromic structure arising from the pole singularities of internal lines, close to that of Feynman integrals in the energy region of interest. (It should also be regular holonomic there.) Moreover, conjectures on discontinuities of individual integrals have been proposed and shown to yield, formally, relations that characterize AC in that region or also other properties (such as discontinuity formulae needed in multiparticle dispersion relations).

4.   Constructive Field Theory

4.1.   Definition of Models and Wightman Axioms

Models are defined in euclidean space-time by functional integrals. E.g., N-point functions are obtained for a $\lambda\varphi^4$ interaction in space-time dimension 2 as the $\Lambda \to \infty$ of the ratio $N_\Lambda(x_1, \ldots, x_N)/Z_\Lambda$ where Λ is a box in euclidean space-time,

$$N_\Lambda(x_1, \ldots, x_N) = \int \left[\prod_{i=1}^{N} \varphi(x_i) \right] e^{-\lambda \int_\Lambda : \varphi^4(x): \, dx} \, d\mu(\varphi), \tag{5}$$

and Z_Λ is defined similarly by removing the product $\prod \varphi(x_i)$. In (5), $d\mu$ is the measure of covariance C, $C(x, y) \equiv C(x - y)$ is the bare propagator, Fourier transform of $\tilde{C}(p) = -(p^2 + m^2)^{-1}$ with, in euclidean energy-momentum space, $p^2 = p_0^2 + \mathbf{p}^2$, p_0, $\mathbf{p}$ real (Note the change of notation); m is the "bare" mass, distinct from the physical mass to be generated. There is also *a priori* an analytic cut-off in euclidean energy–momentum space, i.e., $\tilde{C}(p)$ is replaced by

$$\tilde{C}_\rho(p) = \tilde{C}(p) \exp[-M^{-2\rho}(p^2 + m^2)], \qquad M > 1.$$

Perturbative expansions are easily recovered formally, at Λ and ρ infinite, but are divergent, however small the coupling λ is. Constructive theory makes use of more refined "cluster expansions" of N_Λ which are in some sense minimal expansions, e.g., with respect to a given lattice in Λ. This leads to expansions of connected N-point functions, in euclidean space-time, into sums of contributions associated with sets $\Delta_1, \ldots, \Delta_r$, r arbitrary, of squares of the lattice that vary independently in Λ, except that they must contain the external points $x_1, \ldots, x_N$, and are connected via propagator lines $C(u_l, v_l)$ and "Mayer" lines. (A Mayer line indicates that the two squares must coincide in Λ). For models $\lambda P(\varphi)$ in dimension $d = 2$, where P is a polynomial, this is sufficient (in view of the exponential decay of the propagators $C(u, v)$ as $u - v \to \infty$) to prove convergence of the expansions uniformly in Λ and to define the $\Lambda \to \infty$ limit, at sufficiently small coupling λ. The removal of the momentum cut-off ρ is relatively simple for these models, whereas "phase-space analysis" is needed more generally: e.g., C_ρ is decomposed as a sum $\sum_{i=1}^{\rho} C^{(i)}$, $\tilde{C}^{(1)}(p) = \tilde{C}_1(p)$,

$$\tilde{C}^{(i)}(p) = \{\exp[-M^{-2i}(p^2 + m^2)] - \exp[-M^{-2(i-1)}(p^2 + m^2)]\}\tilde{C}(p)$$

$$\text{for } i > 1.$$

Cluster expansions with respect to conveniently scaled lattices are then considered in each "momentum slice" i. More complicated expansions are obtained, and the $\Lambda \to \infty$ and $\rho \to \infty$ limits can still be treated after "renormalization," i.e., a suitable rearrangement of terms, at sufficiently small renormalized coupling.

Besides $\lambda P(\varphi)$ at $d = 2$, models defined so far include φ^4 at $d = 3$ and models involving more renormalization such as the massive Gross–Neveu model at $d = 2$, which has spin (and colour) indices and is in some sense an "asymptotically free" analogue of $\lambda \varphi^4$ at $d = 4$. There are many reasons to believe that the latter does not exist (i.e., is trivial) and that only more complicated "gauge theories" will exist, and be of physical interest, in dimension 4.

One also proves in euclidean space-time properties ("Osterwalder–Schrader axioms") which are shown, via analytic continuation to (real) Minkowski space-time (from imaginary to real times) to allow the reconstruction of a theory satisfying the Wightman axioms. However, the interest of this result (which includes at that stage no detailed information on spectrum) is limited as appears in Section 3 for our purposes, and there is today no direct approach to spectrum, or to asymptotic completeness.

4.2. Irreducible Kernels and Structure Equations

The approach to scattering developed so far then relies, as in Section 3.4, on the use of irreducible kernels, but analyticity properties of the latter will here be established independently of AC and be used in turn to obtain at the same time results on spectrum, AC and multiparticle analytic structure. Several methods to define irreducible kernels and prove their analyticity properties have been proposed. For conciseness, we only outline the most recent one, which is more general than previous ones.

Cluster expansions "of order p," $p \geq 1$, are introduced. General irreducible kernels are then defined in euclidean space-time by restricting sums involved in the expansions of connected functions to diagrams with corresponding graphical irreducibility properties. Uniform convergence and the existence of the $\Lambda \to \infty$, $\rho \to \infty$ limits are again established at small coupling, as also exponential fall-off (in euclidean space-time) of the form $e^{-(m-\varepsilon)l_{irr.}(x_1,\ldots,x_N)}$ where $l_{irr.}$ is the shortest length of irreducible graphs joining $x_1, \ldots, x_N$ and possibly intermediate points. Via Fourier–Laplace transformation, analyticity properties follow in turn in complex energy–momentum space.

By graphical inspection, structure equations of the type mentioned in Section 3.6 are also obtained in euclidean space, convergence of the expansions being established there at small coupling. The introduction of irreducible kernels and structure equations is thus a way of regrouping terms in the original pth-order expansions of connected functions which allows one to recover the euclidean convergence of the latter. This way is far too complicated if one is merely interested in the topics of Section 4.1 but is well suited to the analysis of scattering by analytic continuation from euclidean to Minkowski space, now in complex energy–momentum space, along the lines outlined in Section 3.6. (The choice of p depends on the energy region one wishes to explore.) In the simplest cases, the 2-point function on internal lines is $H_2(k) =$

$$ \underline{\quad} + -\!\!\textcircled{K}\!\!- + -\!\!\textcircled{K}\!\!-\!\!\textcircled{K}\!\!- + \cdots = [p^2 + m^2 - K(p)]^{-1}, \text{ where } \underline{\quad} $$

is the bare propagator $(p^2 + m^2)^{-1}$ and K is the $1\,p.i.$ 2-point function, shown to be analytic up to $(2m)^2 - \varepsilon$ and bounded in modulus by $\mathrm{Cst}|\lambda|$: hence H_2 has an isolated pole at small λ whose position defines the physical mass.

Remaining problems include the following. First, the analysis of the multisheeted structure of individual Feynman-type (or even Feynman) integrals is still based in general on conjectures. Progress in the derivation of some of these has been made recently, but a general understanding remains a challenge. Secondly, convergence of the structure equations has so far not been established in general away from euclidean space. In fact, even in the simplest case of the Neumann series

$$F = G + \underline{GG} + \ \dots$$

specific kinematical factors in dimension $d = 2$ or 3 yield *divergence*, however small the coupling λ is, in the neighborhood of the 2-particle threshold $s = 4\mu^2$. As studied in the seventies and recently with more general methods, this is linked with the possiblity of new stable particles ("2-particle bound states") in some models. Finally, all results hold only at small coupling. While this type of limitation (or related ones) is standard so far in constructive theory (to ensure convergence properties), coupling constants have moreover in the present discussion to be smaller and smaller as one wishes to explore higher and higher energy regions.

References

For conciseness, we only indicate some recent references, as also some books or review articles, in which original works are indicated and due credit to other authors is given.

Sections 1, 2   G. F. Chew, *The Analytic S Matrix*, Benjamin, New York (1966).

R. J. Eden, P. V. Landshoff, D. I. Olive, and J. C. Polkinghorne, *The Analytic S Matrix*, Cambridge Univ., 1966.

H. P. Stapp, *Structural Analysis of Collision Amplitudes*, R. Balian and D. Iagolnitzer, eds., North-Holland, 1976, p. 191 and 275.

D. Iagolnitzer, *The S Matrix*, North-Holland, 1978, and *New Developments in Mathematical Physics*, H. Mitter and L. Pittner, eds., Springer-Verlag, 1981, p. 235.

M. Kashiwara and T. Kawai, *Complex Analysis, Microlocal Calculus and Relativistic Quantum Theory*, Lecture Notes in Physics 126 (D. Iagolnitzer, ed.), Springer-Verlag, 1980, p. 5.

Section 3.1   R. F. Streater and A. S. Wightman, *PCT, Spin-Statistics and All That*, Benjamin, 1964.

R. Jost, *The General Theory of Quantized Fields*, Am. Math. Soc. (1965).

Section 3.2   H. Epstein, V. Glaser, and R. Stora, *Structural Analysis* . . . (op. cit.), p. 5.

Section 3.3   H. Epstein, V. Glaser, and D. Iagolnitzer, *Commun. Math. Phys.* **84** (1981) 99.

Section 3.4   J. Bros and M. Lassalle, *Structural Analysis* . . . (op. cit.), p. 97.

                   J. Bros, *New Developments* . . . (op. cit.), p. 329, and *Physica* **125A** (1984), 145.

Section 3.5   J. Bros, *Physics Reports* **134**, 5-6, 325 (1986) (in memoriam V. Glaser).

Section 3.6   D. Iagolnitzer, *Fizika* **17**, 3 (1985) 361.

Section 4.1   J. Glimm, A. Jaffe, *Quantum Physics*, Springer, 1981.

                   V. Rivasseau, *VIIth International Conference on Mathematical Physics*, M. Mebkhout and R. Seneor, eds., World Scientific, 1987, p. 257.

Sections 4.1, 4.2 D. Iagolnitzer and J. Magnen, *Commun. Math. Phys.* **110** (1987), 51, **111** (1987) 81.

                   D. Iagolnitzer, *VIIIth International* . . . (op. cit.), p. 257.

b-Functions and p-adic Integrals

Jun-ichi Igusa*

The Johns Hopkins University
Department of Mathematics
Baltimore, Maryland

We shall summarize some results on complex-valued p-adic zeta functions which are related to b-functions. We shall also explain a role of such zeta functions in the adelic situation.

1.

If $s, x_1, \ldots, x_n$ are letters, F is a field of characteristic 0 and $f(x)$ is an element of $F[x_1, \ldots, x_n]$ other than 0, there exists an element P of the associative $F[s]$-algebra $F[s, x_1, \ldots, x_n, \partial/\partial x_1, \ldots, \partial/\partial x_n]$ satisfying

$$Pf(x)^{s+1} = b(s)f(x)^s$$

for some $b(s)$ in $F[s]$ of degree $d \geq 0$, in which $f(x)^s$ is a symbol such that

$$\frac{\partial}{\partial x_i}\{\phi(s, x)f(x)^s\} = \left\{\frac{\partial\phi}{\partial x_i} + s\phi(s, x)f(x)^{-1}\frac{\partial f}{\partial x_i}\right\}f(x)^s, \qquad 1 \leq i \leq n$$

* This work was partially supported by the National Science Foundation.

for every $\phi(s, x)$ in $F[s, x_1, \ldots, x_n, f(x)^{-1}]$; cf. Bernstein [1]. We assume that d is the smallest and also that the coefficient of s^d in $b(s)$ is 1. This is the b-function of $f(x)$; it was introduced in the 1960s by Sato for an important class of $f(x)$. If $f(x)$ is not in F, then $b(s)$ is divisible by $s+1$. If further $f(x)$ and $\partial f/\partial x_i$ for $1 \leq i \leq n$ have no common zeros over an algebraic closure of F, then $b(s) = s + 1$. A basic theorem by Kashiwara [11] states that if

$$b(s) = \prod_\lambda (s + \lambda),$$

then all λ's are positive rational numbers. We keep in mind that $b(s)$ remains the same under any extension of F.

In the case where, e.g., $F = \mathbf{C}$, if s is a complex variable in the right-half plane and Φ is an arbitrary element of the Schwartz space $\mathscr{S}(\mathbf{C}^n)$, then

$$\prod_\lambda \Gamma(s + \lambda)^{-1} \cdot \int_{\mathbf{C}^n} |f(x)|_{\mathbf{C}}^s \Phi(x) \, dx,$$

where $|a|_{\mathbf{C}} = a\bar{a}$ for any a in $\mathbf{C}$, has a holomorphic continuation to the whole s-plane. This follows from

$$\int_{\mathbf{C}^n} |f(x)|_{\mathbf{C}}^s \Phi(x) \, dx = \prod_{1 \leq j \leq m} b(s + j - 1)^{-1} \cdot \int_{\mathbf{C}^n} |f(x)|_{\mathbf{C}}^s f(x)^m (Q_m \Phi)(x) \, dx,$$

in which Q_m is an element such as P for $m = 1, 2, 3, \ldots$. We have recalled this well-known fact because the above argument, which is at the basis of the connection of the b-function and the integral on the left-hand side, breaks down in the p-adic case. Therefore it seems quite remarkable that the b-function plays a similar role in the p-adic case.

2.

In order to motivate the p-adic case we start from an algebraic number field k. We know that k has, in addition to $\mathbf{R}$ and/or $\mathbf{C}$, infinitely many p-adic completions. As we emphasized in one of our early papers, they all have "equal rights" in mathematics. We therefore take any completion K of k as F. We have learned from Weil [20], pp. 157–158, that a generalization of the Schwartz space was given by Bruhat: Every locally compact abelian group $\mathscr{X}$ has the Schwartz–Bruhat space $\mathscr{S}(\mathscr{X})$ and its topological dual $\mathscr{S}(\mathscr{X})'$, called the space of distributions in $\mathscr{X}$. If $\mathscr{X}$ has arbitrarily large and small compact open subgroups, e.g., if $\mathscr{X}$ is a finite-dimensional vector space over a *p*-adic field, then $\mathscr{S}(\mathscr{X})$ consists of complex-valued locally constant functions Φ on $\mathscr{X}$ with compact support. We define $|a|_K$ for any a in K as the rate of Haar measure change in K under the multiplication by a with the understanding that $|0|_K = 0$. As for dx we simply take a Haar measure on K^n. Thus all data in

$$|f|_K^s(\Phi) = \int_{K^n} |f(x)|_K^s \Phi(x)\, dx$$

are explained and the integral clearly defines a holomorphic function on the right-half *s*-plane. A well-known theorem then states that $|f|_K^s$, as a distribution in K^n, has a meromorphic continuation to the whole *s*-plane, i.e., for every Φ in $\mathscr{S}(K^n)$ the complex-valued function $|f|_K^s(\Phi)$ has a meromorphic continuation to the whole *s*-plane with poles among a fixed set such that the order of each pole is bounded, in fact by n in this case. If K is a *p*-adic field, the theorem goes further: Let $\mathcal{O}_K$ denote the maximal compact subring of K and $\pi_K \mathcal{O}_K$ the ideal of nonunits of $\mathcal{O}_K$; then $\mathcal{O}_K / \pi_K \mathcal{O}_K$ is a finite field with say $q = q_K$ elements and for every Φ the above meromorphic function on the *s*-plane is a rational function of $t = q^{-s}$. This basic rationality was first proved in [5] by using Hironaka's theorem on desingularization; later a generalization has been given by Denef [2], in fact by a new method.

3.

We introduce a special $|f|_K^s(\Phi)$ for each K: According as K is **R**, **C** or a *p*-adic field, we take $\exp(-\pi{}^t xx)$, $\exp(-2\pi{}^t \bar{x}x)$ or the characteristic function of $\mathcal{O}_K^n$ as $\Phi(x)$ and denote the corresponding $|f|_K^s(\Phi)$ by $Z_K(s)$; we also normalize dx so that $Z(0) = 1$ and denote the normalized dx by dx_K. We know by experience that they are the exact analogues of each other. At any rate in the *p*-adic case

$$Z_K(s) = \int_{\mathcal{O}_K^n} |f(x)|_K^s\, dx_K,$$

in which $\mathrm{vol}(\mathcal{O}_K^n) = 1$. We now ask about the nature of $Z_k(s)$ as a rational function of $t = q^{-s}$.

If the coefficients of $f(x)$ are in $\mathcal{O}_K$, then $Z_K(s)$ can always be written as

$$Z_K(s) = N(t) \Big/ \prod_{E \in \mathscr{E}} (1 - q^{-n_E} t^{N_E})$$

for a finite set $\mathscr{E}$, in which n_E, N_E are positive integers and $N(t)$ is an element of $\mathbf{Z}[q^{-1}, t]$ satisfying $N(1) = \prod (1 - q^{-n_E})$; this is a part of our theorem. In the case where $f(x)$ is homogeneous, if $f(x)$ has a "good reduction mod π_K," a condition satisfied by almost all K if the coefficients of $f(x)$ are in k, then

$$\deg_t(Z_K(s)) = -\deg(f(x));$$

this remarkable relation has recently been proved by Denef [3]. A simple example is as follows:

If $f(x)$ is homogeneous of degree say d with coefficients in $\mathcal{O}_K$ and if $\tilde{f}(x)$ denotes its reduction mod π_K, then

$$Z_K(s) = \frac{(1)(n)t + (1 - q^{-n}N(0))(1 - t)}{(1 - q^{-1}t)(1 - q^{-n}t^d)},$$

in which $(j) = 1 - q^{-j}$ and $N(0)$ is the number of solutions of $\tilde{f}(x) = 0$ in $\mathbf{F}_q^n$, provided that $\partial \tilde{f}/\partial x_i$ for $1 \le i \le n$ do not have a common zero in $\mathbf{F}_q^n - \{0\}$. Since $(1)(n) \ne 1 - q^{-n}N(0)$, we indeed have $\deg_t(Z_K(s)) = 1 - (1 + d) = -d = -\deg(f(x))$.

If $f(x)$ is not homogeneous, the above relation need not be true. For instance, if $n = 2$ and $f(x) = x_1^3 + x_2^2$, then

$$Z_K(s) = (1) \frac{1 - q^{-2}t(1 - t) - q^{-5}t^5}{(1 - q^{-1}t)(1 - q^{-5}t^6)},$$

hence $\deg_t(Z_K(s)) = -2$ while $\deg(f(x)) = 3$. We might mention that in this case $b(s) = (s + 1)(s + 5/6)(s + 7/6)$.

4.

We shall now specialize $f(x)$ to a relative invariant of Sato's "regular prehomogeneous vector space" (G, X); cf. [18]. We take a connected reductive F-subgroup G of GL_n acting transitively on the complement Y of an absolutely irreducible F-hypersurface $f(x) = 0$ in $X = \text{Aff}^n$, the affine n-space; then $f(x)$ is homogeneous and $f(gx) = \nu(g)f(x)$ for every g in G with a rational character ν of G; and if $b(s)$ is the b-function of $f(x)$, then $\deg(f(x)) = \deg(b(s))$. Therefore if $F = k$, then $\deg_t(Z_K(s)) = -\deg(b(s))$ for almost all p-adic completions K of k by Denef. We further recall the following from Sato's theory:

If G_0 is a connected reductive F-group, ρ_0 is an F-representation of G_0 in Aff^{n_0} and $n_0 = r_1 + r_2$ is a partition of n_0, then $G_0 \times SL_{r_j}$ acts on $X_j = M_{n_0, r_j}$ as

$$\rho_1(g_0, g_1)x_1 = \rho_0(g_0)x_1 {}^t g_1, \qquad \rho_2(g_0, g_2)x_2 = {}^t \rho_0(g_0)^{-1} x_2 {}^t g_2.$$

We put $G_j = \mathrm{Im}(\rho_j)$ and assume that (G_j, X_j) satisfies the above condition for $f_j(x_j)$; then $\deg(f_1(x_1))/r_1 = \deg(f_2(x_2))/r_2 = d_0$. Furthermore if $b_j(s)$ is the b-function of $f_j(x_j)$ and if we assume, by summary, that $r_2 \geq r_1$, then

$$\frac{b_2(s)}{b_1(s)} = \prod_{r_1 < i \leq r_2} \prod_{0 \leq j < d_0} (s + d_0^{-1}(i+j));$$

this relation is quoted in [12], p. 78 as Shintani's theorem.

Now the fact is that if K is a p-adic field and if the coefficients of $f_j(x_j)$ are relatively prime, i.e., they are in $\mathcal{O}_K$ but not all in $\pi_K \mathcal{O}_K$, and if $Z_j(s)$ is the $Z_K(s)$ for $f_j(x_j)$, then

$$\frac{Z_2(s)}{Z_1(s)} = \prod_{r_1 < j \leq r_2} \frac{(1 - q^{-j})}{(1 - q^{-(d_0 s + j)})};$$

cf. [10]. Therefore if $Z_1(s)$ is known, so is $Z_2(s)$.

We consider the set of all (G, X) introduced in the beginning and identify F-isomorphic pairs. We then say that two such pairs are equivalent if they can be included in a finite sequence whose consecutive terms are as (G_1, X_1), (G_2, X_2) above. If for a moment we regard $\dim(G)$ as the height of (G, X), the going-down process by such a sequence is unique, hence every class contains the bottom pair, which is called reduced; cf. [18], pp. 39–40.

5.

In the case where G is irreducible as a matrix group and $F = \mathbf{C}$ the set of all reduced pairs has been divided into 29 types by Sato and Kimura [18] and the corresponding b-functions, except for one, are tabulated in Kimura [12], pp. 76–78. If we restrict to F-split groups, the absolute classification gives a relative classification over F. In the following we shall give a corresponding list of $Z_K(s)$ together with the set $\Lambda = \{\lambda\}$ for $b(s) = \prod (s + \lambda)$ assuming that the groups split over K; the numbering is the same as in op. cit. and $(j) = 1 - q^{-j}$ as before while $(j)_+ = 1 + q^{-j}$. Furthermore the coefficients of $f(x)$ are assumed to be relatively prime and q is odd in (II).

(I) Λ *consists of integers*

$\{j\}_{1\leq j\leq d}$ (Type 1), $\{2j-1\}_{1\leq j\leq d}$ (Type 3),

$\{2j-1, 2(m-j+1)\}_{1\leq j\leq d/2}$, $m\geq d$ and d even (Type 13),

$\{1, n/2\}$, n even (subtype of Type 15), $n=8$ (Type 16), $n=16$ (Type 19),

$\{1, 4, 5, 8\}$ (Type 20), $\{1, 5, 9\}$ (Type 27);

$$Z_K(s) = \prod_\lambda (\lambda)/(1-q^{-\lambda}t).$$

(II) *other types*

$\{(j+1)/2\}_{1\leq j\leq d}$ (Type 2);

$$Z_K(s) = \prod_{1\leq j\leq[d/2]} (2j-1)/(1-q^{-2j-1}t^2)\begin{cases}(d)/(1-q^{-1}t) & d \text{ odd} \\ (1-q^{-d-1}t)/(1-q^{-1}t) & d \text{ even};\end{cases}$$

$\{1, 1, 5/6, 7/6\}$ (Type 4);

$$Z_K(s) = (1)(2)(1-q^{-5}t^5)/(1-q^{-1}t)(1-q^{-2}t^2)(1-q^{-5}t^6);$$

$\{1, (j+3)/2, (2j+3)/2, (3j+4)/2\}$, $j=1$ (Type 14), $j=2$ (Type 5),

$j=4$ (Types 22 and 23), $j=8$ (Type 29);

$$Z_K(s) = (1)(4)(1-q^{-7}t)/(1-q^{-1}t)(1-q^{-4}t^2)(1-q^{-7}t^2) \text{ for } j=1,$$

$$Z_K(s) = (1)(3j/2+2)N_j(t)/D_j(t),$$

where

$$N_j(t) = (3j/2+2)_+ - q^{-j-3}(1+q^{-j/2}+q^{-j}-q^{-2j-2})t$$

$$+ q^{-3j/2-3}(1-q^{-j-2}-q^{-3j/2-2}-q^{-2j-2})t^2 + q^{-3j-6}(3j/2+2)_+t^3,$$

$$D_j(t) = (1-q^{-1}t)(1-q^{-j-3}t^2)(1-q^{-2j-3}t^2)(1-q^{-3j-4}t^2) \quad \text{for } j\neq 1;$$

$\{1, 2, 5/2, 7/2, 3, 4, 5\}$ (Type 6);

$$Z_K(s) = \prod_{j=1,3,5} (j)/(1-q^{-j}t) \cdot (4)(7)/(1-q^{-4}t^2)(1-q^{-7}t^2);$$

$\{(j+1)/2, (m-j+1)/2\}_{1\leq j\leq r}$, $m\geq 2r$, 3 (Type 15),

$m=7$ and $r=1$ (Type 25),

$m=7$ and $r=2$ (Type 26), $m=8$ and $r=2$ (Type 17),

$m=8$ and $r=3$ (Type 18);

$$Z_K(s) = D(1)N(t)/N(1)D(t), \text{ where}$$

$$D(t) = (1-q^{-1}t)\cdot \prod_{1\leq j\leq[r/2]} (1-q^{-2j-1}t^2)\cdot \prod_{1\leq j\leq r} (1-q^{-m+j-1}t^2)$$

in all cases and

$$
N(t) = \begin{cases}
\displaystyle\prod_{1 \le j \le [r/2]} (1 - q^{-m+2j-2} t^2) \begin{cases} (1 - q^{-m+r-1} t) & r \text{ odd} \\ (1 - q^{-r-1} t) & r \text{ even} \end{cases} m \text{ odd} \\[20pt]
\displaystyle\prod_{1 \le j \le [r/2]} (1 - q^{-m+2j-1} t^2) \cdot (1 + q^{-m/2} t) \qquad r \text{ odd and } m \text{ even,}
\end{cases}
$$

while $N(t)$ is of different nature if both m and r are even:

$$
N(t) = \prod_{1 \le j < r/2} (1 - q^{-m+2j-1} t^2) \cdot \{(m/2)_+
$$

$$
- (1 + q^{-m/2+r} + q^{-m+2r} - q^{-m+r}) q^{-r-1} t
$$

$$
+ (1 - q^{-r} - q^{-m/2} - q^{-m+r}) q^{-m/2-1} t^2 + (m/2)_+ q^{-m-2} t^3 \}.
$$

The above results were obtained on different occasions for various purposes. Except for Types 6 and 15, they are stated in [7], [8] with references and at least with indication of proof. Type 6 has recently been settled in [10], and Type 15 follows easily from some results in that paper, of which we shall later make some comments. Also some nonsplit cases were examined in op. cit. and in Robinson [16].

If G splits over K, then $Z_K(s)$ in the general case can be obtained from that in the reduced case by multiplying a finite number of factors of the form $1/(1 - q^{-j} t^{d_0})$. Therefore the numerator of $Z_K(s)$ in the reduced case can at most be cancelled. This never takes place for the cubic polynomials in Types 5, 15, 17, 22, 23 and 29. Therefore each of those bizarre polynomials is invariantly associated with the whole class.

Another remark is that in the above examples real poles of $Z_K(s)$ are zeros of $b(s)$. In the two-variable case this has recently become a theorem by Loeser [14]. Therefore, accepting for a moment that it is true in general, we can ask about the rule, if any, for the disappearance of some zeros of $b(s)$ as real poles of $Z_K(s)$. It is hardly necessary to say that such a rule is not only interesting but will considerably simplify the computation of $Z_K(s)$.

6.

We put $X = M_{m,r}$, where $m \ge r$, and we let $GL_m \times SL_r$ act on X as $\rho(g_1, g_2)x = g_1 x\, {}^t g_2$; further, we denote by $I(X)$ the affine variety with the ring of SL_r-invariants on X as its coordinate ring and by $i_X : X \to I(X)$ the corresponding morphism. Then GL_m is transitive on $I(X)' = I(X) - \{0\}$ and $GL_m(\mathcal{O}_K)$ is transitive on $U = I(X)'(\mathcal{O}_K)$ for any nonarchimedean local field K. Therefore U carries a $GL_m(\mathcal{O}_K)$-invariant measure di, which will be normalized as $\mathrm{vol}(U) = 1$.

Now let $f_0(i)$ denote a homogeneous element of the coordinate ring of $I(X)$ of degree d_0 with coefficients in K and put $f(x) = f_0(i_X(x))$; then

$$\int_{X(\mathcal{O}_K)} |f(x)|_K^s \, dx = \prod_{1 \leq j \leq r} \frac{(m-j+1)}{(1-q^{-m+j-1}t^{d_0})} \cdot \int_U |f_0(i)|_K^s \, di.$$

This is a special case of [10], Lemma 8, and it implies the formula for "$Z_2(s)/Z_1(s)$." The point is that if (G, X) is of Type 15, a case discussed in [17], pp. 168–172, we have the above formula with $d_0 = 2$. Furthermore, if q is odd, the integral over U can easily be computed by the uniform method explained in [7], p. 1025 and supplemented by some general results in [10].

We might mention that among the 29 types of Sato and Kimura there are only 9 types which are not included in our list of $Z_K(s)$. To these, except for Types 7 and 24, the above reduction of $Z_K(s)$ to an integral over U is applicable. Furthermore the computation of that integral in some of those types seems to be easy. However there are certain types for which the computation of $Z_K(s)$ will become enormously long, at least by our method.

7.

Although the fact that real poles of $Z_K(s)$ are zeros of $b(s)$ is still "experimental," we can prove the following weak statement: If every λ in $\Lambda - \{1\}$ is larger than 1, then $(1 - q^{-(s+1)})|f|_K^s$ is holomorphic for $\mathrm{Re}(s) \geq -1$, hence all real poles of $(s+1)Z_K(s)$ are smaller than -1. The proof is as follows.

We denote by F the extension of $\mathbf{Q}$ obtained by adjoining all coefficients of $f(x)$ and construct an F-desingularization $h: \tilde{X} \to X = \mathrm{Aff}^n$ of the hypersurface $f(x) = 0$ by Hironaka [4], p. 176. Then the divisors of $f \circ h$ and $h^*(dx_1 \wedge \cdots \wedge dx_n)$ on $\tilde{X}$ are of the form

$$\sum_{E \in \mathscr{E}} N_E E, \qquad \sum_{E \in \mathscr{E}} (n_E - 1)E,$$

in which E's are irreducible smooth subvarieties of X of codimension 1 meeting transversally on $\tilde{X}$ and N_E, n_E are positive integers. We associate a simplicial complex $\mathscr{N}(\mathscr{E})$ to $\mathscr{E}$, called its "nerve," equipped with the function $E \to (N_E, n_E)$ on the set of its vertices. We embed F in $\mathbf{C}$ and consider the distribution $|f|_{\mathbf{C}}^s$ on $\mathbf{C}^n$. The assumption on Λ then implies that

$(s+1)|f|_C^s$ is holomorphic for $\text{Re}(s) \geq -1$. This implies that $\mathcal{N}(\mathcal{E})$ has the property (P): For every simplex σ of $\mathcal{N}(\mathcal{E})$ we have $n_E \geq N_E$ for all E in σ such that $n_E = N_E$ for at most one E in σ with the consequence $n_E = N_E = 1$ for that E; cf. [6], p. 48 and p. 53. Then $(1 - q^{-(s+1)}) \cdot |f|_K^s$ is holomorphic for $\text{Re}(s) > -\min\{n_E/N_E; E \in \mathcal{E}, n_E > N_E\}$.

In this connection we might mention the following consequence of some results of Milnor and Malgrange in the case where the hypersurface $f(x) = 0$ in $\mathbf{C}^n$ for $n \geq 2$ has 0 as its only singularity: If the intersection of a small $(2n-1)$-sphere in $\mathbf{C}^n$ with center 0 and $f(x) = u$ for $u \neq 0$ very close to 0 in $\mathbf{C}$ is a homology sphere over $\mathbf{Q}$, then $\Lambda - \{1\}$ does not contain any integer. On the other hand, in the *p*-adic case, Langlands has given a formula in [13] for the principal parts of the Laurent expansions of $|f|_K^s$ around its poles on the *t*-plane by using principal-value integrals, in which the integrands are defined in terms of local expressions on $\tilde{X}$ of $f \circ h$ and $h^*(dx_1 \wedge \cdots \wedge dx_n)$.

8.

We shall now consider the adelic situation: We denote by Σ the set of all $|\ |_K$, where K is a completion of a number field k, and use Σ as a set of subscripts, e.g., k_v, $\mathcal{O}_v$, q_v instead of K, $\mathcal{O}_K$, q_K. We also use v (resp. A) to denote the localization (resp. adelization) relative to k, e.g., $X_v = k_v^n$ (resp. $X_A = k_A^n$) for $X = \text{Aff}^n$; we put $|a|_A = \Pi |a_v|_v$ for any $a = (a_v)$ in $(GL_1)_A$. We define Σ_f to be the set of all v in Σ such that k_v is a *p*-adic field and denote by Σ_∞ the complement of Σ_f in Σ. We recall that every Φ in $\mathscr{S}(X_A)$ is of the form

$$\Phi = \Phi_S \otimes \left(\bigotimes_{v \notin S} \Phi_v \right),$$

in which S is a finite subset of Σ containing Σ_∞ such that Φ_S is in the Schwartz–Bruhat space of the product of X_v for all v in S and Φ_v is the characteristic function of $X(\mathcal{O}_v) = \mathcal{O}_v^n$. If $f(x)$ in $k[x_1, \ldots, x_n]$ is absolutely irreducible and $(1)_v = 1 - q_v^{-1}$, the product of $(1)_v^{-1} \cdot Z_v(s)$ for all v not in S is convergent for $\text{Re}(s) > 0$ by Ono [15]. We might recall that this is based on the Lang–Weil estimate on the number of $\mathbf{F}_q$-rational points of an $\mathbf{F}_q$-variety. Therefore

$$\|f\|^s = \left(\bigotimes_{v \in \Sigma_f} (1)_v^{-1} |f|_v^s \right) \otimes \left(\bigotimes_{v \in \Sigma_\infty} |f|_v^s \right)$$

defines a distribution in X_A which is holomorphic on the right-half s-plane. Furthermore, the theorem on $|f|_v^s$ shows that $\|f\|^s(\Phi)$ has a meromorphic continuation for every Φ in $\mathscr{S}(X_A)$ if and only if the product of $(1)_v^{-1} \cdot Z_v(s)$ for almost all v in Σ_f has one. This explains one role of $Z_v(s)$ in the adelic situation.

In the regular prehomogeneous case, i.e., if a connected reductive k-subgroup G of GL_n acts transitively on $Y = X - f^{-1}(0)$ so that $f(gx) = \nu(g)f(x)$ for every g in G, we can introduce

$$Z(s)(\Phi) = \int_{G_A/G_k} \left(\sum_{\xi \in Y_k} \Phi(g\xi) \right) |\nu(g)|_A^s \, dg,$$

in which dg is a Haar measure on G_A. If G^1 is the kernel of ν, $d^1 g$ is a Haar measure on G_A^1 and $\Phi^g(x) = \Phi(gx)$ for every g in G_A, then $Z(s)(\Phi)$ is a Mellin transform of $I'(\Phi^g)$, where

$$I'(\Phi) = \int_{G_A^1/G_k^1} \left(\sum_{\xi \in Y_k} \Phi(g\xi) \right) d^1 g.$$

The importance of this kind of integral in number theory has been emphasized in Weil [21], p. 14. However even if we assume that G is irreducible so that G^1 becomes semisimple, we, or rather I, do not know any criterion for the convergence of $I'(\Phi)$ for every Φ in $\mathscr{S}(X_A)$. Therefore we cannot really talk about $Z(s)(\Phi)$ in general. The situation becomes different if we replace Y in $I'(\Phi)$ by X and denote the corresponding integral by $I(\Phi)$: The sufficient condition in Weil op. cit., p. 20 for the convergence of $I(\Phi)$ for every Φ in $\mathscr{S}(X_A)$ is also necessary and we have a classification of all k-split semisimple groups G^1 for which the criterion is satisfied. Furthermore the convergence remains valid even if we replace G^1 by a k-form of G^1 and $Z(s)$, as a distribution in X_A, has a meromorphic continuation to the whole s-plane satisfying the standard functional equation; the argument is the same as in Tate's case where $G = GL_1$.

Now G is transitive on Y by assumption, hence G_v has a finite number of orbits in Y_v by Serre [19], III-33. However the number of G_A-orbits in Y_A may be infinite. We have shown in [9] that if G_A has a finite number of orbits in Y_A, the above $Z(s)$ not only exists but is related to $\|f\|^s$ as follows:

$$Z(s) = \text{const.} \|f\|^{s-\kappa}, \qquad \kappa = \dim(X)/\deg(f(x));$$

this is in the case where G is irreducible. We can easily show that $Z(s)$ has a meromorphic continuation to the whole s-plane and satisfies the standard functional equation if either G splits over k or (G, X) is reduced.

The above relation is subject to a minor restriction if (G, X) is of Type 1: We may have to restrict the arbitrary connected irreducible k-group which appears in G as an almost direct factor. It is very unlikely that such a relation exists if the number of G_A-orbits in Y_A is infinite. At any rate the Euler factor $(1)_v^{-1} \cdot Z_v(s - \kappa)$ is known if G splits over k_v because (G, X) in the reduced case is either one of the eight types with Λ consisting of integers or Type 6. We might say that this is another role of $Z_v(s)$ in the adelic situation.

References

[1] I. N. Bernstein, The analytic continuation of generalized functions with respect to a parameter, *Functional Analysis and its Applications* **6** (1972) 273–285.

[2] J. Denef, The rationality of the Poincaré series associated to the *p*-adic points on a variety, *Invent. Math.* **77** (1984) 1–23.

[3] J. Denef, On the degree of Igusa's local zeta function, *Amer. J. Math.* **109** (1987).

[4] H. Hironaka, Resolution of singularities of an algebraic variety over a field of characteristic zero. I-II, *Ann. Math.* **79** (1964) 109–326.

[5] J. Igusa, Complex powers and asymptotic expansions. I, *Crelles J. Math.* **268/279** (1974) 110–130; II. *Ibid.* **278/279** (1975) 307–321; or *Forms of Higher Degree*, Tata Inst. Lect. Notes 59, Springer-Verlag (1978).

[6] J. Igusa, Criteria for the validity of a certain Poisson formula, in: *Algebraic number theory*, Jap. Soc. Prom. Sci. (1977) 43–65.

[7] J. Igusa, Some results on *p*-adic complex powers, *Amer. J. Math.* **106** (1984) 1013–1032.

[8] J. Igusa, On functional equations of complex powers, *Invent. Math.* **85** (1986) 1–29.

[9] J. Igusa, Zeta distributions associated with some invariants, *Amer. J. Math.* **109** (1987) 1–34.

[10] J. Igusa, On the arithmetic of a singular invariant, preprint.

[11] M. Kashiwara, *B*-functions and holonomic systems (Rationality of *b*-functions), *Invent. Math.* **38** (1976) 33–53.

[12] T. Kimura, The *b*-functions and holonomy diagrams of irreducible regular prehomogeneous vector spaces, *Nagoya Math. J.* **85** (1982), 1–80.

[13] R. P. Langlands, Orbital integrals on forms of $SL(3)$. I, *Amer. J. Math.* **105** (1983) 465–506.

[14] F. Loeser, Fonctions d'Igusa *p*-adiques et polynômes de Bernstein, *Amer. J. Math.* **109** (1988).

[15] T. Ono, An integral attached to a hypersurface, *Amer. J. Math.* **90** (1968) 1224–1236.

[16] M. M. Robinson, On the complex powers associated with the twisted cases of the determinant and the Pfaffian, Thesis, Johns Hopkins (1986).

[17] M. Sato, M. Kashiwara, T. Kimura, and T. Oshima, Micro-local analysis of prehomogeneous vector spaces, *Invent. Math.* **62** (1980) 117–179.

[18] M. Sato and T. Kimura, A classification of irreducible prehomogeneous vector spaces and their relative invariants, *Nagoya Math. J.* **65** (1977) 1–155.

[19] J.-P. Serre, *Cohomologie Galoisienne*, Lect. Notes in Math. 5, Springer-Verlag (1965).

[20] A. Weil, Sur certains groupes d'opérateurs unitaires, *Acta Math.* **111** (1964) 143–211; *Collected Papers III*, 1–69.

[21] A. Weil, Sur la formule de Siegel dans la théorie des groupes classiques, *Acta Math.* **113** (1965) 1–87; *Collected Papers III*, 71–157.

On the Poles of the Scattering Matrix for Several Convex Bodies

Mitsuru Ikawa

Department of Mathematics
Osaka University
Osaka, Japan

1. Introduction

Let $\mathcal{O}$ be an open bounded set in $\mathbf{R}^3$ with smooth boundary Γ. We set $\Omega = \mathbf{R}^3 - \bar{\mathcal{O}}$. Suppose that Ω is connected. Consider the following acoustic problem with Neumann boundary condition:

$$\Box u(x, t) = \frac{\partial^2 u}{\partial t^2} - \Delta u = 0 \qquad \text{in } \Omega \times (-\infty, \infty),$$

$$B u(x, t) = 0 \qquad \text{on } \Gamma \times (-\infty, \infty),$$

$$u(x, 0) = f_1(x),$$

$$\frac{\partial u}{\partial t}(x, 0) = f_2(x),$$

(1.1)

where $\Delta = \sum_{j=1}^{3} \partial^2 / \partial x_j^2$ and $B = \sum_{j=1}^{3} n_j(x)\, \partial / \partial x_j$, $n(x) = (n_1(x), n_2(x), n_3(x))$ being the unit outer normal of Γ at x.

Copyright © 1988 by Academic Press, Inc.
All rights of reproduction in any form reserved.
ISBN 0-12-400465-2

Question. How is the location of the poles of the scattering matrix $\mathscr{S}(z)$ (see, [LP 1, page 9], [LP 3]) related to the geometry of $\mathcal{O}$?

Concerning the above question, it was proved that, when $\mathcal{O}$ is nontrapping (see [LP 1, page 155] and [MeS]), there exist positive constants a and b such that $\{z \in \mathbf{C}; \operatorname{Im} z < a \log|z| + b\}$ is free for poles ([LP 1], [MoRSt], [Me]). For trapping obstacles, Lax and Phillips conjectured in [LP 1, page 158] (see also [R]) that $\mathscr{S}(z)$ has a sequence of poles converging to the real axis. But a counterexample to the conjecture was found in [Ik 1]. Thus we would like to modify Lax and Phillips's conjecture as follows: *when $\mathcal{O}$ is trapping there exists $\alpha > 0$ such that a slab domain $\{z \in \mathbf{C}; 0 < \operatorname{Im} z < \alpha\}$ contains an infinite number of poles.* If this conjecture is correct, the existence of such α can be a characterization of trapping obstacles.

As to the above conjecture, there are only a few examples for which is shown the existence of such α. To my best knowledge, it is proved only for obstacles composed of two disjoint convex bodies ([Ik 1, 2, 3], [G]).

We shall consider in this paper the case that $\mathcal{O}$ is composed of several disjoint strictly convex bodies. More precisely, let $\mathcal{O}_j, j = 1, 2, \ldots, J$ be open bounded sets in $\mathbf{R}^3$ with smooth boundary Γ_j, and let

$$\mathcal{O} = \bigcup_{j=1}^{J} \mathcal{O}_j. \tag{1.2}$$

We assume the following:

(H.1) Every $\mathcal{O}_j$ is strictly convex. That is, the Gaussian curvature of Γ_j is positive everywhere.

(H.2) For all $\{j_1, j_2, j_3\} \in \{1, 2, \ldots, J\}^3$ such that $j_l \neq j_h$ if $l \neq h$, the convex hull of $\bar{\mathcal{O}}_{j_1}$ and $\bar{\mathcal{O}}_{j_2}$ has no intersection with $\bar{\mathcal{O}}_{j_3}$.

Theorem. *Suppose that $\mathcal{O}$ is of the form (1.2) and satisfies the hypothesis (H.1) and (H.2). Then there exists a positive number α such that a slab domain $\{z \in \mathbf{C}; 0 < \operatorname{Im} z < \alpha\}$ contains an infinite number of the poles of $\mathscr{S}(z)$.*

The essential difference between the case of two bodies and the case of more than three should be mentioned. In the geometric point of view, in the case of two bodies there is only one primitive periodic ray in Ω. On the other hand, in the case of more than three bodies there exists an infinite number of the primitive periodic rays in Ω. This fact causes the behavior of the solution of (1.1) to be very complicated.

Thus the method for two bodies used in [Ik 2, 4] and [G], which could determine precisely the position of the poles, is not applicable to this case.

Therefore we shall use a trace formula due to Bardos, Guillot and Ralston [BGR]. The main point of the proof is an estimate from below of the trace of the distribution kernel of the fundamental solution of (1.1). For this estimate, we shall use essentially the fact that the boundary condition is of Neumann type. In the case of the Dirichlet condition we have not yet succeeded in obtaining such an estimate. Thus, we do not know at the present time how to extend the theorem to the case of Dirichlet condition.

2.   On the Trace Formula and the Reduction of the Problem

We denote by $-A$ the self-adjoint realization in $L^2(\Omega)$ of the Laplacian in Ω with the boundary condition $Bu = 0$ on Γ, and by $-A_0$ the self-adjoint realization in $L^2(\mathbf{R}^3)$ of the Laplacian in $\mathbf{R}^3$. Then a trace formula

$$\mathrm{tr}_{L^2(\mathbf{R}^3)} \int_{\mathbf{R}} \rho(t)(\cos t\sqrt{A}\oplus 0 - \cos t\sqrt{A_0})\, dt = \sum_{j=1}^{\infty} \hat{\rho}(\lambda_j) \qquad (2.1)$$

holds for all $\rho \in C_0^\infty(0, \infty)$, where

$$\hat{\rho}(\lambda) = \int e^{i\lambda t}\rho(t)\, dt,$$

and $\{\lambda_j\}_{j=1}^{\infty}$ is a numbering of all the poles of $\mathscr{S}(z)$. Here $\cos t\sqrt{A}\oplus 0$ is an operator in $L^2(\mathbf{R}^3)$ defined by

$$(\cos t\sqrt{A}\oplus 0)(x) = \begin{cases} (\cos t\sqrt{A}\chi f)(x) & \text{for } x \in \Omega \\ 0 & \text{for } x \in \mathscr{O} \end{cases}$$

for $f \in L^2(\mathbf{R}^3)$, where χ denotes the characteristic function of Ω. In [BGR] (2.1) is proved for the Dirichlet condition, but their argument is also valid for the Neumann condition (see also [Me]).

Let $\rho(t) \in C_0^\infty(-1, 1)$ such that $\rho \geq 0$ for all $t \in \mathbf{R}$ and

$$\rho(0) = 1, \quad \rho(-t) = \rho(t) \qquad \text{for all } t \in \mathbf{R},$$

$$\hat{\rho}(\lambda) \geq 0 \qquad \text{for all } \lambda \in \mathbf{R}.$$

Define $\rho_q(t)$, $q = 1, 2, \ldots$ a sequence of functions in $C_0^\infty(0, \infty)$ by

$$\rho_q(t) = \rho(m_q(t - l_q)),$$

where $\{m_q\}_{q=1}^{\infty}$ and $\{l_q\}_{q=1}^{\infty}$ are sequences of positive numbers such that

$$l_q \to \infty \qquad \text{as } q \to \infty, \tag{2.2}$$

$$m_q \to \infty \qquad \text{as } q \to \infty, \tag{2.3}$$

which will be determined later. Evidently we have

$$\text{supp } \rho_q \subset [l_q - m_q^{-1}, l_q + m_q^{-1}], \qquad \rho_q(l_q) = 1. \tag{2.4}$$

Modifying the proof of Lemma 2.1 of [Ik 2] we have

Lemma 2.1. *Suppose that*

$$\#\{j; \text{ Im } \lambda_j \le \alpha\} = P(\alpha) < \infty.$$

Then it holds that

$$\sum_{j=1}^{\infty} |\hat{\rho}(\lambda_j)| \le C_\alpha m_q^4 \, e^{-\alpha l_q} + CP(\alpha)m_q^{-1}, \tag{2.5}$$

where C is independent of q and α, and C_α is independent of q.

Concerning the estimate of the right-hand side of (2.5) we shall show the following

Proposition 2.2. *There exists a sequence $\{l_q\}_{q=1}^{\infty}$ satisfying (2.2) such that we have an estimate*

$$\left| \text{tr}_{L^2(\mathbf{R}^3)} \int_{\mathbf{R}} \rho_q(t)(\cos t\sqrt{A} \oplus 0 - \cos t\sqrt{A_0}) \, dt \right| \ge e^{-\alpha_0 l_q} - C e^{\alpha_1 l_q} m_q^{-1} \tag{2.6}$$

for all q, where α_0, α_1 and C are positive constants independent of the choice of sequence $\{m_q\}_{q=1}^{\infty}$.

Admit first Proposition 2.2. Then we can deduce immediately the Theorem from Lemma 2.1 and Proposition 2.2. Indeed, let $\{l_q\}_{q=1}^{\infty}$ be the sequence in Proposition 2.2, and take

$$\alpha = 5(\alpha_0 + \alpha_1 + 2), \tag{2.7}$$

$$m_q = e^{\alpha l_q / 5}. \tag{2.8}$$

Suppose that $P(\alpha) < \infty$. Then we have from Lemma 2.1 and (2.7), (2.8)

$$\sum_{j=1}^{\infty} |\hat{\rho}_q(\lambda_j)| \le (C_\alpha + CP(\alpha)) \, e^{-\alpha l_q / 5}. \tag{2.9}$$

On the other hand, Proposition 2.2 gives

$$\left| \mathrm{tr}_{L^2(\mathbf{R}^3)} \int_{\mathbf{R}} \rho_q(t)(\cos t\sqrt{A} \oplus 0 - \cos t\sqrt{A_0}) \, dt \right| \geq (1 - C e^{-l_q}) \, e^{-\alpha_0 l_q}. \quad (2.10)$$

Set $\rho = \rho_q$ in (2.1). Then it follows from (2.9) and (2.10) that

$$(1 - C e^{-l_q}) \, e^{-\alpha_0 l_q} \leq (C_\alpha + CP(\alpha)) \, e^{-\alpha l_q/5}.$$

Since $\alpha/5 > \alpha_0$, the above inequality shows a contradiction when q increases infinitely.

In the next section we shall show the outline of the proof of Proposition 2.2.

3. Proof of Proposition 2.2

3.1. On the Kernel of cos tA

Suppose that $\mathcal{O} \subset \{x; |x| < R\}$. Let $E(x, y, t)$ be the distribution kernel of $\cos tA$. As is shown in [BGR], for any $\rho \in C_0^\infty(2R, \infty)$,

$$F_\rho(x, y) = \int \rho(t) E(x, y, t) \, dt$$

is smooth in $(x, y) \in \bar{\Omega} \times \bar{\Omega}$, and an equality

$$\mathrm{tr}_{L^2(\mathbf{R}^3)} \int \rho(t)(\cos t\sqrt{A} \oplus 0 - \cos t\sqrt{A_0}) \, dt = \int_{\Omega(T+R)} F_\rho(x, x) \, dx, \quad (3.1)$$

holds, where $\Omega(r) = \Omega \cap \{x; |x| < r\}$ and $T = \sup \mathrm{supp} \, \rho$.

Let w and $\tilde{w}$ be functions in $C_0^\infty(\Omega)$ such that

$$\tilde{w} = 1 \qquad \text{on supp } w.$$

Let y, ω and k be parameters moving in Ω, S^2 and $(0, \infty)$ respectively. We denote by $s(x, t; \omega, k)$ the solution to the problem

$$\Box s = 0 \qquad \text{in } \Omega \times \mathbf{R}$$

$$\frac{\partial s}{\partial n} = 0 \qquad \text{on } \Gamma \times \mathbf{R}$$

$$s(x, 0; y, \omega, k) = \tilde{w}(x) \, e^{ikx \cdot \omega} \qquad\qquad (3.2)$$

$$\frac{\partial s}{\partial t}(x, 0; \omega, k) = 0.$$

Then we have an expression

$$F_\rho(x, y)w(y) = O_s - \int_0^\infty k^2 \, dk \int_{S^2} d\omega \int dt \, \rho(t)s(x, t; \omega, k) \, e^{-iky \cdot \omega} w(y). \tag{3.3}$$

To estimate $\int_{\Omega(2l_q)} F_q(x, x)w(x) \, dx$ we have to know the behavior of $s(x, t; \omega, k)$ for large t. To this end, we shall use an asymptotic solution u, which approximates s.

3.2. Asymptotic Solutions

Following the procedure in [Ik 4] we explicitly construct asymptotic solutions for the initial-boundary value problem (3.2). We shall use freely the notations and the results in it.

Suppose that

$$\text{supp } \tilde{w} \subset \text{convex hull of } \mathcal{O}_j \text{ and } \mathcal{O}_l. \tag{3.4}$$

Then we have an asymptotic solution $u(x, t; \omega, k)$ with the following properties:

(i) u is of the form

$$u = \sum_{j \in I} (u_j^+ + u_j^-),$$

$$\tag{3.5}$$

$$u_j^\pm(x, t; \omega, k) = e^{ik(\varphi_j^\pm(x, \omega) - t)} \sum_{h=0}^N v_{j,h}^\pm(x, t; \omega)(ik)^{-h},$$

where $\{\varphi_j^\pm; j \in I\}$ are the sequences of phase functions starting from $\varphi = x \cdot \omega$ ([Ik 4, Section 2]).

(ii) Amplitude functions $v_{j,h}^\pm(x, t; \omega)$ satisfy an estimate for $(x, t) \in (\text{supp } \tilde{w}) \times \mathbf{R}$

$$|v_{j,h}^\pm(x, t; \omega)|_p \le C_{p,l} \, e^{-\alpha_2 |\mathbf{j}|}(t+1)^h M_{N,p},$$

$$M_{N,p} = |\tilde{w}|_{2N+p},$$

where α_2 is a positive constant independent of $\mathbf{j}$ and h.

(iii) If $x \in \text{supp } \tilde{w}$ and $v_{j,0}^\pm(x, t; \omega) \ne 0$, we have an expression

$$v_{j,0}^\pm(x, t; \omega) = \Lambda_j^\pm(x, \omega)\tilde{w}(Y_j^\pm(x, t; \omega)), \tag{3.6}$$

$$\Lambda_j^\pm(x, \omega) = \Lambda_{\varphi_{(j,j_1,\dots,j_n)}^\pm}(x) \cdot \Lambda_{\varphi_{(j,j_1,\dots,j_{n-1})}^\pm}(X^{-1}(x, \nabla\varphi_j^\pm))$$

$$\times \cdots \Lambda_\varphi(X^{-(n+1)}(x, \nabla\varphi_j^\pm))$$

$$\times \left[\frac{G_\varphi(X^{-(n+1)}(x, \nabla\varphi_j^\pm))}{G_\varphi(Y_j^\pm(x, t; \omega))}\right]^{1/2}, \tag{3.7}$$

where we set $\mathbf{j} = (j, j_1, \ldots, j_n)$, and $Y_{\mathbf{j}}^{\pm}(x, t; \omega)$ denotes the point on the segment $X^{-n}(x, \nabla \varphi_{\mathbf{j}}^{\pm}) X^{-(n+1)}(x, \nabla \varphi_{\mathbf{j}}^{\pm})$ such that

$$|x - X^{-q}(x, \nabla \varphi_{\mathbf{j}}^{\pm})| + \sum_{q=1}^{n} |X^{-q}(x, \nabla \varphi_{\mathbf{j}}^{\pm}) - X^{-q-1}(x, \nabla \varphi_{\mathbf{j}}^{\pm})|$$

$$+ |X^{-(n+1)}(x, \nabla \varphi_{\mathbf{j}}^{\pm}) - Y_{\mathbf{j}}^{\pm}(x, t; \omega)| = t.$$

(*iv*)   For $(x, t) \in (\text{supp } \tilde{w}) \times \mathbf{R}$ we have

$$\sum_{|\beta| \leq p} |\partial_{x,t}^{\beta}(s(x, t; \omega, k) - u(x, t; \omega, k))| \leq C_{p,l} e^{-\alpha_2 |\mathbf{j}|} (t+1)^l M_{N,p}.$$

$$(3.8)$$

Now we consider the following integral:

$$L_{\mathbf{j}}^{\pm}(t, k) = \int_{S^2} d\omega \int_{\Omega} dx \, u_{\mathbf{j}}^{\pm}(x, t; \omega, k) \, e^{-ikx \cdot \omega} w(x). \qquad (3.9)$$

Substituting (3.5) into (3.9) we have

$$L_{\mathbf{j}}^{\pm}(t, k) = \sum_{h=0}^{N} (ik)^{-h} L_{\mathbf{j},h}^{\pm}(t, k) \, e^{-ikt},$$

$$L_{\mathbf{j},h}^{\pm}(t, k) = \int_{S^2} d\omega \int_{\Omega} dx \, e^{ik\Phi_{\mathbf{j}}^{\pm}(x,\omega)} v_{\mathbf{j},h}^{\pm}(x, t) w(x)$$

where

$$\Phi_{\mathbf{j}}^{\pm}(x, \omega) = \varphi_{\mathbf{j}}^{\pm}(x, \omega) - x \cdot \omega.$$

Note that

$$\{(x, \omega) \in \text{supp } \tilde{w} \times S^2; |d_x \Phi_{\mathbf{j}}^{\pm}| + |d_{\omega} \Phi_{\mathbf{j}}^{\pm}| = 0\}$$

$$= (\text{supp } \tilde{w} \cap \gamma) \times \{a_{n+1} a_1 / |a_{n+1} a_0|\},$$

where $\gamma = \bigcup_{q=1}^{n+1} a_q a_{q+1}$, $a_q \in \Gamma_{j_q}$, $a_{n+2} = a_1$, is the periodic ray such that $\mathbf{j} \in \mathscr{J}(\gamma)$. We may suppose that a_1 and a_{n+1} are on the x_1-axis and that a_1 is on the right-hand side of a_{n+1}. Then for each x_1, the stationary point of $\Phi_{\mathbf{j}}^{+}(x_1, \cdot, \cdot)$ with respect to x_2, x_3, ω variables is unique and $x_2 = x_3 = 0$ and $\omega = (1, 0, 0)$, which is non-degenerate. Applying the method of stationary phase to each $L_{\mathbf{j},h}^{\pm}(t, k)$, we have

$$L_{\mathbf{j}}^{\pm}(t, k) \sim e^{\pm ik(l(\gamma)-t)} k^{-2} \sum_{q=0}^{\infty} (ik)^{-q} \left(\sum_{h=0}^{N} c_{\mathbf{j},h,q}^{\pm}(t)(ik)^{-h} \right). \qquad (3.10)$$

The coefficients $c_{\mathbf{j},h,q}^{\pm}(t)$ satisfy

$$|c_{\mathbf{j},h,q}^{\pm}(t)|_p \leq C_{N,h,q} e^{-\alpha_2 |\mathbf{j}|} (t+1)^N M_{N,p+2h+2q}.$$

Especially

$$c_{\mathbf{j},0,0}^{\pm}(t) = \tfrac{1}{2}(\det(I - P_\gamma))^{-1/2} \int w(x_1 - (t - l(\gamma)), 0, 0)\, dx_1,$$

where P_γ denotes the Poincaré map of γ.

3.3. Proof of Proposition 2.2

It follows from (3.8) that

$$\left| \int_{\Omega(2l_q)} F_{\rho_q}(x, x) w(x)\, dx - \sum_{\mathbf{j} \in I} \int (L_{\mathbf{j}}^+(t, k) + L_{\mathbf{j}}^-(t, k)) \rho_q(t)\, dt\, dk \right|$$

$$\leq C e^{-\alpha_3 l_q} \# \left\{ \mathbf{j}; |\mathbf{j}| \leq \frac{2l_q}{d} \right\} l_q^{2N+2} m_q^{-1}.$$

By using (3.10) and (3.11) we have

$$\left| \int_{\Omega(2l_q)} F_{\rho_q}(x, x) w(x)\, dx - \sum_\gamma \#\{\mathbf{j}; \mathbf{j} \in \mathscr{J}(\gamma)\} \right.$$

$$\left. \times (\det(I - P_\gamma))^{-1/2} \cdot \rho_q l(\gamma) \int_\gamma w(x)\, ds \right|$$

$$\leq C e^{-\alpha_3 l_q} \# \left\{ \mathbf{j}; |\mathbf{j}| \leq \frac{2l_q}{d} \right\} l_q^{2N+2} m_q^{-1}. \tag{3.11}$$

We can extend (3.11) for any $w \in C_0^\infty(\bar{\Omega})$ (when $\overline{\operatorname{supp} w} \cap L \neq \varnothing$, we shall use the argument in [Ik 2, Section 6]). Then by a partition of the unity over $\overline{\Omega(ql_q)}$ we have

$$\left| \int_{\Omega(2l_q)} F_{\rho_q}(x, x)\, dx - \sum_\gamma T(\gamma)(\det(I - P_\gamma))^{-1/2} \cdot \rho_q(l(\gamma)) \right|$$

$$\leq C e^{-\alpha_3 l_q} \# \left\{ \mathbf{j}; |\mathbf{j}| \leq \frac{2l_q}{d} \right\} l_q^{2N+2} m_q^{-1}. \tag{3.12}$$

For each positive integer q choose a $\mathbf{j} \in I$ such that $|\mathbf{j}| = q$. We set $l_q = l(\gamma)$ for γ such that $\mathbf{j} \in \mathscr{J}(\gamma)$. Note that there exists $\alpha_0 > 0$ such that

$$(\det(I - P_\gamma))^{-1/2} \geq e^{-\alpha_0 l(\gamma)} \qquad \text{for all } \gamma.$$

On the other hand

$$\# \left\{ \mathbf{j}; |\mathbf{j}| \leq \frac{2l_q}{d} \right\} \leq e^{\alpha_4 l_q} \qquad \text{for all } q$$

holds for some $\alpha_4 > 0$. Thus Proposition 2.2 is proved.

References

[BGR]   C. Bardos, J. C. Guillot and J. Ralston, La relation de Poisson pour l'équation des ondes dans un ouvert non borné. Application à la theorie de la diffusion, *Comm. Partial Diff. Equ.* **7** (1982) 905–958.

[G]   C. Gérard, Asymptotique des poles de la matrice de scattering pour deux obstacles strictement convexes, preprint, Univ. Paris-Sud.

[Ik 1]   M. Ikawa, On the poles of the scattering matrix for two strictly convex obstacles, *J. Math. Kyoto Univ.* **23** (1983) 127–194.

[Ik 2]   M. Ikawa, Trapping obstacles with a sequence of poles of the scattering matrix converging to the real axis, *Osaka J. Math.* **22** (1985) 657–689.

[Ik 3]   M. Ikawa, Precise information on the poles of the scattering matrix for two strictly convex obstacles, *J. Math. Kyoto Univ.* **27** (1987) 69–102.

[Ik 4]   M. Ikawa, Decay of solutions of the wave equation in the exterior of several convex bodies, To appear in *Ann. Inst. Fourier.*

[LP 1]   P. D. Lax and R. S. Phillips, *Scattering Theory*, Academic Press, 1967.

[LP 2]   P. D. Lax and R. S. Phillips, A logarithmic bound on the location of the poles of the scattering matrix, *Arch. Rat. Mech. Anal.* **40** (1971) 268–280.

[LP 3]   P. D. Lax and R. S. Phillips, Scattering theory for dissipative hyperbolic systems, *J. Funct. Anal.* **14** (1973) 172–235.

[Me 1]   R. Melrose, Singularities and energy decay in acoustical scattering, *Duke Math. J.* **46** (1979) 43–59.

[Me 2]   R. Melrose, Polynomial bound on the number of scattering poles, *J. Funct. Anal.* **53** (1983) 287–303.

[MeS]   R. Melrose and J. Sjöstrand, Singularities of boundary value problems, I and II, *Comm. Pure Appl. Math.* **31** (1979) 593–617; **35** (1982) 129–168.

[MoRSt] C. S. Morawetz, J. Ralston, and W. A. Strauss, Decay of solutions of the wave equation outside nontrapping obstacles, *Comm. Pure Appl. Math.* **30** (1977) 447–508.

[P]   V. M. Petkov, La distribution des poles de la matrice de diffusion, *Seminaire Goulaouic-Meyer-Schwartz*, 1982–1983.

[R]   J. Ralston, The first variation of the scattering matrix: An addendum, *J. Differential Equations* **28** (1978) 155–162.

Symmetric Tensors of the $A_{n-1}^{(1)}$ Family

Michio Jimbo, Tetsuji Miwa, and Masato Okado

Research Institute for Mathematical Sciences
Kyoto University
Kyoto, Japan

1. Introduction

The mathematical contents of nonlinear phenomena have been attracting hot attention of many mathematicians. The latest achievement concerning solvable models is particularly important. Sato's theory [1] of the universal Grassmann manifold revealed the intrinsic mechanism of exact solvability of the soliton equations. Exact solvability in statistical mechanics and quantum field theory is also of primary importance. As for lattice models an extrinsic criterion is known for exact solvability. That is the commutativity of transfer matrices, or its local version called the *Yang–Baxter equation* or the *star-triangle relation.* Its role is similar to that of the Lax pair in the soliton theory. Until several years ago only a few solutions of the Yang-Baxter equation or the star-triangle relation were known. Recently there has been a lot of progress in this respect. Now the list of solutions is quantitatively no less than that of soliton equations.

As for the class of lattice models called the vertex models, a line of investigations by Baxter, Belavin, Sklyanin and Cherednik is notable. They

Algebraic Analysis, Volume I

deal with the models with elliptic parametrizations in connection with the representation theory of A_{n-1}. Andrews–Baxter–Forrester [2] studied a series of IRF models corresponding to the eight-vertex model. In [3] the symmetric tensors of the ABF models were introduced, while in [4] the higher-rank version was given. The purpose of this paper is to give the symmetric tensor construction of the higher-rank case as announced in [5].

We hope that our understanding of the exact solvability of lattice statistical mechanics will be raised to an intrinsic level in the near future.

2. Elementary Blocks

2.1. Local States

Fix an integer $n \geq 2$. We denote by $\mathfrak{h}^*$ an $(n+1)$-dimensional vector space with the distinguished basis $\Lambda_0, \ldots, \Lambda_{n-1}, \delta$. We define a symmetric bilinear form on $\mathfrak{h}^*$ by

$$\langle \Lambda_\mu, \Lambda_\nu \rangle = \min(\mu, \nu) - \frac{\mu\nu}{n}, \qquad \langle \Lambda_\mu, \delta \rangle = 1, \qquad \langle \delta, \delta \rangle = 0.$$

The space $\mathfrak{h}^*$ is isomorphic to the dual space of the Cartan subalgebra of $A_{n-1}^{(1)}$, in which Λ_μ ($\mu = 0, \ldots, n-1$) are called the *fundamental weights* and δ is called the *null root*.

We denote by $\bar{\mathfrak{h}}^*$ the orthogonal complement of $\mathbf{C}\Lambda_0 \oplus \mathbf{C}\delta$ and call it the *classical* part of $\mathfrak{h}^*$. Consider an n-dimensional orthogonal vector space with the distinguished basis $\varepsilon_0, \ldots, \varepsilon_{n-1}$. The classical part $\bar{\mathfrak{h}}^*$ is identified with the orthogonal complement of $\mathbf{C}(\varepsilon_0 + \cdots + \varepsilon_{n-1})$ by

$$\Lambda_\mu - \Lambda_0 = \varepsilon_0 + \cdots + \varepsilon_{\mu-1} - \frac{\mu}{n}(\varepsilon_0 + \cdots + \varepsilon_{n-1}).$$

An element $a = (\Lambda(a), m(a))$ of $\mathfrak{h}^* \times \mathbf{Z}$ is called a *local state*. For a local state a we define a^μ, $a_{\mu\nu}$ and $l(a)$ as follows:

$$\Lambda(a) \equiv \sum_{\mu=0}^{n-1} a^\mu \Lambda_\mu \bmod \mathbf{C}\delta,$$

$$a_{\mu\nu} = \langle \Lambda(a) + \rho, \varepsilon_\mu - \varepsilon_\nu \rangle, \qquad (\rho = \Lambda_0 + \Lambda_1 + \cdots + \Lambda_{n-1}),$$

$$l(a) = \langle \Lambda(a), \delta \rangle = \sum_{\mu=0}^{n-1} a^\mu.$$

A local state a is called *admissible* if $a^\mu \in \mathbf{N} = \{0, 1, 2, \ldots\}$. The $l(a)$ is called the *level* of a. We often abbreviate $l(a)$ to l if the specification of a is obvious.

We set

$$\hat{\mu} = (\Lambda_{\mu+1} - \Lambda_{\mu}, 1) \in \mathfrak{h}^* \times \mathbf{Z}, \qquad (\mu = 0, \ldots, n-1; \Lambda_{\mu} = \Lambda_{\mu+n}).$$

They are called *elementary vectors*. We consider $\mathfrak{h}^* \times \mathbf{Z}$ as an additive group. A sequence of local states $(a_0, a_1, \ldots, a_N)$ is called a *path* from a_0 to a_N if each $a_j - a_{j-1}$ $(j = 1, \ldots, N)$ is an elementary vector. An ordered pair (a, b) is called *weakly admissible* (resp. *N-weakly admissible*) if there exists a path (resp. a path of length N) from a to b. If (a, b) is weakly admissible, then $l(a) = l(b)$ and $p = b - a$ is written as $p = \sum_{\mu=0}^{n-1} p_{\mu} \hat{\mu}$, $p_{\mu} \in \mathbf{N}$. We shall keep this notation: (p_{μ}) means the coordinates of p with respect to $\hat{\mu}$. An ordered pair is called *admissible* (resp. *N-admissible*) if the following are valid:

(1) (a, b) is weakly admissible (resp. N-weakly admissible).
(2) Any path from a to b contains only admissible local states.

If (a, b) is admissible, the states a and b are admissible.

2.2. Theta Functions

We use the following elliptic theta function with variables $u \in \mathbf{C}$ and p $(-1 < p < 1)$.

$$\theta_1(u, p) = 2|p|^{1/8} \sin u \prod_{k=1}^{\infty} (1 - 2p^k \cos 2u + p^{2k})(1 - p^k).$$

We use an abbreviated notation,

$$[u] = \theta_1\left(\frac{\pi u}{L}, p\right), \qquad [A, A+k] = [A][A+1] \cdots [A+k].$$

The L is a non-zero complex number. The choice of L is arbitrary unless otherwise stated.

Lemma 2.1.   *If a is admissible and $L = l(a) + n$, then we have $[a_{\mu\nu}] \neq 0$* $(\mu \neq \nu)$.

Proof.   If $\mu < \nu$, we have $1 \leq a_{\mu\nu} = a^{\mu+1} + \cdots + a^{\nu} + \nu - \mu \leq L - 1$.

2.3. Elementary Blocks

A square $\left(\begin{smallmatrix} a & b \\ d & c \end{smallmatrix}\right)$ consisting of local states a, b, c, d is called an (M, N) block if (a, d), (b, c) are M-weakly admissible and (a, b), (d, c) are N-weakly admissible. A block is called *admissible* if (a, d), (b, c), (a, b), (d, c) are admissible. A $(1, 1)$ block is also called an *elementary block*. We define the weights of elementary blocks as follows:

$$W\begin{pmatrix} a & a+\hat{\mu} \\ a+\hat{\mu} & a+2\hat{\mu} \end{pmatrix} = \frac{[u+1]}{[1]}.$$

For $\mu \neq \nu$,

$$W\begin{pmatrix} a & a+\hat{\mu} \\ a+\hat{\mu} & a+\hat{\mu}+\hat{\nu} \end{pmatrix} = \frac{[a_{\mu\nu}-u]}{[a_{\mu\nu}]},$$

$$W\begin{pmatrix} a & a+\hat{\nu} \\ a+\hat{\mu} & a+\hat{\mu}+\hat{\nu} \end{pmatrix} = \frac{[u][a_{\mu\nu}+1]}{[1][a_{\mu\nu}]}.$$

The weights satisfy the following symmetries.

Proposition 2.2.

$$W\begin{pmatrix} -2\rho-c & -2\rho-b \\ -2\rho-d & -2\rho-a \end{pmatrix} u = W\begin{pmatrix} a & b \\ d & c \end{pmatrix} u.$$

Let σ denote the cyclic automorphism on $\mathfrak{h}^*$ such that

$$\sigma(\Lambda_\mu) = \Lambda_{\mu+1}, \qquad \sigma(\delta) = \delta.$$

Proposition 2.3. *If we choose $L = l + n$, then we have*

$$W\begin{pmatrix} \sigma(a) & \sigma(b) \\ \sigma(d) & \sigma(c) \end{pmatrix} u = W\begin{pmatrix} a & b \\ d & c \end{pmatrix} u.$$

The proofs of these Propositions are left to the reader. In the later sections we define the weights for general (M, N) blocks. Propositions 2.3 and 2.4 are equally valid for the general case. (Note that Proposition 2.3 is valid when we restrict L in the left- (resp. right-) hand side to $l(-2a-\rho)+n$ (resp. $l(a)+n$) because $l(-2\rho-a)+n = -(l(a)+n)$.) Therefore we shall present the statements and the proofs in reduced form by these symmetries.

The weights for elementary blocks satisfy the following star-triangle relation.

Theorem 2.4.

$$\sum_g W\begin{pmatrix} a & b \\ f & g \end{pmatrix} u+v \end{pmatrix} W\begin{pmatrix} f & g \\ e & d \end{pmatrix} u \end{pmatrix} W\begin{pmatrix} b & c \\ g & d \end{pmatrix} v \end{pmatrix}$$

$$= \sum_g W\begin{pmatrix} a & g \\ f & e \end{pmatrix} v \end{pmatrix} W\begin{pmatrix} a & b \\ g & c \end{pmatrix} u \end{pmatrix} W\begin{pmatrix} g & c \\ e & d \end{pmatrix} u+v \end{pmatrix}.$$

3. Fusion

3.1. (M, N) Blocks

We now define the weights of an (M, N) block $\begin{pmatrix} a & b \\ d & c \end{pmatrix}$ by fusing those of elementary blocks. Choose a path $(a_0, \ldots, a_{M+N})$ from a to c in such a way that $a_M = d$. We define

$$W\begin{pmatrix} a & b \\ d & c \end{pmatrix} u \end{pmatrix} = \sum_\sigma \prod_{i=0}^{M-1} \prod_{j=0}^{N-1} W\begin{pmatrix} \sigma_{i+1j} & \sigma_{i+1j+1} \\ \sigma_{ij} & \sigma_{ij+1} \end{pmatrix} u+i+j-M+1 \end{pmatrix}.$$

The sum extends over $\sigma = (\sigma_{ij})_{0 \le i \le M, 0 \le j \le N}$ such that

(i) $\sigma_{M-i0} = a_i \ (0 \le i \le M)$,

(ii) $\sigma_{0j} = a_{M+j} \ (0 \le j \le N)$,

(iii) $\sigma_{MN} = b$,

(iv) $\begin{pmatrix} \sigma_{i+1j} & \sigma_{i+1j+1} \\ \sigma_{ij} & \sigma_{ij+1} \end{pmatrix} (0 \le i \le M-1, 0 \le j \le N-1)$ are elementary blocks.

Theorem 3.1. *The weight $W\begin{pmatrix} a & b \\ d & c \end{pmatrix} u\end{pmatrix}$ is independent of the choice of the path $(a_0, \ldots, a_{M+N})$ and satisfies the star-triangle relation.*

The proof follows from

Proposition 3.2. *Let $\begin{pmatrix} a & b \\ d & c \end{pmatrix}$ be a $(1, N)$ block. We set $p = b - a$, $\hat{\mu} = d - a$, $\hat{\nu} = c - b$. Then we have*

$$W\begin{pmatrix} a & b \\ d & c \end{pmatrix} u \end{pmatrix} = \prod_{j=1}^{N-1} \frac{[u+j]}{[1]} \frac{[u+a_{\nu\mu}+p_\nu] \prod_{\lambda(\ne\mu)} [a_{\nu\lambda}+p_\nu+1]}{\prod_\lambda [c_{\nu\lambda}+\delta_{\nu\lambda}]}.$$

Proof. We use an induction on N. In Fig. 1 we have shown how to divide a $(1, N)$ block into $(1, 1)$ and $(1, N-1)$ blocks. The details are left to the reader.

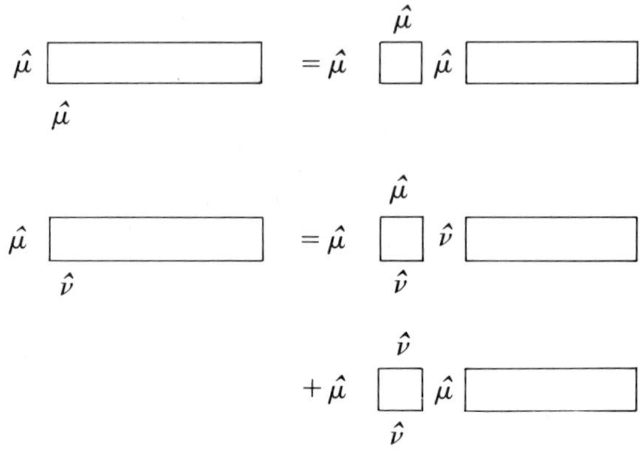

FIGURE 1　Induction step for $(1, N)$ weights.

3.2.　Symmetrization

Let (a, b) be an N-weakly admissible pair. We set $p = b - a$. We define

$$s(a, b) = \prod_{\mu < \nu} \frac{[a_{\mu\nu} - p_\nu, a_{\mu\nu} + p_\mu]}{\sqrt{[a_{\mu\nu}]}\sqrt{[b_{\mu\nu}]}} \prod_\mu [1, p_\mu].$$

Proposition 3.3.　*Let $\begin{pmatrix} a & b \\ d & c \end{pmatrix}$ be an (M, N) block. Then we have*

$$W\begin{pmatrix} a & b \\ d & c \end{pmatrix} u\bigg) = \frac{s(a, d)s(d, c)}{s(a, b)s(b, c)} W\begin{pmatrix} a & d \\ b & c \end{pmatrix} u - M + N\bigg).$$

Proof.　The case of $(1, N)$ block can be proved by using Proposition 3.2. The general case follows from this case. The argument for the $(2, 3)$ block is illustrated in Fig. 2. The general case is left to the reader.

4.　Restriction

4.1.　Restriction

Take an integer $l \geq 2$. The l restriction means to restrict local states and blocks to be admissible of level l and the parameter L to $l + n$.

Theorem 4.1.　*The l restricted weights $W\begin{pmatrix} a & b \\ d & c \end{pmatrix} u)$ are finite and satisfy the star-triangle relation among themselves.*

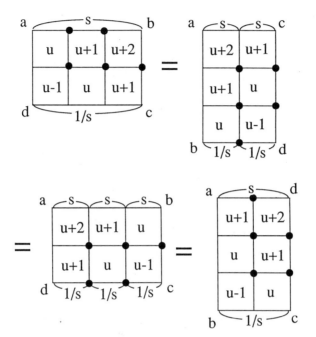

FIGURE 2   The symmetry of a $(2, 3)$ weight. Dots mean the summations. The factors $s(a, b)$ $1/s(d, c)$, etc. are abbreviated to s or $1/s$.

The finiteness follows from

Proposition 4.2.   *Let $\begin{pmatrix} a & b \\ d & c \end{pmatrix}$ be an (M, N) block. If (b, c) is admissible, then the weight $W(\begin{smallmatrix} a & b \\ d & c \end{smallmatrix}|u)$ with $L = l + n$ is finite.*

Proof.   We use the expression of $W(\begin{smallmatrix} a & b \\ d & c \end{smallmatrix}|u)$ in terms of the $(1, N)$ weights. We choose a path $(a_0, \ldots, a_M)$ from a to d. Then we have

$$W\begin{pmatrix} a & b \\ d & c \end{pmatrix}u \Bigg) = \sum_{(b_0, \ldots, b_M)} X(a_0, \ldots, a_M; b_0, \ldots, b_M),$$

$$X(a_0, \ldots, a_M; b_0, \ldots, b_M) = \prod_{i=0}^{M-1} W\begin{pmatrix} a_i & b_i \\ a_{i+1} & b_{i+1} \end{pmatrix}u - i \Bigg),$$

where the sum is over the paths from b to c. Since (b, c) is admissible, the finiteness of each $X = X(a_0, \ldots, a_M; b_0, \ldots, b_M)$ follows from Proposition 3.2.

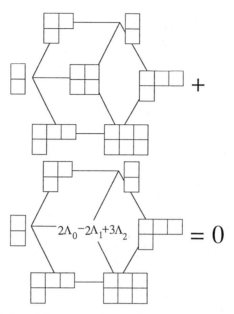

FIGURE 3   Cancellation of the unwanted terms. ($N = 2$, $l = n = 3$.) A Young tableau with signature (f_1, f_2) signifies the weight $(3 - f_1)\Lambda_0 + (f_1 - f_2)\Lambda_1 + f_2\Lambda_2$.

The key to the proof of the star-triangle relation is to show the cancellation of the terms containing non-admissible blocks. In Fig. 3 we have shown an example of such cancellation. We need some preliminaries for the general proof.

4.2.   Reflection

We denote by r_μ the reflection of $\mathfrak{h}^* \times \mathbf{Z}$,

$$r_\mu(a) = (\Lambda(a) - (a^\mu + 1)(\widehat{(\mu - 1)} - \hat{\mu}), m(a)).$$

An N-weakly admissible pair (a, b) is called μ-*reflexible* if and only if $(a, r_\mu(b))$ is also N-weakly admissible. This is equivalent to saying that there is a path $(a_0, \ldots, a_N)$ from a to b such that $a_i^\mu = -1$ for some i. From this remark follows

Lemma 4.3.   *Assume that a is admissible and (a, b) is weakly admissible. We make the restriction $L = l + n$. We set $p = b - a$. Then we have*

$$\prod_{\kappa < \lambda} [a_{\kappa\lambda} - p_\lambda, \, a_{\kappa\lambda} + p_\kappa] = 0$$

if and only if (a, b) is μ-reflexible for some μ.

Proposition 4.4. *Assume that (a, b) and (b, c) are admissible, (a, b') and (b', c) are weakly admissible and $m(b) = m(b')$. If (a, b') and (b', c) are μ-reflexible for $\mu = \mu_1, \mu_2$, then $\mu_1 \neq \mu_2 \pm 1$.*

Proof. It is sufficient to show that the following are contradictory.

(1) (a, b') is μ-reflexible.
(2) (b', c) is $(\mu + 1)$-reflexible.

Let $b' - b = p - q$ where $p_\mu, q_\mu \geq 0$ and $p_\mu \cdot q_\mu = 0$. Because (a, b) is admissible, any path from a to b stays in the admissible weights. The condition (1) means there exists a path from a to b' which crosses the hyperplane $\{x \in \mathfrak{h}^* \times \mathbf{Z} | x^\mu = -1\}$. Therefore we have $p_\mu > 0$. Similarly the condition (2) implies $q_\mu > 0$. This is a contradiction.

Lemma 4.5. *We make the restriction $L = l + n$. Assume that a local state x satisfies $x^\mu = -1$. We set $\bar{g} = r_\mu(g)$. If $\left(\begin{smallmatrix} a & b \\ x & g \end{smallmatrix}\right)$ (resp. $\left(\begin{smallmatrix} g & b \\ x & a \end{smallmatrix}\right)$) is a block such that (b, g) is μ-reflexible, we have*

$$\frac{W\left(\begin{matrix} a & b \\ x & g \end{matrix} \middle| u\right)}{W\left(\begin{matrix} a & b \\ x & \bar{g} \end{matrix} \middle| u\right)} = 1 \qquad \left(resp. \quad \frac{W\left(\begin{matrix} g & b \\ x & a \end{matrix} \middle| u\right)}{W\left(\begin{matrix} \bar{g} & b \\ x & a \end{matrix} \middle| u\right)} = 1 \right).$$

If $\left(\begin{smallmatrix} x & c \\ g & d \end{smallmatrix}\right)$ is a block such that (g, d) is ν-reflexible, we have

$$\frac{W\left(\begin{matrix} x & c \\ g & d \end{matrix} \middle| u\right)}{W\left(\begin{matrix} x & c \\ \bar{g} & d \end{matrix} \middle| u\right)} = 1.$$

Proof. The assertions follow from the expressions of the weights given in the proof of Proposition 4.2.

4.3. Cancellations

We start the proof of the star-triangle relation by showing the finiteness of the unwanted terms.

Proposition 4.6. *Let $\left(\begin{smallmatrix} a & b \\ d & c \end{smallmatrix}\right)$ be an (M, N) block. If both (a, b) and (b, c) are admissible, then $W\left(\begin{smallmatrix} a & b \\ d & c \end{smallmatrix} \middle| u\right)/s(d, c)$ with $L = l + n$ is finite.*

Proof. We use the expression of $W(\begin{smallmatrix} a & b \\ d & c \end{smallmatrix}|u)$ given in the proof of Proposition 4.2. We already know that each X is finite. Set $p = b - a$ and $r = c - d$. For $\kappa < \lambda$ we shall show the following:

(1) $[a_{\kappa\lambda} - p_\lambda, a_{\kappa\lambda} + p_\kappa] \neq 0.$

(2)   If $d_{\kappa\lambda} - r_\lambda \leq a_{\kappa\lambda} - p_\lambda - 1$, then $[d_{\kappa\lambda} - r_\lambda], \ldots, [a_{\kappa\lambda} - p_\lambda - 1]$

   appears in the numerator of X.

(3)   If $a_{\kappa\lambda} + p_\kappa + 1 \leq d_{\kappa\lambda} + r_\kappa$, then $[a_{\kappa\lambda} + p_\kappa + 1], \ldots, [d_{\kappa\lambda} + r_\kappa]$ appears in the numerator of X.

Then (1) follows from Lemma 2.1 and the assumption that (a, b) is admissible. Let us prove (2). We set $\Delta_i = (a_i)_{\kappa\lambda} - (p_i)_\lambda$ where $p_i = b_i - a_i$ $(i = 0, \ldots, M)$. We have $\Delta_i - \Delta_{i-1} = -1$ if and only if $a_i - a_{i-1} \neq \hat{\kappa}$ and $b_i - b_{i-1} = \hat{\lambda}$. Moreover the numerator of X contains $[\Delta_i]$ in this case. Since $\Delta_i - \Delta_{i-1} = 0, \pm 1$, we have

$$\{\Delta_i | \Delta_i - \Delta_{i-1} = -1\} \ni d_{\kappa\lambda} - r_\lambda, \ldots, a_{\kappa\lambda} - p_\lambda - 1.$$

This implies (2). The proof of (3) is similar. Since the numerator of $s(d, c)$ is $\prod_{\kappa < \lambda} [d_{\kappa\lambda} - r_\lambda, d_{\kappa\lambda} + r_\kappa]$, (1)-(3) are enough for the finiteness of X.

Proposition 4.7.   *Consider three blocks $(\begin{smallmatrix} a & b \\ f & g \end{smallmatrix})$, $(\begin{smallmatrix} f & g \\ e & d \end{smallmatrix})$, $(\begin{smallmatrix} b & c \\ g & d \end{smallmatrix})$ as in Fig. 4, in which M, N, P denotes the lengths of the paths. The summand in the star-triangle relation corresponding to these blocks is*

$$S = W\begin{pmatrix} a & b \\ f & g \end{pmatrix} u + v \end{pmatrix} W\begin{pmatrix} f & g \\ e & d \end{pmatrix} u \end{pmatrix} W\begin{pmatrix} b & c \\ g & d \end{pmatrix} v \end{pmatrix}.$$

We assume that (a, b), (b, c), (c, d), (a, f), (f, e), (e, d) are all admissible, but not all of (b, g), (f, g), (g, d) are admissible. We make the restriction

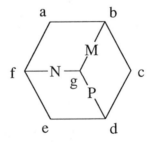

FIGURE 4   A summand in the star-triangle relation.

$L = l + n$. Then we have the following alternatives:

(1) $S = 0$.

(2) There exists v such that (b, g), (f, g), (g, d) are v-reflexible.

Proof. For a fixed v there are 2^3 possibilities: (A) or not (A), (B) or not (B), and (C) or not (C), where

(A) (b, g) is v-reflexible.

(B) (f, g) is v-reflexible.

(C) (g, d) is v-reflexible.

We shall show that except for the two cases, (A), (B), (C) and not(A), not(B), not(C), the product S is zero. If not(A), not(B), not(C) are valid for any v, (b, g), (f, g), (g, d) are all admissible. Therefore, the assertion follows.

Consider the case (A), (B), not (C). Using Proposition 3.3 we write

$$S = W\!\left(\begin{matrix} a & b \\ f & g \end{matrix}\middle| u+v\right) W\!\left(\begin{matrix} f & e \\ g & d \end{matrix}\middle| u-P+N\right) W\!\left(\begin{matrix} b & c \\ g & d \end{matrix}\middle| v\right) \frac{s(f, e)s(e, d)}{s(f, g)s(g, d)}.$$

By Proposition 4.2 $W(\begin{smallmatrix} a & b \\ f & g \end{smallmatrix}|u+v)$ is finite. By Proposition 4.6 $W(\begin{smallmatrix} f & e \\ g & d \end{smallmatrix}|u-P+N)/s(g, d)$ and $W(\begin{smallmatrix} b & c \\ g & d \end{smallmatrix}|v)/s(g, d)$ are both finite. By Proposition 4.3 $s(f, g)^{-1}$ is finite, while $s(g, d)$ is zero. Therefore we have $S = 0$ as desired. The rest of the proof is left to the reader.

We keep the assumption of Proposition 4.7. We now focus on the case in which (A), (B), (C) holds for some v. We denote $\bar{g} = r_v(g)$ for short.

Proposition 4.8.

$$W\!\left(\begin{matrix} a & b \\ f & g \end{matrix}\middle| u+v\right) W\!\left(\begin{matrix} f & g \\ e & d \end{matrix}\middle| u\right) W\!\left(\begin{matrix} b & c \\ g & d \end{matrix}\middle| v\right)$$

$$+ W\!\left(\begin{matrix} a & b \\ f & \bar{g} \end{matrix}\middle| u+v\right) W\!\left(\begin{matrix} f & \bar{g} \\ e & d \end{matrix}\middle| u\right) W\!\left(\begin{matrix} b & c \\ \bar{g} & d \end{matrix}\middle| v\right) = 0.$$

We divide the proof into three parts.

Lemma 4.9.

$$\frac{s(f, g)s(g, d)}{s(f, \bar{g})s(\bar{g}, d)} = -1.$$

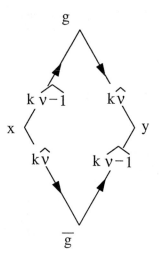

FIGURE 5   The ν reflection.

Proof.   There exist $k \in \mathbf{N}$ and a local state x, y such that (see Fig. 5)

$$g = x + k(\widehat{\nu - 1}) = y - k\hat{\nu}, \qquad \bar{g} = x + k\hat{\nu} = y - k(\widehat{\nu - 1}), \qquad x^\nu = y^\nu = -1.$$

We have

$$\prod_{\kappa < \lambda} \frac{[g_{\kappa\lambda}]}{[\bar{g}_{\kappa\lambda}]} = \prod_{\kappa < \lambda} \frac{[(x + k(\widehat{\nu - 1}))_{\kappa\lambda}]}{[(x + k\hat{\nu})_{\kappa\lambda}]} = -1.$$

We set $p = x - f, q = d - y$ and

$$Y = \frac{[1, p_\nu][1, p_{\nu-1} + k][1, q_{\nu-1}][1, q_\nu + k]}{[1, p_{\nu-1}][1, p_\nu + k][1, q_\nu][1, q_{\nu-1} + k]}.$$

Then we have

$$\prod_\lambda \frac{[1, (p + k(\widehat{\nu - 1}))_\lambda][1, (k\hat{\nu} + q)_\lambda]}{[1, (p + k\hat{\nu})_\lambda][1, (k(\widehat{\nu - 1}) + q)_\lambda]} = Y,$$

and

$$\prod_{\kappa < \lambda} \frac{[f_{\kappa\lambda} - (p + k(\widehat{\nu - 1}))_\lambda, f_{\kappa\lambda} + (p + k(\widehat{\nu - 1}))_\lambda]}{[f_{\kappa\lambda} - (p + k\hat{\nu})_\lambda, f_{\kappa\lambda} + (p + k\hat{\nu})_\lambda]}$$
$$\times \frac{[d_{\kappa\lambda} - (k\hat{\nu} + q)_\kappa, d_{\kappa\lambda} + (k\hat{\nu} + q)_\lambda]}{[d_{\kappa\lambda} - (k(\widehat{\nu - 1}) + q)_\kappa, d_{\kappa\lambda} + (k(\widehat{\nu - 1}) + q)_\lambda]} = Y^{-1}.$$

Lemma 4.10.

$$W\begin{pmatrix} a & b \\ f & g \end{pmatrix}u+v\end{pmatrix} = W\begin{pmatrix} a & b \\ f & \bar{g} \end{pmatrix}u+v\end{pmatrix}.$$

Proof. Let us denote $w = u + v$ for short. We keep the notations in the proof of Proposition 4.8. We use the following expression:

$$W\begin{pmatrix} a & b \\ f & g \end{pmatrix}w\end{pmatrix} = \sum_h W\begin{pmatrix} a & h \\ f & x \end{pmatrix}w\end{pmatrix} W\begin{pmatrix} h & b \\ x & g \end{pmatrix}w+N-k\end{pmatrix}.$$

Here the sum extends over h such that (a, h) is $(N-k)$-admissible and (h, b) is k-admissible. The assertion follows from Lemma 4.5.

The final step is to show that

$$\frac{W\begin{pmatrix} f & e \\ g & d \end{pmatrix}u\end{pmatrix} W\begin{pmatrix} b & c \\ g & d \end{pmatrix}v\end{pmatrix}}{s(f, g)s(g, d)} + \frac{W\begin{pmatrix} f & e \\ \bar{g} & d \end{pmatrix}u\end{pmatrix} W\begin{pmatrix} b & c \\ \bar{g} & d \end{pmatrix}v\end{pmatrix}}{s(f, \bar{g})s(\bar{g}, d)} = 0.$$

We use the expressions

$$W\begin{pmatrix} f & e \\ g & d \end{pmatrix}u\end{pmatrix} = \sum_h W\begin{pmatrix} f & h \\ g & y \end{pmatrix}u\end{pmatrix} W\begin{pmatrix} h & e \\ y & d \end{pmatrix}u+N-k\end{pmatrix},$$

$$W\begin{pmatrix} b & c \\ g & d \end{pmatrix}v\end{pmatrix} = \sum_{h'} W\begin{pmatrix} b & c \\ x & h' \end{pmatrix}v\end{pmatrix} W\begin{pmatrix} x & h' \\ g & d \end{pmatrix}v+M-k\end{pmatrix},$$

and the similar expressions with g replaced by $\bar{g}$. Note that the proof of Proposition 4.6 shows that each term

$$\frac{W\begin{pmatrix} f & h \\ g & y \end{pmatrix}u\end{pmatrix} W\begin{pmatrix} h & e \\ y & d \end{pmatrix}u+N-k\end{pmatrix} W\begin{pmatrix} b & c \\ x & h' \end{pmatrix}v\end{pmatrix} W\begin{pmatrix} x & h' \\ g & d \end{pmatrix}v+M-k\end{pmatrix}}{s(f, g)s(g, d)}$$

is finite. Termwise cancellations follow from Lemma 4.5 and Lemma 4.9.

References

[1] M. Sato and Y. Sato, Soliton equations as dynamical systems on infinite dimensional Grassmann manifold. Nonlinear partial differential equations in applied sciences, *Proceedings of the U.S.–Japan Seminar, Tokyo* (1982) 259–271, Kinokuniya/North-Holland.

[2] G. E. Andrews, R. J. Baxter, and P. J. Forrester, Eight-vertex SOS model and generalized Rogers–Ramanujan's identities, *J. Stat. Phys.* **35** (1984) 3908–3915.

[3] E. Date, M. Jimbo, T. Miwa, and M. Okado, Fusion of the eight-vertex SOS model, *Lett. Math. Phys.* **12** (1986) 209–215.

[4] M. Jimbo, T. Miwa, and M. Okado, Solvable lattice models whose states are dominant integral weights of $A_{n-1}^{(1)}$, *Lett. Math. Phys.* **14** (1987) 123–131.

[5] M. Jimbo, T. Miwa, and M. Okado, An $A_{n-1}^{(1)}$ family of solvable lattice models, *Mod. Phys. Lett.* **B1** (1987) 73–79.

On Hyperfunctions with Analytic Parameters

Akira Kaneko

Department of Mathematics
College of General Education
University of Tokyo
Komaba, Meguro-ku
Tokyo, Japan

1. Introduction

The notion of hyperfunctions with analytic parameters is important in various applications. Here we shall survey their fundamental properties, stressing the unique continuation properties. First we briefly recall their definitions.

A hyperfunction $f(x, t)$ in the variables $(x, t) \in \Omega \times T \subset \mathbf{R}^{n+m}$ is said to contain t as real analytic parameters if S.S. f (the *singular spectrum* of f) does not contain elements with direction components of the form $\theta \, dt$, or

$$\text{S.S.} \, f(x, t) \cap i\mathbf{S}^{m-1} \, dt = \varnothing. \tag{1.1}$$

We remark that this notion was already introduced in Sato [12] before the foundation of the theory of S.S. as follows: Let $\mathcal{O}_{z,\tau}$ denote the sheaf of germs of holomorphic functions on $\mathbf{C}^{n+m}$ with the coordinates (z, τ). Then the sheaf of hyperfunctions in (x, t) containing t as real analytic parameters is defined by $\mathscr{BA} := \mathscr{H}^n_{\mathbf{R}^{n+m}}(\mathcal{O}_{z,\tau}|_{\mathbf{C}^n \times \mathbf{R}^m})$. (We are using symbols a little different from the original ones.) This is the pure component just as in the definition of the sheaf $\mathscr{B}$ of usual hyperfunctions. A section of this sheaf admits a boundary value representation of the form

$$f(x, t) = \sum_{j=1}^{N} F_j(x + i\Gamma_{j0}, t), \qquad (1.2)$$

where, by denoting the closed dual cone of Γ_j by Γ_j, S.S. $f(x, t) \subset \bigcup_{j=1}^{N} \Gamma_j^\circ$. By Kashiwara's lemma each $F_j(z, \tau)$ is in fact holomorphic on a wedge-like domain whose opening is an $(n+m)$-dimensional cone containing $\Gamma_j \times \{0\}$. From this we can easily see the equivalence of the two definitions.

Another important notion is the hyperfunction with complex holomorphic parameters: A hyperfunction $f(x, \tau)$ is said to contain $\tau = t + is$ as complex holomorphic parameters if it satisfies the partial Cauchy–Riemann system $\bar{\partial}_\tau f(x, \tau) = 0$. In [12] the sheaf of such hyperfunctions was introduced by $\mathscr{BO} := \mathscr{H}^n_{\mathbf{R}^n \times \mathbf{C}^m}(\mathcal{O}_{z,\tau})$. For a time the restriction of this sheaf to the real domain $\mathscr{BO}|_{\mathbf{R}^{n+m}}$ was also called by the same name. Such a hyperfunction $f(x, t)$ of course contains t as real analytic parameters and moreover satisfies

$$\text{S.S. } f(x, t) \subset \mathbf{R}^{n+m} \times i\mathbf{S}^{n-1} \, dx. \qquad (1.3)$$

However, this estimate is weaker than the fact $f \in \mathscr{BO}|_{\mathbf{R}^{n+m}}$. This sheaf may be used to formulate a necessary and sufficient condition of some existence problems for linear partial differential equations with constant coefficients (see [5]).

There are a few generalizations of these notions. Also, for those hyperfunctions whose supports are compact with respect to x, we can paraphrase the definition of analytic parameters by means of topological terms (see [7]).

2. Holmgren-Type Theorem and Watermelon Theorem

These two theorems are not necessarily concerned with the hyperfunctions with analytic parameters. But they are fundamental to understand the meaning of unique continuation properties.

Theorem 2.1 (Holmgren-type theorem of Kawai–Kashiwara). *Let $f(x, t)$ be a germ of hyperfunction on $\mathbf{R}^{n+1}$ such that $O \in \text{supp} f \subset \{t \geq \varphi(x)\}$, where*

O denotes the origin of $\mathbf{R}^{n+1}$ and $t = \varphi(x)$ is a hypersurface of class C^1 satisfying $\varphi(0) = \nabla\varphi(0) = 0$. Then we have $(O, \pm i\,dt) \in \text{S.S.}\,f$.

This theorem was first proved by Kawai in the form of Theorem 3.1 below (hence with the weaker conclusion that either of $(O, \pm i\,dt)$ must belong to S.S. f). This final form is due to Kashiwara. We introduce here a sketch of an elementary proof employing the twisted Radon decomposition of $\delta(x)$, in order to prepare the related notation which will be repeatedly used in the sequel. (This proof is given in [4]. Its original form, employing the classical Radon transform of f, is found in [8], the manuscript of which I consulted in writing [4].) Let $W(x, \omega)$ be the component of Kashiwara's twisted Radon decomposition of $\delta(x)$:

$$W(x, \omega) = \frac{(n-1)!}{(-2\pi i)^n} \frac{(1 - ix\omega)^{n-1} - (1 - ix\omega)^{n-2}(x^2 - (x\omega)^2)}{\{x\omega + i(x^2 - (x\omega)^2) + i0\}^n},$$

and put

$$W(x, \Gamma_j^\circ) = \int_{\Gamma_j^\circ \cap \mathbf{S}^{n-1}} W(x, \omega)\,d\omega, \tag{2.1}$$

where $\mathbf{S}^{n-1} = \bigcup \Gamma_j^\circ$ is a partition by closed convex proper cones. First assuming $\varphi(x) = x^2$, we consider

$$f_j(x, t) = \int_{\mathbf{R}^n} f(y, t)\,W(x - y, \Gamma_j^\circ)\,dy.$$

If either of $(O, \pm i\,dt)$ is absent from S.S. f, then S.S. $f_j(x, t)$ is easily seen to be contained in one closed convex proper cone Δ_j°. Hence it can be written as $F_j((x, t) + i\Delta_j O)$. By the injectivity of boundary operation, the vanishing of f_j on $t < 0$ implies $F_j(z, \tau) \equiv 0$. Thus

$$f(x, t) = \sum_{j=1}^{N} f_j(x, t) \equiv 0.$$

By a real analytic local coordinate transformation fixing the conormal elements $(O, \pm i\,dt)$, any hypersurface $t = \varphi(x)$, of class C^2 with $\varphi(0) = \nabla\varphi(0) = 0$, can be brought into the region $t \geq x^2$. Note however that this is not true for a hypersurface of class C^1. But then we can employ the method of sweeping out by a family of real analytic hypersurfaces with the conormal elements $(O, \pm i\,dt)$ to obtain the same conclusion. (This remark is due to T. Ôaku.)

This theorem has various microlocalized versions. Since they are rather results on propagation of micro-analyticity, we omit to introduce them (see, e.g., Laurent [10]).

Theorem 2.2 (Watermelon theorem). *Let $f(x, t)$ be a germ of hyperfunction satisfying $O \in \operatorname{supp} f(x, t) \subset \{t \geq \varphi(x)\}$, where $\varphi(x)$ is a function of class C^2 satisying $\varphi(0) = \nabla\varphi(0) = 0$. Then the fiber $(S.S. f)_O$ of S.S. f at O has the form of the union of slices of watermelons*:

$$(S.S. f)_O = \{\pm i\, dt\} \cup \rho^{-1}(E),$$

where ρ denotes the projection from $\mathbf{S}^n \backslash \{\pm i\, dt\}$ to the equator $\mathbf{S}^{n-1} = \{i\xi\, dx;\ \xi \neq 0\}$ and $E \subset \mathbf{S}^{n-1}$ is a closed subset.

Proofs of this were given by various authors: Morimoto-Kashiwara (unpublished), Kataoka [9], and myself [3]; Sjöstrand [13] gave a proof which simultaneously implies the Holmgren-type theorem, too. We shall sketch here a very elementary proof which is a simplification of mine by M. Noro.

It suffices to show that $(S.S. f)_O \cap \rho^{-1}(i\xi_0\, dx) = \varnothing$ from the assumption $i\xi_0\, dx + i\theta\, dt \notin (S.S. f)_O$ for some θ. By a real analytic coordinate transformation fixing the conormal $(O, \pm i\, dt)$, we can assume that $\operatorname{supp} f \subset \{t \geq x^2\}$. Hence we can use the convolution by (2.1) and therefore assume that

$$(S.S. f)_O \subset \{i\xi_0\, dx + i\theta\, dt;\ |\xi - \xi_0| < \varepsilon,\ -\infty \leq \theta \leq c_1 \text{ or } c_2 \leq \theta \leq +\infty\}.$$

Choose $\xi_0 = (1, 0, \ldots, 0)$ to fix the discussion. Then by a linear coordinate transformation of the form

$$\tilde{x}_1 = x_1 + \lambda t, \qquad \tilde{x}_2 = x_2, \qquad \ldots, \qquad \tilde{x}_n = x_n, \qquad \tilde{t} = t,$$

which preserves the conormal direction dt, we can reduce the situation to the case $c_1 < 0 < c_2$. Then f admits a boundary value representation of the form

$$f(x, t) = F_+(x + i\Gamma 0, t) - F_-(x - i\Gamma 0, t),$$

where $F_\pm(z, t)$ agree to a common holomorphic function $F(z, \tau)$ through $t < 0$. Then $G(z, \tau) = F(z, \tau^2)$ is defined and holomorphic on a neighborhood of $\Omega \times T + \{(0, \ldots, 0, is);\ 0 < s < \varepsilon\}$, and hence by Kashiwara's lemma it is holomorphic in (z, τ) on a wedge-like neighborhood. An elementary calculation on the domain of holomorphy of $F(z, \tau) = G(z, \sqrt{\tau})$ then shows that the opening of the wedge on which F is defined approaches the half-space $\pm \operatorname{Im} \tau > 0$ as $t = \operatorname{Re} \tau$ approaches 0. Thus $(S.S. f)_O = \{\pm i\, dt\}$.

As far as I know, it is not yet determined if the assumption of C^2 regularity for $\varphi(x)$ can be replaced by that of C^1 here, too.

3. Restriction Data and Uniqueness

If $f(x, t)$ is a hyperfunction with real analytic parameters t, then we can consider the restriction data with respect to t:

$$\partial_t^\alpha f(x, t)|_{t=0} = \partial_{t_1}^{\alpha_1}(\cdots(\partial_{t_m}^{\alpha_m} f(x, t)|_{t_m=0})\cdots)|_{t_1=0}, \qquad \alpha \in (\overline{\mathbf{Z}^+})^m.$$

Without loss of generality we shall assume that the parameter t is in one variable. From the naming of "real analytic parameter" one may imagine that we will have the unique continuation property with respect to the Taylor coefficients

$$\partial_t^k f(x, t)|_{t=0}, \qquad k = 0, 1, 2, \ldots. \tag{3.1}$$

But Sato showed by an example that this is not true in general. In this section we collect results related with this problem. The following is most fundamental:

Theorem 3.1. *Let* $f(x, t) \in \mathscr{B}\mathscr{A}(\mathbf{R}^n \times T)$. *Assume that* $\operatorname{supp} f \subset K \times T$, *where* $K \subset \mathbf{R}^n$ *is a compact subset. Then the data* (3.1) *determines* f *completely.*

In view of the decomposition formula

$$f(x, t) = \sum_{j=1}^N F_j(x + i\Gamma_{j0}, t),$$

$$F_j(z, t) = \int_{\mathbf{R}^n} f(x, t) W(z - x, \Gamma_j^\circ) \, dx,$$

this theorem can be reduced to the following:

Lemma 3.2. *Let* $F(x + i\Gamma 0, t) \in \mathscr{B}\mathscr{A}(\Omega \times T)$. *Then the data* (3.1) *determines* $F(z, \tau)$ *completely.*

This lemma is easily obtained from the injectivity of the boundary value operation. In fact, the vanishing of (3.1) will imply the vanishing of all $\partial_\tau^k F(z, \tau)|_{\tau=0}$, hence $F(z, \tau) \equiv 0$.

From this theorem we can see especially $f(x, t) \equiv 0$ from the assumption $f(x, t)|_{t=c} = 0$ for every c. However, this form of uniqueness holds for general hyperfunctions with real analytic parameter:

Theorem 3.3.   *Let* $f(x, t) \in \mathcal{BA}(\Omega \times T)$. *If* $f(x, t)|_{t=c} = 0$ *for every* $c \in T$, *then* $f(x, t) \equiv 0$ *as a hyperfunction of* $n + 1$ *variables.*

This theorem was first found by T. Oshima, and then rediscovered by K. Kataoka. But both neglected to write the proof, of which the outline is as follows: Let $\Omega_1 \subset\subset \Omega_2 \subset\subset \Omega_3 \subset\subset \Omega$. We first choose a modification $\tilde{f}(x, t)$ such that $\mathrm{supp}\, \tilde{f} \subset \overline{\Omega}_3 \times T$, $\tilde{f} = f$ on $\Omega_1 \times T$ and such that $\tilde{f}$ still contains t as real analytic parameter. (We can construct such an $\tilde{f}$ as follows: Employing the flabbiness of the sheaf $\mathcal{C}$ of microfunctions, we first take a hyperfunction g such that S.S. $g \subset \overline{\text{S.S.}\, f}|_{\Omega_2 \times T}$ and that $h = g - f$ is real analytic in $\Omega_2 \times T$, then $\tilde{f}(x, t) := g(x, t)\chi_{\Omega_3}(x) - h(x, t)\chi_{\Omega_1}(x)$ has the desired properties. A similar technique shows that $\mathcal{BA}|_{t=0}$ is a soft sheaf on $\mathbf{R}^n$.)
Next we consider

$$F_j(z, t) = \int_{\mathbf{R}^n} \tilde{f}(x, t)\, W(z - x, \Gamma_j^{\circ})\, dx.$$

Originally, this is holomorphic in (z, τ) on a neighborhood of $(\Omega_3 + i\Gamma_j 0) \times T$. But since $\tilde{f}(x, t) \equiv 0$ in Ω_1 for each fixed t, the domain of the integral is in fact reduced to the annulus-like domain $\overline{\Omega}_3 \backslash \Omega_1$. Hence for each fixed $t \in T$, $F_j(z, t)$ can be continued in z to a fixed complex neighborhood of Ω_1 determined by the domain of holomorphy of $W(z, \Gamma_j^{\circ})$. Hence we can apply the Malgrange-Zerner theorem to conclude that $F_j(z, t)$ can be analytically continued with respect to the combined variables (z, t) to a neighborhood of $\Omega_1 \times T$. Thus $\tilde{f}(x, t) = \sum_{j=1}^{N} F_j(x + i\Gamma_j 0, t)$ and hence $f(x, t)$ become real analytic on $\Omega_1 \times T$, and the assertion of uniqueness is now obvious.
The above proof shows that we have the following generalization:

Theorem 3.4.   *Let* $f(x, t) \in \mathcal{BA}(\Omega \times T)$. *Assume that there exists a fixed complex neighborhood U of Ω such that for each fixed t, $f(x, t)$ can be continued to a holomorphic function of x on U. Then $f(x, t)$ is real analytic in the joint variables* (x, t).

Note however that the assumption that $f(x, t)$ is real analytic on Ω for each fixed t implies nothing. As an example, consider $f(x, t) = t \exp(1/(x + it^2 + i0))$. For each fixed $t \in \mathbf{R}$, $f(x, t)$ is obviously real analytic (especially $f(x, 0) \equiv 0$). But $f(x, t)$ is even beyond a distribution.
Returning to the case of restriction data to one point $t = 0$, we have the following

Theorem 3.5.   *Let $f(x, t)$ be a germ of $\mathcal{BA}$ at $O \in \mathbf{R}^{n+1}$. If the data $J(\partial_t)f(x, t)|_{t=0}$ vanish (resp. are real analytic) as germs of $\mathcal{B}$ at 0 for all the local operators with constant coefficients $J(\partial_t)$, then $f(x, t)$ vanishes (resp. is real analytic) as a germ of $\mathcal{BA}$ at O.*

The proof is technical so we omit it (see [2]). Further, Chou–Marti [1] showed that we can diminish the normal derivations $J(\partial_t)$ up to all the non-quasi-analytic type ones.

Sato's counterexample is given in [2]. In general whether the countable data (3.1) suffice or not to determine $f(x, t)$ on a neighborhood of $t = 0$, depends on the regularity of $f(x, t)$ as a generalized function. See [7] for further such discussion.

4.   Analyticity of the Discrete Loci

Consider now a hyperfunction $f(x, t)$ with real analytic parameters t whose support is limited to a submanifold of the form $x = \varphi(t)$. It is natural to expect that in this case φ should undergo a severe restriction. In fact we have the following

Theorem 4.1.   *Let $f(x, t)$ be a germ of $\mathcal{BA}$ possessing a non-trivial support contained in the graph of a germ of continuous mapping $x = \varphi(t)$. Then $\varphi(t)$ is necessarily a real analytic germ.*

This theorem can be reduced to the following function-theoretic assertion generalizing a classical theorem by Hartogs:

Lemma 4.2.   *Let $F(z, \tau)$ be a function holomorphic on $\mathbf{C} \times \Delta \setminus B$, where $\Delta = \Delta_1 \times \cdots \times \Delta_m$ is a polydisc and B is a closed subset of $\mathbf{C} \times \Delta$ such that for each $\tau \in \Delta$ the section $B_\tau := \{z \in \mathbf{C}; (z, \tau) \in B\}$ is compact, and moreover it reduces to one point $\{\varphi(t)\}$ for real t. Then $F(z, \tau)$ can be continued to another domain of similar form such that B_τ always reduces to one point $\{\varphi(\tau)\}$, and it is the locus of a holomorphic function $z = \varphi(\tau)$ (unless B is completely removable).*

Recall that Hartogs's classical theorem asserts the analyticity of $\varphi(\tau)$ under the assumption that $B_\tau = \{\varphi(\tau)\}$ is always one point. Recall also that in the case when $C \times \Delta \setminus B$ is a Stein domain it is already shown by Oka [11] that B_τ is in fact the set of zeros of a holomorphic function.

We shall show the analyticity of the coordinate functions $x_j = \varphi_j(t)$, $j = 1, \ldots, n$, componentwise. To fix the discussion, set $j = 1$. Then in view of Theorem 3.1 we can find some real analytic function $\psi(x)$ such that

$$g(x_1, t) = \int_{\mathbf{R}^{n-1}} f(x, t)\psi(x)\, dx_2 \cdots dx_n$$

becomes a hyperfunction with nontrivial support contained in $x_1 = \varphi_1(t)$. Thus we can reduce n to 1. In that case $f(x, t)$ admits a boundary value representation with two terms,

$$f(x, t) = F_+(x + i0, t) - F_-(x - i0, t).$$

Here $F_\pm(z, \tau)$ are holomorphic on a wedge with the opening $\pm\Gamma$ containing $(\pm 1, 0, \ldots, 0)$, respectively. Since $\operatorname{supp} f \subset C := \{x = \varphi(t)\}$, in view of the edge-of-the-wedge theorem they are patched to one holomorphic function $F(z, \tau)$ through the real domain $\mathbf{R}^{1+m} \setminus C$. If we choose the component at infinity of the Laurent expansion of $F(z, \tau)$ with respect to z if necessary, we obtain a holomorphic function in (z, τ) which is defined on a domain as just described in Lemma 4.2.

I could only prove this lemma for some restricted cases including the case when $f(x, t)$ is a distribution. Then T. Ohsawa informed me that the above lemma is a particular case of the following result of H. Yamaguchi in geometric function theory, which extends the work of T. Nishino in this direction:

Theorem Y. *Let $f: M \to \Delta$ be a holomorphic mapping from a two-dimensional Stein manifold to the unit disc $\{\tau \in \mathbf{C}; |\tau| < 1\}$. Assume that the fibers $M_\tau = f^{-1}(\tau)$ contain Riemann surfaces of parabolic type in their irreducible components for τ belonging to a subset of Δ of positive logarithmic capacity. Then all the fibers in fact consist of irreducible components of parabolic type. Further, if M_τ are planar domains for τ belonging to a subset of Δ of positive logarithmic capacity, then all M_τ become planar, and the part M' of M lying on a little shrunk $\Delta' \subset\subset \Delta$ admits a uniformization of the form*

$$
\begin{array}{ccc}
M' & \longrightarrow & \mathbf{CP}^1 \times \Delta' \\
\downarrow & & \downarrow \\
\Delta' & \xrightarrow{\ id\ } & \Delta'.
\end{array}
$$

The deduction of Lemma 4.2 from Theorem Y is as follows: Since it suffices to show that B_τ is one point for τ with one-by-one complexified τ_j, we can assume that τ is also one variable. Then let M be the natural domain of definition of $F(z, \tau)$. It is a Stein Riemannian domain extended over $\mathbf{C} \times \Delta$. Let $M \to \Delta$ be the natural projection. Its fiber M_t for real t is the original one $\mathbf{C}\setminus\{\varphi(t)\}$ unless F can be extended univalently to the whole $\mathbf{C} \times \Delta$. Since the real interval has positive logarithmic capacity, we can thus apply Theorem Y to M to obtain a uniformization in $\mathbf{CP}^1 \times \Delta$. (We neglect to shrink Δ.) Thus by the above-mentioned classical theorem of Oka we conclude that M has a planar fiber of the form $\mathbf{C}\setminus\{\varphi(\tau)\}$ for every τ, where $\varphi(\tau)$ is a holomorphic function.

Theorem Y is found in Yamaguchi [14], where the uniformization assertion is proved under the assumption that all the fibers M_τ are planar. The above definitive form was given in Yamaguchi [15]. Ignoring the latter when writing [6], we only cited the results of [14] and reproved a part of the uniformization theorem for our special case with the "general" fiber isomorphic to $\mathbf{C}\setminus\{0\}$, by introducing the fiberwise universal covering of M.

The same argument also gives the following somewhat general form:

Theorem 4.1′. *Let $f(x, t)$ be a germ of $\mathcal{B}\mathcal{A}$ possessing a nontrivial support contained in a closed subset B with discrete fibers B_t. Then B_t agrees with the common zeros of germs of some real analytic functions $\varphi_j(x_j, t) = 0$, $j = 1, \ldots, n$.*

Note however that analytic subsets with full singularities never appear in this way. In fact, the assumption $f \in \mathcal{B}\mathcal{A}$ together with the Holmgren-type theorem requires that the set cannot have conormal directions in generalized sense of the form $i\theta\, dt$.

As an application of this theorem we can prove the following

Example 4.3. Let $f(x, t)$ be a germ of real analytic solution of the wave equation

$$(\partial_t^2 - \Delta_x)f(x, t) = 0 \qquad \text{on } \mathbf{R}^{n+1},$$

defined at O outside a continuous curve $C = \{x = \varphi(t)\}$ which is contained in a non-spacelike, non-characteristic, real analytic hypersurface S. Unless $f(x, t)$ can be continued as a (hyperfunction) solution to a whole neighborhood of O, C must be a germ of a real analytic curve. In fact, assume $S = \{x_1 = 0\}$ for simplicity. Then we can apply the above theorem to the differences of the boundary values $f_k(x', t) = \partial_1^k f(+0, x', t) - \partial_1^k f(-0, x', t)$, $k = 0, 1$, which obviously have supports in C, and which contain t as real analytic parameter by a result on the micro-analyticity of the boundary values of real analytic solutions. On the other hand, by a general theory of non-characteristic boundary value problem we know that the triviality of all $f_k(x', t)$ implies that $f(x, t)$ can be continued to C as a hyperfunction solution. This argument applies to prove the real analyticity of minimal dimensional singularities of real analytic solutions of more general equations. See [6].

References

[1] C. C. Chou and J.-A. Marti, Un théorème d'unicité sur les hyperfonctions régulières par rapport à une des variables, *C.R. Acad. Sc. Paris* **288** (1979) 493-495.

[2] A. Kaneko, Remarks on hyperfunctions with analytic parameters, *J. Fac. Sci. Univ. Tokyo Sec. 1A* **22** (1975) 371-407; II, *ibid.* **25** (1978) 67-73.

[3] A. Kaneko, Introductions aux Hyperfonctions, Cours de DEA, Univ. de Grenoble, 1978.

[4] A. Kaneko, *Introduction to Hyperfunctions*, Reidel, to appear (original Japanese in two volumes 1980-1982).

[5] A. Kaneko, Estimation of singular spectrum of boundary values for some semihyperbolic operators, *J. Fac. Sci. Univ. Tokyo Sec. 1A* **27** (1980) 401-461.

[6] A. Kaneko, On the analyticity of the locus of singularity of real analytic solutions with minimal dimension, *Nagoya Math. J.* **104** (1986) 63-84.

[7] A. Kaneko, Remarks on hyperfunctions with analytic parameters III, in preparation.

[8] M. Kashiwara, T. Kawai, and T. Kimura, *Foundation of Algebraic Analysis*, Princeton, 1986 (original Japanese 1980).

[9] K. Kataoka, On the theory of Radon transformations of hyperfunctions, *J. Fac. Sci. Univ. Tokyo Sec. 1A* **28** (1981) 331-413.

[10] Y. Laurent, *Théorie de la Deuxième Microlocalisation dans le Domaine Complexe*, Birkhäuser, 1985.

[11] K. Oka, Note sur les familles de fonctions analytiques multiformes etc., *J. Sci. Hiroshima Univ.* **4** (1934) 93-98.

[12] M. Sato, Theory of hyperfunctions II, *J. Fac. Sci. Univ. Sec. I* **8** (1960) 387-437.

[13] J. Sjöstrand, Singularités analytiques microlocales, *Astérisque, Soc. Math. France* **95** (1982) 1-166.

[14] H. Yamaguchi, Parabolicité d'une fonction entière, *J. Math. Kyoto Univ.* **16** (1976) 71-92.

[15] H. Yamaguchi, Famille holomorphe de surfaces de Riemann ouvertes, *ibid.* **16** (1976) 497-530.

[16] A. Kaneko, Nishino-Yamaguchi theory and its application to the theory of hyperfunctions, Proceedings of Workshop in Pure Math. Vol. 7, Part II (Ed. K. S. Chang and H. O. Kim), Seoul, 1987, pp. 240-261.

[17] A. Kaneko, A topological characterization of hyperfunctions with real analytic parameters, *Sci. Pap. Coll. Gen. Educ. Univ. of Tokyo* **38** (1988) 1-6.

The Invariant Holonomic System on a Semisimple Lie Group

Masaki Kashiwara

Research Institute for Mathematical Sciences
Kyoto University
Kyoto, Japan

0. Introduction

0.1.

Let G be a connected reductive algebraic group defined over $\mathbf{C}$ and $\mathfrak{G}$ its Lie algebra. The center $\mathfrak{Z}(\mathfrak{G})$ of the universal enveloping algebra $U(\mathfrak{G})$ is identified with the ring of bi-invariant differential operators on G. Let θ_G be the sheaf of vector fields on G. The adjoint action of G on G induces the Lie algebra homomorphism

$$\mathrm{Ad}: \mathfrak{G} \to \Gamma(G, \theta_G). \tag{0.1}$$

We shall denote by $\mathscr{D}_G$ the ring of differential operators on G. For any character $\chi: \mathfrak{Z}(\mathfrak{G}) \to \mathbf{C}$, let $\mathscr{M}_\chi$ be the $\mathscr{D}_G$-module $\mathscr{D}_G/(\mathscr{D}_G \, \mathrm{Ad}(\mathfrak{G}) + \sum_{P \in \mathfrak{Z}(\mathfrak{G})} \mathscr{D}_G(P - \chi(P)))$.

If $G_{\mathbf{R}}$ is a real form of G, any invariant eigendistribution satisfies the system of differential equations $\mathcal{M}_\chi$. The property of $\mathcal{M}_\chi$ is deeply investigated by Harish-Chandra [H-C]). In this article, we shall give the proof of the following theorem on $\mathcal{M}_\chi$.

Theorem 1. (i) $\mathcal{M}_\chi$ is a regular holonomic $\mathcal{D}_G$-module. (ii) $\mathcal{M}_\chi$ is the minimal extension (see Section 2.2) of $\mathcal{M}_\chi|G_{\mathrm{reg}}$. Here G_{reg} is the open set of semisimple regular elements of G.

The version of this theorem in the Lie algebra case is already proven in [H-K], and we shall use this result to prove Theorem 1. When χ is a trivial infinitesimal character, this theorem is shown in [H-K'] by a completely different method.

The author acknowledges R. Hotta for some interesting discussions.

0.2.

Let T^*G be the cotangent bundle of G and we identify T^*G with $G \times \mathfrak{G}^*$. Let $N(\mathfrak{G}^*)$ be the set of nilpotent elements of $\mathfrak{G}^*$. Let V be the common zeroes of the principal symbol of $\mathrm{Ad}(\mathfrak{G})$. Hence we have

$$V \cong \{(g, \xi) \in G \times \mathfrak{G}^*; \ g \cdot \xi = \xi\}. \tag{0.2}$$

Here $g \cdot \xi$ is the coadjoint action on $\mathfrak{G}^*$. We shall also prove the following theorem.

Theorem 2. (i) (Richardson [R]) V is irreducible with dimension $\dim G + \mathrm{rank}\ G$.
　(ii) $\dim(V \cap (G \times N(\mathfrak{G}^*))) = \dim G$.
　(iii) The characteristic cycle of $\mathcal{M}_\chi$ coincides with $V \cdot (G \times N(\mathfrak{G}^*))$.

The statement (i) is proven by Richardson ([R]).

0.3.

We define a $\mathcal{D}_G$-module $\mathcal{N}_G$ by $\mathcal{N}_G = \mathcal{D}_G / \mathcal{D}_G \ \mathrm{Ad}(\mathfrak{G})$. Then $\mathfrak{z}(\mathfrak{G})$ acts on $\mathcal{N}_G$ by the right multiplication.

For a $\mathfrak{z}(\mathfrak{G})$-module M, we set

$$\mathcal{N}_G(M) = \mathcal{N}_G \underset{\mathfrak{z}(\mathfrak{G})}{\otimes} M. \tag{0.3}$$

Hence if M is the one-dimensional $\mathfrak{Z}(\mathfrak{G})$-module corresponding to a character χ of $\mathfrak{Z}(\mathfrak{G})$, $\mathscr{N}_G(M)$ is nothing but $\mathscr{M}_\chi$. We shall also prove the following theorem.

Theorem 3. (i) $\mathscr{N}_G$ *is flat over* $\mathfrak{Z}(\mathfrak{G})$.

(ii) $\mathscr{E}xt^j_{\mathscr{D}_G}(\mathscr{N}_G, \mathscr{D}_G) = 0$ *for* $j \neq \dim G - \mathrm{rank}\, G$.

(iii) *Assume further* G *is semisimple. Then for any* $\mathfrak{Z}(\mathfrak{G})$-*module* M, *we have* $\mathscr{H}^0_{N(G)}(\mathscr{N}_G(M)) = 0$, *where* $N(G)$ *is the set of unipotent elements of* G.

1.   Proof of Theorems 1 and 3

1.1.

We shall prove these theorems by induction on $\dim G$. We can reduce to the case where G has a trivial center. The following lemma is proven by Harish-Chandra.

Lemma 1.1 ([H-C]).   *Let* S *be a nonempty closed subset of* G *invariant by the adjoint action. If* S *contains no semisimple element other than* 1, *then* S *is contained in* $N(G)$.

1.2.

Hereafter we assume that G has a trivial center. Let us take a semisimple element $a \neq 1$, and let G' be the centralizer of a and $\mathfrak{G}'$ its Lie algebra. Then by the hypothesis of induction, Theorems 1 and 3 are true for G'.

We can easily prove

Lemma 1.2. $\pi^{-1}(a) \cap T^*_{G'}G \cap V \subset T^*_G G$. *Here* π *is the projection* $T^*G \to G$ *and* $T^*_G G$ *is the conormal bundle.*

Corollary 1.3. $\mathscr{N}_G|_{G'}$ *is generated by* $u_G|_{G'}$ *as a* $\mathscr{D}_{G'}$-*module on a neighborhood of* a. *Here* u_G *is the canonical generator of* $\mathscr{N}_G$.

1.3.

Lemma 1.4.   *If Theorem 3 is true for* G, *then* $\mathscr{H}^0_S(\mathscr{N}_G) = 0$ *for any closed nowhere dense subset* S *of* G.

In fact, we have ch $\mathcal{H}^0_S(\mathcal{N}_G) \subset \pi^{-1}(S) \cap V$, and hence codim ch $\mathcal{H}^0_S(\mathcal{N}_G) >$ dim $G -$ rank G. Since $\mathcal{E}xt^j(\mathcal{N}_G, \mathcal{D}_G) = 0$ for $j \neq$ dim $G -$ rank G, $\mathcal{H}^0_S(\mathcal{N}_G) = 0$ (Theorem 2.12 [K']).

We can also prove easily by induction on dim M.

Lemma 1.5. *If Theorem 1 is true for G, then for any finite-dimensional* $\mathfrak{Z}(\mathfrak{G})$*-module M, $\mathcal{N}_G(M)$ is regular holonomic and it is the minimal extension of $\mathcal{N}_G(M)|_{G_{\mathrm{reg}}}$.*

1.4.

Set $\nu(g) = \det(\mathrm{Ad}(g) - 1; \mathfrak{G}/\mathfrak{G}')$ for $g \in G'$. Then $u_{G'} \mapsto \nu^{1/2}(u_G|_{G'})$ defines a $\mathcal{D}_{G'}$-linear homomorphism on a neighborhood of a:

$$\mathcal{N}_{G'} \to \mathcal{N}_G|_{G'}. \tag{1.1}$$

The hypothesis of induction along with Lemma 1.4 implies

$$\mathcal{H}^0_{G' \backslash G_{\mathrm{reg}}}(\mathcal{N}_{G'}) = 0. \tag{1.2}$$

Since (1.1) is surjective by Corollary 1.3 and bijective on $G' \cap G_{\mathrm{reg}}$, (1.2) implies that (1.1) is an isomorphism on a neighborhood of a.

Let us embed $\mathfrak{Z}(\mathfrak{G})$ into $\mathfrak{Z}(\mathfrak{G}')$. Then $\mathfrak{Z}(\mathfrak{G}')$ is a free $\mathfrak{Z}(\mathfrak{G})$-module of finite rank. By Harish-Chandra [H-C], (1.1) is $\mathfrak{Z}(\mathfrak{G})$-linear on $G' \cap G_{\mathrm{reg}}$. Hence (1.2) implies the following lemma.

Lemma 1.6. *$\mathcal{N}_{G'}$ and $\mathcal{N}_G|_{G'}$ are isomorphic as $(\mathcal{D}_G, \mathfrak{Z}(\mathfrak{G}))$-bimodules on a neighborhood of a.*

1.5.

Now, we shall show Theorem 3 and

$$\mathcal{H}^0_{G \backslash G_{\mathrm{reg}}}(\mathcal{M}_\chi) = 0 \tag{1.3}$$

on a neighborhood of a.

Lemma 1.7. *If $\mathcal{L}$ is a coherent $\mathcal{D}_G$-module such that ch $\mathcal{L} \subset V$ and $\mathcal{L}|_{G'} = 0$, then $\mathcal{L} = 0$ on a neighborhood of a.*

This follows immediately from Lemma 1.4 and Theorem 2.6.17 in [K].

For a $\mathfrak{Z}(\mathfrak{G})$-module M, set $M' = M \underset{\mathfrak{Z}(\mathfrak{G})}{\otimes} \mathfrak{Z}(\mathfrak{G}')$. Then we have

$$\mathcal{T}or_j^{\mathfrak{Z}(\mathfrak{G}')}(\mathcal{N}_{G'}, M') \simeq \mathcal{T}or_j^{\mathfrak{Z}(\mathfrak{G})}(\mathcal{N}_G, M)|_{G'}.$$

Hence we have $\mathcal{T}or_j(\mathcal{N}_G, M) = 0$ for $j \neq 0$. The other statements follow in a similar way.

1.6.

By using Lemma 1.1, Theorem 3 and (1.3) is true outside $N(G)$.

2. Proof of Theorem 1 (Continued)

2.1.

In order to describe $\mathcal{M}_\chi|_{G_{\mathrm{reg}}}$, let us choose a Cartan subgroup T of G. Let t be its Lie algebra, Δ the root system for $(\mathfrak{G}, t)$ and W the Weyl group. For $\alpha \in \Delta$, let ξ_α be the corresponding character of T. Set $T_{\mathrm{reg}} = T \cap G_{\mathrm{reg}}$. Let $\varphi : \mathfrak{Z}(\mathfrak{G}) \to U(t)^W$ be the canonical isomorphism and we identify $U(t)$ with the ring of invariant differential operators on T. Then by Harish-Chandra ([H-C]), $\mathcal{M}_\chi|_{T_{\mathrm{reg}}}$ is equal to the system of differential equations

$$D^{-1/2}(\varphi(P) - \chi(P))D^{1/2}u = 0 \qquad \text{for } P \in \mathfrak{Z}(\mathfrak{G}), \tag{2.1}$$

where

$$D = \prod_{\alpha \in \Delta} (\xi_\alpha^{1/2} - \xi_\alpha^{-1/2}).$$

2.2.

Since (2.1) is regular holonomic, $\mathcal{M}_\chi|_{G_{\mathrm{reg}}}$ is regular holonomic. Let $^\pi(\mathcal{M}_\chi|_{G_{\mathrm{reg}}})$ be its minimal extension, i.e., a regular holonomic $\mathscr{D}_G$-module such that $^\pi(\mathcal{M}_\chi|_{G_{\mathrm{reg}}})|_{G_{\mathrm{reg}}} \cong \mathcal{M}_\chi|_{G_{\mathrm{reg}}}$ and such that it has neither non-zero submodule nor quotient whose support is contained in G_{reg}. By Harish-Chandra ([H-C]), we have

$\mathcal{M}_\chi$ has no non-zero quotient whose support is contained in $G \backslash G_{\mathrm{reg}}$.

$$\tag{2.2}$$

Hence we have a canonical homomorphism

$$\mathcal{M}_\chi \to {}^\pi(\mathcal{M}_\chi|_{G_{\mathrm{reg}}}). \tag{2.3}$$

This is evidently surjective.

2.3.

By the result of Section 1, $\operatorname{supp}(\mathcal{H}^0_{G\setminus G_{\mathrm{reg}}}(\mathcal{M}_\chi))$ is contained in the set $N(G)$ of unipotent elements, and hence (2.3) is an isomorphism outside $N(G)$. Thus it is enough to show that (2.3) is an isomorphism on a neighborhood of 1.

2.4.

Let us take a small neighborhood U of 0 in $\mathfrak{G}$ such that $\exp : \mathfrak{G} \to G$ is an isomorphism from U onto $V = \exp(U)$. Let $\tilde{\mathcal{M}}_\chi$ be the $\mathcal{D}_\mathfrak{G}$-module $\mathcal{D}_\mathfrak{G}/(\mathcal{D}_\mathfrak{G} \operatorname{ad} \mathfrak{G} + \sum_{P \in S(\mathfrak{G})^G} \mathcal{D}_\mathfrak{G}(P - \chi(P)))$. Here ad is the homomorphism $\mathfrak{G} \to \Gamma(\mathfrak{G}; \theta_g)$ given by the adjoint action. We identify $\mathfrak{Z}(\mathfrak{G})$ with $S(\mathfrak{G})^G$ and $S(\mathfrak{G})^G$ with the ring of G-invariant constant-coefficient differential operators on $\mathfrak{G}$. By [H-K], $\tilde{\mathcal{M}}_\chi$ is the minimal extension of $\tilde{\mathcal{M}}_\chi|_{\mathfrak{G}_{\mathrm{reg}}}$. Here $\mathfrak{G}_{\mathrm{reg}}$ is the set of regular semisimple elements of $\mathfrak{G}$. Moreover by [H-C], we have

$$(\exp)_* \operatorname{Hom}(\tilde{\mathcal{M}}_\chi \mathcal{O}_{\mathfrak{G}_{\mathrm{an}}}) \cong \operatorname{Hom}_{\mathcal{D}_G}(\mathcal{M}_\chi, \mathcal{O}_{G_{\mathrm{an}}}).$$

on $V \cap G_{\mathrm{reg}}$. Here G_{an} and $\mathfrak{G}_{\mathrm{an}}$ are the underlying complex manifolds. This implies

$$({}^\pi \mathcal{M}_\chi|_{G_{\mathrm{reg}}})|_V \cong \exp_*(\tilde{\mathcal{M}}_\chi|_U). \tag{2.4}$$

2.5.

Now, we shall use the same argument as in [H-K].

Let $\mathcal{L}$ be the kernel of (2.3) and S the support of $\mathcal{L}$. Then one can easily show (by a similar argument as in [H-K], Section 6.6) that codim $S \geq 2$. Assume $S \ni e$. We take a generic point g of $S \cap V$. Since $\mathcal{E}xt^1_{\mathcal{D}_\mathfrak{G}}(\tilde{\mathcal{M}}_\chi, \mathcal{B}_{(\exp)^{-1}S|\mathfrak{G}})_{\log(g)} = 0$ by Lemma 6.7.1 [H-K],

$$0 \to \mathcal{L} \to \mathcal{M}_\chi \to {}^\pi(\mathcal{M}_\chi|_{\mathfrak{G}_{\mathrm{reg}}}) \to 0 \tag{2.5}$$

splits on a neighborhood of g. Hence $\mathcal{M}_\chi$ has a non-zero quotient supported in $G\setminus G_{\mathrm{reg}}$. This contradicts Harish-Chandra's result (2.2).

3. Proof of Theorem 3 (Continued)

3.1.

We proved already Theorem 3 outside the set $N(G)$ of unipotent elements. In the sequel, we use the following simple result.

Lemma 3.1. *Let $\mathcal{M}$ be a coherent $\mathcal{D}_G$-module such that* $\text{ch}\,\mathcal{M} \subset \pi^{-1}(N(g)) \cap V$. *Then $\mathcal{M}$ is holonomic.*

In fact $\pi^{-1}N(G) \cap V$ is Lagrangean because $N(G)$ has finitely many G-orbits.

3.2.

We shall first prove Theorem 3(iii) by induction on dim M. If dim $M = 0$, this is true by Lemma 1.5. Hence we shall assume dim $M > 0$. Let M' be the largest submodule of M with dim $M' = 0$, and let $M'' = M/M'$. Since $0 \to \mathcal{N}_G(M') \to \mathcal{N}_G(M) \to \mathcal{N}_G(M'') \to 0$ is exact outside $N(G)$ and $\mathcal{H}^0_{N(G)}(\mathcal{N}_G(M')) = 0$, this is exact on the whole of G. Hence in order to prove $\mathcal{H}^0_{N(G)}(\mathcal{N}_G(M)) = 0$, it is enough to show $\mathcal{H}^0_{N(G)}(\mathcal{N}_G(M'')) = 0$. Replacing M with M'', we shall assume $M' = 0$ from the beginning. Then there exists $P \in \mathfrak{Z}(\mathfrak{G})$ such that $P - c$ acts injectively on M for any $c \in C$. Then for any non-zero polynomial $b(P)$, $\dim(M/b(P)M) < \dim M$. Set $\mathcal{L} = \mathcal{H}^0_{N(G)}(\mathcal{N}_G(M))$. Then by Lemma 3.1, $\mathcal{L}$ is holonomic and hence $\text{End}(\mathcal{L})$ is finite-dimensional. Therefore there exists a non-zero polynomial $b(P)$ such that $b(P)\mathcal{L} = 0$. Since $b(P)$ acts injectively on $\mathcal{N}_G(M)$ outside $N(G)$, $b(P)$ acts injectively in $\mathcal{N}_G(M)/\mathcal{L}$. Moreover the kernel of $b(P)$ in $\mathcal{N}_G(M)$ is $\mathcal{L}$. Hence $\mathcal{L} \to \mathcal{N}_G(M/b(P)M)$ is injective. Since $\mathcal{H}^0_{N(G)}(\mathcal{N}_G(M/b(P)M)) = 0$ by the hypothesis of induction, we have $\mathcal{L} = 0$.

3.3.

In order to prove Theorem 3(ii), we shall prove the following generalized statement.

Lemma 3.2. *If G is semisimple, $\mathcal{E}xt^j(\mathcal{N}_G(M), \mathcal{D}_G) = 0$ for $j \neq \dim G - \dim M$ for any finitely generated Cohen-Macaulay $\mathfrak{Z}(\mathfrak{G})$-module M.*

Proof. We may assume that the center of G is trivial. Since $\text{codim ch}(\mathcal{N}_G(M)) \geq \dim G - \dim M$, we have $\mathcal{E}xt^j(\mathcal{N}_G(M), \mathcal{D}_G) = 0$ for $j < \dim G - \dim M$. Since $\dim \text{proj}\, M = \text{rank}\, G - \dim M$, M has a free resolution of length $m = \text{rank}\, G - \dim M$: $0 \leftarrow M \leftarrow L_0 \leftarrow L_1 \leftarrow \cdots \leftarrow L_M \leftarrow 0$. Hence we have a resolution outside $N(G)$: $0 \leftarrow \mathcal{N}_G(M) \leftarrow \mathcal{N}_G(L_0) \leftarrow \cdots \leftarrow \mathcal{N}_G(L_M) \leftarrow 0$. Since $\mathcal{E}xt^j(\mathcal{N}_G, \mathcal{D}_G) = 0$ outside $N(G)$ for $j > \dim G - \text{rank}\, G$, we have

$$\text{supp } \mathscr{E}xt^j(\mathscr{N}_G(M), \mathscr{D}_G) \subset N(G)$$

for $j > \dim G - \text{rank } G + m = \dim G - \dim M.$

Hence $\mathscr{E}xt^j(\mathscr{N}_G(M), \mathscr{D}_G)$ is holonomic by Lemma 3.1. Now we shall proceed with the proof by induction on $\dim M$. If $\dim M = 0$, this is trivial. If $\dim M > 0$, take $P \in \mathfrak{Z}(\mathfrak{G})$ such that $\dim(\text{supp } M \cap \{P = c\}) = \dim M - 1$ for any $c \in \mathbf{C}$. Then for any non-zero polynomial $b(P)$, $\mathscr{E}xt^j(\mathscr{N}_G(M/b(P)M), \mathscr{D}_G) = 0$ for $j > \dim G - \dim M + 1$ by the hypothesis of induction. Hence $b(P)$ acts surjectively on $\mathscr{L} = \mathscr{E}xt^j(\mathscr{N}_G(M), \mathscr{D}_G)$ for $j > \dim G - \dim M$. Since $\mathscr{L}$ is holonomic, there exists a non-zero $b(P)$ such that $b(P)\mathscr{L} = 0$. This implies $\mathscr{L} = 0$.

3.4.

We shall prove Theorem 3(i). In order to see this, it is enough to show that for an injective morphism $M' \to M$ of finitely generated $\mathbf{Z}(\mathfrak{G})$-modules, $\mathscr{N}_G(M') \to \mathscr{N}_G(M)$ is injective. Since this is injective outside $N(G)$, this follows from Theorem 3(iii).

4. The Proof of Theorem 2

4.1.

Theorem 2(i) is a result of Richardson [R]. Let us prove Theorem 2(ii). Let q be the projection from V to $\mathfrak{G}^*$. Let S be a G-orbit of $N(\mathfrak{G}^*)$. Let $\xi \in S$. Then $\pi(V \cap q^{-1}(\xi)) = G_\xi$. Hence $\dim G_\xi + \dim S = \dim G$. This shows that $V \cap q^{-1}(S)$ is a non-singular manifold of $\dim G$. Since $N(\mathfrak{G}^*)$ has finitely many G-orbits, $V \cap q^{-1}(N(\mathfrak{G}^*))$ has pure dimension $\dim G$.

4.2.

Let us prove Theorem 2(iii). We may assume that G has a trivial center. By Proposition 4.8.3 and Theorem 6.1 in [H-K], the characteristic cycle of $\exp_*(\tilde{\mathcal{M}}_x)$ is $V \cdot (G \times N(\mathfrak{G}^*))$ on a neighborhood of 1 (with the notation in Section 2.4). Hence, by the result in Section 2, Theorem 2(iii) is true on a neighborhood of 1.

4.3.

If Theorem 2 is true, then for a finite-dimensional $\mathfrak{Z}(\mathfrak{G})$-module M, the characteristic cycle of $\mathscr{N}_G(M)$ is $(\dim M)V \cdot (G \times N(\mathfrak{G}^*))$.

4.4.

By Lemma 1.1, it is enough to show $\underline{\mathrm{ch}}\, \mathscr{M}_\chi = V \cdot (G \times N)$ on a neighborhood of a semisimple element a. Here $\underline{\mathrm{ch}}$ denotes the characteristic cycle. Let G' be the centralizer of a. Let ρ be the projection $G' \times_G T^*G \to T^*G'$ and $\tilde{\omega}$ the embedding $G' \times_G T^*G \to T^*G$. Then since $\mathscr{M}'$ is non-characteristic, we have $\underline{\mathrm{ch}}(\mathscr{M}_\chi|_{G'}) = \rho_* \tilde{\omega}^{-1}(\underline{\mathrm{ch}}\, \mathscr{M}_\chi)$ (see Chapter II, Section 6 [K]). Hence it is enough to show

$$\underline{\mathrm{ch}}(\mathscr{M}_\chi|_{G'}) = \rho_* \tilde{\omega}^{-1}(V \cdot G \times N).$$

Let V' and N' be the sets defined as V and N replacing G with G'. Since $\mathscr{M}_\chi|_{G'}$ is isomorphic to $\mathscr{M}_{G'}(M)$ with $M = \mathbf{C} \underset{\mathfrak{Z}(\mathfrak{G}')}{\otimes} \mathfrak{Z}(\mathfrak{G})$ by Lemma 1.4, we have

$$\underline{\mathrm{ch}}(\mathscr{M}_\chi|_{G'}) = \underline{\mathrm{ch}}(\mathscr{M}_{G'}(M))$$
$$= (\dim_{\mathbf{C}} M) \cdot V' \cdot (G' \times N'). \tag{4.1}$$

Note that $\dim_{\mathbf{C}} M = \#(W/W')$. Here W and W' are the Weyl group of G and G' respectively. Hence it is enough to show

Lemma 4.1. *On a neighborhood of a,*

$$\rho_* \tilde{\omega}^{-1}(V \cdot G \times N) = \#(W/W')(V' \cdot G' \times N').$$

Proof. Set $\mathfrak{u} = \{\xi \in \mathfrak{G}^*; a\xi = \xi\}$. Then, $\mathfrak{u} \to \mathfrak{G}^*$ is an isomorphism. We shall show that

$$(G' \times \mathfrak{G}^*) \cap V \subset G' \times \mathfrak{u} \text{ on a neighborhood of } a. \tag{4.2}$$

In fact, it is enough to show that $g: \mathfrak{G}^*/\mathfrak{u} \to \mathfrak{G}^*/\mathfrak{u}$ has no eigenvalue 1 if $g \in G'$ is sufficiently near a. This is evident. Hence $(G' \times \mathfrak{G}^*) \cap V$ is isomorphic to V'. Take homogeneous functions $f_1, \ldots, f_r$ ($r = \mathrm{rank}\, G$) on $\mathfrak{G}^*$ such that $\mathbf{C}[f_1, \ldots, f_r] = S(\mathfrak{G})^G$. Then

$$\tilde{\omega}^{-1}(V \cdot N) = ((G' \times \mathfrak{G}^*) \cap V) \cap (f_1 = \cdots = f_r = 0).$$

Hence $\rho_* \tilde{\omega}^{-1}(V \cdot N) = V' \cap (f_1 = \cdots = f_r = 0)$. Since $S(\mathfrak{G}')^{G'}$ is a free module over $S(\mathfrak{G})^G$ of rank $\#(W/W')$, $\mathfrak{G}'^* \cap (f_1 = \cdots = f_r = 0) = \#(W/W')N'$.   Q.E.D.

References

[H-C]   Harish-Chandra, Invariant eigendistributions on a semisimple Lie group, *Trans. A.M.S.* **119** (1965) 457–508.

[H-K]   R. Hotta and M. Kashiwara, The invariant holonomic system on a semisimple Lie algebra, *Invent. Math.* **75** (1984) 327–358.

[H-K']   R. Hotta and M. Kashiwara, Quotients of Harish-Chandra system by primitive ideal, *Geometry of Today*, Giornate di Geometria, Roma 1984, Progress in Math., Birkhäuser (1985) 185–205.

[K]   M. Kashiwara, Systems of Microdifferential Equations, *Progress in Math.* **34** (1983) Birkhäuser.

[K']   M. Kashiwara, *B*-functions and holonomic systems, *Invent. Math.* **38** (1976) 33–53.

[R]   R. W. Richardson, Commuting varieties of semisimple Lie algebras and algebraic groups, *Compositio Math.* **38** (1979) 311–322.

Some Applications of Microlocal Energy Methods to Analytic Hypoellipticity

Kiyômi Kataoka

Department of Mathematics
Faculty of Science
The University of Tokyo
Hongo, Bunkyo-Ku
Tokyo, Japan

Introduction

We extend our "microlocal energy methods" in microfunction theory ([10, 11]) to system-cases admitting some degeneracy. That is, we show the "quasi-positivity" of sesquilinear forms of microfunctions with coefficients in pseudo-differential operators under a weaker condition than in [10]. As applications, we give new proofs of analytic hypoellipticity for parabolic boundary value problems, and for Trèves-type operators with double characteristics. Further we give some new results for more degenerate operators.

The author would like to express his hearty gratitude to Prof. C. Parenti and Prof. A. Bove for their kind advice and encouragement.

Algebraic Analysis, Volume I

1. Preliminaries

We recall our microlocal energy methods developed in [10] (for a brief survey and easy applications, see [11]). Let X be an abstract set. Then, a C-valued function $K(x, u)$ defined on $X \times X$ is said to be a *positive Hermite kernel* if $(K(x_j, x_k))_{j,k}$ is a positive semi-definite Hermite matrix for every $N = 1, 2, \ldots$ and every $x_1, \ldots, x_N \in X$. We employ the notation "$K \gg 0$" for a positive Hermite kernel K. A simple but basic fact is the following:

$$K_1 \gg 0 \text{ and } K_2 \gg 0 \implies K_1 + K_2 \gg 0 \text{ and } K_1 K_2 \gg 0.$$

Here, the product $K_1 K_2$ is defined by the product as scalar functions on $X \times X$. Hence, the Hermite positivity induces an order structure in the R-algebra of Hermite kernels. In our theory, analytic Hermite kernels play an important role. Let X be an open set in $\mathbf{C}^n$ and X^c be the complex conjugate of X. Then, a holomorphic function $G(z, w)$ on $X \times X^c$ is said to be an analytic Hermite kernel on $X \times X$ if $G(z, \bar{w})$ is Hermitian on $X \times X$. It is the most important and well known fact for analytic Hermite kernels that "positivity" extends "analytically" (see Section 1 of [10]). By using these facts, we can formulate the microlocalization of Hermite positivity. That is, we can introduce an order structure for "Hermite microkernels," which constitute a subsheaf of $\mathscr{C}_{\mathbf{R}^{n+n}}|_{\Delta^a(iS^*\mathbf{R}^{n+n})}$. Here, $\Delta^a(iS^*\mathbf{R}^{n+n}) = \{(x, u; i\xi, i\eta); x = u, \xi = -\eta\}$.

2. The Quasi-Positivity and Some Inequalities

In this section we develop some tools. Our main result here is Theorem 2.4, which is a generalization of Theorem 3.14 of [10]; that is, we treat sesquilinear forms of microfunctions with degenerate coefficients in $\mathscr{E}^{\mathbf{R}}$. By using this result, we introduce the notion "quasi-positivity" explicitly as an order relation "$\gg_q$, or $\ll_q$". Further we give some inequalities of Sobolev's type, which will be applied in Section 3.

From now on, we use the notation

$$f^* = \overline{f(t, u)}, \qquad P^* = \overline{P(\bar{t}, \bar{w}, \bar{\eta})} \tag{2.1}$$

for the conjugates of a microfunction $f(t, x)$ or a symbol $P(t, z, \xi)$, respectively. Further we inherit some terminologies: "a formal symbol of product

Hermite type," "depending real analytically on $t \in T$ at (t_0, x_0)," etc. from Sections 2 and 3 of [10].

Lemma 2.1. *Let S be an abstract set, and N be a positive integer. Let $K = (K^{jk}(x, u))_{j,k=1,\ldots,N}$ be a Hermite kernel on $(\{1, 2, \ldots, N\} \times S)^2$; that is, the value at $(j, x) \times (k, u)$ is given by $K^{jk}(x, u)$ for any $j, k = 1, \ldots, N$ and any $x, u \in S$. Then $K \gg 0$ on $(\{1, \ldots, N\} \times S)^2$ iff*

$$\sum_{s,t=1}^{L} \sum_{j,k=1}^{N} K^{jk}(x_s, x_t) \cdot \lambda_{js}\overline{\lambda_{kt}} \ge 0 \tag{2.2}$$

holds for any $L = 1, 2, \ldots$, any $x_1, \ldots, x_L \in S$, and any $\lambda_{js} \in \mathbf{C}$ ($j = 1, \ldots, N$, $s = 1, \ldots, L$).

Proof. The necessity of (2.2) is clear. So we show the sufficiency of (2.2). Let $\{\lambda_{j's}; 1 \le j' \le N, 1 \le s \le L\}$, $\{x_{j's}; 1 \le j' \le N, 1 \le s \le L\}$ be arbitrary finite subsets respectively of $\mathbf{C}$ and S indexed by j' and s. Set $\mu_{j,j's} = \delta_{jj'}\lambda_{js}$. Therefore by (2.2), we obtain

$$0 \le \sum_{s,t=1}^{L} \sum_{j',k'=1}^{N} \sum_{j,k=1}^{N} K^{jk}(x_{j's}, x_{k't})\mu_{j,j's}\overline{\mu_{k,k't}}$$

$$= \sum_{s,t=1}^{N} \sum_{j,k=1}^{N} K^{jk}(x_{js}, x_{kt})\lambda_{js}\overline{\lambda_{kt}}. \tag{2.3}$$

Since L, $\{x_{js}\}_{js}$, $\{\lambda_{js}\}_{js}$ are arbitrary, this proves the Hermite positivity of K.

Lemma 2.2. *Let $p_0 = (z_0, \bar{z}_0; \xi_0 dz + \bar{\xi}_0 dw)$ be a point of $T^*\mathbf{C}^{n+n}$ with $|\xi_0| = 1$. Let $\{\sum_{j,k=0}^{\infty} Q_{jk}^{\lambda}(z, w, \xi, \eta); \lambda \in \Lambda\}$ be a set of formal symbols at p_0 of product Hermite type. Suppose that there exist some positive constants r, A ($0 < A < 1$), C and α independent of $\lambda \in \Lambda$, and that each Q_{jk}^{λ} is holomorphic in $V_j(r) \times V_k(r)^c$ with*

$$V_j(r) = \{(z, \xi) \in \mathbf{C}^{n+n}; |z - z_0| < r, |(\xi/|\xi|) - \xi_0| < r, |\xi| > (j+1)r^{-1}\} \tag{2.4}$$

satisfying the estimate

$$|Q_{jk}^{\lambda}(z, w, \xi, \eta)| \le CA^{j+k}(|\xi| \cdot |\eta|)^{\alpha/2} \quad \text{on } V_j(r) \times V_k(r)^c. \tag{2.5}$$

Then for any $\beta \in \mathbf{R}$ ($\alpha < \beta < 1$), there exist a positive number r' ($< r$), and a formal symbol R of form

$$R = \sum_{j,k=0}^{\infty} \delta_{jk} \cdot R_j(z, w, \xi, \eta), \tag{2.6}$$

with R_j defined in $V_j(r') \times V_j(r')^c$ for every $j \geq 0$ such that the inequality

$$- \sum_{j,k=0}^{\infty} \delta_{jk} R_j(z, \bar{w}, \xi, \bar{\eta}) \ll \sum_{j,k=0}^{\infty} Q_{jk}(z, \bar{w}, \xi, \bar{\eta})$$

$$\ll \sum_{j,k=0}^{\infty} \delta_{jk} R_j(z, \bar{w}, \xi, \bar{\eta}) \qquad (2.7)$$

holds for every $\lambda \in \Lambda$ as a kernel on $(\coprod_{j=0}^{\infty}\{j\} \times V_j(r'))^2$. Further, $\{R_j\}_j$ satisfy the following estimates

$$|R_j| \leq \begin{cases} C'A'^{2j} \min\{|\xi|^{\beta}, |\eta|^{\beta}\} & \text{if } \beta \geq 0, \\ C'A'^{2j}(|\xi|+|\eta|)^{\beta} & \text{if } \beta \leq 0, \end{cases} \qquad (2.8)$$

$$|\text{grad}_{\xi,\eta} R_j| \leq C'A'^{2j}(|\xi|+|\eta|)^{\beta-1} \qquad (2.9)$$

on $V_j(r') \times V_j(r')^c$ for every $j \geq 0$ with some positive constants C', A' $(0 < A' < 1)$.

Proof. This is the extension of Theorem 3.1 of [10] to the case of formal symbols. So the proof goes in the same way with some modifications. Put

$$\Omega_j = \left\{z \in \mathbb{C}^n; |z - z_0| < r\right\} \times \left\{\xi \in \mathbb{C}^n; \frac{\xi}{j+1} \in U_N\right\} \qquad (2.10)$$

for $j = 0, 1, \ldots$, where U_N is an open set defined in (3.38) of [10]. Then, replace Ω by a disconnected complex manifold $\Omega' = \coprod_{j=0}^{\infty} \{j\} \times \Omega_j$. We define an L^2-norm for holomorphic functions on Ω' by

$$\|f\|_{\Omega'}^2 = \sum_{j=0}^{\infty} \frac{A^{-2j}}{(j+1)^2} \int_{\Omega_j} |f_j(z, \xi)|^2 \cdot |j+1+\langle \xi, \bar{\xi}_0\rangle|^{-2n-\beta} \, dv(z, \xi), \quad (2.11)$$

where $f_j(z, \xi)$ is the restriction of $f(z, \xi)$ to $\{j\} \times \Omega_j$ for every j, $\langle \xi, \eta \rangle = \xi_1\eta_1 + \cdots + \xi_n\eta_n$ for any vectors $\xi, \eta \in \mathbb{C}^n$, and $dv(z, \xi)$ is the Lebesgue measure on $\mathbb{R}^{2n} \times \mathbb{R}^{2n}$. Hence we have only to replace K_{Ω}^{ρ} in (3.39) of [10] by the Bergman kernel K' of Ω' for the norm (2.11). It is easy to see that

$$K'|_{\Omega_j \times \Omega_k} = \delta_{jk}(j+1)^{\beta+2}A^{2j} \cdot K_{\Omega}^{\rho}\left(z, \bar{w}, \frac{\xi}{j+1}, \frac{\bar{\eta}}{j+1}\right) \qquad (2.12)$$

(see (3.39) of [10]). This implies the estimates (2.8), (2.9) for a little larger A' (<1) than A.

Theorem 2.3. *Let $p_0 = (z_0, \bar{z}_0; \xi_0 \, dz + \bar{\xi}_0 \, dw)$ be a point of $T^*\mathbb{C}^{n+n}$ with $|\xi_0| = 1$. Let $\{Q_\lambda(z, w, \xi, \eta); \lambda \in \Lambda\}$ be a set of symbols at p_0 of product Hermite type satisfying the following (i)-(iii):*

(i) *Each $Q_\lambda(z, w, \xi, \eta)$ is holomorphic in $V_0(r) \times V_0(r)^c$ (see (2.4)) for some positive r independent of λ.*

(ii) *Quasi-positivity: $Q_\lambda(z, w, \xi, \eta) \notin \{x \in \mathbf{R}; x \le 0\}$ on $V_0(r) \times V_0(r)^c$.*

(iii) *Ellipticity: There exist some positive constants C, σ $(0 < \sigma < 1)$ independent of $\lambda \in \Lambda$, and a number $\gamma_\lambda \in \mathbf{R}$ depending on $\lambda \in \Lambda$ such that*

$$\sup\{|\gamma_\lambda|; \lambda \in \Lambda\} \le C, \tag{2.13}$$

$$C^{-1} \le |Q_\lambda(z, w, \xi, \eta)/(|\xi| + |\eta|)^{\gamma_\lambda}| \le C, \tag{2.14}$$

$$|\mathrm{grad}_{\xi,\eta} Q_\lambda| \le C(|\xi| + |\eta|)^{\gamma_\lambda + \sigma - 1} \tag{2.15}$$

hold for every $\lambda \in \Lambda$ and every $(z, w, \xi, \eta) \in V_0(r) \times V_0(r)^c$. Then, there exist a positive number r' $(<r)$ and a formal symbol $P = \sum_{j,k=0}^\infty P_{jk}(z, w, \xi, \eta)$ at p_0 of positive Hermite type such that P_{jk} is holomorphic on $V_j(r') \times V_k(r')^c$ for every j, $k \ge 0$. $:P:$ is elliptic at p_0, and $:E^\lambda: = :P: :Q_\lambda:$ has a positive formal symbol for every $\lambda \in \Lambda$. Exactly speaking, the Leibniz composition $\sum_{j,k=0}^\infty E_{jk}^\lambda(z, w, \xi, \eta)$ of $\sum_{j,k=0}^\infty P_{jk}$ and $\sum_{j,k=0}^\infty \delta_{j,0}\delta_{k,0} \cdot Q_\lambda$ defined in (3.13) of [10] *becomes a positive Hermite kernel on $(\coprod_{j=0}^\infty \{j\} \times V_j(r'))^2$ for every $\lambda \in \Lambda$.*

Remark. We obtained a similar theorem in [10] (Theorem 3.12). In that, there is some ambiguity for E^λ coming from the simplification of formal symbols. In this theorem, we have no such ambiguity though P cannot be chosen as a simple symbol.

Proof. The proof goes in the same way as that of Theorem 3.12 and Remark 3.13 of [10]. We have only to avoid the simplification of formal symbols in the context. Instead, at the last step ((3.47) of [10]), we apply Lemma 2.2. to the construction of positive formal symbol "T."

Let $p_0 = (x_0, x_0; i\xi_0(dx - du))$ be a point of $\Delta^a(iS^*\mathbf{R}^{n+n})$ with $|\xi_0| = 1$, and T_l be a bounded open set in $\mathbf{R}^{m_l}(m_l \ge 0)$ with real analytic boundary for $l = 1, \ldots, L$. Further, let $\rho_l(t_l)$ be a real analytic function on a neighborhood of $\bar{T}_l$ for every $l = 1, \ldots, L$ satisfying $\rho_l(t_l) \ge 0$ on T_l and $\rho_l \not\equiv 0$ on any connected component of T_l. Then we have the following theorem as an extension of Theorem 3.14 of [10].

Theorem 2.4. *We inherit the notation from above. Let $(Q_l^{jk}(t_l, z, \xi))_{jk}$ be an $N_l \times N_l$ matrix of simple symbols of pseudo-differential operators with real analytic parameters $t_l \in T_l$ for every $l = 1, \ldots, L$ such that $Q_l^{jk}(t_l, z, \xi)$ is* *holomorphic in*

$$X_l = \{(t_l, z, \xi) \in \mathbf{C}^{m_l + n + n}; \, t_l \in U_l, \, |z - x_0| < r, \, |(\xi/|\xi|) - i\xi_0| < r, \, |\xi| > r^{-1}\},$$

(2.16)

where U_l is a complex neighborhood of $\bar{T}_l$, and $r > 0$. Further, let $\{f_l^j(t_l, x); j = 1, \ldots, N_l\}$ be hyperfunctions defined in $T_l \times \{x \in \mathbf{R}^n; \, |x - x_0| < r\}$ for $l = 1, \ldots, L$ such that each f_l^j is depending real analytically on $t_l \in T_l$ at every point of $\bar{T}_l \times \{x_0\}$. Suppose that there exist positive constants C, $\gamma_1, \ldots, \gamma_L, \theta$ and continuous functions $q_l^{jk}(t_l)$ $(j, k = 1, \ldots, N_l, \, l = 1, \ldots, L)$ such that, for $l = 1, \ldots, L$ and $j, k = 1, \ldots, N_l$,

$$|Q_l^{jk}(t_l, z, \xi)| \le C |\xi|^{\gamma_l} \rho_l(t_l) \qquad on \, X_l \cap \{\mathrm{Im} \, t_l = 0\}, \tag{2.17}$$

$$|Q_l^{jk}(t_l, x_0, is\xi_0) - s^{\gamma_l} q_l^{jk}(t_l)| \le C s^{\gamma_l - \theta} \cdot \rho_l(t_l) \qquad for \, every \, s > \frac{1}{r}, \, t_l \in T_l,$$

(2.18)

any eigenvalue of $(q_l^{jk}(t_l) + \overline{q_l^{kj}(t_l)})_{jk} > C^{-1} \rho_l(t_l) \qquad for \, t_l \in T_l.$ (2.19)

Further we suppose the microlocal inequality

$$\sum_{l=1}^{L} \int_{\mathbf{R}^{m_l}} \sum_{j,k=1}^{N_l} (Q_l^{jk}(t_l, x, D_x) + Q_l^{kj*}(t_l, u, D_u)) v_l^{jk}(t_l, x, u) \, dt_l \ll 0 \quad (2.20)$$

at p_0, where $(D_x, D_u) = (\partial/\partial x, \partial/\partial u)$, and $\{v_l^{jk}\}$ are microfunctions:

$$v_l^{jk}(t_l, x, u) = [\mathrm{ext}(f_l^j(t_l, x) f_l^{k*}(t_l, u))] (= f_l^j(t_l, x) f_l^{k*}(t_l, u) \chi_{T_l}(t_l)) \quad (2.21)$$

for every j, k, l. Then we have

$$\mathrm{SS}(\mathrm{ext}(f_l^j(t_l, x))) \cap \{(t_l, x_0; i\tau_l \, dt_l + i\xi_0 \, dx; t_l, \tau_l \in \mathbf{R}^{m_l}\} = \varnothing \quad (2.22)$$

for every $j = 1, \ldots, N_l$ and every $l = 1, \ldots, L$.

Proof. We prove this by constructing an elliptic and positive pseudo-differential operator $P(x, u, D_x, D_u)$ at p_0, which makes every integrand of (2.20) a positive microkernel. Then the theorem directly follows because we can decompose the lth term of (2.20) into two parts corresponding to the decomposition

$$Q_l^{jk}(t_l, z, \xi) = \delta_{jk} \frac{\rho_l(t_l)}{4C} \langle \xi, -i\xi_0 \rangle^{\gamma_l} + \left(Q_l^{jk} - \delta_{jk} \frac{\rho_l(t_l)}{4C} \langle \xi, -i\xi_0 \rangle^{\gamma_l} \right). \quad (2.23)$$

In fact, both terms satisfy the conditions similar to (2.17)–(2.19), and so we may assume that P makes all these $2L$ terms positive. Therefore from (2.20) we know

$$(\langle D_x, -i\xi_0\rangle^{\gamma_l} + \langle D_u, i\xi_0\rangle^{\gamma_l}) \int \sum_{j=1}^{N_l} v_l^{jj}(t_l, x, u)\rho_l(t_l)\,dt_l = 0 \qquad \text{at } p_0. \quad (2.24)$$

Hence by ellipticity,

$$\sum_{j=1}^{N_l} \int v_l^{jj}(t_l, x, u)\rho_l(t_l)\,dt_l = 0 \qquad \text{at } p_0. \quad (2.25)$$

Now we can directly apply Theorem 3.14 (or Theorem 2.13) of [10] to conclude (2.22).

Construction of P. To avoid the complexity of notation, we treat only the case $L = 1$ (so we can drop "l") because the proof goes in the same way also in general cases. Take an $N \dot{\times} N$ unitary matrix $(U^{jk}(t))_{jk}$ for every $t \in T$ such that

$$\sum_{j',k'=1}^{N} (q^{j'k'}(t) + \overline{q^{k'j'}(t)}) U^{jj'}(t)\overline{U^{kk'}(t)} = \delta_{jk}q^j(t) \qquad \text{for every } j,k, \quad (2.26)$$

where the $q^j(t)$'s are real-valued functions, and these functions of t may be discontinuous. Put a simple symbol

$$R^{jk}(t, z, \xi) = \sum_{j',k'=1}^{N} \rho(t)^{-1} Q^{j'k'}(t, z, \xi) U^{jj'}(t)\overline{U^{kk'}(t)}, \quad (2.27)$$

and simple symbols of product Hermite type

$$Y^{jk} = R^{jk}(t, z, \xi) + R^{jk*}(t, w, \eta), \qquad Z^{jk} = \frac{R^{jk}(t, z, \xi) - R^{jk*}(t, w, \eta)}{i},$$

$$(2.28)$$

$$A(\xi, \eta, \gamma) = \langle \xi, -i\xi_0\rangle^{\gamma} + \langle \eta, i\xi_0\rangle^{\gamma}, \quad (2.29)$$

for $j, k = 1, \ldots, N$, $t \in \{t \in T; \rho(t) > 0\}$. Consider a formal symbol $P = \sum_{\alpha,\beta=0}^{\infty} P_{\alpha\beta}(z, w, \xi, \eta)$ at p_0 of positive Hermite type, which will be determined later. Let $\sum_{\alpha,\beta} \hat{R}_{\alpha\beta}^{jk}$, $\sum_{\alpha,\beta} \hat{Y}_{\alpha\beta}^{jk}$, $\sum_{\alpha,\beta} \hat{Z}_{\alpha\beta}^{jk}$ be the Leibniz compositions of P and $R^{jk}(t, z, \xi) + R^{kj*}(t, w, \eta)$, P and Y^{jk}, or P and Z^{jk} respectively. We fix t for the moment, and introduce infinitely many complex variables $\lambda_{j\alpha}$, $\lambda_{j\alpha}^* \ (j = 1, \ldots, N, \ \alpha = 0, 1, \ldots)$, which should be zero except for some finitely many indices (j, α). Then we have

$$\sum_{j,k=1}^{N} \sum_{\alpha,\beta=0}^{\infty} \hat{R}_{\alpha\beta}^{jk}(t, z, w, \xi, \eta)\lambda_{j\alpha}\lambda_{k\beta}^{*}$$

$$= \sum_{j,k=1}^{N} \sum_{\alpha,\beta=0}^{\infty} \tfrac{1}{2}(\hat{Y}_{\alpha\beta}^{jk} + i\hat{Z}_{\alpha\beta}^{jk} + \hat{Y}_{\alpha\beta}^{kj} - i\hat{Z}_{\alpha\beta}^{kj})\lambda_{j\alpha}\lambda_{k\beta}^{*}$$

$$= \sum_{j=1}^{N} \sum_{\alpha,\beta=0}^{\infty} \left(\hat{Y}_{\alpha\beta}^{jj} - \frac{P_{\alpha\beta}}{4C}A\right)\lambda_{j\alpha}\lambda_{j\beta}^{*}$$

$$+ \tfrac{1}{4} \sum_{\varepsilon=\pm 1} \sum_{1 \le j < k \le N} \sum_{\alpha,\beta=0}^{\infty} [(\mu P_{\alpha\beta}A + \varepsilon(\hat{Y}_{\alpha\beta}^{jk} + \hat{Y}_{\alpha\beta}^{kj}))$$

$$\times (\lambda_{j\alpha} + \varepsilon\lambda_{k\alpha})(\lambda_{j\beta}^{*} + \varepsilon\lambda_{k\beta}^{*})$$

$$+ (\mu P_{\alpha\beta}A - \varepsilon(\hat{Z}_{\alpha\beta}^{jk} - \hat{Z}_{\alpha\beta}^{kj}))(\lambda_{j\alpha} + i\varepsilon\lambda_{k\alpha})(\lambda_{j\beta}^{*} - i\varepsilon\lambda_{k\beta}^{*})], \qquad (2.30)$$

where $\mu = 1/\{4C(N-1)\}$. Let $\mathscr{F}$ be the set of simple symbols at p_0 of product Hermite type:

$$\mathscr{F} = \left\{ Y^{hh} - \frac{A}{4C}, \mu A \pm (Y^{jk} + Y^{kj}), \mu A \pm (Z^{jk} - Z^{kj}); \right.$$

$$\left. 1 \le h \le N, 1 \le j < k \le N, t \in T, \rho(t) > 0 \right\}$$

$$= \left\{ \tfrac{1}{2}E^{hh}(t, z, \xi) - \frac{1}{4C}\langle \xi, -i\xi_0 \rangle^{\gamma} + \tfrac{1}{2}E^{hh*}(t, w, \eta) - \frac{1}{4C}\langle \eta, i\xi_0 \rangle^{\gamma}, \right.$$

$$\mu\langle \xi, -i\xi_0 \rangle^{\gamma} + \varepsilon E^{jk}(t, z, \xi) + \mu\langle \eta, i\xi_0 \rangle^{\gamma} + \varepsilon E^{jk*}(t, w, \eta),$$

$$\mu\langle \xi, -i\xi_0 \rangle^{\gamma} + \varepsilon F^{jk}(t, z, \xi) + \mu\langle \eta, i\xi_0 \rangle^{\gamma} + \varepsilon F^{jk*}(t, w, \eta);$$

$$\left. \varepsilon = \pm 1, 1 \le h \le N, 1 \le j < k < N, t \in T, \rho(t) > 0 \right\}. \qquad (2.31)$$

Here $E^{jk}(t, z, \xi) = R^{jk} + R^{kj}$, $F^{jk}(t, z, \xi) = (R^{jk} - R^{kj})/i$. Now by using condition (2.17) we have

$$|R^{jk}(t, z, \xi)| \le CN^2|\xi|^{\gamma} \qquad (2.32)$$

in $\{|z - x_0| < r, |(\xi/|\xi|) - i\xi_0| < r, |\xi| > r^{-1}\}$ for every j, k and every $t \in T$. The combination of (2.18) and the Cauchy estimate for $\mathrm{grad}_{z,\xi}R^{jk}$ implies

$$\left|\mathrm{Re}\, E^{jk}(t, z, \xi) - \delta_{jk}\frac{q^j(t)}{\rho(t)}|\xi|^{\gamma}\right| + |\mathrm{Re}\, F^{jk}(t, z, \xi)|$$

$$\le C'|\xi|^{\gamma}(|z - x_0| + |(\xi/|\xi|) - i\xi_0| + |\xi|^{-\theta}) \qquad (2.33)$$

on $\{|z - x_0| < r', |(\xi/|\xi|) - i\xi_0| < r', |\xi| > r'^{-1}, t \in T, \rho(t) > 0\}$ for every j, k with some positive constants C', r'. Hence $\mathscr{F}$ satisfies the conditions of Theorem 2.3 with $\sigma = \max\{0, 1 - \gamma\} < 1$ if we choose r' as sufficiently small, because $q^j(t)/\rho(t) > C^{-1}$ for every j and $t \in T$. Let $P = \sum_{\alpha,\beta} P_{\alpha\beta}(z, w, \xi, \eta)$ be the formal symbol at p_0 of positive Hermite type obtained in Theorem 2.3 for $\mathscr{F}$. Then by the calculation (2.30) and Lemma 2.1 we know that, for every $t \in \{t \in T; \rho(t) > 0\}$, the Leibniz composition of P and $R^{jk} + R^{kj}*$ is a positive kernel on $(\{1, \ldots, N\} \times \coprod_{j=0}^{\infty} V_j(r'))^2$. Therefore the Leibniz composition $\sum_{\alpha,\beta=0}^{\infty} \hat{Q}_{\alpha\beta}^{jk}(t, z, w, \xi, \eta)$ of P and $Q^{jk} + Q^{kj}*$ is also positive on $(\{1, \ldots, N\} \times \coprod_{j=0}^{\infty} V_j(r'))^2$ for the same t. On the other hand, since $\{t \in T; \rho(t) > 0\}$ is dense in T, such positivity holds everywhere on T (recall that $\hat{Q}_{\alpha\beta}^{jk}(t, z, w, \xi, \eta)$ is analytic in t). This concludes our first claim. Thus the proof of the theorem is completed.

Remark 2.5. We excluded the case $\gamma_l = 0$ (the order of $Q_l^{jk} = 0$) in Theorem 2.4. However we can treat this case in the same way if the Q_l^{jk}'s are all only multiplication-operators for such l. In fact "$\mathscr{F}$" in the proof satisfies (2.15) of Theorem 2.3 trivially. The author believes that the theorem is also true for every "$\gamma_l = 0$"-case (at least, some approach is possible for the case that $\{Q_l^{jk}\}$ are homogeneous of degree 0 with respect to ξ).

Definition 2.6. Let p_0 be a point of $\Delta^a(iS^*\mathbf{R}^{n+n})$. Then, a Hermite microkernel v at p_0 ($v \in \mathscr{C}_{\mathbf{R}^{n+n}}|_{p_0}$ and $v(x, u) = \overline{v(u, x)}$) is said to be "quasi-positive" if $v(x, u)$ is written as a finite sum of positive Hermite microkernels at p_0 and of the microkernels defined by the left-hand side of (2.20) for some $\{f_l^j\}$ and $\{Q_l^{jk}\}$ satisfying the conditions (including the exceptional cases in Remark 2.5). Then,

Theorem 2.7. *"Quasi-positivity" also gives an order structure to Hermite microkernels at p_0, which is weaker than "$\ll$."*

This is a direct corollary of Theorem 2.4. Hereafter, we denote by "$\ll_q$" or "$\gg_q$" this order relation.

We give inequalities of Sobolev's type, which correspond to the $\frac{1}{2}$-differentiability of traces of 1-differentiable functions in the L^2 sense. Let T be a bounded open set in $\mathbf{R}^m$ with C^ω-boundary ∂T, and $\nu = \sum_{j=1}^m \nu_j(t)D_{t_j}, d\sigma(t)$ be the outer unit normal vector field to ∂T, and the volume element of ∂T, respectively.

Proposition 2.8. *Set $\Omega = T \times U$ with $U = \{x \in \mathbf{R}^n; |x - x_0| < r\}$. Let $f(t, x)$ be a hyperfunction on Ω depending real analytically on $t \in T$ at every point of $\bar{T} \times U$, and let $B(t, x, D_t) = \sum_{j=0}^m \alpha_j(t, x)D_{t_j} + \beta(t, x)$ be a differential operator with real analytic coefficients defined on $\partial T \times U$. Assume that B is tangential to $\partial T \times U$; that is,*

$$\sum_{j=1}^m \alpha_j(t, x)\nu_j(t) = 0 \qquad on \ \partial T \times U. \tag{2.34}$$

Then,

$$I = \int_{\partial T} \{(B(t, x, D_t)f(t, x))f^*(t, u) + f(t, x) \cdot B^*(t, u, D_t)f^*(t, u)\} \, d\sigma(t)$$

$$\ll_q \sum_{j,k=1}^m \int_T (\mu_{jk}(t, x) + \mu_{kj}^*(t, u) + \varepsilon\delta_{jk})f_{t_j}f_{t_k}^* \, dt + \left(C + \frac{C}{\varepsilon}\right) \int_T ff^* \, dt \tag{2.35}$$

at $(x_0, x_0; i\xi_0(dx - du))$ for every $\varepsilon > 0$ and every $\xi_0 \in \mathbf{S}^{n-1}$ with some constant $C > 0$ independent of $f(t, x)$. Here,

$$\mu_{jk}(t, x) = \tilde{\nu}_k(t)\tilde{\alpha}_j(t, x) - \tilde{\nu}_j(t)\tilde{\alpha}_k(t, x), \tag{2.36}$$

and $\{\tilde{\alpha}_j\}$, $\{\tilde{\nu}_k\}$ are any real analytic extensions of $\{\alpha_j\}$, $\{\nu_k\}$ to $\bar{T} \times U$ ($\{\tilde{\nu}_k\}$ are real-valued on $\bar{T}$).

Proof. It follows from the Gauss formula that

$$I = \int_T \sum_{j=1}^m \frac{\partial}{\partial t_j} \{\tilde{\nu}_j(t)((\tilde{B}f)f^* + f(\tilde{B}^*f^*))\} \, dt. \tag{2.37}$$

On the other hand, by (2.34), we have

$$0 = \int_{\partial T} \sum_{j,k=1}^m (\tilde{\nu}_k\tilde{\alpha}_k\tilde{\nu}_j f_{t_j}f^* + \tilde{\nu}_k\tilde{\alpha}_k^*\tilde{\nu}_j ff_{t_j}^*) \, d\sigma(t)$$

$$= \int_T \sum_{j,k=1}^m \frac{\partial}{\partial t_k} (\tilde{\alpha}_k\tilde{\nu}_j f_{t_j}f^* + \tilde{\alpha}_k^*\tilde{\nu}_j ff_{t_j}^*) \, dt. \tag{2.38}$$

So $I =$ the difference of the right-hand side of (2.37) and (2.38). Then the terms containing $f_{t_j t_k}$ or $f_{t_j t_k}^*$ cancel each other. Further we can directly apply Theorem 2.4 to the estimates of the other terms by the right-hand side of (2.35); set $\{f_1^j\} = \{f, f_{t_1}, \ldots, f_{t_m}\}$ and see Remark 2.5.

3. Applications to Analytic Hypoellipticity

The first application concerns parabolic problems. The easiest case was treated in [11]. The analyticity of solutions for parabolic boundary value problems can be proven also by constructing fundamental solutions with estimates for analytic wave front sets. For example, if the coefficients do not depend on the time variable, the theory of analytic semi-groups will be applicable. Our method is not constructive but direct, and the spirit is similar to classical energy methods. We refer to [7, 3] concerning general boundary conditions, and to [14, 16] concerning parabolic interior problems in microfunction theory. Before stating the theorem, we stress here that $x \in \mathbf{R}$ is the "time variable," $t \in \mathbf{R}^m$ are the space variables, and that $D_x = \partial/\partial x$, $D_t = \partial/\partial t$.

Theorem 3.1. *Let T be a bounded domain of $\mathbf{R}_t^m$ with C^ω-boundary ∂T, and I be an open interval of $\mathbf{R}_x$. Let $a_{jk}(t, x)$ $(= a_{kj}; j, k = 1, \ldots, m)$, $b_j(t, x)$ $(j = 1, \ldots, m)$, $c(t, x)$, $g(t, x)$ be C^ω-functions in a neighborhood of $\bar{T} \times I$, and $\alpha_j(t, x)$ $(j = 1, \ldots, m)$, $\beta(t, x)$, $h(t, x)$ be C^ω-functions on the boundary $\partial T \times I$. Suppose that a hyperfunction $f(t, x)$ on $T \times I$ satisfies the following:*

$$\left(\sum_{j,k=1}^{m} a_{jk}(t, x) D_{t_j} D_{t_k} + \sum_{j=1}^{m} b_j(t, x) D_{t_j} - D_x + c(t, x) \right) f = g(t, x) \qquad in \ T \times I$$

$$(3.1)$$

$$\left(\sum_{j=1}^{m} \alpha_j(t, x) D_{t_j} + \beta(t, x) \right) f = h(t, x) \qquad as \ a \ trace \ on \ \partial T \times I. \quad (3.2)$$

Then $f(t, x)$ is analytic in a neighborhood of $\bar{T} \times I$ if the following conditions are fulfilled:

  (i) *$(a_{jk}(t, x))_{jk}$ is a real, symmetric and positive matrix for every $(t, x) \in \bar{T} \times I$.*

  (ii) *$\{\alpha_j(t, x); j = 1, \ldots, m\}$ are real-valued, and transversal to $\partial T \times I$, that is,*

$$\sum_{j=1}^{m} \alpha_j(t, x) \nu_j(t) \neq 0 \qquad on \ \partial T \times I, \qquad (3.3)$$

where $\nu = \sum_{j=1}^{m} \nu_j(t) D_{t_j}$ is the outer unit normal vector field to ∂T.

Remark. (*i*) The trace in (3.2) is well-defined because $\partial T \times I$ is non-characteristic for (3.1). (*ii*) We can treat some other boundary conditions in the same way; the Dirichlet condition (easier), or the case $\mathrm{Im}\,\alpha_j \neq 0$ under some additional assumptions on $\{a_{jk}\}$, $\{\mathrm{Im}\,\alpha_j\}$ and b_0.

Proof. First, we remark that $f(t, x)$ depends real analytically on $t \in T$ at every point of $\bar{T} \times I$. Because (3.1) is elliptic with respect to t, and the boundary condition (3.2) satisfies the Lopatinski condition microlocally on $\{(t, x; i\tau\,dt + i\xi\,dx); t \in \partial T, \xi \neq 0\}$ (note that the α_j's are real-valued; see [14, 9, 15]). Hence we can apply our energy method to $f(t, x)f^*(t, u)$. Fix a point $(x_0; i\xi_0\,dx) \in iS^*\mathbf{R}$ ($\xi_0 = \pm 1$), and put $p_0 = (x_0, x_0; i\xi_0(dx - du))$. For the sake of simplicity, we consider only the case $\xi_0 = 1$. Then we have the following equality at p_0:

$$
0 = \int_T g(t, x)f^*(t, u)\,dt
$$

$$
= \int_{\partial T} \sum_{j,k=1}^m \nu_k a_{jk} f_{t_j} f^*\,d\sigma(t)
$$

$$
+ \int_T \left\{ -\sum_{j,k=1}^m a_{jk} f_{t_j} f_{t_k}^* + \sum_{j=1}^m b_j' f_{t_j} f^* - (D_x - c)ff^* \right\}\,dt, \qquad (3.4)
$$

where $b_j' = b_j - \sum_{k=1}^m \partial a_{jk}/\partial t_k$ for $j = 1, \ldots, m$. On the other hand, considering the condition (*ii*), we can rewrite (3.2) as follows:

$$
\sum_{j,k=1}^m \nu_k(t)a_{jk}(t, x)D_{t_j}f = \sum_{j=1}^m \alpha_j'(t, x)D_{t_j}f + \beta'(t, x)f + h'(t, x) \qquad (3.5)
$$

on $\partial T \times I$, where $\{\alpha_j'\}_j$ are real-valued and $\sum_{j=1}^m \alpha_j' D_{t_j}$ is tangential to $\partial T \times I$; that is,

$$
\sum_{j=1}^m \alpha_j'(t, x)\nu_j(t) = 0 \qquad \text{on } \partial T \times I. \qquad (3.6)
$$

Further by the cohomological triviality of the sheaf $\mathscr{A}$ we can extend $\nu_k(t)$, $\alpha_j'(t, x)$, $\beta'(t, x)$ real analytically to $\bar{T} \times I$. We use the same notation. Then, by the condition (*ii*), we may assume that the ν_k's and the α_j''s are real-valued on $\bar{T} \times I$. Thus, by Proposition 2.8 and Theorem 2.4 (Remark 2.5), we know that, for $\omega = 1, -i$, and every $\varepsilon > 0$,

$$\int_T \sum_{j,k=1}^m (\omega a_{jk} + \bar{\omega} a_{kj}^*) f_{t_j} f_{t_k}^* \, dt + \int_T \{\omega(D_x - c) + \bar{\omega}(D_u - c^*)\} ff^* \, dt$$

$$= \int_T \sum_{j=1}^m (\omega b_j' f_{t_j} f^* + \bar{\omega} b_j'^* ff_{t_j}^*) \, dt$$

$$+ \int_{\partial T} \sum_{j=1}^m \{\omega(\alpha_j' f_{t_j} + \beta' f) f^* + \bar{\omega} f(\alpha_j'^* f_{t_j}^* + \beta'^* f^*)\} \, d\sigma(t) \qquad (3.7)$$

$$\ll_q \int_T \sum_{j,k=1}^m (\omega \mu_{jk} + \bar{\omega} \mu_{kj}^* + \varepsilon \delta_{jk}) f_{t_j} f_{t_k}^* \, dt + \left(M + \frac{M}{\varepsilon}\right) \int_T ff^* \, dt$$

holds at p_0 with a large constant $M > 0$ independent of ε, where

$$\mu_{jk} = v_k(t)\alpha_j'(t, x) - v_j(t)\alpha_k'(t, x). \qquad (3.8)$$

Set $a_{jk}'(t, x) = a_{jk} - \mu_{jk}$. Then, since the a_{jk}''s are real, any eigenvalue of the matrix $(a_{jk}' + \overline{a_{kj}'})_{jk} = (2a_{jk})_{jk}$ is positive for $(t, x) \in \bar{T} \times \{x_0\}$. Taking $\omega = 1$, we know by Theorem 2.4 (Remark 2.5) that

$$(\lambda - \varepsilon) \int_T \sum_{j=1}^m f_{t_j} f_{t_j}^* \, dt \ll_q \int_T \sum_{j,k=1}^m (a_{jk}' + a_{kj}'^* - \varepsilon \delta_{jk}) f_{t_j} f_{t_k}^* \, dt$$

$$\ll_q \int_T \left(-D_x + c - D_u + c^* + M + \frac{M}{\varepsilon}\right) ff^* \, dt, \qquad (3.9)$$

at p_0, where λ is some positive constant independent of ε. Similarly, from (3.7) with $\omega = -i$ we obtain

$$\int_T \left\{-i(D_x - c) + i(D_u - c^*) - M - \frac{M}{\varepsilon}\right\} ff^* \, dt$$

$$\ll_q \int_T \sum_{j,k=1}^m (ia_{jk}' - ia_{kj}'^* + \varepsilon \delta_{jk}) f_{t_j} f_{t_k}^* \, dt$$

$$\ll_q (\lambda' + \varepsilon) \int_T \sum_{j=1}^m f_{t_j} f_{t_j}^* \, dt \qquad (3.10)$$

at p_0. Hence, putting $\varepsilon = \lambda/2$, the combination of (3.9) and (3.10) implies

$$\int_T \left\{\left(-i + \frac{2\lambda' + \lambda}{\lambda}\right)(D_x - c) + \left(i + \frac{2\lambda' + \lambda}{\lambda}\right)(D_u - c^*) - C_0\right\} ff^* \, dt \ll_q 0$$

$$(3.11)$$

at p_0, where C_0 is a real constant. It is clear that the left-hand side is quasi-positive at $p_0 = (x_0, x_0; i(dx - du))$. Thus, by Theorem 2.4 we conclude

$$\text{SS}(\text{ext}(f(t, x))) \cap \{(t, x_0; i\tau dt + idx); t, \tau \in \mathbf{R}^m\} = \varnothing.$$

The same conclusion is proven in a similar manner for $\xi_0 = -1$. Therefore f is analytic on $\bar{T} \times \{x_0\}$. This completes the proof.

The next problem concerns operators of Trèves type. There are many proofs of analytic hypoellipticity ([19, 18, 12, 17]). We treat here only the easiest case. However, the effects by lower-order terms to analytic hypoellipticity will be clarified.

First, we give a new result on a special, but a more degenerate case, which is not contained in Okaji's result [(13)]; for example, $P = D_{t_1}^2 + D_{t_2}^2 + (t_1^2 + t_2^4 + xt_1 t_2^2)D_x^2$ at $(0; i\, dx)$.

Theorem 3.2 (cf. Theorem 17 of [11]).   *Let P be a pseudo-differential operator at $x_0^* = (0, x_0; i\xi_0\, dx) \in iS^*(\mathbf{R}_t^m \times \mathbf{R}_x^n)$ $(|\xi_0| = 1)$ given by*

$$P = \sum_{j,k=1}^{m} D_{t_j} a_{jk}(t, x) D_{t_k} + Q(t, x, D_x), \qquad (3.12)$$

where $(a_{jk}(t, x))_{jk}$ is an $m \times m$ matrix of C^ω-functions at $(0, x_0)$ satisfying $(a_{jk}(0, x_0) + \overline{a_{kj}(0, x_0)})_{jk} \gg (\lambda \delta_{jk})_{jk}$ for some $\lambda > 0$, and $Q(t, z, \xi)$ is a simple symbol of a second-order microdifferential operator defined in

$$V = \{(t, z, \xi) \in \mathbf{C}^{m+n+n}; |t| < r, |z - x_0| < r, |(\xi/|\xi|) - i\xi_0| < r, |\xi| > r^{-1}\}$$

and Q satisfies the inequality

$$C^{-1}|\xi|^2 \rho(t) \le -\text{Re } Q(t, z, \xi) \le |Q(t, z, \xi)| \le C|\xi|^2 \rho(t) \qquad (3.13)$$

on $\{\text{Im } t = 0\} \cap V$ with a constant $C > 0$ and a C^ω-function $\rho(t)$ such that $\rho(t) \ge 0$ in $\{t \in \mathbf{R}^m; |t| < r\}$ and $\rho^{-1}(0) = \{0\}$. Then, P is analytically hypoelliptic at x_0^*.

This is a corollary of Theorem 2.4. The proof goes in a similar way as that of Theorem 17 of [11].

Consider the following operator with double characteristics on $\mathbf{R}_t^m \times \mathbf{R}_x$:

$$P = \sum_{j=1}^{m} \alpha_j(D_{t_j}^2 + t_j^2 D_x^2) + \sum_{j=1}^{m} \beta_j D_{t_j} + \gamma D_x + \delta, \qquad (3.14)$$

where $\alpha_j, \beta_j, \gamma, \delta$ are complex numbers for $j = 1, \ldots, m$ such that $\text{Re } \alpha_j > 0$ for every j. This is the simplest case of operators treated by Trèves [19].

We show the analytic hypoellipticity of P at $(0; i\,dx)$ under the following well known condition (so-called "sub-ellipticity"):

$$i\gamma \neq (2n_1+1)\alpha_1 + \cdots + (2n_m+1)\alpha_m \tag{3.15}$$

for any non-negative integers $n_1, \ldots, n_m$.

Our proof. Let $f(t, x)$ be any microfunction at $x_0^* = (0; i\,dx)$ satisfying $Pf = 0$ at x_0^*. Since P is elliptic off $\Sigma = \{t = 0, \xi_t = 0\}$, the support of f is contained in Σ. Therefore we can represent f by a hyperfunction in $V = \{|t| < 2r, |x| < r\}$ with some $r > 0$, whose singular spectrum is contained in Σ. We use the same notation f for such a hyperfunction. Hence, $f(t, x)$ depends real analytically on t. Set $p_0 = (0, 0; i(dx - du)) \in iS^*\mathbf{R}^{1+1}$ and $T = \{|t| < r\}$. Since f is analytic on $|t| = r$, we obtain the following equality at p_0:

$$0 = \int_T (Pf \cdot f^* + fP^*f^*)\, dt = \int_T (\gamma D_x + \delta + \bar{\gamma}D_u + \bar{\delta})ff^*\, dt$$

$$+ \sum_{j=1}^m \int_T \{-2\,\mathrm{Re}\ \alpha_j \cdot f_{t_j}f_{t_j}^* + \beta_j f_{t_j}f^*$$

$$+ \bar{\beta}_j ff_{t_j}^* + t_j^2(\alpha_j D_x^2 + \bar{\alpha}_j D_u^2)ff^*\}\, dt. \tag{3.16}$$

Put $\lambda = \min\{\mathrm{Re}\ \alpha_j;\ j = 1, \ldots, m\} > 0$. Noting that $\alpha_j D_x^2 + \bar{\alpha}_j D_u^2$ is quasi-negative at p_0, we have

$$2\lambda \int_T \sum_{j=1}^m f_{t_j}f_{t_j}^*\, dt \ll_q \varepsilon \int_T \sum_{j=1}^m f_{t_j}f_{t_j}^*\, dt$$

$$+ \int_T \left(\gamma D_x + \bar{\gamma}D_u + \delta + \bar{\delta} + \frac{C}{\varepsilon}\right)ff^*\, dt$$

$$\tag{3.17}$$

at p_0 for any $\varepsilon > 0$, where $C > 0$ is some constant independent of ε. Hence, setting $\varepsilon = \lambda$, we have

$$\int_T \left(\gamma D_x + \bar{\gamma}D_u + \delta + \bar{\delta} + \frac{C}{\varepsilon}\right)ff^*\, dt \gg_q 0. \tag{3.18}$$

If $\mathrm{Re}(i\gamma) < 0$, the left-hand side becomes quasi-negative. So f must vanish at x_0^*. That is, "$\mathrm{Re}(i\gamma) < 0$" is a sufficient condition for analytic hypoellipticity. To weaken this condition, we employ the well known method of "concatenations" ([19]). Put

$$Z_j = D_{t_j} + it_jD_x, \quad Z_j^* = D_{t_j} - it_jD_x \quad \text{for } j = 1, \ldots, m. \tag{3.19}$$

Then we have

$$P = \sum_{j=1}^{m} (\alpha_j Z_j Z_j^* + \tfrac{1}{2}\beta_j (Z_j + Z_j^*)) + (\gamma + i\alpha_1 + \cdots + i\alpha_m) D_x + \delta, \quad (3.20)$$

$$Z_j^* P = (P + 2i\alpha_j D_x) Z_j^* + i\beta_j D_x. \tag{3.21}$$

For the sake of simplicity, we suppose $\beta_j = 0$ for every j. Then "$Pf = 0$" implies "$(P + 2i\alpha_j D_x) Z_j^* f = 0$" for every j. Assume that "$\mathrm{Re}(i\gamma) < 2\lambda$, and $i\gamma \neq \alpha_1 + \cdots + \alpha_m$." Then, $P + 2i\alpha_j D_x$ is analytically hypoelliptic at x_0^* for every j. Hence "$Pf = 0$" implies "$Z_j^* f = 0$" for every j at x_0^*. By using (3.20), we have $\{(\gamma + i\alpha_1 + \cdots + i\alpha_m) D_x + \delta\} f = 0$. This concludes "$f = 0$ at x_0^*." That is, "$\mathrm{Re}(i\gamma) < 2\lambda$, and $i\gamma \neq \alpha_1 + \cdots + \alpha_m$" is a weakened condition. Repeating this procedure, then we arrive at the final condition (3.15).

Remark 3.3. To prove general cases for the Trèves operator, we need a technique concerning a series of *a priori* estimates. The details will be published elsewhere.

Remark 3.4. "Sub-ellipticity" (3.15) is not necessary for analytic hypoellipticity. In fact, we know from the proof above that the degenerate case ($m = 1$, $\alpha = 1$, $\beta = 0$, $\gamma = -i$) is analytically hypoelliptic if $\delta \neq 0$.

References

[1] T. Aoki, Invertibility for microdifferential operators of infinite order. *Publ. Res. Inst. Math. Sci.* **18** (1982) 1–29.
[2] T. Aoki, The exponential calculus of microdifferential operators of infinite order. II. *Proc. Jap. Acad.* **58** (1982) 154–157.
[3] R. Arima, On general boundary value problem for parabolic equations. *J. Math. Kyoto Univ.* **4-1** (1964) 207–243.
[4] J. M. Bony and P. Schapira, Propagation des singularités analytiques pour les solutions des équations aux derivées partielles. *Ann. Inst. Fourier* **26** (1976) 81–140.
[5] L. Boutet de Monvel, Opérateurs pseudo-différentiels analytiques et opérateurs d'ordre infini. *Ann. Inst. Fourier, Grenoble* **22** (1972) 229–268.
[6] L. Boutet de Monvel and P. Krée, Pseudo-differential operators and Gevrey classes. *Ann. Inst. Fourier* **17** (1967) 295–323.
[7] S. Itô, Fundamental solutions of parabolic differential equations and boundary value problems. *Japan J. Math.* **27** (1957) 55–102.
[8] K. Kataoka, Micro-local theory of boundary value problems I. *J. Fac. Sci. Univ. Tokyo* **27** (1980) 355–399.
[9] K. Kataoka, Micro-local theory of boundary value problems II. *J. Fac. Sci. Univ. Tokyo* **28** (1981) 31–56.

[10] K. Kataoka, Microlocal energy methods and pseudo-differential operators. *Invent. Math.* **81** (1985) 305-340.

[11] K. Kataoka, Quasi-positivity for pseudo-differential operators and microlocal energy methods. In proceedings "Taniguchi Symp. HERT Katata," 1984, pp. 125-141

[12] G. Métivier, Analytic hypoellipticity for operators with multiple characteristics. *Comm. Partial Differential Equations,* **6** (1981) 1-90.

[13] T. Ōkaji, Analytic hypoellipticity for operators with symplectic characteristics. *J. Math. Kyoto Univ.* **25**-3 (1985) 489-514.

[14] M. Sato, T. Kawai and M. Kashiwara, Microfunctions and Pseudo-differential Equations. *Lecture Notes in Math,* Vol. 287, Springer, Berlin-Heidelberg-New York, 1973, pp. 265-529.

[15] P. Schapira, Propagation au bord et reflexion des singularités analytiques des solutions des équations aux derivées partielles. *Sém. Goulaouic-Schwartz 1975-1976.*

[16] P. Schapira, Propagation au bord et reflexion des singularités analytiques des solutions des équations aux derivées partielles, II. *Sém. Goulaouic-Schwartz 1976-1977.*

[17] J. Sjöstrand, Analytic wavefront sets and operators with multiple characteristics. *Hokkaido Math. J.* **12** (1983) 392-433.

[18] D. S. Tartakoff, The local real analyticity of solutions to $\Box_b$ and the $\bar{\partial}$-Neumann problem. *Acta Math.* **145** (1980) 177-204.

[19] F. Trèves, Analytic hypoellipticity of a class of pseudo-differential operators with double characteristics. *Comm. Partial Differential Equations* **3** (1978) 475-642.

A Proof of the Transformation Formula of the Theta-Functions

Yukiyosi Kawada

University of Tokyo
Hongô, Bunkyo-ku
Tokyo, Japan

The purpose of this short note is to give a simple proof of the classical transformation formula of the theta-function ϑ_3:

$$\vartheta_3\left(\frac{v}{\tau}\bigg|\frac{-1}{\tau}\right) = \sqrt{\frac{\tau}{i}}\, e^{(i\pi/\tau)v^2}\vartheta_3(v\,|\,\tau), \qquad \tau, v \in \mathbf{C}, \qquad \mathrm{Im}\,\tau > 0 \qquad (1)$$

after the simple proof of

$$\eta\left(\frac{-1}{\tau}\right) = \sqrt{\frac{\tau}{i}}\, \eta(\tau) \qquad (2)$$

given by C. L. Siegel[1] for

$$\eta(\tau) = q^{1/12} \prod_{n=1}^{\infty} (1 - q^{2n}), \qquad q = e^{\pi i \tau}, \qquad \mathrm{Im}\,\tau > 0, |q| < 1.$$

---

[1] C. L. Siegel, A Simple Proof of $\eta(-1/\tau) = \eta(\tau)\sqrt{\tau/i}$, *Mathematica* **1** (1954), 4 (*Gesammelte Abhandlungen*, Band **III**, 188).

Algebraic Analysis, Volume I

His proof is a straightforward application of the theorem of residues. Namely, taking the logarithm of the both sides of (2), it suffices to prove

$$\pi i \frac{\tau + \tau^{-1}}{12} + \frac{1}{2} \log \frac{\tau}{i} = \sum_{k=1}^{\infty} \left(\frac{1}{e^{-2\pi i k \tau} - 1} - \frac{1}{e^{2\pi i k / \tau} - 1} \right). \tag{3}$$

Put

$$f(z) = \cot z \cot \frac{z}{\tau}, \qquad \nu = (n + \tfrac{1}{2})\pi, \qquad (n = 0, 1, \ldots). \tag{4}$$

Then the meromorphic function $z^{-1}f(\nu z)$ has simple poles at $z = \pm k\pi/\nu$ and $z = \pm k\pi\tau/\nu$ with the residues $(1/k\pi)\cos(k\pi/\tau)$ and $(1/k\pi)\cot(k\pi\tau)$ respectively for $k = 1, 2, \ldots$; in addition there is a pole of the third order at $z = 0$ with the residue $-\frac{1}{3}(\tau + \tau^{-1})$. Consider the integral of the function $f(\nu z)/8z$ along the contour C of the parallelogram with vertices at 1, τ, -1, $-\tau$. We obtain by the theorem of residues

$$\pi i \frac{\tau + \tau^{-1}}{12} + \int_C f(\nu z) \frac{dz}{8z} = \frac{i}{2} \sum_{k=1}^{n} \left(\cot k\pi\tau + \cot \frac{k\pi}{\tau} \right)$$

$$= \sum_{k=1}^{n} \frac{1}{k} \left(\frac{1}{e^{-2\pi i k \tau} - 1} - \frac{1}{e^{2\pi i k / \tau} - 1} \right). \tag{5}$$

As $n \to \infty$ the function $f(\nu z)$ is uniformly bounded on C and has on the four sides, excluding the vertices, the limit values 1, -1, 1, -1. Hence

$$\lim_{n \to \infty} \int_C f(\nu z) \frac{dz}{z} = \left(\int_1^\tau - \int_\tau^{-1} + \int_{-1}^{-\tau} - \int_{-\tau}^1 \right) \frac{dz}{z}$$

$$= 4 \log \frac{\tau}{i}, \tag{6}$$

which gives the formula (3).

Now we start from the infinite product representation of $\vartheta_3(v \mid \tau)$:

$$\vartheta_3(v \mid \tau) = \prod_{n=1}^{\infty} (1 - q^{2n}) \prod_{n=1}^{\infty} (1 + q^{2n-1}z^2)(1 + q^{2n-1}z^{-2}) \qquad q = e^{\pi i \tau}, z = e^{\pi i v}.$$

$$\tag{7}$$

Our purpose is to prove

$$\prod_{n=1}^{\infty} (1 - q^{*2n}) \prod_{n=1}^{\infty} (1 + q^{*2n-1}z^{*2})(1 + q^{*2n-1}z^{*-2})$$

$$= \sqrt{\frac{\tau}{i}} e^{i\pi v^2/\tau} \prod_{n=1}^{\infty} (1 - q^{2n}) \prod_{n=1}^{\infty} (1 + q^{2n-1}z^2)(1 + q^{2n-1}z^{-2}),$$

$$q^* = e^{-\pi i / \tau}, \, z^* = e^{\pi i v / \tau}. \tag{8}$$

Since $\vartheta_3(v|\tau)$ is analytic, it is sufficient to prove (8) for (τ, v) with Im $\tau > 0$ and $|v| < \varepsilon(\tau)$ $(\varepsilon(\tau) > 0)$ such that $|q^*| < 1$, $|qz^{\pm 2}| < 1$, and $|q^*z^{*\pm 2}| < 1$ hold.

Taking the logarithm of the both sides of (8) we obtain as (3)

$$
\frac{1}{2}\log\frac{\tau}{i} + i\pi\frac{v^2}{\tau} = \sum_{k=1}^{\infty}\frac{1}{k}\left(\frac{1}{e^{-2\pi ik\tau}-1} - \frac{1}{e^{2\pi ik/\tau}-1}\right)
$$

$$
+ \sum_{k=1}^{\infty}\frac{(-1)^k}{k}\left(e^{-\pi ik\tau}\frac{e^{2\pi ikv}+e^{-2\pi ikv}}{e^{-2\pi ik\tau}-1} - e^{\pi ik/\tau}\frac{e^{2\pi ikv/\tau}+e^{-2\pi ikv/\tau}}{e^{2\pi ik/\tau}-1}\right).
$$

(9)

Since we have already the formula (3) by Siegel, it suffices, in order to obtain (9), to prove

$$
-\pi i\frac{\tau+\tau^{-1}}{12} + i\pi\frac{v^2}{\tau}
$$

$$
= \sum_{k=1}^{\infty}\frac{(-1)^k}{k}\left(e^{-\pi ik\tau}\frac{e^{2\pi ikv}+e^{-2\pi ikv}}{e^{-2\pi ik\tau}-1} - e^{\pi ik/\tau}\frac{e^{2\pi ikv/\tau}+e^{-2\pi ikv/\tau}}{e^{2\pi ik/\tau}-1}\right). \quad (10)
$$

For the proof of (10) put

$$
g(z) = \operatorname{cosec} z \operatorname{cosec}\frac{z}{\tau}e^{2ivz/\tau}, \qquad \nu = (n+\tfrac{1}{2})\pi, \ (n=0, 1, \ldots) \quad (11)
$$

and use the meromorphic function $z^{-1}g(\nu z)$ instead of $z^{-1}f(\nu z)$. The function $z^{-1}g(\nu z)$ has simple poles at $z = \pm k\pi/\nu$ and $z = \pm k\pi\tau/\nu$ with the residues

$$
\frac{(-1)^k}{k\pi}\operatorname{cosec}\frac{k\pi}{\tau}e^{\pm 2ik\pi v/\tau} \quad \text{and} \quad \frac{(-1)^k}{k\pi}\operatorname{cosec} k\pi\tau\, e^{\pm 2ik\pi v}
$$

respectively for $k=1, 2, \ldots$. In addition there is a pole of order 3 at $z=0$ with the residue $\frac{1}{6}(\tau+\tau^{-1})$. Consider the integral of the function $g(\nu z)/4z$ along the contour C as above. Then we obtain

$$
-\frac{\pi i}{12}(\tau+\tau^{-1}) + i\pi\frac{v^2}{\tau} + \int_C\frac{g(\nu z)}{4z}\,dz
$$

$$
= \frac{i}{2}\sum_{\substack{k=-n\\k\neq 0}}^{n}\frac{(-1)^k}{k}\left\{\operatorname{cosec}\frac{\pi k}{\tau}e^{2\pi ikv/\tau} + \operatorname{cosec}\pi k\tau\, e^{2\pi ikv}\right\}
$$

$$
= \sum_{k=1}^{n}\frac{(-1)^k}{k}\left(e^{-\pi ik\tau}\frac{e^{2\pi ikv}+e^{-2\pi ikv}}{e^{-2\pi ik\tau}-1} - e^{\pi ik/\tau}\frac{e^{2\pi ikv/\tau}+e^{-2\pi ikv/\tau}}{e^{2\pi ik/\tau}-1}\right). \quad (12)
$$

As $n\to\infty$ the function $g(\nu z)/4z$ is uniformly bounded on C and $\operatorname{cosec}\nu z$

tends to zero except $z = \pm 1$ and $\operatorname{cosec} (\nu z / \tau) \, e^{2i\nu\nu z/\tau}$ tends to zero except $z = \pm \tau$. Hence we have

$$\lim_{n \to \infty} \int_C \frac{g(\nu z)}{4z} \, dz = 0.$$

Hence from (12) follows the formula (10), which was to be proved.

Microlocal Analysis of Infrared Singularities

Takahiro Kawai

Research Institute for Mathematical Sciences
Kyoto University
Kyoto, Japan

and

Henry P. Stapp

Lawrence Berkeley Laboratory
University of California
Berkeley, California

0. Introduction

Probably the most secure feature of relativistic quantum theory is the physical-region analytic structure: the singularities of any scattering function are confined to certain surfaces, called *Landau-Nakanishi surfaces* [1], or *Landau surfaces* for short, and in a neighborhood of any regular point of such a surface the scattering function is, normally, the boundary value of a function holomorphic on a certain specified kind of domain. Moreover, if all the relevant particles have strictly positive mass then the singularity on the Landau surface specified by $\varphi = 0$, at a regular point with grad $\varphi \neq 0$, normally has the form $A\varphi^{\lambda}(\log \varphi)^{\nu}$ where A is nonsingular, $\lambda = \frac{3}{2}N - 2n + \frac{3}{2}$,

Algebraic Analysis, Volume I

$\nu = 1$ or 0 according as λ is a non-negative integer or not, and N and n are the numbers of internal lines and vertices in the "Landau diagram" associated with this Landau surface [2]. These analytic properties seem quite secure because they follow from general properties that do not depend on the short-range (i.e., high-energy) contributions that can lead to ultraviolet divergences. Moreover, this physical-region singularity structure is the momentum-space form of an asymptotic spacetime structure that corresponds to classical physics in just the way demanded by the correspondence principle: the physical-region momentum-space singularity structure corresponds to the classical property that a stable particle, left undisturbed, moves with constant velocity, and neither disappears nor generates clones of itself, in the course of an undisturbed motion [4].

This correspondence-principle connection evidently requires that this same form $\varphi^{\lambda}(\log \varphi)^{\nu}$ be maintained modulo weaker singularities in the presence of radiative corrections coming from massless particles. For an electron appears neither to disappear nor to generate clones of itself during its asymptotic motion, as it would be effectively forced to do if λ or ν were altered by the effects of the massless photons.

Several difficulties block the direct extension of the earlier calculations of the values of λ and ν to cases in which (radiative) corrections due to exchange of photons, assumed massless, are included. The first is the infrared divergence: straightforward perturbative calculations lead to divergent integrals, when they are confined to the mass-shell manifold. The essential ideas needed to circumvent these divergences are well known [5]: one must separate out for special treatment the classical part of the problem, and use the fact that photons of sufficiently small energy are not observed. These ideas allow the calculations to be reorganized according to the principle that the classical part must be treated as a whole.

These ideas have recently been applied to the problem of the elucidation of the effects of radiative corrections on the singularity structure of scattering functions, and on the asymptotic behavior in spacetime [6, 7]. In [6], the photon interaction was separated into a "classical" part and a "quantum" part. The contributions from "classical photons" which by definition have "classical" couplings at each end, can then be treated exactly, without using any limiting procedure involving a fictitious nonzero photon mass, and these classical contributions sum to a unitary operator. The expected asymptotic behavior in spacetime demanded by the correspondence principle was then deduced, under the condition that the radiative corrections that arise from (massless) photons coupled via the residual "quantum" couplings lead to no infrared divergences, and to no disruption of the dominant contributions to the physical-region singularity structure described above.

The dominant term arising from the quantum coupling has in place of the usual photon coupling factor γ_μ rather the factor $\gamma_\mu k$. Here $k = k^\nu \gamma_\nu$, the quantities k^ν are components of the photon momentum-energy vector, and the γ_μ are Dirac matrices. The extra factor k, which vanishes at $k = 0$, is expected to remove all infrared divergencies. But if one wishes to examine the asymptotic behavior in spacetime, or the corresponding momentum-space analytic structure, then difficulties still remain. These arise from the singular character of the photon propagator $(k^2 + i0)^{-1}$ at $k = 0$.

In the present context the usual procedure for avoiding this problem by introducing a fictitious photon mass μ, which is set to zero at the end of the computation, is not efficient. For the character of the singularity surface $k^2 - \mu^2 = 0$ changes abruptly when the photon mass μ is set to zero, and this change induces an abrupt change in the character of the singularity on or near the Landau surface $\varphi = 0$, when the photon mass is set to zero. For example, if we consider the (renormalized) Feynman function for the

self-energy diagram , the exponent λ at the threshold

$p^2 = (m + \mu)^2$ is equal to $\frac{1}{2}$ for $\mu \neq 0$ and equal to 1 (and $\nu = 1$) for $\mu = 0$. Thus one is left with the highly nontrivial problem of proving that the $\mu \to 0$ limiting procedure yields precisely the answer that would be obtained if one used the true propagator $(k^2 + i0)^{-1}$.

Our approach is to deal directly with the photon propagator $(k^2 + i0)^{-1}$, without introducing any limiting procedure. This makes the mathematical situation different in principle from the familiar one associated with the propagators of massive particles. The main content of this paper consists, then, basically of a rigorous treatment of the problems associated with the singular character of the photon propagator at $k = 0$, in the context of the simplest nontrivial example.

The initial problem in setting up such a formulation is to give well-defined meaning to the photon propagator $(k^2 + i0)^{-1}$. This is done in Section 1, where a rigorous meaning is given to the formal expression

$$(k^2 + i0)^{-1} = -2\pi i \delta(k_0 - \sqrt{\mathbf{k}^2})(2k_0)^{-1} + ((k_0 - i0)^2 - \mathbf{k}^2)^{-1} \qquad (0.1)$$

and its $k \mapsto -k$ transform

$$(k^2 + i0)^{-1} = -2\pi i \delta(k_0 + \sqrt{\mathbf{k}^2})(2k_0)^{-1} + ((k_0 + i0)^2 - \mathbf{k}^2)^{-1}. \qquad (0.2)$$

Here, and in what follows, $\mathbf{k}$ denotes (k_1, k_2, k_3), and $\mathbf{k}^2 = \sum_{j=1}^{3} k_j^2$. The apparent ambiguity of $\delta(k_0 \pm \sqrt{\mathbf{k}^2})(2k_0)^{-1}$ at $k = 0$ will be removed in Section 1, and $(k^2 + i0)^{-1}$ will then be expressed (in two different ways) as a sum of two well-defined hyperfunctions (cf. (1.10)). Each of the two terms in either of those expressions has important properties which we will exploit. But in none of these terms is there a separation of the singularities into two disjoint parts, one confined to the region of positive photon energy and the other confined to the region of negative photon energy. This separation holds for the massive particle case, and it plays an important role in the usual derivations of singularity properties. The failure of these properties forces us to devise new methods, in order to deal with the $k \doteq 0$ contributions.

A specific aim of our analysis is to show that in our simple case the discontinuity around the singularity surface under consideration is given by Cutkosky-type formulas [8]. This property does not hold for the original Feynman function, with photon couplings γ_μ, because of infrared divergences. In particular, for our case in which the external particles are all neutral the original Feynman function is well-defined (i.e., it is infrared finite), and so is the discontinuity. But the expression for the discontinuity obtained from the $\mu \to 0$ limiting procedure is a sum of four Cutkosky-type functions each of which is infrared divergent. In our case, with couplings $\gamma_\mu k$, we find, in the end, that the discontinuity is given by the same Cutkosky-type formula that was formally obtained from the $\mu \to 0$ limiting procedure. Now, however, the four discontinuities are, as expected, infrared finite, due to the extra power of k in each quantum coupling. The formula allows one to exhibit explicitly the character of the singularity at $\varphi = 0$, and confirms that the character of the dominant singularity is not altered by the radiative corrections, as was demanded in [6].

The scattering functions, considered as functions of real momentum vectors, are hyperfunctions: in a sufficiently small real open neighborhood Ω of any point p, the scattering function $f(p')$ can be represented as a sum of terms, each of which is a boundary value $f_j(p')$ of a function $\tilde{f}_j(p' + iq')$ that is holomorphic on an open region $p' \in \Omega$, $q' \in \Gamma_j$, where Γ_j is a domain that tends to an open cone with vertex at $q' = 0$ as q' tends to zero. We use the framework of microlocal analysis, which is described in [9, 10]. Very briefly, the formalism rests on the concept of the "singularity spectrum" of a hyperfunction. For each point $x = (x_1, \ldots, x_n)$ in a real manifold M where the hyperfunction is *not* analytic there is a set of pairs $(x; \sqrt{-1}w)$, where the "cotangential" components w ($\neq 0$), which are defined up to a strictly positive scalar multiple, are determined as dual cones of the set of allowed directions along which the real point x is to be approached if the hyperfunction is to be represented as a sum of boundary values of holomorphic functions. The totality of such pairs $(x; \sqrt{-1}w)$ is called the singularity spectrum of f, and it is denoted as S.S. f. The singularity spectrum is, by its definition, a subset of the (pure imaginary) spherical cotangent bundle, which is usually denoted $\sqrt{-1}S^*M$.

Two important properties of hyperfunctions used repeatedly in this work are:

(a)   The product $f = \prod_{l \in L} f_l$ of finitely many hyperfunctions f_l is a well-defined hyperfunction at x if there is no solution to $\sum_{l \in L(x)} \alpha_l w_l = w = 0$, $\alpha_l \geq 0$, $\sum_{l \in L(x)} a_l \neq 0$, $\alpha_l = 0$ $(l \notin L(x))$, where $L(x) = \{l \in L; f_l$ is not analytic at $x\}$ and each w_l $(l \in L(x))$ belongs to a pair $(x; \sqrt{-1}w_l)$ associated with f_l (i.e., $(x; \sqrt{-1}w_l)$ belongs to the singularity spectrum of f_l). If there is no such solution $w = 0$ then the singularity spectrum of f over the base point x is contained in the set of pairs $(x; \sqrt{-1}w)$ such that w is a solution of the above set of equations with the condition $w = 0$ excluded (cf. [10], p. 109).

(b)   If $f(x, y)$ is a hyperfunction then the integral $F(x) = \int dy f(x, y)$ is a well-defined hyperfunction provided the support of $f(x, y)$ in y, as x ranges over a compact set, is compact. Further the singularity spectrum of $F(x)$ is confined to the set of pairs $(x; \sqrt{-1}u)$ such that $((x, y); (\sqrt{-1}u, \sqrt{-1} \cdot 0))$ is in the singularity spectrum of $f(x, y)$. In particular, $F(x)$ is analytic if there is no point of the form $(x, y; (\sqrt{-1}u, \sqrt{-1} \cdot 0))$ $(u \neq 0)$ in the singularity spectrum of $f(x, y)$, provided that the condition on the support of $f(x, y)$ is satisfied.

In Section 1, we define $(k^2 + i0)^{-1}$ as a hyperfunction in accordance with (0.1) and (0.2), and give conditions on the singularity spectra of the terms appearing on the right-hand side of these equations. In Section 2, we define our problem, which is to determine the effects of a radiative correction to a charged-particle triangular closed loop. In Section 3, it is noticed that the relevant scattering function is not infrared divergent, and has a singularity on the Landau surface $\varphi = 0$ associated with the triangle diagram. In Sections 4, 5, 6 and 7, we derive the discontinuity around this surface. The method is the same as was employed in [3] and [11]: the scattering function f^+ is transformed into $\Delta + f^-$, where Δ is zero in $\varphi < 0$, and f^+ and f^- have plus $i0$ and minus $i0$ continuations around $\varphi = 0$. Then Δ is the discontinuity. The character of the singularity at $\varphi = 0$ is then discussed in Section 8.

1. The Photon Propagator

Our first task is to make the Feynman photon propagator $(k^2 + i0)^{-1}$ well-defined within our framework. Here k is the energy-momentum four-vector of the photon, $(k_0, k_1, k_2, k_3) = (k_0, \mathbf{k})$ and k^2 means $k_0^2 - \mathbf{k}^2$. At $k = 0$ the meaning of the symbol $+i0$ is not clear, because $\text{grad}_k\, k^2$ vanishes there. This is in contrast to the case of massive particle, for which the propagator is $(p^2 - m^2 + i0)^{-1}$ $(m > 0)$. Then $\text{grad}_p(p^2 - m^2) \neq 0$ if $p^2 - m^2 = 0$, and the symbol $+i0$ acquires unambiguous meaning (cf. [10], p. 89).

One way, natural both from the mathematical and physical viewpoints, is to start from the retarded propagator $R(k)$ and the advanced propagator $A(k)$; they are, by definition, given respectively by

$$R(k) = ((k_0 + i0)^2 - \mathbf{k}^2)^{-1} \tag{1.1}$$

and

$$A(k) = ((k_0 - i0)^2 - \mathbf{k}^2)^{-1}. \tag{1.2}$$

It is known that each of these two functions, $R(k)$ and $A(k)$, can be realized as the boundary value of a holomorphic function ([10], p. 90).

More specifically, their singularity spectra are given as follows:

S.S. $R(k) \subset \{(k; \sqrt{-1}w) \in \sqrt{-1}S^*\mathbf{R}^4;\ k^2 = 0,\ k \neq 0,$

$$w = \tfrac{1}{2}c\,\text{sgn}(k_0)\,\text{grad}_k\,k^2 = c\,\text{sgn}(k_0)k\ (c > 0)\}$$

$$\cup \left\{(k; \sqrt{-1}w);\ k = 0,\ w^2 \underset{\text{def}}{=} w_0^2 - \sum_{j=1}^{3} w_j^2 \geq 0 \text{ and } w_0 > 0\right\}, \tag{1.3}$$

S.S. $A(k) \subset \{(k; \sqrt{-1}w) \in \sqrt{-1}S^*\mathbf{R}^4;\ k^2 = 0,\ k \neq 0,\ w = -c\,\text{sgn}(k_0)k\ (c > 0)\}$

$$\cup \{(k; \sqrt{-1}w);\ k = 0,\ w^2 \geq 0,\ w_0 < 0\}. \tag{1.4}$$

Here, and in what follows, we identify $\text{grad}_k\,k^2$ with $2k$, using the Minkowsky metric ([10], p. 90).

Furthermore, thanks to the four-dimensionality, we can verify that the d'Alembertian $\Box = \partial^2/\partial k_0^2 - \sum_{j=1}^{3} \partial^2/\partial k_j^2$ annihilates both $R(k)$ and $A(k)$ ([10], p. 150). Hence, Sato's lemma on microlocal ellipticity ([10], p. 140) entails

$$\text{S.S. } R(k) \subset \{(k; \sqrt{-1}w);\ w^2 = 0\} \tag{1.5}$$

and

$$\text{S.S. } A(k) \subset \{(k; \sqrt{-1}w);\ w^2 = 0\}. \tag{1.6}$$

That is, the cotangential component of the singularity spectra of $R(k)$ and $A(k)$ are confined to the light cone, provided we restrict our considerations to the four-dimensional world. These extra relations (1.5) and (1.6) will be used effectively in later sections. The same result holds, in fact, for all even-dimensional cases.

Now, by the result (A) stated in Section 0, we find that $\theta(k_0)R(k)$, $\theta(k_0)A(k)$ and $\theta(k_0)(R(k)-A(k))/(-2\pi i)$ are all well-defined. Here, and in what follows, $\theta(k_0)$ denotes the Heaviside function. Since $\theta(k_0)(R(k)-A(k))/(-2\pi i)$ coincides with $\delta(k^2)$ for $k_0 > 0$), and vanishes identically for $k_0 < 0$, we denote it by $\delta^+(k^2)$. Then, the massless propagator $(k^2+i0)^{-1}$ is, by definition,

$$(-2\pi i)\delta^+(k^2)+A(k). \tag{1.7}$$

Note that it can be expressed also in the following form:

$$(+2\pi i)\delta^-(k^2)+R(k). \tag{1.8}$$

Here $\delta^-(k^2)$ is, by definition,

$$\frac{1}{2\pi i}\,\theta(-k_0)(A(k)-R(k)). \tag{1.9}$$

Summing up, we have

$$\frac{1}{k^2+i0}=(-2\pi i)\delta^+(k^2)+A(k)$$
$$=(+2\pi i)\delta^-(k^2)+R(k). \tag{1.10}$$

This relation will be used frequently in the ensuing discussion.

2. Definition of the Problem

Throughout this article, we denote by D the following triangle graph:

D:

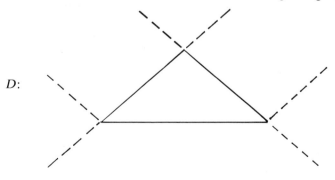

Each solid line is associated with a charged particle of mass $m > 0$. Each dashed line is a neutral particle. Let D' denote the following graph, which corresponds to an electromagnetic correction:

D':

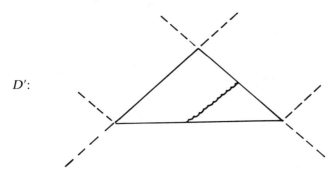

Here the wiggly line is associated with the photon propagator $g^{\mu\nu}/(k^2 + i0)$, where $g^{\mu\nu}$ is the Minkowsky metric tensor. We call each end point of the wiggly line a photon vertex.

The modified Feynman function, which is the quantum coupled function associated with the graph D', is given by the following formula (2.1). There, and in the sequel, we denote by e a constant (electric charge). We also use the symbol $\not{p}$ to denote $p^\mu \gamma_\mu = g^{\mu\nu} \gamma_\mu p_\nu$, where γ_μ $(\mu = 0, 1, 2, 3)$ denotes the Dirac gamma matrix.

$$F_{QQ}(q) = \int \frac{d^4k}{(2\pi)^4} \frac{g^{\mu\nu}}{k^2 + i0} \int \frac{d^4p_1}{(2\pi)^4}$$
$$\mathrm{Tr}\left[F_{Q\mu}(p_1, k) V_1 F'_{Q\nu}(-p_1 + q_1, k) V_2 \right.$$
$$\left. \times \frac{(-\not{p}_1 + \not{q}_1 + \not{q}_2 + m)}{(-p_1 + q_1 + q_2)^2 - m^2 + i0} V_3 \right]$$
$$\times (2\pi)^4 \delta^4(q_1 + q_2 + q_3), \tag{2.1}$$

where V_j $(j = 1, 2, 3)$ is some Dirac matrix, $F_{Q\mu}(p_1, k)$ is, by definition, given by

$$G_\mu(p_1, k) - \int_0^1 d\lambda\, G_\mu(p_1 + \lambda k, 0) \tag{2.2}$$

with

$$G_\mu(p_1, k) = \frac{(\not{p}_1 + m)(-ie\gamma_\mu)(\not{p}_1 + \not{k} + m)}{(p_1^2 - m^2 + i0)((p_1 + k)^2 - m^2 + i0)},$$

and $F'_{Q\nu}$ is the similar function for the other photon vertex.

As our concern in this article is the infrared problem, i.e., the problems arising from points near $k = 0$, we neglect the ultraviolet problem, i.e., the contribution to the integral from large k.

By performing the λ-integration in (2.2) explicitly, we find that $F_{Q\mu}(p_1, k)$ has the form

$$\Phi_\mu(p_1, k) + \rho_\mu(p_1, k), \tag{2.3}$$

where $\Phi_\mu(p_1, k)$ has a pole on both $\{p_1^2 = m^2\}$ and $\{(p_1 + k)^2 = m^2\}$, and it does not vanish when $k^2 = 0$, whereas $\rho_\mu(p_1, k)$ is the sum of terms that do not have these properties. The explicit form of $\Phi_\mu(p, k)$ is

$$\frac{(\not{p}_1 + m)\gamma_\mu \not{k}}{(p_1^2 - m^2 + i0)((p_1 + k)^2 - m^2 - i0)}, \tag{2.4}$$

and the actual computation shows that $\rho_\mu(p_1, k)$ contributes singularities to $F_{QQ}(q)$ along the leading Landau-Nakanishi surface $L_0^+(D)$ that are weaker than those coming from $\Phi_\mu(p_1, k)$ (see [12]).

Hence the most singular part of $F_{QQ}(q)$ is given by the combination of $\Phi_\mu(p_1, k)$ and the similar function $\Phi_\nu(p_1 + q_1, -k)$ determined by $F'_{Q\nu}(p_1 + q_1, -k)$. This fact confirms a formula given in [6] obtained by studying the asymptotic behavior of the corresponding function (i.e., the inverse Fourier transform) in position space.

By virtue of these results we may focus our attention on the integral $\Phi(q)$ defined by replacing in (2.1) $F_{Q\mu}$ by Φ_μ and $F_{Q\nu}$ by Φ_ν.

3. Properties of $\Phi(q)$

It is an easy consequence of the results (A) and (B) in Section 0 that $\Phi(q)$ is a well-defined hyperfunction, provided we ignore any possible problem associated with the contributions at large k. Moreover, near $L_0^+(D)$ its singularities are confined to $L_0^+(D)$. It is also a $+i0$ boundary value along $L_0^+(D)$. (See [12], Appendix B.) Hence we can consider its discontinuity $\Delta(q)$ around $L_0^+(D)$. This latter function coincides with $\Phi(q)$ in the $+i0$-direction (i.e., its difference with $\Phi(q)$ has no singularity associated with the $+i0$-direction), and it vanishes below the threshold $L_0^+(D)$).

Our specific task here is to show that the discontinuity is represented as a sum of functions that are determined diagrammatically, are easily calculable, and, in particular, are of the kind proposed by Cutkosky for the massive-particle case. From the formula we may obtain the singularity structure of $\Phi(q)$ near $L_0^+(D)$. In deriving the discontinuity formula, the factor k in the numerators of Φ_μ and Φ_ν plays a decisively important role; this factor, which the ordinary Feynman function lacks, makes the integral convergent near $k = 0$.

4. The Discontinuity around $L_0^+(D)$

We now turn to the derivation of the discontinuity formula for $\Phi(q)$ around the surface $L_0^+(D)$. Our method is based on a graphical analysis combined with microlocal analysis.

In what follows we identify a graph with an associated function. The symbols $\xrightarrow{\;+\;}$, $\xrightarrow{\;-\;}$ and $\longmapsto$ respectively denote $(p^2 - m^2 + i0)^{-1}$, $(p^2 - m^2 - i0)^{-1}$, and $(-2\pi i)\delta^+(p^2 - m^2) = (-2\pi i)\theta(p_0)\delta(p^2 - m^2)$ (sometimes multiplied by $p\!\!\!/ + m$). Moreover, the symbols

$$\overset{\nu\;\;+\;\;\mu}{\wwave}, \quad \overset{\nu\;\;\;\;\;\mu}{\wwavearrow}, \quad \overset{\nu\;\;+\;\;\mu}{\cdots\cdots\!\!\blacktriangleright\!\cdots} \quad \text{and} \quad \overset{\nu\;\;-\;\;\mu}{\cdots\cdots\!\!\blacktriangleright\!\cdots}$$

respectively denote $g^{\mu\nu}/(k^2 + i0)$, $(-2\pi i)g^{\mu\nu}\delta^+(k^2)$, $g^{\mu\nu}R(k)$ and $g^{\mu\nu}A(k)$. An arrow on a line indicates the direction in which the momentum-energy vector p or k flows along that line. In what follows, we often suppress μ and ν at photon vertices (and hence $g^{\mu\nu}$ also) for the sake of simplicity. Note that the formula (1.10) (or (0.1) and (0.2)) can be rewritten diagrammatically as follows:

$$(4.1)$$

The two forms differ only in regard to which direction one takes the vector called k to be flowing along the line.

For any two graphs D_1 and D_2 the symbol $D_1 \cong D_2$ means that the function associated with D_1 minus the function associated with D_2 is a $(-i0)$-boundary value at a generic point of $L_0^+(D)$. For the sake of the simplicity of notations, we will omit the external lines of each graph in the sequel.

The first step of our graphical analysis is to decompose

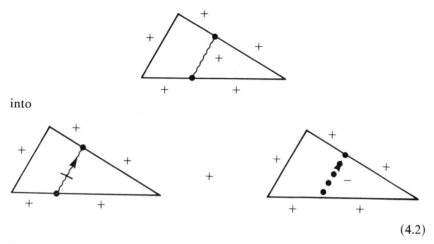

into

$$(4.2)$$

We call the first (resp., second) graph D_0 (resp., D_1). To manipulate D_0 and D_1, we will in the next two sections first prepare some auxiliary results. Then the required decompositions of D_0 and D_1 will be given in Section 7.

5. Polar Coordinates

As noted earlier, our reasoning relies heavily on the extra factor k at each photon vertex. To make full use of this factor, we sometimes use the polar coordinate system (r, Ω) in k-space; $k = r\Omega$ $(r \geq 0)$ with $\Omega\tilde{\Omega} = 1$, where $\tilde{\Omega} = (\Omega_0, -\Omega_1, -\Omega_2, -\Omega_3)$. Then we put a symbol r on a photon vertex if the vertex is associated with a factor k. Using this symbol we obtain the following relation:

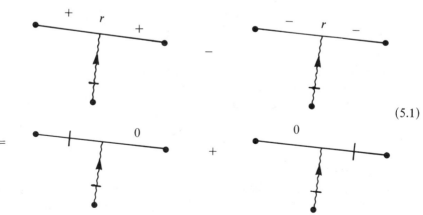

$$(5.1)$$

The line segments with a symbol zero represent the factor $r((p')^2 - m^2)^{-1}$, where p' is the energy-momentum vector associated with the line, subjected to the constraints $\Omega^2 = 0$ and $(p'') = m^2$ that arise from the slashes on the other two lines. Each of these factors associated with the zero line is nonsingular in (r, Ω)-variables, since $p^2 = m^2 > 0$ and $\Omega^2 = 0$ entail $p\Omega \neq 0$.

The result (5.1) is an immediate consequence of

$$\frac{r}{p^2 - m^2 \pm i0} \cdot \frac{1}{p^2 - m^2 + 2rp\Omega \pm i0}$$

$$= \frac{1}{p^2 - m^2 \pm i0} \cdot \frac{1}{2p\Omega} - \frac{1}{2p\Omega} \cdot \frac{1}{p^2 - m^2 + 2rp\Omega \pm i0} \qquad (5.2)$$

(the signs $\pm i0$ should be uniformly used).

6. Some Auxiliary Results

Another needed auxiliary result pertains to the analyticity of the function associated with the following graph:

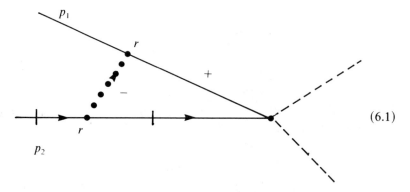

$$(6.1)$$

The associated function $I(p_1, p_2)$ takes the form (6.2) below in the (r, Ω)-variable. There $\delta \ll m$ is a strictly positive constant that is introduced in order to restrict our considerations to contributions coming from a neighborhood of $k = 0$, and $P^{\mu\nu}$ is a polynomial in p and $r\Omega$:

$$\int_0^\delta r^2 \, dr \int d^4\Omega \, \delta(\Omega\tilde{\Omega} - 1) P^{\mu\nu} \frac{\Omega_\mu \Omega_\nu \delta(-2p_2\Omega + r\Omega^2)}{((\Omega_0 - i0)^2 - \Omega^2)((p_1 + r\Omega)^2 - m^2 + i0)}. \qquad (6.2)$$

Using the results (A) and (B) in Section 0, we can easily verify that this integral defines a well-defined hyperfunction. However, we cannot see immediately that there is no net contribution to the singularity from the endpoint $r = 0$. To obtain that result we use the following trick: Let us consider the coordinate transformation $(r, \Omega) \mapsto (-r, -\Omega)$. Then $I(p_1, p_2)$ can be expressed in the form

$$\frac{1}{2} \int_{-\delta}^{\delta} r^2 \, dr \int d^4\Omega \, \delta(\Omega\tilde{\Omega} - 1) P^{\mu\nu} \frac{\Omega_\mu \Omega_\nu \delta(-2p_2\Omega + r\Omega^2)}{((\Omega_0 - i0)^2 - \Omega^2)((p_1 + r\Omega)^2 - m^2 + i0)}.$$
(6.3)

This form has no endpoint contribution from $r = 0$. The integral given in (6.3) is analytic except at points where contributions from the endpoints $r = \pm\delta$ are relevant. To convert $I(p_1, p_2)$ to the form of (6.3), it suffices to note that the $(-i0)$ in the first denominator should be understood as $-i0/\operatorname{sgn} r$ if $r \neq 0$. This fact, combined with the trivial fact that $r\Omega$ is invariant under the coordinate transformation in question, guarantees that the integrand of the integral I is invariant under the coordinate transformation $(r, \Omega) \mapsto (-r, -\Omega)$. The required result then follows immediately. The same reasoning also gives this same analyticity property for the function specified by the following graph:

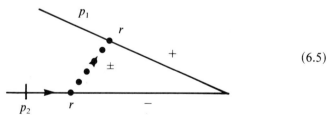

(6.4)

The relations (1.3)–(1.6) guarantee the analyticity of the functions specified by either

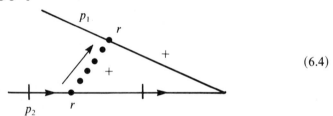

(6.5)

or

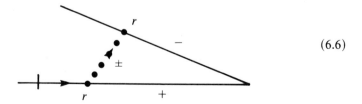

(6.6)

In fact, using the results (A) and (B) in Section 0, we deduce from (1.3)–(1.6) that the analyticity (at a generic point of $L_0^+(D)$) of these functions fails only when $\alpha_1 p_1 + \alpha_2 p_2 = w$ holds for a non-zero light cone vector w with $\alpha_1, \alpha_2 \geq 0$, $\alpha_1 + \alpha_2 \neq 0$ and $\alpha_1(p_1^2 - m^2) = \alpha_2(p_2^2 - m^2) = 0$. Therefore $\alpha_1 \alpha_2 \neq 0$, and hence $p_1^2 = p_2^2 = m^2$. But this cannot happen, as $p_{1,0}$ and $p_{2,0}$ are of the same sign at the relevant points.

It follows from the definitions that

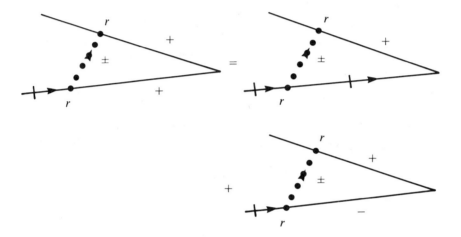

Thus we have the same analyticity property also for the graphs

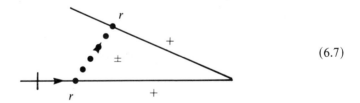

(6.7)

Combining the graph (6.6) and (6.7) we also find this same analyticity for the graphs

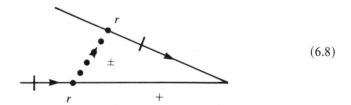

(6.8)

7. Derivation of a Discontinuity Formula

Let us now transform 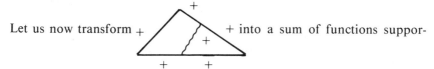 + into a sum of functions suppor-

ted above the threshold $L_0^+(D)$, modulo some $(-i0)$-boundary value along generic points of $L_0^+(D)$. The transformation given below relies on the results (A) and (B) in Section 0 supplemented by the results obtained in the previous two sections.

Let us first consider D_0. Then, using the result in Section 5, we find the following:

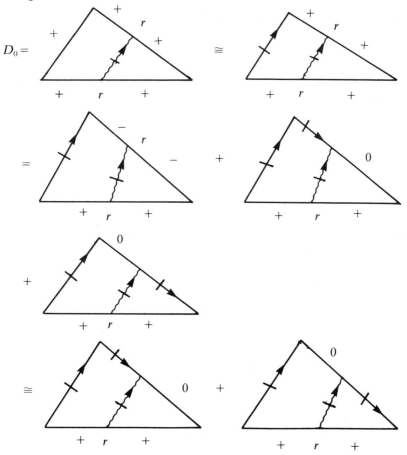

$$(7.1)$$

Let D_j ($j = 2, 3, 4, 5$) denote the $(j-1)^{\text{th}}$ graph in the last expression. Observing the direction of the energy flows, we immediately see that D_2, D_3 and D_5 vanish below the threshold $L_0^+(D)$. Hence they are a part of the discontinuity function as they stand. To manipulate D_4 further we again apply (4.1):

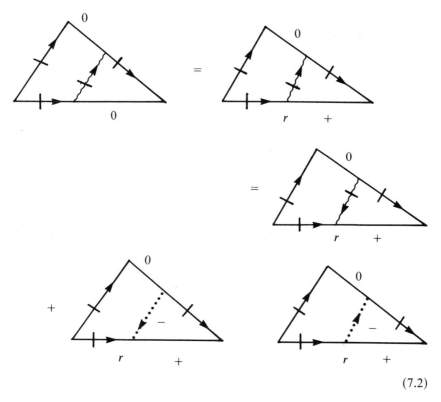

$$(7.2)$$

Then, in view of the analyticity established in Section 6 (which lets the small triangle involving the dotted minus line to be contracted to a point), we obtain

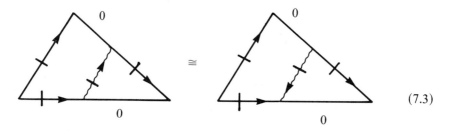

$$(7.3)$$

The right-hand side of (7.3), which we call D_d, vanishes below the threshold $L_0^+(D)$, and hence it is a part of the discontinuity function.

Combining (7.1) and (7.3), we find that the discontinuity function for D_0 is a sum of the terms D_2, D_3, D_5 and D_d.

Next let us consider D_1. For D_1, we find the following:

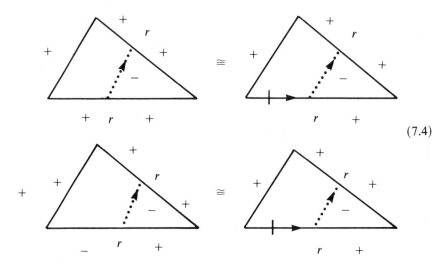

$$(7.4)$$

Here we have used (1.4) and (1.6) to obtain

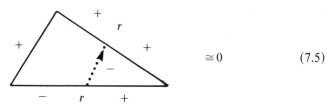

$$\cong 0 \qquad (7.5)$$

In view of the analyticity for the graph in (6.7), we find the discontinuity function for the last term in (7.4) is given by

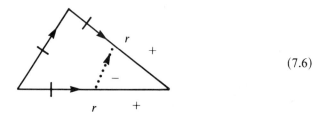

$$(7.6)$$

Combining this contribution with the one corresponding to D_2 we obtain, using (4.1),

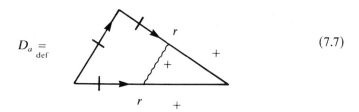

$$D_a \underset{\text{def}}{=} \qquad (7.7)$$

Summing up the results obtained so far, we find the following formula:

$$(7.8)$$

where each term on the right-hand side is supported in the region lying above $L_0^+(D)$.

In what follows, we denote by D_b, D_c and D_d respectively the last three terms in the right-hand side of (7.8).

8. Explicit Form of D_a, D_b, D_c and D_d

Let us now show that

(I) The character of the singularity of D_a along $L_0^+(D)$ is the same as

that of  and that

(II) The singularities of D_b, D_c and D_d along generic points of $L_0^+(D)$ are strictly weaker than that of D_b.

The assertion (I) is an immediate consequence of the analyticity of

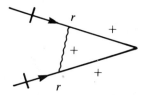

which follows from (6.7) and the analyticity of

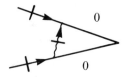

The analysis of D_c and that of D_d are similar. Let us consider D_c. The function associated with

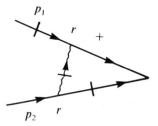

is, by definition, of the following form:

$$I_c(p_1, p_2) = \int d^4\Omega \, A(p_1, p_2, \Omega) \delta^+(\Omega^2) \, \delta(\Omega\tilde{\Omega} - 1)$$

$$\times \int_0^\delta \delta^+(p_2^2 - m^2 - 2rp_2\Omega)r^2 \, dr, \qquad (8.1)$$

where $A(p_1, p_2, \Omega)$ is nonsingular. At the endpoint contribution from $r = \delta$ does not give any singularities at the generic point of $L_0^+(D)$, we neglect it. Then we find

$$I_c = \int d^4\Omega A(p_1, p_2, \Omega) \frac{(p_2^2 - m^2)^2 \theta(p_2^2 - m^2)}{(2p_2\Omega)^3} \delta^+(\Omega^2)\delta(\Omega\tilde{\Omega} - 1). \quad (8.2)$$

Since $p_2\Omega \neq 0$ at relevant points, the integral I_c has the form

$$B(p_1, p_2)(p_2^2 - m^2)^2 \theta(p_2^2 - m^2), \quad (8.3)$$

where B is nonsingular.

Thus the function corresponding to D_c has the form

$$\int C(p, q_1, q_2)\delta^+(p^2 - m^2)\delta^+((p + q_1)^2 - m^2)$$

$$\times ((q_2 - p)^2 - m^2)\theta((q_2 - p)^2 - m^2)\, d^4p,$$

where C is nonsingular. Hence it follows from a result in [10], p. 422 that the above integral has the form

$$A_c(q_1, q_2)\varphi(q_1, q_2)^2 \theta(\varphi(q_1, q_2)), \quad (8.4)$$

where A_c is analytic and φ is a defining function of $L_0^+(D)$ (with its sign suitably chosen). This is much weaker than D_a, which has the form $A_a(q_1, q_2)\theta(\varphi(q_1, q_2))$.

Finally, let us consider D_b. Taking into account the extra r^2 in the numerator, we find that it assumes the following form:

$$\int d^4\Omega \delta^+(\Omega^2)\delta(\Omega\tilde{\Omega} - 1) \int_0^\delta r\, dr \int d^4pA(p, q_1, q_2, \Omega, r)$$

$$\times \delta^+(p^2 - m^2)\delta^+((q_1 + p + r\Omega)^2 - m^2)\delta^+((q_2 - p - r\Omega)^2 - m^2), \quad (8.5)$$

where A is nonsingular. By performing the p-integration, we find the p-integral has the form

$$B(q_1, q_2, \Omega, r)\theta(\varphi(q_1 + r\Omega, q_2 - r\Omega)), \quad (8.6)$$

where B is again nonsingular. Let $\psi(q_1, q_2, \Omega, r)$ denote $\varphi(q_1 + r\Omega, q_2 - r\Omega)$. Then we find

$$\left.\frac{\partial\psi}{\partial r}\right|_{r=0} = \Omega \cdot \left(\left.\frac{\partial\varphi}{\partial q_1}\right|_{r=0} - \left.\frac{\partial\varphi}{\partial q_2}\right|_{r=0}\right). \quad (8.7)$$

On the other hand, the Landau-Nakanishi equations tell us

$$\left.\frac{\partial \varphi}{\partial q_1}\right|_{r=0} - \left.\frac{\partial \varphi}{\partial q_2}\right|_{r=0} = -\alpha p \tag{8.8}$$

with $\alpha > 0$, $p^2 = m^2$, if (q_1, q_2) lies in $L_0^+(D)$. Hence if $\Omega^2 = 0$ and $\Omega_0 > 0$, the right-hand side of (8.7) is strictly negative. Since $\psi|_{r=0} = \varphi(q_1, q_2)$ holds, the r-integration should be done from φ/c to δ with some nonzero analytic function c. Hence, again neglecting the contribution from the endpoint $r = \delta$, we find that D_b has the form

$$\int d^4\Omega \, \delta^+(\Omega^2) \delta(\Omega\tilde{\Omega} - 1) C(q_1, q_2, \Omega) \varphi(q_1, q_2)^2 \theta(\varphi(q_1, q_2)), \tag{8.9}$$

where C is nonsingular. Hence the Ω-integration can be trivially done to give the form

$$A_b(q_1, q_2) \varphi(q_1, q_2)^2 \theta(\varphi(q_1, q_2)). \tag{8.10}$$

This is again much weaker than D_a. Thus we have verified both (I) and (II).

References

[1] L. D. Landau, in *Proc. Kiev Conference on High-Energy Physics* (1959); *Nuclear Physics* **13** (1959) 181. N. Nakanishi, *Prog. Theor. Phys.* **21** (1959) 135.

[2] R. J. Eden, P. V. Landshoff, D. I. Olive, and J. C. Polkinghorne, *The Analytic S-matrix*, Cambridge University Press, p. 57. See also T. Kawai and H. P. Stapp, *Publ. RIMS, Kyoto Univ.* **12** Suppl. (1977) 155.

[3] H. P. Stapp, *J. Math. Phys.* **9** (1968) 1548. J. Coster and H. P. Stapp, *J. Math. Phys.* **10** (1969) 371; **11** (1970) 1441; **11** (1970) 2743; **16** (1975) 1288. H. P. Stapp, in *Structural Analysis of Collision Amplitudes*, (R. Balian and D. Iagolnitzer, ed.), North-Holland, 1976. D. Iagolnitzer, *The S-matrix*, North-Holland, 1978.

[4] C. Chandler and H. P. Stapp, *J. Math. Phys.* **10** (1969) 826. D. Iagolnitzer and H. P. Stapp, *Commun. Math. Phys.* **14** (1969) 15. H. P. Stapp, in *Structural Analysis of Collision Amplitudes* (*ibid.*). D. Iagolnitzer, *The S-matrix* (*ibid.*).

[5] F. Bloch and A. Nordsieck, *Phys. Rev.* **52** (1937) 54. D. Yennie, S. Frautchi, and H. Suura, *Ann. Phys.* (N.Y.) **13** (1961) 379.

[6] H. P. Stapp, *Phys. Rev.* **28** (1983) 1386.

[7] E. D'Emilio and M. Mintchev, *Phys. Rev.* **27** (1983) 1840. M. V. Bergere and L. Syzmanowski, *Phys. Rev.* **26** (1982) 3550.

[8] R. E. Cutkosky, *J. Math. Phys.* **1** (1960) 429.

[9] M. Sato, T. Kawai, and M. Kashiwara, *Lect. Notes in Math.*, No. 287 (Springer) (1973) 265.

[10] M. Kashiwara, T. Kawai, and T. Kimura, *Foundations of Algebraic Analysis*, Princeton University Press (1986).

[11] T. Kawai and H. P. Stapp, *Publ. RIMS, Kyoto Univ.* **12** Suppl. (1977) 155.

[12] T. Kawai and H. P. Stapp, Infrared finiteness and analytic structure, in preparation.

On the Global Existence of Real Analytic Solutions of Systems of Linear Differential Equations

Takahiro Kawai

Research Institute for Mathematical Sciences
Kyoto University
Kyoto, Japan

and

Yoshitsugu Takei

Department of Mathematics
Faculty of Science
Kyoto University
Kyoto, Japan

0. Introduction

For general systems of linear differential equations not necessarily with constant coefficients, almost no global existence theorems are known, with the exception of the pioneering work of Hörmander [H] for the first-order systems with one unknown function. The purpose of this article is to present some global existence theorems for real analytic solutions on a compact set K satisfying some convexity condition with respect to the system in question.

Throughout this article we always assume that the linear differential operators in question are with analytic coefficients.

We hope this article will accelerate the current trend that not only single equations but also systems should be studied in the theory of differential equations. This is a viewpoint which Sato has been emphasizing since 1960 (or most likely some day prior to this date, although the reference cannot be traced back before 1960).

1.   Uniform $\mathcal{M}$-Convexity

Let us first prepare some notations. Throughout this article M denotes an open subset of $\mathbf{R}^n$, X denotes a Stein open neighborhood of M in $\mathbf{C}^n$, T^*X denotes the cotangent bundle of X and ω_X denotes the canonical 1-form of T^*X, i.e., $\sum_{j=1}^{n} \zeta_j \, dz_j$. We denote by $\mathcal{A}$ (resp., $\mathcal{O}$) the sheaf of real analytic functions on M (resp., holomorphic functions on X). Let $\mathcal{D}_X$ denote the sheaf of linear differential operators on X, and let $\mathcal{M}$ denote a system, i.e., a coherent left $\mathcal{D}_X$-module. Hereafter we suppose that $\mathcal{M}$ satisfies the following conditions:

(1.1) $\mathcal{M}$ is with one unknown function, that is, $\mathcal{M}$ has the form $\mathcal{D}_X / \mathcal{I}$, where $\mathcal{I} = \mathcal{D}_X P_1 + \cdots + \mathcal{D}_X P_d$ for some linear differential operators P_α $(1 \le \alpha \le d)$ defined on X.

(1.2) $[P_\alpha, P_\beta]$, the commutator of P_α and P_β, identically vanishes for $\alpha, \beta = 1, \ldots, d$.

Denoting by $p_\alpha(z, \zeta)$ the principal symbol of the operator $P_\alpha(z, D)$, we further suppose

(1.3) $dp_1, \ldots, dp_d$ and ω_X are linearly independent on $\{(z, \zeta) \in T^*X;$ $\zeta \ne 0, \, p_1(z, \zeta) = \cdots = p_d(z, \zeta) = 0\}$.

The condition (1.2) guarantees that we can construct a Koszul complex using $P_1, \ldots, P_d$, and the condition (1.3) guarantees that it is a free resolution of $\mathcal{M}$ with length d (cf. [SKK], p. 389), and further it entails

$$\mathcal{E}xt^j_{\mathcal{D}_X}(\mathcal{M}, \mathcal{D}_X) = 0 \qquad \text{for } j \ne d. \tag{1.4}$$

In order to define a convexity condition on a compact subset K of M with respect to the system $\mathcal{M}$, we further introduce the following notations. First we suppose that there exists a real-valued real analytic function $\psi(x)$

defined on M for which

$$K = \{x \in M; \psi(x) \le 0\} \tag{1.5}$$

holds. We suppose for simplicity that grad ψ never vanishes on the boundary of K, although we can slightly weaken this condition in what follows. Let A be a positive constant, and let z and $\zeta(z; A)$ respectively denote $x + \sqrt{-1}y$ ($x \in M$, $y \in \mathbf{R}^n$) and $\frac{1}{2}$ grad $\psi(x) - \sqrt{-1}Ay$. Define a function $g_{\alpha,\beta}(z; A)$ ($1 \le \alpha, \beta \le d$) by the following:

$$
\begin{aligned}
g_{\alpha,\beta}(z; A) = &\sum_{1 \le j,k \le n} \frac{\partial^2 \psi}{\partial x_j \, \partial x_k}(x) \overline{p_\alpha^{(j)}(z, \zeta(z; A))} p_\beta^{(k)}(z, \zeta(z; A)) \\
&+ \sum_{1 \le j \le n} (\overline{p_\alpha^{(j)}(z, \zeta(z; A))} p_{\beta(j)}(z, \zeta(z; A)) \\
&+ \overline{p_{\alpha(j)}(z, \zeta(z; A))} p_\beta^{(j)}(z, \zeta(z; A))) \\
&- \frac{2}{A} \sum_{1 \le j \le n} \overline{p_{\alpha(j)}(z, \zeta(z; A))} p_{\beta(j)}(z, \zeta(z; A)).
\end{aligned}
\tag{1.6}
$$

Here, and in what follows, $p^{(j)}(z, \zeta)$ (resp., $p_{(j)}(z, \zeta)$) denotes $\partial p(z, \zeta)/\partial \zeta_j$ (resp., $\partial p(z, \zeta)/\partial z_j$).

It immediately follows from the definition that $G(z; A) = (g_{\alpha,\beta}(z; A))_{1 \le \alpha, \beta \le d}$ is a Hermitian matrix. Throughout this article we use the symbol "$G_1 \ge G_2$" for a pair of Hermitian matrices G_1 and G_2 to mean that $G_1 - G_2$ is positive semi-definite.

Using these notations we can now state the convexity condition on K with respect to $\mathcal{M}$, which we call the uniform $\mathcal{M}$-convexity. In the definition m_α ($1 \le \alpha \le d$) denotes the degree of $p_\alpha(z, \zeta)$ with respect to ζ.

Definition 1.1. A compact set K having the form $\{x \in M; \psi(x) \le 0\}$ is said to be *uniformly $\mathcal{M}$-convex*, if there exist strictly positive constants A_0 and C for which the following inequality (1.7) holds for $A > A_0$ whenever $(z, \zeta(z; A))$ satisfies the conditions (1.8) and (1.9) below:

$$G(z; A) \ge C \begin{pmatrix} |\zeta(z; A)|^{2(m_1 - 1)} & & 0 \\ & \ddots & \\ 0 & & |\zeta(z; A)|^{2(m_d - 1)} \end{pmatrix}. \tag{1.7}$$

$$p_1(z, \zeta(z; A)) = \cdots = p_d(z, \zeta(z; A)) = 0. \tag{1.8}$$

$$A\psi(x) + A^2|y|^2 = 1. \tag{1.9}$$

Remark 1.2. The above definition does not depend on the choice of generators P_α's of $\mathscr{I}$. To see this, let us choose another set of generators $\{Q_\alpha\}$ of $\mathscr{I}$ that satisfies conditions (1.2) and (1.3). Let $q_\alpha(z, \zeta)$ denote the principal symbol of $Q_\alpha(z, \zeta)$, and denote by n_α its degree with respect to ζ. Then there exist $s_{\alpha,\beta}(z, \zeta)$ which satisfy

$$q_\alpha(z, \zeta) = \sum_{\beta=1}^{d} s_{\alpha,\beta}(z, \zeta) p_\beta(z, \zeta), \tag{1.10}$$

$$s_{\alpha,\beta}(z, \zeta) \text{ is homogeneous of degree } n_\alpha - m_\beta, \tag{1.11}$$

and

$$S \underset{\text{def}}{=} (s_{\alpha,\beta}(z, \zeta))_{1 \le \alpha, \beta \le d} \text{ is invertible (for } \zeta \ne 0). \tag{1.12}$$

Let $\tilde{g}_{\alpha,\beta}(z; A)$ denote the function given by (1.6) with p_α being replaced by q_α, and let $\tilde{G}(z; A)$ denote the matrix $(\tilde{g}_{\alpha,\beta}(z; A))$. It is then clear that

$$\tilde{g}_{\alpha,\beta}(z; \zeta(z, A)) = \sum_{1 \le \mu, \nu \le d} \overline{s_{\alpha,\mu}(z, \zeta(z; A))} s_{\beta,\nu}(z, \zeta(z; A)) g_{\mu,\nu}(z; \zeta(z; A)) \tag{1.13}$$

holds if $p_\alpha(z, \zeta(z; A)) = 0$ $(1 \le \alpha \le d)$, that is,

$$\tilde{G}(z; \zeta(z; A)) = \overline{S(z)} G(z)^t S(z). \tag{1.14}$$

Hence (1.7) entails

$$\begin{pmatrix} |\zeta(z; A)|^{1-n_1} & & 0 \\ & \ddots & \\ 0 & & |\zeta(z; A)|^{1-n_d} \end{pmatrix} \tilde{G}(z; \zeta(z; A)) \begin{pmatrix} |\zeta(z; A)|^{1-n_1} & & 0 \\ & \ddots & \\ 0 & & |\zeta(z; A)|^{1-n_d} \end{pmatrix}$$

$$\ge C \begin{pmatrix} |\zeta(z; A)|^{1-n_1} & & 0 \\ & \ddots & \\ 0 & & |\zeta(z; A)|^{1-n_d} \end{pmatrix} \overline{S(z, \zeta(z; A))}$$

$$\times \begin{pmatrix} |\zeta(z; A)|^{2(m_1-1)} & & 0 \\ & \ddots & \\ 0 & & |\zeta(z; A)|^{2(m_d-1)} \end{pmatrix}$$

$$\times {}^t S(z, \zeta(z; A)) \begin{pmatrix} |\zeta(z; A)|^{1-n_1} & & 0 \\ & \ddots & \\ 0 & & |\zeta(z; A)|^{1-n_d} \end{pmatrix}. \tag{1.15}$$

Let $T(z, \zeta)$ denote

$$\begin{pmatrix} |\zeta|^{1-n_1} & & 0 \\ & \ddots & \\ 0 & & |\zeta|^{1-n_d} \end{pmatrix} S(z, \zeta) \begin{pmatrix} |\zeta|^{m_1-1} & & 0 \\ & \ddots & \\ 0 & & |\zeta|^{m_d-1} \end{pmatrix}. \tag{1.16}$$

Then $T(z, \zeta)$ is homogeneous of degree 0 with respect to ζ. Furthermore it is invertible (if $\zeta \neq 0$). Since $\overline{T(z, \zeta)}' T(z, \zeta)$ is clearly non-negative, this means $\overline{T(z, \zeta(z; A))}' T(z, \zeta(z; A))$ is positive–definite for (z, A) satisfying (1.9). Hence the required result follows from (1.7).

Remark 1.3. In case $d = 1$ (i.e., the case where $\mathcal{M}$ reduces to a single equation $\mathcal{D}/\mathcal{D}P$), the uniform $\mathcal{M}$-convexity of K entails the uniform P-convexity of K (in the sense of [KT]).

Remark 1.4. If $p_\alpha(x, \xi)$ is real for any (x, ξ) in $M \times \mathbf{R}^n$, then the uniform $\mathcal{M}$-convexity of K implies that $K_\varepsilon = \{x \in M; \psi(x) \leq \varepsilon\}$ $(0 < \varepsilon \ll 1)$ is bicharacteristically convex in the following sense: For any x in the boundary of K_ε such that $p_\alpha(x, \mathrm{grad}_x \psi(x)) = 0$ $(1 \leq \alpha \leq d)$, the bicharacteristic manifold passing through $(x, \mathrm{grad}_x \psi(x))$ does not intersect the interior of K in a small neighborhood of the point x.

2.   Main Results

One of the most important analytic implications of the uniform $\mathcal{M}$-convexity is the following

Theorem 2.1. *Let K be a compact subset of M that has the form (1.5), and let $\mathcal{M}$ be a system on M satisfying conditions (1.1), (1.2) and (1.3). Suppose that K is uniformly $\mathcal{M}$-convex. Then we find that $\mathrm{Ext}^j_{\mathcal{D}_X}(K; \mathcal{M}, \mathcal{A})$ is of countable dimension for each $j \geq 1$.*

It follows from Schwartz's lemma (cf. [Ko] for example) that the countable-dimensionality entails the space in question is actually a DFS-space, and in particular, Hausdorff. Putting this fact differently we find the operator

$$\delta_j(\mathcal{M}) : \mathcal{A}(K)^{\binom{n}{j}} \to \mathcal{A}(K)^{\binom{n}{j+1}} \tag{2.1}$$

appearing in the Koszul complex determined by $\{P_\alpha\}_{1 \leq \alpha \leq d}$ is of closed range. This means that the obstruction $\text{Ext}^j_{\mathscr{D}_X}(K;\mathscr{M},\mathscr{A})$ $(j \geq 1)$, if it exists, is tied up with the global geometry of K, rather than the analytic complexity of the system $\mathscr{M}$. In order to make this manifest, let us introduce the following variant of the notion of the uniform $\mathscr{M}$-convexity.

Definition 2.2. For a compact set $K = \{x \in M; \psi(x) \leq 0\}$, where $\psi(x)$ is a real-valued real analytic function, we say K is *completely $\mathscr{M}$-convex* if the following four conditions are satisfied for some point x_1 in K:

$$\psi(x) \geq \psi(x_1) \quad \text{holds for any } x \text{ in } M. \tag{2.2}$$

$$\bigcap_{t > \psi(x_1)} \{x \in M; \psi(x) < t\} = \{x_1\}. \tag{2.3}$$

$$\text{grad } \psi(x) \neq 0 \quad \text{on } \{x \in M; x \neq x_1\}. \tag{2.4}$$

$$\text{The inequality (1.7) holds, if } A\psi(x) + A^2|y|^2 \leq 1 \text{ and}$$
$$\text{if the condition (1.8) is satisfied.} \tag{2.5}$$

The complete $\mathscr{M}$-convexity of K is intended to enable us to make use of the stability of the cohomology groups in question under the perturbation of the domain (cf. [K2]). Actually, using a result in [K2], we have the following vanishing theorem.

Theorem 2.2. *Let K and $\mathscr{M}$ be the same as in Theorem 2.1. Suppose further that K is completely $\mathscr{M}$-convex. Then we find*

$$\text{Ext}^j_{\mathscr{D}_X}(K;\mathscr{M},\mathscr{A}) \text{ vanishes for any } j \geq 1. \tag{2.6}$$

As an immediate consequence of Theorem 2.2, we obtain the following Hartogs-type result.

Theorem 2.3. *Let K and $\mathscr{M}$ be the same as in the preceding theorem, and suppose $d \geq 2$. Then any hyperfunction solution of $\mathscr{M}$ on $M - K$ can be uniquely extended as a hyperfunction solution of $\mathscr{M}$ on M.*

In fact, using the pairing between $\mathscr{A}(K)$ and the space $\mathscr{B}_K$ of hyperfunctions supported by K, we find $\text{Ext}^j_K(M;\mathscr{M},\mathscr{B}) = 0$ $(j = 0, \ldots, d-1)$ from Theorem 2.2 adapted to the adjoint system $\mathscr{M}^*$ of $\mathscr{M}$. Note that $\mathscr{M}^*$ also satisfies conditions (1.1)–(1.3), and that K is completely $\mathscr{M}^*$-convex if K is completely $\mathscr{M}$-convex. Since $d \geq 2$ by the assumption, this means the vanishing of $\text{Ext}^1_K(M;\mathscr{M},\mathscr{B})$, from which Theorem 2.3 immediately follows.

3. Examples

Before proving Theorem 2.1 and Theorem 2.2, let us examine some simple examples to get some idea about the convexity conditions we are using.

Example 3.1. Let $\mathcal{M}$ be the partial de Rham system on $\mathbf{R}^n$, that is,

$$\mathcal{M} = \mathcal{D}_{\mathbf{C}^n} \Big/ \left(\sum_{\alpha=1}^{d} \mathcal{D}_{\mathbf{C}^n} \frac{\partial}{\partial x_\alpha} \right)$$

$(1 \le d \le n)$ and let K be a compact subset of $\mathbf{R}^n$ given by $K = \{x \in \mathbf{R}^n; \psi(x) \le 0\}$. We suppose that K satisfies the conditions (2.2)-(2.4). In this case it follows from the definition that

$$g_{\alpha,\beta}(z; A) = \frac{\partial^2 \psi}{\partial x_\alpha \, \partial x_\beta}(x).$$

Since $m_\alpha = 1 \; (1 \le \alpha \le d)$ in our case, the complete $\mathcal{M}$-convexity condition on K reads as follows:

$$G = \left(\frac{\partial^2 \psi}{\partial x_\alpha \, \partial x_\beta}(x) \right)_{1 \le \alpha, \beta \le d} \ge C \quad \text{holds}$$

if $A\psi(x) + A^2|y|^2 \le 1$ and

$$\frac{1}{2} \frac{\partial \psi}{\partial x_\alpha}(x) - \sqrt{-1} A y_\alpha = 0 \quad (1 \le \alpha \le d) \text{ are satisfied.} \tag{3.1}$$

Since the matrix G and the constant C are independent of A and y, we find that the above condition (3.1) is equivalent to the following:

There exists a positive constant δ such that

$$G = \left(\frac{\partial^2 \psi}{\partial x_\alpha \, \partial x_\beta}(x) \right)_{1 \le \alpha, \beta \le d} \ge C \quad \text{holds}$$

for every $x \in \mathbf{R}^n$ satisfying $\psi(x) \le \delta$ and $\dfrac{\partial \psi}{\partial x_1}(x) = \cdots = \dfrac{\partial \psi}{\partial x_d}(x) = 0$.

$$\tag{3.2}$$

This means that for each $t \le \delta$ the compact set $K_t = \{x \in \mathbf{R}^n; \psi(x) \le t\}$ is bicharacteristically convex in the sense of Remark 1.4. That is, the complete $\mathcal{M}$-convexity in this case is equivalent to the bicharacteristical convexity of K_t for each $t < \delta \; (\delta > 0)$.

Example 3.2. Let us now regard $\mathbf{C}^l_{(z_1,\ldots,z_l)}$ as the $2l$-dimensional Euclidean space $\mathbf{R}^{2l}_{(x_1,\ldots,x_{2l})}$ $(z_j = x_j + \sqrt{-1}x_{l+j})$, and consider the Cauchy–Riemann equations

$$\mathcal{M} = \mathcal{D}_{\mathbf{C}^{2l}} \bigg/ \bigg(\sum_{j=1}^{l} \mathcal{D}_{\mathbf{C}^{2l}} \bar{\partial}_j \bigg)$$

on $\mathbf{C}^l \cong \mathbf{R}^{2l}$. Here $\mathbf{C}^{2l}$ is a complexification of $\mathbf{R}^{2l}$, and $\bar{\partial}_j$ denotes $\partial/\partial \bar{z}_j$, i.e., $\frac{1}{2}(\partial/\partial x_j + \sqrt{-1}\,\partial/\partial x_{l+j})$ for $1 \le j \le l$. Now, suppose that a compact subset $K = \{z \in \mathbf{C}^l; \psi(z) \le 0\}$ of $\mathbf{C}^l \cong \mathbf{R}^{2l}$ satisfies the conditions (2.2)–(2.4) and that it is strongly pseudo-convex (in the sense of the theory of holomorphic functions). Then K satisfies the condition of complete $\mathcal{M}$-convexity.

In fact, by an easy computation, we find

$$g_{\alpha,\beta} = \frac{1}{4}\bigg(\frac{\partial^2 \psi}{\partial x_\alpha\,\partial x_\beta} + \sqrt{-1}\,\frac{\partial^2 \psi}{\partial x_\alpha\,\partial x_{l+\beta}} - \sqrt{-1}\,\frac{\partial^2 \psi}{\partial x_{l+\alpha}\,\partial x_\beta} + \frac{\partial^2 \psi}{\partial x_{l+\alpha}\,\partial x_{l+\beta}} \bigg)$$

$$= \frac{\partial^2 \psi}{\partial z_\alpha\,\partial \bar{z}_\beta}.$$

Therefore, if K is strongly pseudo-convex, then $G = (g_{\alpha,\beta})$ is positive-definite. Since $m_\alpha = 1$ for $\alpha = 1, \ldots, l$ in our case, this implies that K satisfies the condition of complete $\mathcal{M}$-convexity.

4.   Proof of the Theorems

Let us now prove Theorem 2.1. The strategy of our proof is the same as in the case of single equations treated in [KT]. Let us first sketch it briefly.

Let A be a sufficiently large positive number, and let $\varphi_A(z)$ $(z = x + \sqrt{-1}y \in \mathbf{C}^n)$ denote $\psi(x) + A|y|^2$. Set $\Omega_A = \{z \in M \times \sqrt{-1}\mathbf{R}^n_y;\ \varphi_A(z) < 1/A\}$. Let X' be a complexification of $\mathbf{R}^n_x \times \mathbf{R}^n_y$. Since $\{\Omega_A\}_{A>0}$ is a fundamental system of neighborhoods of K and Ω_A is contained in X for A sufficiently large, we may regard $\mathcal{M}$ as a $\mathcal{D}_{X'}$-module defined on Ω_A, and we find

$$\mathrm{Ext}^j_{\mathcal{D}_X}(K; \mathcal{M}, \mathcal{A}) = \varinjlim_{A \to \infty} \mathrm{Ext}^j_{\mathcal{D}_{X'}}(\Omega_A; \mathcal{M}, \mathcal{O}_{X'}). \tag{4.1}$$

Let us now define a $\mathcal{D}_{X'}$-module $\mathcal{M}'$ by

$$\mathcal{D}_{X'} \bigg/ \bigg(\sum_{\alpha=1}^{d} \mathcal{D}_{X'} P_\alpha + \sum_{k=1}^{n} \mathcal{D}_{X'} \bar{\partial}_k \bigg), \tag{4.2}$$

where $\bar{\partial}_k$ denotes $(\partial/\partial x_k + \sqrt{-1}\partial/\partial y_k)/2$. Note that P_α and $\bar{\partial}_k$ commute. Since Ω_A is Stein for A sufficiently large, we find

$$\text{Ext}^j_{\mathscr{D}_X}(\Omega_A; \mathscr{M}, \mathscr{O}_{X'}) \xrightarrow{\sim} \text{Ext}^j_{\mathscr{D}_X}(\Omega_A; \mathscr{M}', \mathscr{B}_{\mathbf{R}^{2n}_{(x,y)}}). \tag{4.3}$$

Therefore, if we can prove that the right-hand side of (4.3) is finite-dimensional, it follows from (4.1) that $\text{Ext}^j(K; \mathscr{M}, \mathscr{A})$ is countable-dimensional. Here we have used the fact that we can choose countably many Ω_A's to constitute a fundamental system of Stein neighborhoods of K. This is a sketch of our strategy.

To prove the finite-dimensionality of the right-hand side of (4.3), we use the ellipticity of the system $\mathscr{M}'$. The ellipticity enables us to use the results in [K1] on the finite-dimensionality of cohomology groups; we calculate the generalized Levi form of the "positive" tangential system $\mathscr{N}_{A,+}$ on the boundary of Ω_A induced from $\mathscr{M}'$. By the assumptions (1.1)–(1.3) on the system $\mathscr{M}$, we find that $\mathscr{N}_{A,+}$ admits a free resolution of length $n + d - 1$ by the sheaf $\mathscr{D}_{Y'}$, where Y' is a complexification of the boundary of Ω_A. Hence, if we can verify that the generalized Levi form is positive–definite at each characteristic point of $\mathscr{N}_{A,+}$, then the right-hand side of (4.3) is finite-dimensional.

Now, by straightforward calculations we find that the characteristic variety of $\mathscr{N}_{A,+}$ is given as follows:

$$\{(z, n(z)); \varphi_A(z) = 1/A \text{ and } p_\alpha(z, n(z)) = 0 \text{ for } 1 \leq \alpha \leq d\},$$

where

$$n(z) = \left(\frac{1}{\sqrt{-1}} \frac{\partial \varphi_A}{\partial z_j}(z)\right)_{j=1,\ldots,n}.$$

Hence the cotangential component of a characteristic point of $\mathscr{N}_{A,+}$ is determined by its base point z. We denote by L_z the generalized Levi form calculated at the characteristic point $(z, n(z))$. The definition of the generalized Levi form is given in [SKK], Chapter III, Definition 2.3.1, and an explicit form is given in [P] in the case of single equations. In our case, L_{z_0} is positive-definite if and only if the Hermitian form $Q_{z_0}(\tau) = \sum_{1 \leq j, k \leq n+d} q_{j,k}(z_0)\tau_j\bar{\tau}_k$ is positive-definite on $\{\tau = (\tau_1, \ldots, \tau_{n+d}) \in \mathbf{C}^{n+d}; \sum_{j=1}^n (\partial \varphi_A/\partial z_j)(z_0)\tau_j = 0\}$, where $q_{j,k}(z_0)$ are given as follows:

$$q_{j,k}(z_0) = \frac{\partial^2 \varphi_A}{\partial z_j \, \partial \bar{z}_k}(z_0) \qquad (1 \leq j, k \leq n). \tag{4.4}$$

$$q_{j,n+\alpha}(z_0) = \overline{q_{n+\alpha,j}(z_0)}$$

$$= p_{\alpha(j)}(z_0, \text{grad}_z\, \varphi_A(z_0))$$

$$+ \sum_{k=1}^{n} p_{\alpha}^{(k)}(z_0, \text{grad}_z\, \varphi_A(z_0)) \frac{\partial^2 \varphi_A}{\partial z_j\, \partial z_k}(z_0)$$

$$(1 \le j \le n, 1 \le \alpha \le d). \qquad (4.5)$$

$$q_{n+\alpha, n+\beta}(z_0)$$

$$= \sum_{1 \le j, k \le n} \overline{p_{\alpha}^{(j)}(z_0, \text{grad}_z\, \varphi_A(z_0))} p_{\beta}^{(k)}(z_0, \text{grad}_z\, \varphi_A(z_0)) \frac{\partial^2 \varphi_A}{\partial z_j\, \partial z_k}(z_0)$$

$$(1 \le \alpha, \beta \le d). \qquad (4.6)$$

Here z_0 is supposed to satisfy $\varphi_A(z_0) = 1/A$ and $p_\alpha(z_0, \text{grad}_z\, \varphi_A(z_0)) = 0$ for $1 \le \alpha \le d$.

To show the positive-definiteness of $Q_{z_0}(\tau)$, let us first note that the matrix

$$H(z_0) \underset{\text{def}}{=} (q_{j,k}(z_0))_{1 \le j,k \le n} = \left(\frac{\partial^2 \varphi_A}{\partial z_j\, \partial z_k}(z_0) \right)_{1 \le j,k \le n}$$

is positive-definite for A sufficiently large, because

$$\frac{\partial^2 \varphi_A}{\partial z_j\, \partial z_k}(z_0) = \frac{1}{4} \frac{\partial^2 \psi}{\partial x_j\, \partial x_k}(\text{Re } z_0) + \tfrac{1}{2} A \delta_{jk}$$

holds. In particular, $H(z_0)$ is nonsingular. Using this fact, we define σ by the following:

$$\begin{pmatrix} \tau_1 \\ \vdots \\ \tau_n \\ \hline \tau_{n+1} \\ \vdots \\ \tau_{n+d} \end{pmatrix} = \left(\begin{array}{ccc|c} 1 & & 0 & \\ & \ddots & & -H(z_0)^{-1} \Lambda(z_0) \\ 0 & & 1 & \\ \hline & & & 1 \\ 0 & & 0 & \ddots \\ & & & & 1 \end{array} \right) \begin{pmatrix} \sigma_1 \\ \vdots \\ \sigma_n \\ \hline \sigma_{n+1} \\ \vdots \\ \sigma_{n+d} \end{pmatrix},$$

where

$$\Lambda(z_0) = \begin{pmatrix} q_{1,n+1}(z_0) & \cdots & q_{1,n+d}(z_0) \\ \vdots & & \vdots \\ q_{n,n+1}(z_0) & \cdots & q_{n,n+d}(z_0) \end{pmatrix}.$$

Then we obtain

$$Q_{z_0}(\tau) = \sum_{1 \le j, k \le n} q_{j,k}(z_0) \sigma_j \overline{\sigma_k} + \sum_{1 \le \alpha, \beta \le d} a_{\alpha,\beta}(z_0) \sigma_{n+\alpha} \overline{\sigma_{n+\beta}}, \qquad (4.7)$$

where the matrix $K(z_0) \underset{\text{def}}{=} (a_{\alpha,\beta}(z_0))_{1 \le \alpha, \beta \le d}$ is given as follows:

$$K(z_0) = (q_{n+\alpha, n+\beta}(z_0))_{1 \le \alpha, \beta \le d} - {}^t\overline{\Lambda(z_0)} H(z_0)^{-1} \Lambda(z_0).$$

Since $H(z_0)$ is positive-definite as stated above, it suffices for us to verify the positive-definiteness of $K(z_0)$.

Next we note the fact that the uniform $\mathcal{M}$-convexity is invariant under a real orthogonal transformation. That is, if we define $\tilde{z} = \tilde{x} + \sqrt{-1}\tilde{y}$ by

$$\tilde{z} = M^{-1}(x - \operatorname{Re} z_0 + \sqrt{-1}y) \qquad (4.8)$$

for a real orthogonal matrix M, the domain Ω_A assumes the form

$$\left\{ \tilde{z} = \tilde{x} + \sqrt{-1}\tilde{y} \in \mathbf{C}^n; \ \psi(\operatorname{Re} z_0 + M\tilde{x}) + A|\tilde{y}|^2 < \frac{1}{A} \right\}.$$

On the other hand, $g_{\alpha,\beta}(z; A)$ defined by (1.5) is invariant under any real orthogonal transformation, if we rewrite ζ as $\tilde{\zeta} = \operatorname{grad}_{\tilde{z}} \varphi_A(\tilde{z})$. Hence the uniform $\mathcal{M}$-convexity holds for the new variable $\tilde{z}$.

Now, to calculate $a_{\alpha,\beta}(z_0)$, let us choose an orthogonal matrix M that brings $((\partial^2\psi/\partial x_j \, \partial x_k)(\operatorname{Re} z_0))_{1 \le j, k \le n}$ to a diagonal matrix in the coordinate system $(\tilde{x}, \tilde{y})$ defined by (4.8). Then we find

$a_{\alpha,\beta}(z_0)$

$$= \sum_{j=1}^{n} (c_j + \tilde{A}) \overline{p_\alpha^{(j)}(\tilde{z}_0, \operatorname{grad}_{\tilde{z}} \varphi_A(\tilde{z}_0))} p_\beta^{(j)}(\tilde{z}_0, \operatorname{grad}_{\tilde{z}} \varphi_A(\tilde{z}_0))$$

$$- \sum_{j=1}^{n} (c_j + \tilde{A})^{-1} \overline{(p_{\alpha(j)}(\tilde{z}_0, \operatorname{grad}_{\tilde{z}} \varphi_A(\tilde{z}_0)) + (c_j - \tilde{A}) p_\alpha^{(j)}(\tilde{z}_0, \operatorname{grad}_{\tilde{z}} \varphi_A(\tilde{z}_0)))}$$

$$\times (p_{\beta(j)}(\tilde{z}_0, \operatorname{grad}_{\tilde{z}} \varphi_A(\tilde{z}_0)) + (c_j - \tilde{A}) p_\beta^{(j)}(\tilde{z}_0, \operatorname{grad}_{\tilde{z}} \varphi_A(\tilde{z}_0))), \qquad (4.9)$$

where

$$c_j = \frac{1}{4} \frac{\partial^2 \psi}{\partial \tilde{x}_j^2}(\tilde{x}_0), \qquad \tilde{A} = \tfrac{1}{2}A \quad \text{and} \quad \tilde{z}_0 = \sqrt{-1} M^{-1}(\operatorname{Im} z_0).$$

Set $\rho = |\zeta(\tilde{z}_0; A)|$. Since $\text{grad}_{\tilde{z}}\, \varphi_A(\tilde{z}_0) = \zeta(\tilde{z}_0; A)$ holds, we then obtain

$$
\begin{aligned}
a_{\alpha,\beta}(z_0) = 4 \sum_{j=1}^{n} & \overline{c_j p_\alpha^{(j)}(\tilde{z}_0, \text{grad}_{\tilde{z}}\, \varphi_A(\tilde{z}_0))} p_\beta^{(j)}(\tilde{z}_0, \text{grad}_{\tilde{z}}\, \varphi_A(\tilde{z}_0)) \\
& + \sum_{j=1}^{n} \overline{(p_\alpha^{(j)}(\tilde{z}_0, \text{grad}_{\tilde{z}}\, \varphi_A(\tilde{z}_0))} p_{\beta(j)}(\tilde{z}_0, \text{grad}_{\tilde{z}}\, \varphi_A(\tilde{z}_0)) \\
& + \overline{p_{\alpha(j)}(\tilde{z}_0, \text{grad}_{\tilde{z}}\, \varphi_A(\tilde{z}_0))} p_\beta^{(j)}(\tilde{z}_0, \text{grad}_{\tilde{z}}\, \varphi_A(\tilde{z}_0))) \\
& - \frac{1}{\tilde{A}} \sum_{j=1}^{n} \overline{p_{\alpha(j)}(\tilde{z}_0, \text{grad}_{\tilde{z}}\, \varphi_A(\tilde{z}_0))} p_{\beta(j)}(\tilde{z}_0, \text{grad}_{\tilde{z}}\, \varphi_A(\tilde{z}_0)) + R_{\alpha,\beta}(\tilde{z}_0) \\
= \; & g_{\alpha,\beta}(\tilde{z}_0; A) + R_{\alpha,\beta}(\tilde{z}_0),
\end{aligned} \tag{4.10}
$$

where

$$
|R_{\alpha,\beta}(\tilde{z}_0)| \le C_1 \rho^{m_\alpha + m_\beta - 2}(1 + \rho A^{-1}) \rho A^{-1} \tag{4.11}
$$

holds for a constant C_1. The first term on the right-hand side of (4.10) is positive-definite by the uniform $\mathcal{M}$-convexity assumption, i.e.,

$$
G = (g_{\alpha,\beta}(z_0; A)) \ge C \begin{pmatrix} \rho^{2(m_1-1)} & & 0 \\ & \ddots & \\ 0 & & \rho^{2(m_d-1)} \end{pmatrix}.
$$

Hence, in order to prove the positive-definiteness of $K(\tilde{z}_0) = (a_{\alpha,\beta}(\tilde{z}_0))$, it suffices to show that

$$
K'(\tilde{z}_0) \underset{\text{def}}{=} C \begin{pmatrix} \rho^{2(m_1-1)} & & 0 \\ & \ddots & \\ 0 & & \rho^{2(m_d-1)} \end{pmatrix} + (R_{\alpha,\beta}(\tilde{z}_0))
$$

is positive-definite. We can easily verify the positive-definiteness of the matrix $K'(\tilde{z}_0)$. In fact, let

$$
W = \begin{pmatrix} \rho^{1-m_1} & & 0 \\ & \ddots & \\ 0 & & \rho^{1-m_d} \end{pmatrix}.
$$

Then we find

$$
{}^tW K'(\tilde{z}_0) \bar{W} = C \begin{pmatrix} 1 & & 0 \\ & \ddots & \\ 0 & & 1 \end{pmatrix} + (R'_{\alpha,\beta}(\tilde{z}_0)),
$$

where

$$\left| R'_{\alpha,\beta}(\tilde{z}_0) \right| = \left| R_{\alpha,\beta}(\tilde{z}_0) \rho^{2-(m_\alpha + m_\beta)} \right| \leq C_1 (1 + \rho A^{-1}) \rho A^{-1}$$

holds by (4.11). Since $\varphi_A(\tilde{z}_0) = 1/A$ by the assumption, ρA^{-1} tends to zero as A tends to infinity. This implies that $K'(\tilde{z}_0)$ is positive-definite for sufficiently large A, completing the proof of Theorem 2.1.

Using a result on the stability of the cohomology groups under the perturbation of the domain ([K2], Theorem 2), we can prove Theorem 2.2 in a similar manner as above, that is, by calculating the generalized Levi form. We need, however, the following three auxiliary results besides the positivity of the generalized Levi form.

The first one is an algebraic one; by using the assumptions (1.1), (1.2) and (1.3), we find

$$\mathscr{E}\!xt^j_{\mathscr{D}_{X'}}(\mathscr{M}', \mathscr{D}_{X'}) = 0 \qquad \text{for } j \neq n + d. \tag{4.12}$$

Here $\mathscr{M}'$ is the $\mathscr{D}_{X'}$-module given by (4.2).

The second one is a rather trivial geometric one; the function $\varphi_A(z)$ given by $\psi(x) + A|y|^2$ satisfies conditions similar to (2.2)-(2.4), that is,

$$\varphi_A(z) \geq \varphi_A(x_1) \ (= \psi(x_1)) \text{ holds for any } z \text{ in } \Omega_A = \{z; \varphi_A(z) < 1/A\}. \tag{4.13}$$

$$\bigcap_{t > \varphi_A(x_1)} \{z \in \Omega_A; \varphi_A(z) < t\} = \{x_1\}. \tag{4.14}$$

$$\operatorname{grad}_z \varphi_A(z) \neq 0 \text{ on } \{z \in \Omega_A; z \neq x_1\}. \tag{4.15}$$

The third one is an analytic one; we have

$$\mathscr{E}\!xt^j(\mathscr{M}, \mathscr{A})_{x_1}, \text{ the germ of the sheaf } \mathscr{E}\!xt^j(\mathscr{M}, \mathscr{A}) \text{ at } x_1,$$

vanishes for each $j \geq 1$. $\tag{4.16}$

To verify (4.16), we first note that the space $\mathscr{E}\!xt^j(\mathscr{M}, \mathscr{A})_{x_1}$ is seen to be Hausdorff by the same reasoning as in the proof of Theorem 2.1. Hence the operator $\delta_j(\mathscr{M})$ given by (2.1) is of closed range (in $\mathscr{A}(\{x_0\})^{(n/j-1)}$). Therefore the vanishing theorem for relative cohomology groups (for the adjoint system $\mathscr{M}^*$ of $\mathscr{M}$) ([K3], Theorem 1) entails (4.16).

Combining these observations with Theorem 2 of [K2], we find that the right-hand side of (4.3) vanishes. This proves Theorem 2.2.

References

[H]   L. Hörmander, The Frobenius–Nirenberg theorem, *Ark. för Mat.* **5** (1964) 425–432.

[K1]   T. Kawai, Theorems on the finite-dimensionality of cohomology groups. III, *Proc. Japan Acad.* **49** (1973) 243–246.

[K2]   T. Kawai, Theorems on the finite-dimensionality of cohomology groups. V, *Proc. Japan Acad.* **49** (1973) 782–784.

[K3]   T. Kawai, Extension of solutions of systems of linear differential equations, *Publ. RIMS, Kyoto Univ.* **12** (1976) 215–227.

[KT]   T. Kawai and Y. Takei, On a closed range property of a linear differential operator, *Proc. Japan Acad. Ser. A* **62** (1986) 386–388.

[Ko]   H. Komatsu, Projective and injective limits of weakly compact sequences of locally convex spaces, *J. Math. Soc. Japan* **19** (1967) 366–383.

[P]   P. Pallu de La Barrière, Existence et prolongement des solutions holomorphes des équations aux dérivées partielles, I, *J. Math. Pures et Appl.* **55** (1976) 21–46.

[SKK]   M. Sato, T. Kawai, and K. Kashiwara, Microfunctions and pseudo-differential equations, *Lecture Notes in Math.* No. 287, Springer, Berlin–Heidelberg–New York, 1973, pp. 265–529.

Complex Powers on p-adic Fields and a Resolution of Singularities

Tatsuo Kimura

The Institute of Mathematics
University of Tsukuba
Ibaraki, Japan

Introduction

Let K be a p-adic field, i.e., a finite algebraic extension of $\mathbf{Q}_p$ for some prime number p. We shall denote by O_K the maximal compact subring of K, by πO_K its ideal of nonunits and by q the number of elements of $O_K/\pi O_K (\approx \mathbf{F}_q)$. We shall denote the normalized absolute value on K by $|a|_K$ so that $|\pi|_K = q^{-1}$. We normalize the Haar measure dx on K^n by $\mathrm{vol}(O_K^n) = \int_{O_K^n} dx = 1$. We denote by $O_K[x_1, \ldots, x_n]$ the polynomial ring of n variables over O_K. We shall consider the calculation of complex powers $Z(s) = \int_{O_K^n} |f(x)|_K^s \, dx$ for $f(x) \in O_K[x_1, \ldots, x_n]$. Professor J. Igusa proved that $Z(s)$ is a rational function of $t = q^{-s}$ and sometimes it is called an Igusa's local zeta function (named by J. Serre). When $f(x)$ is a relative invariant of (irreducible) regular prehomogeneous vector spaces (Sato-Kumura [4]), $Z(s)$ has interesting properties, and in some cases, it is explicitly calculated (Igusa [3]). The n-dimensional affine space over K will be denoted by $\mathbf{A}_K^n$. We shall denote by $V(f)$ the zeros of f in $\mathbf{A}_K^n$. By

Algebraic Analysis, Volume I

(Hironaka [2] Cor. 3. p. 146), there exists an embedded resolution of singularities of $V(f) \subset A_K^n$, i.e., there exist a nonsingular absolutely irreducible variety Y defined over K and a regular map $h: Y \to A_K^n$, also defined over K, such that the following three conditions are satisfied:

(1) h is biregular on $Y \setminus h^{-1}(V(f))$,
(2) h is the composition of finitely many blowing-up maps each with a nonsingular center defined over K, and
(3) the K-irreducible components of $h^{-1}(V(f))$ are nonsingular and intersect transversally, i.e., the reduced K-subscheme associated to $h^{-1}(V(f))$ has only normal crossing in the sense of (Hironaka [2], Def 2. p. 141).

In view of (2), we shall give an integration formula corresponding to the blowing up at the origin in Section 1. In view of (3), we shall consider the complex power of $f(x) = f_1(x)^{N_1} \cdots f_t(x)^{N_t}$ where $V(f_i)$ $(i = 1, \ldots, t)$ are irreducible, nonsingular and intersect transversally. In general, $f(x)$ modulo πO_K will be denoted by $\bar{f}$. When $V(\bar{f}_i) \subset A_{F_q}^n$ $(i = 1, \ldots, t)$ satisfy the additional condition (4) that the $V(\bar{f}_i)$ are irreducible, nonsingular and intersect transversally, we can calculate the complex power of $f(x)$ (see Theorem 2.1, or, Denef [1]). Still in the general case, i.e., without assumption of (4), we shall prove in Section 2 that $Z(s) = \int_{O_K^n} |f(x)|_K^s \, dx$ can be calculated by applying an integration formula $\int_{O_K^n} = \sum_{\xi \in F_q^m} \int_{(\xi + \pi O_K^m) \times O_K^{n-m}}$ finitely many times. Namely, applying this integration formula, we can reduce the general case to the case with the condition (4). This is a kind of desingularization. In other words, the result of Section 2 guarantees that one can calculate the p-adic complex power of f when one has an explicit embedded resolution.

This paper consists of the following two sections:

1. The Integration Formula Corresponding to the Blowing-up at the Origin.
2. The resolution of singularities modulo π.

The author would like to express his hearty thanks to Professor M. Sato, who made me a researcher of mathematics by his severe and kind teaching about 16 years ago. The author also expresses his hearty thanks to Professor J. Igusa, who invited me to The Johns Hopkins University in 1986 and gave me a lot of mathematical stimulation. The discussion there with Professor G. Kempf was very helpful.

1. The Integration Formula Corresponding to the Blowing-up at the Origin

First we shall review the blowing-up Y of $\mathbf{A}^n$ at the origin. A subvariety Y of $\mathbf{A}^n \times \mathbf{P}^{n-1}$ is defined by $Y = \{(x_1, \ldots, x_n; z_1 : \ldots : z_n); x_i z_j - x_j z_i = 0 \text{ for } i, j = 1, \ldots, n\} = (\mathbf{A}^n - \{0\}) \cup \mathbf{P}^{n-1}$ (disjoint union). We have an affine open covering $Y = Y_1 \cup \cdots \cup Y_n$ where $Y_i = \{(x_1, \ldots, x_n; z_1 : \ldots : z_n); z_i \neq 0, x_j = (z_j/z_i)x_i \text{ for } j = 1, \ldots, n\}$ $(i = 1, \ldots, n)$. Put $\lambda = x_i$ and $y_j = z_j/z_i$ for $j = 1, \ldots, n$ and $j \neq i$. Then $(y_1, \ldots, y_{i-1}, \lambda, y_{i+1}, \ldots, y_n)$ is a local coordinate of Y_i and we have

$$(x_1, \ldots, x_n) = (\lambda y_1, \ldots, \lambda y_{i-1}, \lambda, \lambda y_{i+1}, \ldots, \lambda y_n). \tag{1.1}$$

Now put $D'_i = \{(x_1^{(i)}, \ldots, x_{i-1}^{(i)}, \lambda_i, x_{i+1}^{(i)}, \ldots, x_n^{(i)}) \in O_K^n; (x_1^{(i)}, \ldots, x_{i-1}^{(i)}) \in \pi O_K^{i-1}, \lambda_i \neq 0\}$ $(i = 1, \ldots, n)$.

Lemma 1.1. *We have a bijection Ψ between $O_K^n - \{0\}$, and $D'_1 \cup \cdots \cup D'_n$* (*disjoint union*).

Proof. For any element $x = (x_1, \ldots, x_n)$ in $O_K^n - \{0\}$, there exists uniquely an index i satisfying $\min\{\text{ord } x_1, \ldots, \text{ord } x_n\} = \text{ord } x_i (\neq \text{ord } x_j$ for all $j < i)$. In this case, put

$$\Psi(x_1, \ldots, x_n) = (x_1^{(i)}, \ldots, x_{i-1}^{(i)}, \lambda_i, x_{i+1}^{(i)}, \ldots, x_n^{(i)})$$

where

$$(x_1, \ldots, x_i, \ldots, x_n) = (\lambda_i x_1^{(i)}, \ldots, \lambda_i x_{i-1}^{(i)}, \lambda_i, \lambda_i x_{i+1}^{(i)}, \ldots, \lambda_i x_n^{(i)})$$

(cf. (1.1)). Since ord $x_i < $ ord x_j for all $j < i$, we have $(x_1^{(i)}, \ldots, x_{i-1}^{(i)}) \in \pi O_K^{i-1}$ and hence $\Psi(x_1, \ldots, x_n) \in D'_i$. Note that ord $0 = +\infty$. The bijectivity of Ψ is clear.   Q.E.D.

Theorem 1.2 (*An integration formula*). *For $1 \leq m \leq n$, put $D_i = \{(x_1^{(i)}, \ldots, x_{i-1}^{(i)}, \lambda_i, x_{i+1}^{(i)}, \ldots, x_m^{(i)}) \in O_K^m; (x_1^{(i)}, \ldots, x_{i-1}^{(i)}) \in \pi O_K^{i-1}\}$. Then we have*

$$\int_{O_K^n} |f(x)|_K^s \, dx = \sum_{i=1}^m \int_{D_i \times O_K^{n-m}} |f(\lambda_i x_1^{(i)}, \ldots, \lambda_i x_{i-1}^{(i)}, \lambda_i, \lambda_i x_{i+1}^{(i)}, \ldots,$$

$$\lambda_i x_m^{(i)}, x_{m+1}, \ldots, x_n)|_K^s \lambda_i^{m-1} \, d\lambda_i \, dx_1^{(i)}$$

$$\cdots dx_{i-1}^{(i)} \, dx_{i+1}^{(i)} \cdots dx_m^{(i)} \, dx_{m+1} \cdots dx_n.$$

Proof. It is clear by Lemma 1.1 and

$$\int_{O_K^n} = \int_{O_K^n - \{0\}}, \qquad \int_{D_i \times O_K^{n-m}} = \int_{D_i' \times O_K^{n-m}}. \qquad \text{Q.E.D.}$$

Example 1.3. We shall prove the well-known fact

$$Z_n(s) = \int_{M_n(O_K)} |\det X|_K^s \cdot dX = \prod_{i=1}^n \frac{(1 - q^{-i})}{(1 - q^{-(s+i)})}$$

by using Theorem 1.2. When $n = 1$, we have

$$\int_{O_K} |x|_K^s \, dx = \sum_{m=0}^\infty \int_{\pi^m O_K} |x|_K^s \, dx = \sum_{m=0}^\infty q^{-sm-m} \int_{U_K} dx = \frac{(1 - q^{-1})}{(1 - q^{-(s+1)})}.$$

We use an induction on n. We have

$$Z_n(s) = \sum_{i=1}^n \int_{D_i \times O_K^{n^2-n}} \left| \det \begin{pmatrix} \lambda x_{11}' & \cdots & \lambda & \cdots & \lambda x_{1n}' \\ x_{21} & \cdots & x_{2i} & \cdots & x_{2n} \\ \vdots & & \vdots & & \vdots \\ x_{n1} & \cdots & x_{ni} & \cdots & x_{nn} \end{pmatrix} \right|_K^s$$

$$\times \lambda^{n-1} \, d\lambda \, dx_{11}' \cdots dx_{nn}$$

$$\overset{(*)}{=} \sum_{i=1}^n \int_{\substack{x_{11}',\ldots,x_{1,i-1}' \in \pi O_K \\ \lambda, x_{kl} \in O_K}} |\lambda|_K^{s+n-1} \cdot \left| \det \begin{pmatrix} 0 & \cdots & 010 & \cdots & 0 \\ x_{21} & \cdots & 0 & \cdots & x_{2n} \\ \vdots & & \vdots & & \vdots \\ x_{n1} & \cdots & 0 & \cdots & x_{nn} \end{pmatrix} \right|_K^s$$

$$\times d\lambda \, dx_{11}' \cdots dx_{nn}$$

$$= \sum_{i=1}^n q^{-(i-1)} \int_{O_K} |\lambda|_K^{s+n-1} \, d\lambda \cdot Z_{n-1}(s) = Z_{n-1}(s) \frac{(1 - q^{-n})}{(1 - q^{-(s+n)})}.$$

Hence we obtain our result. Note that we changed the variables at $(*)$.

2. The Resolution of Singularities Modulo π

We shall consider the complex power $Z(s) = \int_{O_K^n} |f(x)|_K^s \, dx$ of $f(x) = f_1(x)^{N_1} \cdots f_t(x)^{N_t}$ where $C_i = \{f_i = 0\}$ ($i \in T = \{1, \ldots, t\}$) are K-irreducible, nonsingular and intersect transversally. The following result is a very special case of (Denef [1]) and one can prove directly by change of variables and by a formula $\int_{O_K^n} = \sum_\xi \int_{\xi + \pi O_K^n}$ where ξ runs over a complete set of representatives of O_K^n modulo πO_K^n.

Theorem 2.1 (cf. Denef [1] Theorem 3.1). *Assume that* $\bar{C}_i = \{\bar{f}_i = 0\}$ *($i \in T$) are also irreducible, nonsingular and interesect transversally. Then we have*

$$\int_{O_K^n} |f_1(x)^{N_1} \cdots f_t(x)^{N_t}|^s \, dx = q^{-n} \sum_{I \subset T} C_I \prod_{i \in I} \frac{(q-1)q^{-N_i s - 1}}{(1 - q^{-N_i s - 1})}$$

where

$$C_I = \# \left\{ \bar{\alpha} \in \mathbf{F}_q^n; \ \bar{\alpha} \in \bigcap_{i \in I} \bar{C}_i, \ \bar{\alpha} \notin \bigcup_{j \notin I} \bar{C}_j \right\}.$$

However, in general, $\bar{C}_i$ ($i \in T$) do not satisfy the conditions in Theorem 2.1. In this section, we shall deal with the general case.

Definition 2.2. Put $A = O_K[x_1, \ldots, x_n] - \pi O_K[x_1, \ldots, x_n]$ and $A' = \{f \in A; \{f = 0\}$ is nonsingular$\}$. For any $\xi \in O_K^n$, we define two maps $R_\xi : A \to A$ and $L_\xi : A \to A$ as follows. For $f \in A$, put $(R_\xi f)(x) = \pi^{-r} f(\xi + \pi x)$ (resp. $(L_\xi f)(x) = \pi^r f((x - \xi)/\pi))$ where r is uniquely determined by the condition that $R_\xi f \in A$ (resp. $L_\xi f \in A$).

When I showed the first version of this paper to Professor M. Harada, he pointed out to me that this procedure $R\xi$ is known as Néron's p-desingularization or blowing-up to p-direction (cf. A. Néron, Pub. Math. I.H.E.S. No. 21 (1964)).

Since

$$\int_{\xi + \pi O_K^n} |f(x)|^s \, dx = q^{-n} \int_{O_K^n} |f(\xi + \pi x)|^s \, dx = q^{-rs-n} \int_{O_K^n} |R_\xi f(x)|^s \, dx,$$

we have

$$\int_{O_K^n} |f(x)|^s \, dx = \sum_\xi q^{-rs-n} \int_{O_K^n} |R_\xi f(x)|^s \, dx,$$

where ξ runs O_K^n modulo πO_K^n ($= \mathbf{F}_q^n$).

Thus the calculation of $\int_{O_K^n} |f(x)|^s \, dx$ reduces to that of $\int_{O_K^n} |R_\xi f(x)|^s \, dx$ ($\bar{\xi} \in \mathbf{F}_q^n$). The following proposition is clear.

Proposition 2.3. *For any $\xi \in O_K^n$, we have $R_\xi L_\xi = L_\xi R_\xi = \mathrm{id}_A$. In particular, $R_\xi : A \to A$ is bijective.*

Proposition 2.4. *For any $\xi \in O_K^n$, $R_\xi f_1, \ldots, R_\xi f_t$ are (i) K-irreducible, (ii) nonsingular, and (iii) they intersect transverally.*

Proof. (i) If $R_\xi f = gh$, then $f = (L_\xi g)(L_\xi h)$. (ii) If η is a singular point of $R_\xi f$, then $\xi + \pi \eta$ is a singular point of f. (iii) We have $f(\xi + \pi x) = \pi^r \cdot (R_\xi f)(x)$ for some r. For any $\eta \in O_K^n$, put $\zeta = \xi + \pi \eta$. If $f_1(\zeta) = \cdots = f_l(\zeta) = 0$, $f_{l+1}(\zeta) \neq 0, \ldots, f_t(\zeta) \neq 0$, then we have $f_k(x) = a_{k1}(x_1 - \zeta_1) + \cdots + a_{kn}(x_n - \zeta_n) + F_k(x)$ where $F_k = \sum_d F_k^d (d \geq 2)$ and F_k^d is homogeneous of degree d with respect to $x - \zeta$ $(k = 1, \ldots, l)$. By the assumption, we have rank $(a_{ij}) = l(1 \leq i \leq l, 1 \leq j \leq n)$.

Since $(R_\xi f_k)(x) = a_{k1}(x_1 - \eta_1) + \cdots + a_{kn}(x_n - \eta_n) + \{\text{degree} \geq 2$ with respect to $x - \eta\}$ $(k = 1, \ldots, l)$ and $R_\xi f_k(\eta) \neq 0$ $(k = l+1, \ldots, t)$, we have (iii).   Q.E.D.

Now, for any $f(x) \in A$ and $\xi \in O_K^n$, we can write $f(x) = \sum_d \pi^{a_d} f_d(x - \xi)$ where $f_d(x - \xi) \in A$ is homogeneous of degree d with respect to $x - \xi = (x_1 - \xi_1, \ldots, x_n - \xi_n)$. Then the following proposition is clear.

Proposition 2.5. *When $\deg \bar{f} = d_0$, the following conditions are equivalent.*

(1) $\deg \overline{R_\xi f} < \deg \bar{f}$
(2)   *There exists $d < d_0$ satisfying $a_d + d < d_0$.*

Lemma 2.6. *Let $F(x)$ be an element of A with $\deg \bar{F} = d_0$. Then the following conditions are equivalent.*

(1)   *There exist an infinite sequence $\xi_1, \ldots, \xi_m, \ldots$ of O_K^n satisfying $\deg \overline{R_{\xi_m} \cdots R_{\xi_1} F} = d_0$ for all m.*
(2)   *There exists $\eta \in O_K^n$ satisfying $F(x) = \sum_{d \geq d_0} F_d(x - \eta)$ where $F_d(x - \eta)$ is homogeneous of degree d with respect to $x - \eta = (x_1 - \eta_1, \ldots, x_n - \eta_n)$.*

Proof. (2)$\Rightarrow$(1): We may take $\xi_1 = \eta$, $\xi_k = 0$ for all $k \geq 2$. (1)$\Rightarrow$(2): By Proposition 2.5, we may express $R_{\xi_m} \cdots R_{\xi_1} F(x) = \sum_d \pi^{a_{m,d}} f_{m,d}(x - \xi_{m+1})$ where $f_{m,d} \in A$, $a_{m,d} > 0$ (all $d > d_0$), $a_{m,d_0} = 0$, and $a_{m,d} + d \geq d_0$ for all $d < d_0$. We shall calculate $R_{\xi_{m-1}} \cdots R_{\xi_1} F = L_{\xi_m}(R_{\xi_m} \cdots R_{\xi_1} F)$ (cf. Proposition 2.3).

We have

$$\sum_d \pi^{a_{m,d}} f_{m,d}((x-\xi_m)/\pi - \xi_{m+1}) = \sum_d \pi^{a_{m,d}-d} f_{m,d}(x - \xi_m - \pi\xi_{m+1}).$$

Since $\deg \overline{R_{\xi_{m-1}} \cdots R_{\xi_1} F} = d_0$ and $a_{m,d_0} = 0$, we have

$$L_{\xi_m}(R_{\xi_m} \cdots R_{\xi_1} F) = \sum_d \pi^{a_{m,d}+(d_0-d)} f_{m,d}(x - \xi_m - \pi\xi_{m+1}).$$

Repeating this procedure, we obtain that

$$F(x) = \sum_d \pi^{a_{m,d}+m(d_0-d)} f_{m,d}(x - \xi_1 - \pi\xi_2 - \cdots - \pi^m \xi_{m+1})$$

for all m. Put $\eta = \lim_{m\to\infty} (\xi_1 + \pi\xi_2 + \cdots + \pi^m \xi_{m+1}) \in O_K^n$, and express

$$F(x) = \sum_d F_d(x - \eta)$$

as in (2). Since

$$F_d(x - \eta) = \lim_{m\to\infty} \pi^{a_{m,d}+m(d_0-d)} f_{m,d}(x - \xi_1 - \pi\xi_2 - \cdots - \pi^m \xi_{m+1})$$

and $a_{m,d} \geq 0$, we have $F_d(x - \eta) = 0$ for all $d < d_0$, i.e., $F(x) = \sum_{d \geq d_0} F_d(x - \eta)$.   Q.E.D.

Proposition 2.7. *Assume that $F(x) \in A'$ and $\deg \bar{F} = d_0 \geq 2$. Then there exists a natural number N satisfying $\deg \overline{R_{\xi_N} \cdots R_{\xi_1} F} < \deg \bar{F} = d_0$ for all $\xi_1, \ldots, \xi_N \in O_K^n$.*

Proof. If such N does not exist, then we may take an infinite sequence $\xi_1, \ldots, \xi_m, \ldots$ as in (1) of Lemma 2.6. Hence, by Lemma 2.6, we have $F(x) = \sum_{d \geq d_0} F_d(x - \eta)(d_0 \geq 2)$. Then η is a singular point of $\{F = 0\}$ which contradicts the assumption $F \in A'$.   Q.E.D.

Theorem 2.8. *Assume that $F(x) \in A$ is irreducible and nonsingular. Then there exists a natural number N such that*

(1) $|R_{\xi_N} \cdots R_{\xi_1} F| = 1$, *or* (2) $\overline{R_{\xi_N} \cdots R_{\xi_1} F}$ *is irreducible and nonsingular for all $\xi_1, \ldots, \xi_N \in O_K^n$.*

Proof. By Proposition 2.7, we may assume that $\deg \overline{R_{\xi_N} \cdots R_{\xi_1} F} \leq 1$. If $\deg \overline{R_{\xi_N} \cdots R_{\xi_1} F} = 0$, then $|R_{\xi_N} \cdots R_{\xi_1} F|_K = 1$. If $\deg \overline{R_{\xi_N} \cdots R_{\xi_1} F} = 1$, it is clearly irreducible and nonsingular.   Q.E.D.

By Theorem 2.8, we may assume that $f_k(x) = a_k + \sum_{i=1}^{n} a_{ki}x_i + \pi F_k(x)$ where $F_k(x) = \sum_{d \geq 2} F_{k,d}(x)$ and $F_{k,d}(x)$ is homogeneous of degree d with respect to $x = (x_1, \ldots, x_n)$ for $k = 1, \ldots, t$. We shall prepare some tools.

Proposition 2.9. *Assume that* $f(x) \equiv a + \sum_{k=1}^{n} a_k x_k \bmod \pi^m$ *(m $\geq$ 1). Then, for any* $\xi \in O_K^n$, *we have* (1) $|R_\xi f|_K = 1$, *or* (2) $R_\xi f \equiv a' + \sum_{i=1}^{n} a_k' x_i \bmod \pi^{m+1}$, *where* $a_k' \equiv a_k \bmod \pi^m$ $(k = 1, \ldots, n)$.

Proof. Put $f(x) = a + \sum_{k=1}^{n} a_k x_k + \pi^m F(x)$. Since $f(\xi + \pi x) = (a + \sum_{k=1}^{n} a_k \xi_k + \pi^m F(\xi)) + \pi \sum_{k=1}^{n} (a_k + \pi^m \partial F/\partial x_k(\xi))x_k + \pi^{m+2} G(x)$ for some $G(x) = \sum_{d \geq 2} G_d(x)$. If $a + \sum_k a_k \xi_k \not\equiv 0(\pi)$, we have (1). If $a + \sum_k a_k \xi_k \equiv 0(\pi)$, we have (2).   Q.E.D.

Proposition 2.10. *Assume that there exists an infinite sequence* $\xi_1, \ldots, \xi_m, \ldots$ *of* O_K^n *satisfying*

$$R_{\xi_{m-1}} \cdots R_{\xi_1} f = a^{(m)} + \sum_{i=1}^{k} a_i^{(m)} x_i + \sum_{i=k+1}^{n} a_i^{(m)} x_i + \pi^m F_{m-1}(x)$$

with $a_i^{(m)} \equiv 0(\pi^m)$ *and* $|R_{\xi_m} \cdots R_{\xi_1} f| \neq 1$ *for all m and* $i = k+1, \ldots, n$. *Put* $\eta = \xi_1 + \pi \xi_2 + \cdots \in O_K^n$. *Then we have* $R_\eta f(x) = b + \sum_{i=1}^{k} b_i x_i + \pi G(x)$ *with* $G(x) = \sum G_d(x)(d \geq 2)$.

Proof. Put $a_i^{(m)} = \pi^m a_{i,m}$ for all m and $i = k+1, \ldots, n$. Then we have $f(x) = a^{(1)} + \sum_{i=1}^{k} a_i^{(1)} x_i + \sum_{i=k+1}^{n} \pi a_{i,1} x_i + \pi F_0(x)$. First we shall show that $a_{i,1} = -\partial F_0/\partial x_i(\eta)$ for $i = k+1, \ldots, n$. Since $\pi(R_{\xi_m} \cdots R_{\xi_1} f)(x) = (R_{\xi_{m-1}} \cdots R_{\xi_1} f)(\xi_m + \pi x)$, we have $a_{i,m} + \partial F_{m-1}/\partial x_i(\xi_m) = \pi a_{i,m+1}$ for $m \geq 1$. Since $\pi^{1-m} f(\xi_1 + \pi \xi_2 + \cdots + \pi^{m-2} \xi_{m-1} + \pi^{m-1} x) = R_{\xi_{m-1}} \cdots R_{\xi_1} f(x) = a^{(m)} + \sum_{i=1}^{k} a_i^{(m)} x_i + \sum_{i=k+1}^{n} \pi^m a_{i,m} x_i + \pi^m F_{m-1}(x)$, we have

$$\frac{\partial f}{\partial x_i}(\xi_1 + \pi \xi_2 + \cdots + \pi^{m-2} \xi_{m-1} + \pi^{m-1} x) = \frac{\partial}{\partial x_i} \pi^{1-m} f(\xi_1 + \pi \xi_2 + \cdots$$
$$+ \pi^{m-2} \xi_{m-1} + \pi^{m-1} x)$$
$$= \pi^m a_{i,m} + \pi^m \frac{\partial F_{m-1}}{\partial x_i}(x),$$

and hence we have $\partial f/\partial x_i(\xi_1 + \pi \xi_2 + \cdots + \pi^{m-1} \xi_m) = \pi^{m+1} a_{i,m+1} \in \pi^{m+1} O_K$. This implies $\partial f/\partial x_i(\eta) = \pi a_{i,1} + \pi \partial F_0/\partial x_i(\eta) = 0$, i.e., $a_{i,1} = -\partial F_0/\partial x_i(\eta)$ for

$i = k+1, \ldots, n$. Since $\eta \equiv \xi_1(\pi)$, we have $f(\eta) \equiv f(\xi_1) \equiv 0$ and

$$R_\eta f = \frac{1}{n} f(\eta + \pi x)$$

$$= \frac{f(\eta)}{\pi} + \sum_{i=1}^{k} \left(a_i^{(1)} + \pi \frac{\partial F_0}{\partial x_i}(\eta) \right) x_i + \sum_{i=k+1}^{n} \left(a_{i,1} + \frac{\partial F_0}{\partial x_i}(\eta) \right) \pi x_i + \pi G(x)$$

$$= b + \sum_{i=1}^{k} b_i x_i + \pi G(x). \hspace{3cm} \text{Q.E.D.}$$

Proposition 2.11. (1) For any $\xi_1, \ldots, \xi_{m+1} \in O_K^n$, put $\eta = \xi_1 + \pi \xi_2 + \cdots + \pi^m \xi_{m+1}$. Then we have $R_{\xi_m} \cdots R_{\xi_1} f = R_0^m R_\eta f$ for all $f \in A$.
(2) For an infinite sequence $\xi_1, \ldots, \xi_m, \ldots$ of O_K^n, put $\eta = \xi_1 + \pi \xi_2 + \cdots + \pi^m \xi_{m+1} + \cdots$. Then, for any m and $f \in A$, we have $R_0^m R_\eta f = R_\xi R_{\xi_m} \cdots R_{\xi_1} f$ with $\xi \equiv \xi_{m+1}(\pi)$.

Proof. (2) is obtained from (1) by putting $\xi = \xi_{m+1} + \pi \xi_{m+2} + \cdots$ so that $\eta = \xi_1 + \pi \xi_2 + \cdots + \pi^m \xi$. (1) Since $f(\xi_1 + \pi x) = \pi^{r_1}(R_{\xi_1} f)(x)$ and $(R_{\xi_1} f)(\xi_2 + \pi x) = \pi^{r_2}(R_{\xi_2} R_{\xi_1} f)(x)$, we have $f(\xi_1 + \pi \xi_2 + \pi^2 x) = \pi^{r_1 + r_2}(R_{\xi_2} R_{\xi_1} f)(x) = \pi^{r_1 + r_2}(R_0 R_{\xi_1 + \pi \xi_2} f)(x)$. Repeating this procedure, we obtain our result.
$\hspace{10cm} \text{Q.E.D.}$

Definition 2.12. We shall generalize R_ξ as follows. For $1 \le m \le n$ and $\eta = (\eta_1, \ldots, \eta_m) \in O_K^m$, define $R_\eta f$ by $f(\eta + \pi x', x'') = \pi^r (R_\eta f)(x)$ where r is uniquely determined by $R_\eta f \in A$. Here $x = (x', x'')$ with $x' = (x_1, \ldots, x_m)$, $x'' = (x_{m+1}, \ldots, x_n)$. A similar assertion as in Proposition 2.4 holds for this generalized R_η.

Theorem 2.13. *Assume that $f_i(x) = a_i + \sum_{j=1}^{n} a_{ij} x_j + \pi^m F_i(x) \in A'$ ($i = 1, \ldots, t$) are irreducible, nonsingular and they intersect transversally. Here $F_i(x) = \sum F_{i,d}(x) (d \ge 2)$ and $F_{i,d}(x)$ is homogeneous of degree d with respect to x. Then there exists a natural number N such that, for $\xi_1, \ldots, \xi_N$, (1) $|R_{\xi_N} \cdots R_{\xi_1} f_i|_K = 1$ for some i, or (2) $\overline{R_{\xi_N} \cdots R_{\xi_1} f_i}$ intersect transversally ($1 \le i \le t$). Here, for each k ($1 \le k \le N$), a natural number m_k with $1 \le m_k \le n$ is determined and ξ_k is any element of $O_K^{m_k}$.*

The remaining part of this section is devoted to proving Theorem 2.13. We shall prove this by induction on t. If $t = 1$, transversality is obvious. Assume that the theorem holds for any $t \le T - 1$. We shall prove the theorem when $t = T$.

Lemma 2.14. *We may assume without loss of generality that* $a_1 = \cdots = a_l = 0$ *and*

$$(\overline{a_{ij}}) = \begin{pmatrix} I_l & 0 \\ * & 0 \end{pmatrix}.$$

Proof. If rank $(\overline{a_{ij}}) = l$, then by a linear change of variables and renumbering f_i if necessary, we may assume that $(\overline{a_{ij}}) = (\begin{smallmatrix} I_l & 0 \\ * & 0 \end{smallmatrix})$. For any given $\xi \in O_K^n$, if $f_i(\xi) \neq 0(\pi)$ for some i, we have $|R_\xi f_i|_K = 1$ and it reduces to the case $t \leq T-1$. Hence $f_i(\xi) \equiv 0(\pi)$ $(i = 1, \ldots, t)$. By Hensel's lemma, there exists $\xi' \in O_K^n$ satisfying $f_i(\xi') = 0$ $(1 \leq i \leq l)$ and $\xi' \equiv \xi$. Then $R'_\xi f_i$ $(i = 1, \ldots, t)$ satisfies the condition in this lemma.   Q.E.D.

Lemma 2.15. *Assume that* rank$(a_{ij}) > l$, $a_1 = \cdots = a_l = 0$, $(\overline{a_{ij}}) = (\begin{smallmatrix} I_l & 0 \\ * & 0 \end{smallmatrix})$ *and that we may take m sufficiently large compared to* $\min_i\{\text{ord } a_{ij}\}$ *for some* $j \geq l+1$. *Then there exists a natural number N such that* (1) $|R_{\eta_N} \cdots R_{\eta_1} f_i| = 1$, *for some* i, *or* (2) rank$(\overline{a_{ij}}) > l$ *with respect to* $R_{\eta_N} \cdots R_{\eta_1} f_i$, $(1 \leq i \leq t)$.

Proof. If $\eta = (\eta_1, \ldots, \eta_l) \neq 0(\pi)$, e.g., $\eta_i \neq 0(\pi)$, then $|R_\eta f_i| = 1$ $(1 \leq i \leq l)$. Hence we may take $\eta = (0, \ldots, 0) \in O_K^l$. If $a_i \neq 0(\pi)$ for some $i = l+1, \ldots, n$, then $|R_0 f_i| = 1$. Hence we can express as $f_i = \pi a_i' + \sum_{j=1}^l a_{ij} x_j + \sum_{j=l+1}^n \pi a_{ij}' x_j + \pi^m F_i(x)$ with $a_1' = \cdots = a_l' = 0$ $(1 \leq i \leq t)$. Then we have $R_\eta f_i = a_i' + \sum_{j=1}^l a_{ij} x_j + \sum_{j=l+1}^n a_{ij}' x_j + \pi^{m-1} F_i(x)$. Repeating this procedure, we have our result. Note that m is sufficiently large.   Q.E.D.

Now there are two cases. Case (I) We can apply Lemma 2.15 repeatedly so that we may assume that rank$(a_{ij}) = $ rank$(\overline{a_{ij}})$. Case (II) We cannot apply Lemma 2.15 only when the situation in Proposition 2.10 happens. The case I is solved by the following proposition.

Proposition 2.16. *Assume that* $f_k(x) = a_k = a_{k1} x_1 + \cdots + a_{kn} x_n + \pi^m F_k(x)$ $(F_k = \sum F_{k,d}(d \geq 2))$ *where* $k = 1, \ldots, t$; $a_1 = \cdots = a_l = 0$, rank(a_{ij}) $(1 \leq i \leq t, 1 \leq j \leq n) = l = $ rank$(\overline{a_{ij}})$ $(1 \leq i \leq l, 1 \leq j \leq n)$. *Then there exists a natural number N such that, for any* $\xi_1, \ldots, \xi_N \in O_K^n$, (1) $|R_{\xi_N} \cdots R_{\xi_1} f_k|_K = 1$ *for some* k, *or* (2) $\overline{R_{\xi_N} \cdots R_{\xi_1} f_k}$ $(1 \leq k \leq t)$ *intersect transversally.*

Proof. Assume that there exists an infinite sequence $\xi_1, \ldots, \xi_m, \ldots$ satisfying $|R_{\xi_m} \cdots R_{\xi_1} f_k|_K \neq 1$ for any $k = 1, \ldots, t$, and any m. Then $f_1(\xi_1) \equiv \cdots \equiv f_t(\xi_1) \equiv 0 \mod \pi$. Hence, by Hensel's lemma, there exists ξ_1' satisfying $f_1(\xi_1') = \cdots = f_t(\xi_1') = 0$, and $\xi_1' \equiv \xi_1(\pi)$. We may assume that $\xi_1' = \xi_1$. Similarly we may assume that $(R_{\xi_m} \cdots R_{\xi_1} f_k)(\xi_{m+1}) = 0$ for all m

and $1 \leq k \leq l$. This implies that $f_k(\xi_1 + \pi\xi_2 + \cdots + \pi^m\xi_{m+1}) = 0$. Put $\eta = \sum_{i=0}^{\infty} \pi^i \xi_{i+1} \in O_K^n$. Then we have $f_k(\eta) = 0$ $(1 \leq k \leq l)$. By the transversality, we have $f_k(\eta) \neq 0$ for $l+1 \leq k \leq t$. Since $\eta \equiv \xi_1(\pi)$, we have $f_k(\eta) \equiv 0(\pi)$. Hence $R_n f_k(x) = a'_k + a'_{k1}x_1 + \cdots + a'_{kn}x_n + \pi^{m+1}F'_k(x)$ $(1 \leq k \leq t)$ with $a'_k = 0$ $(1 \leq k \leq l)$, $a'_k \neq 0$ $(l+1 \leq k \leq t)$ and $\overline{(a_{ij})} = \overline{(a'_{ij})}$. If $t = l$, clearly $R_n f_k$ intersect transversally. If $t > l$, let $a'_t = \pi^r u$ $(u \in U_K = O_K - \pi O_K)$. Then we have $|R_0^{r+1} R_n f_t|_K = 1$. By Proposition 2.11, this implies that $|R_\xi R_{\xi_{r+1}} \cdots R_{\xi_1} f_t|_K = 1$ with $\xi \equiv \xi_{r+2}(\pi)$ and hence $|R_{\xi_{r+2}} \cdots R_{\xi_1} f_t|_K = 1$, which is a contradiction.   Q.E.D.

Finally we shall deal with Case II. Namely there exists an infinite sequence $\xi_1, \cdots, \xi_m, \ldots$ of O_K^n such that the coefficient of x_i $(l+1 \leq i \leq n)$ of $R_{\xi_m} \cdots R_{\xi_1} f_j$ $(j = 1, \ldots, t)$ is $\equiv 0$ modulo π^m for all m. Put $\eta = \xi_1 + \pi\xi_2 + \cdots \in O_K^n$. By Proposition 2.10, we have $R_n f_i = b_i + \sum_{j=1}^{l} b_{ij}x_j + \pi^m G_i$ $(i = 1, \ldots, t)$. If some $b_i \neq 0$, then $b_i = \pi^r u$ for some r and $u \in U_K = O_K - \pi O_K$. Clearly we have $|R_0^{r+1} R_n f_i|_K = 1$. By Proposition 2.11, this implies $|R_\xi R_{\xi_{r+1}} \cdots R_{\xi_1} f_i| = 1$ with $\xi \equiv \xi_{r+2}(\pi)$, and hence it reduces to the case $t \leq T - 1$. If all $b_i = 0$ $(i = 1, \ldots, t)$, then $R_n f_i$ intersect transversally at 0 and hence $\mathrm{rank}(b_{ij}) = t(\leq l$ since $1 \leq j \leq l)$. Since $l = \mathrm{rank}(\overline{a_{ij}}) = \mathrm{rank}(\overline{b_{ij}}) \leq \mathrm{rank}(b_{ij}) = t \leq l$, we have $\mathrm{rank}(\overline{b_{ij}}) = t$. This implies that $\overline{R_n f_i}$ $(i = 1, \ldots, t = l)$ intersect transversally (at any point). This completes the proof of Case II, and hence the proof of Theorem 2.13. Thus we could obtain the following theorem.

Theorem 2.17 (*a resolution of singularities modulo* π). *Assume that* $f_i(x) \in A = O_K[x_1, \ldots, x_n] - \pi O_K[x_1, \ldots, x_n]$ $(i = 1, \ldots, t)$ *are irreducible, nonsingular and they intersect transversally. Then there exists a natural number* N *such that* $\overline{R_{\xi_N} \cdots R_{\xi_1} f_i}$ (*if not constant*) *are irreducible, nonsingular and they intersect transversally. Here, for each* k $(1 \leq k \leq N)$, *a natural number* m_k *with* $1 \leq m_k \leq n$ *is determined and* ξ_k *is any element of* $O_K^{m_k}$.

Thus once we know the embedded resolution of f, we can calculate the Igusa's local zeta function.

References

[1] J. Denef, On the degree of Igusa's local zeta function, *Amer. J. Math.* **109** (1987) 991–1008.

[2] H. Hironaka, Resolution of singularities of an algebraic variety over a field of characteristic zero, *Annals of Math.* (1964) 109–326.

[3] J. Igusa, Some results on p-adic complex powers, *Amer. J. Math.* **106** (1984) 1013–1032.

[4] M. Sato and T. Kimura, A classification of irreducible prehomogeneous vector spaces and their relative invariants, *Nagoya Math. J.* **65** (1977) 1–155.

Operational Calculus, Hyperfunctions and Ultradistributions

Hikosaburo Komatsu

Department of Mathematics
Faculty of Science
University of Tokyo
Tokyo, Japan

A review is given on a new foundation of the Heaviside operational calculus based on the Laplace transformation of hyperfunctions. Then the hyperfunctions with support bounded from below and the Mikusiński operators are compared. Neither of them is included in the other. The ultradistributions (in a wide sense) with support bounded from below are shown to form an important subclass of the intersection.

1. Introduction

One of the origins of modern analysis seems to be in Heaviside's operational calculus. He gave us a way to consider the derivatives of non-differentiable functions and the operations on them of functions $P(d/dx)$ of the differentiation d/dx.

Copyright © 1988 by Academic Press, Inc.

The typical problem he dealt with was that of finding the solution $u(x)$ of the linear ordinary differential equation

$$P\left(\frac{d}{dx}\right)u(x) = f(x), \qquad x > 0, \tag{1}$$

such that $u(x) = 0$ for $x < 0$. Here

$$P\left(\frac{d}{dx}\right) = a_m\left(\frac{d}{dx}\right)^m + \cdots + a_0 \tag{2}$$

is a differential operator with constant coefficients $a_i \in \mathbf{C}$ and he tacitly identifies the function $f(x)$ with the product $\theta(x)f(x)$ with the Heaviside function

$$\theta(x) = \begin{cases} 1, & x > 0, \\ 0, & x < 0, \end{cases} \tag{3}$$

(cf. Berg [2]).

In view of the Green formula

$$P\left(\frac{d}{dx}\right)(\theta(x)u(x)) = \theta(x)\left(P\left(\frac{d}{dx}\right)u(x)\right)$$

$$+ (a_m u^{(m-1)}(0) + \cdots + a_1 u(0))\delta(x) + \cdots$$

$$+ a_m u(0)\delta^{(m-1)}(x), \tag{4}$$

the equation (1) with the general initial conditions

$$u^{(j)}(0) = g_j, \qquad j = 0, \ldots, m-1, \tag{5}$$

may also be reduced to the equation

$$P\left(\frac{d}{dx}\right)u(x) = f(x) \tag{6}$$

for distributions $f(x)$ and $u(x) \in \mathscr{D}'_{[0,\infty)}$ on $\mathbf{R}$ with support in $[0, \infty)$.

Let

$$p(x) = P\left(\frac{d}{dx}\right)\delta(x). \tag{7}$$

Then the equation (6) becomes the convolution equation

$$p * u(x) = f(x) \tag{8}$$

in $\mathscr{D}'_{[0,\infty)}$.

Heaviside claimed that there was a function

$$q(x) = P\left(\frac{d}{dx}\right)^{-1}\delta(x) \tag{9}$$

and the solution of (8) was obtained as

$$u(x) = q * f(x). \tag{10}$$

(Actually he used the Heavisde $\theta(x)$ in (9) in place of the Dirac $\delta(x)$, so that the right-hand side of (10) had to be differentiated once.) His construction of the resolvent $q(x)$ was by no means absurd but his argument looked quite unsatisfactory (cf. Berg [2]).

The first mathematical justification was accomplished by the Laplace transformation (cf. Doetsch [11]). Suppose that $u(x)$ and its derivatives $u^{(j)}(x)$ up to order m satisfy the exponential type condition

$$|u^{(j)}(x)| \le C\,e^{Hx}, \qquad x > 0, \tag{11}$$

with constants H and C. Then taking the Laplace transforms

$$\hat{f}(\lambda) = \int_0^\infty e^{-\lambda x} f(x)\,dx \tag{12}$$

of both sides of (6) or (8), we have

$$P(\lambda)\hat{u}(\lambda) = \hat{f}(\lambda). \tag{13}$$

Hence the solution $u(x)$ of (6) is obtained by the inversion formula

$$u(x) = \frac{1}{2\pi i}\int_{\Lambda - i\infty}^{\Lambda + i\infty} e^{\lambda x}\hat{u}(\lambda)\,d\lambda, \qquad x > 0, \tag{14}$$

applied to

$$\hat{u}(\lambda) = P(\lambda)^{-1}\hat{f}(\lambda). \tag{15}$$

Or, in other words, the resolvent $q(x)$ is obtained as

$$q(x) = \frac{1}{2\pi i}\int_{\Lambda - i\infty}^{\Lambda + i\infty} e^{\lambda x} P(\lambda)^{-1}\,d\lambda. \tag{16}$$

The solution by formulae (14) and (15) has been criticized because at least $f(x)$ must satisfy the exponential type condition (11) in order that (15) make sense.

To overcome this difficulty Mikusiński [22] invented a new foundation based on the Titchmarsh theorem saying that the convolution algebra

$C([0, \infty))$ of continuous functions on $[0, \infty)$ has no divisors of zero. The quotient field $\mathcal{M}$ is defined to be the space of the *Mikusiński operators*.

The Mikusiński operators $\mathcal{M}$ include the distributions $\mathcal{D}'_{[a,\infty)}$ with support in $[a, \infty]$ for any $a > -\infty$, because if $f(x) \in \mathcal{D}'_{[a,\infty)}$, then the regularization $f * \varphi$ by a non-zero function $\varphi \in \mathcal{D}_{[-a,b]}$ is in the space $C_{[0,\infty)}$ of continuous functions on **R** with support in $[0, \infty)$. In particular, the quotient $\delta = \varphi / \varphi$ in $\mathcal{M}$ belongs to $\mathcal{M}$. If $q = \delta / p$ in $\mathcal{M}$ is a function, as in the case of (7), equation (8) is solved by (10). This is also the case with difference equations.

Mikusiński constructs the resolvent functions $q(x)$ by ad hoc methods. He also defines the fractional powers

$$\left(\frac{d}{dx}\right)^{\alpha} \delta(x) = \frac{1}{\Gamma(-\alpha)} x_+^{-\alpha-1} \tag{17}$$

and their exponentials $\exp(-t(d/dx)^{\alpha})\delta(x)$, but there is no unified method. He is even reluctant to admit that the shift operator

$$\delta(x-a) = \exp\left(-a\frac{d}{dx}\right)\delta(x). \tag{18}$$

We remark that those formulae are most naturally derived from (16) interpreted in the sense of hyperfunction.

We review in Section 2 our theory of Laplace transforms of hyperfunctions ([18], [19]), by which we can consider Laplace transforms of (hyper-)functions having an arbitrary growth order. The elementary proof of Theorem 1 is based on an old result of Sato [28].

In Section 3 we show that the solvability of (8) in the hyperfunctions is easily checked by the location of zeros of the Laplace transform of p.

We consider in Section 4 the relationship of the hyperfunctions $\mathcal{B}_+$ with support bounded below, the Mikusiński operators $\mathcal{M}$, and the ultradistributions $\mathcal{D}_+^{*\prime}$ with support bounded from below.

Of course, there are many investigations on Laplace transforms of generalized functions. Schwartz [30] and Sebastião e Silva [31] for distributions of exponential type, Ditkin [10] and Berg [3] for continuous functions of arbitrary growth, and Kawai [13] and Morimoto [24] for hyperfunctions are among them. Actually, Morimoto's theory is general enough to use for the foundation of the standard operational calculus discussed above. However, our formulation is more convenient in the vector-valued case as we will briefly remark in Section 5.

2. Laplace Hyperfunctions and their Laplace Transforms

Recall that the hyperfunctions $\mathscr{B}_{[a,\infty)}$ on **R** with support in $[a, \infty)$ is defined by

$$\mathscr{B}_{[a,\infty)} = \mathcal{O}(\mathbf{C}\backslash[a, \infty))/\mathcal{O}(\mathbf{C}), \tag{19}$$

where $\mathcal{O}(V)$ denotes the space of all holomorphic functions on V (Sato [28], [29], Komatsu [14]). The hyperfunction $f(x)$ represented by $F(z) \in \mathcal{O}(\mathbf{C}\backslash[a, \infty))$ is written

$$f(x) = F(x + i0) - F(x - i0), \tag{20}$$

and $F(z)$ is called a *defining function* of $f(x)$.

Modifying this definition, we have introduced a class of hyperfunctions, called the Laplace hyperfunctions on the extended real line $[-\infty, +\infty]$ ([18], [19]).

Let

$$\mathbf{O} = \mathbf{C} \cup S^1\infty \tag{21}$$

be the radial compactification of the complex plane and let $\mathcal{O}^{\exp}$ be the sheaf on **O** of *holomorphic functions of exponential type*, i.e., for each open set V in **O** the section space $\mathcal{O}^{\exp}(V)$ is the space of all holomorphic functions $F(z)$ on $V \cap \mathbf{C}$ such that on each closed sector

$$\Sigma = \{z \in \mathbf{C}; \alpha \leq \arg(z - c) \leq \beta\} \Subset V, \tag{22}$$

the estimate

$$|F(z)| \leq C\,e^{H|z|}, \qquad z \in \Sigma, \tag{23}$$

holds with constants H and C.

Definition 1. We define the space $\mathscr{B}_{[a,\infty]}^{\exp}$ of *Laplace hyperfunctions* with support in $[a, \infty]$ by

$$\mathscr{B}_{[a,\infty]}^{\exp} = \mathcal{O}^{\exp}(\mathbf{O}\backslash[a, \infty])/\mathcal{O}^{\exp}(\mathbf{O}). \tag{24}$$

An $F(z) \in \mathcal{O}^{\exp}(\mathbf{O}\backslash[a, \infty])$ is called a *defining function* of the Laplace hyperfunction $f(x)$ it represents and we also write (20).

Theorem 1. The inclusion mappings $\mathcal{O}^{\exp}(\mathbf{O}\backslash[a, \infty]) \to \mathcal{O}(\mathbf{C}\backslash[a, \infty))$ and $\mathcal{O}^{\exp}(\mathbf{O}) \to \mathcal{O}(\mathbf{C})$ induce the surjection

$$\rho : \mathscr{B}_{[a,\infty]}^{\exp} \to \mathscr{B}_{[a,\infty)}. \tag{25}$$

Proof. The proofs we gave in [18] and [19] relied on the solution of the $\bar{\partial}$ equation with bounds. Here we note that the theorem is an immediate consequence of the following lemma stated in Sato [28] without proof. The lemma shows also that no natural Hausdorff locally convex topology can be introduced in the space $\mathscr{B}_{[a,\infty)}$ of hyperfunctions.

Lemma. *Let U be any neighborhood of* $[a, \infty)$ *in* **C**. *For each hyperfunction* $f \in \mathscr{B}_{[a,\infty)}$ *and* $\varepsilon > 0$ *we can find a defining function* $F \in \mathcal{O}(\mathbf{C}\backslash[a, \infty))$ *of f such that*

$$|F(z)| \le \varepsilon, \qquad z \in \mathbf{C}\backslash U. \tag{26}$$

Proof. We reproduce the proof of Kaneko [12] because the proof of Theorem 1 has been the only non-elementary part of our theory.

Divide $[a, \infty)$ into the sequence of compact intervals $I_n = [a_n, a_{n+1}]$, $a = a_1 < a_2 < \cdots$, and take a locally finite system of neighborhoods U_n of I_n in U.

We will prove that the hyperfunction f is decomposed into the sum

$$f = \sum_{n=1}^{\infty} f_n$$

of hyperfunctions $f_n \in \mathscr{B}_{I_n}$ with support in I_n so that their standard defining functions

$$F_n(z) = \frac{-1}{2\pi i} \int_{-\infty}^{\infty} \frac{1}{z-x} f_n(x)\, dx \tag{27}$$

satisfy

$$|F_n(z)| \le \varepsilon 2^{-n}, \qquad z \in \mathbf{C}\backslash U_n. \tag{28}$$

Then it will follow that

$$F(z) = \sum_{n=1}^{\infty} F_n(z)$$

is a desired defining function of f.

By the flabbiness of the hyperfunctions the hyperfunction g_1 on $\mathbf{R}\backslash\{a_2\}$ which is equal to f on $(-\infty, a_2)$ and to 0 on (a_2, ∞) can be continued to a hyperfunction h_1 with support in $I_1 = [a_1, a_2]$. Then its standard defining function $H_1(z)$ is a holomorphic function on $\mathbf{P}^1\backslash I_1$ vanishing at ∞. Since $\mathbf{P}^1\backslash I_1$ is simply connected, it follows from the Runge theorem that there is a polynomial $P_1(z)$ in $1/(z-a_2)$ without constant term such that $F_1(z) = H_1(z) - P_1(z)$ satisfies (28). Since $P_1(z)$ is the standard defining function of a distribution $p_1(x)$ with support at a_2, $f_1 = h_1 - p_1$ has the required property. Since $f - f_1$ is a hyperfunction in $\mathcal{B}_{[a_2,\infty)}$, we can continue a similar construction, and so on.

A Laplace hyperfunction f is said to vanish near $b \in [-\infty, \infty]$, if its defining function F can be continued to a function in $\mathcal{O}^{\exp}$ near b. The *support* of f, supp f, is by definition the complement of the set of all points b near which f vanishes.

The mapping ρ of (25) is not injective. The kernel is composed of all Laplace hyperfunctions with support at most at ∞. In finite region ρ preserves the support:

$$\text{supp } \rho(f) = \text{supp } f \cap \mathbf{R}. \tag{29}$$

When K is a closed set in $[a, \infty]$, we denote by $\mathcal{B}_K^{\exp}$ the subspace of all $f \in \mathcal{B}_{[a,\infty]}^{\exp}$ with support in K. If $a < b < \infty$, $\mathcal{B}_{[b,\infty]}^{\exp}$ as a subspace of $\mathcal{B}_{[a,\infty]}^{\exp}$ is the same as the original one.

Theorem 1 says

$$\mathcal{B}_{[a,\infty)} \cong \mathcal{B}_{[a,\infty]}^{\exp} / \mathcal{B}_{\{\infty\}}^{\exp}. \tag{30}$$

More generally we have for any $-\infty < a < b \leq \infty$

$$\mathcal{B}_{[a,b)} \cong \mathcal{B}_{[a,\infty]}^{\exp} / \mathcal{B}_{[b,\infty]}^{\exp}, \tag{31}$$

where the left-hand side denotes the space of all hyperfunctions on $(-\infty, b)$ with support in $[a, b)$.

Definition 2. The *Laplace transform* $\hat{f}(\lambda)$ of a Laplace hyperfunction $f(x) = F(x + i0) - F(x - i0) \in \mathcal{B}_{[a,\infty]}^{\exp}$ is defined by the integral

$$\hat{f}(\lambda) = \int_\Gamma e^{-\lambda z} F(z) \, dz, \tag{32}$$

where Γ is a path composed of the ray from $e^{i\beta}\infty$ to a point $c < a$ and the ray from c to $e^{i\alpha}\infty$ with $0 < \alpha < \pi/2$ and $3\pi/2 < \beta < 2\pi$.

Let

$$\hat{F}_c(\lambda) = \int_c^{e^{i\theta}\infty} e^{-\lambda z} F(z) \, dz \tag{33}$$

be the Laplace transform with origin at c of the defining function $F(z)$.

Then we have

$$\hat{f}(\lambda) = \hat{F}_c^+(\lambda) - \hat{F}_c^-(\lambda), \tag{34}$$

where $\hat{F}_c^+(\lambda)$ (resp. $\hat{F}_c^-(\lambda)$) is the branch of $\hat{F}_c(\lambda)$ in the sector of opening $(-\pi/2, \pi/2)$ (resp. $(-5\pi/2, -3\pi/2)$).

The classical theorem of Borel [7] and Pólya [26] (see Boas [5]) says that if $F(z)$ is an entire function of exponential type, then $\hat{F}_c(\lambda)$ is a holomorphic function outside a compact set in **C**, so that the difference vanishes. Hence the Laplace transform $\hat{f}(\lambda)$ does not depend on the defining function $F(z)$, though the domain of definition may.

Theorem 2. *The Laplace transformation $\mathcal{L}$ is a one-to-one correspondence between $f(x) \in \mathcal{B}_{[a,\infty]}^{\exp}$ and $\hat{f}(\lambda) \in \mathcal{O}^{\exp}(S_{-\pi/2,\pi/2}\infty)$ such that*

$$\varlimsup_{r \to \infty} \frac{\log|\hat{f}(re^{i\theta})|}{r} \leq -a \cos \theta, \qquad |\theta| < \frac{\pi}{2}, \tag{35}$$

*where $\mathcal{O}^{\exp}(S_{\alpha,\beta}\infty)$ denotes the inductive limit $\varinjlim \mathcal{O}^{\exp}(V)$ as the open set V in **O** tends to the arc $S_{\alpha,\beta}\infty = \{e^{i\theta}\infty; \alpha < \theta < \beta\}$ at infinity.*

Choose an arbitrary point Λ in a ray in the domain Ω of $\hat{f}(\lambda)$. Then the absolutely convergent integral

$$F(z) = \frac{1}{2\pi i} \int_{\Lambda}^{\infty} e^{\lambda z} \hat{f}(\lambda) \, d\lambda \tag{36}$$

is a defining function of the inverse image of $\hat{f}(\lambda)$, where the path of integral is a convex curve in Ω which is eventually a ray tending to $e^{i\theta}\infty$ for a θ with $|\theta| < \pi/2$.

The theorem is essentially the same as Macintyre's result on Laplace transforms of holomorphic functions of exponential type on sectors ([21], [5]), which is proved only by the change of order of integrations and Cauchy's integral formula. See [18] for details.

Let $\alpha < 0 < \beta$. If an $\hat{f}(\lambda) \in \mathcal{LB}_{[a,\infty]}^{\exp}$ has an analytic continuation into a function in $\mathcal{O}^{\exp}(S_{-\pi/2-\beta, \pi/2-\alpha}\infty)$, then the defining function $F(z)$ defined by (36) is a function in $\mathcal{O}^{\exp}(S_{\alpha, 2\pi+\beta}\infty)$, so that the hyperfunction $f(x) = F(x+i0) - F(x-i0)$ is a holomorphic function of exponential type on a sector of opening $(\alpha + \varepsilon, \beta - \varepsilon)$ for any $\varepsilon > 0$ if x is sufficiently large. Conversely if a hyperfunction $f(x) \in \mathcal{B}_{[a,\infty)}$ satisfies this property, then the Laplace transform $\hat{f}(\lambda)$ of $f(x)$ defined naturally is in $\mathcal{O}^{\exp}(S_{-\pi/2-\beta, \pi/2-\alpha}\infty)$.

Hence we have a canonical extension mapping defined for such hyperfunctions into Laplace hyperfunctions.

In particular, hyperfunctions with compact support are naturally continued to Laplace hyperfunctions with the same support.

In general there seems to be no natural extension mapping $\mathcal{B}_{[a,\infty)} \to \mathcal{B}_{[a,\infty]}^{exp}$ which is a right inverse of the mapping ρ of (25).

3. Convolution Algebras of Hyperfunctions

The pointwise multiplication is clearly a bilinear mapping

$$\mathcal{L}\mathcal{B}_{[a,\infty]}^{exp} \times \mathcal{L}\mathcal{B}_{[b,\infty]}^{exp} \to \mathcal{L}\mathcal{B}_{[a+b,\infty]}^{exp}. \tag{37}$$

Definition 3. Let $f \in \mathcal{B}_{[a,\infty]}^{exp}$ and $g \in \mathcal{B}_{[b,\infty]}^{exp}$ be Laplace hyperfunctions. We define their *convolution* $f * g \in \mathcal{B}_{[a+b,\infty]}^{exp}$ by

$$(f * g)^\wedge = \hat{f} \cdot \hat{g}. \tag{38}$$

The totality of Laplace hyperfunctions

$$\mathcal{B}_+^{exp} = \bigcup_{a=-\infty}^{\infty} \mathcal{B}_{[a,\infty]}^{exp} \tag{39}$$

and $\mathcal{B}_{[0,\infty]}^{exp}$ form commutative algebras under addition and convolution.

Clearly $\mathcal{B}_+^{exp}$ and $\mathcal{B}_{[0,\infty]}^{exp}$ have no divisors of zero. However, the precise form of the Titchmarsh theorem

$$\inf \operatorname{supp} f * g = \inf \operatorname{supp} f + \inf \operatorname{supp} g \tag{40}$$

does not hold for Laplace hyperfunctions.

Counterexample. Pólya [26] and Cartwright [9] have shown the following. Let x_n be a sequence of non-zero real numbers such that the number $n(r)$ of x_n with $|x_n| \le r$ is of order $o(r)$ as $r \to \infty$ and $\sum_{n=1}^{N} 1/x_n$ is bounded. Then the entire function

$$\hat{f}(z) = \prod_{n=1}^{\infty} \left(1 - \frac{z}{x_n}\right) e^{z/x_n} \tag{41}$$

is of exponential type and we have

$$\varlimsup_{r\to\infty} \frac{\log|\hat{f}(r\,e^{i\theta})|}{r} = \varlimsup_{r\to\infty} S(r)\cos\theta, \qquad 0 \le \theta < 2\pi, \tag{42}$$

where

$$S(r) = \sum_{|x_n| \le r} \frac{1}{x_n}. \tag{43}$$

Given $a < 0 < b$ we can find such a sequence x_n for which $a = -\overline{\lim}_{r \to \infty} S(r)$ and $b = \underline{\lim}_{r \to \infty} S(r)$. Then $\hat{f}$ is the Laplace transform of a (Laplace) hyperfunction f with support containing $\{a, b\}$.

On the other hand, it is easy to see that

$$\hat{f}(z)\hat{f}(-z) = \prod_{n=1}^{\infty} \left(1 - \left(\frac{z}{x_n}\right)^2\right) \tag{44}$$

satisfies

$$\overline{\lim_{r \to \infty}} \frac{\log|\hat{f}(r\,e^{i\theta})\hat{f}(-r\,e^{i\theta})|}{r} = 0, \qquad 0 \le \theta < 2\pi. \tag{45}$$

Hence $f * \check{f}$ has support only at 0, where $\check{f}(x) = f(-x)$.

Definition 4. Let $f \in \mathscr{B}_{[a,a+c)}$ and $g \in \mathscr{B}_{[b,b+c)}$, with $0 < c \le \infty$. We define the *convolution* $f * g \in \mathscr{B}_{[a+b,a+b+c)}$ by the restriction to $(-\infty, a+b+c)$ of the convolution $\tilde{f} * \tilde{g}$ of extensions $\tilde{f} \in \mathscr{B}_{[a,\infty]}^{\exp}$ of f and $\tilde{g} \in \mathscr{B}_{[b,\infty]}^{\exp}$ of g. It is easy to see that $f * g$ does not depend on the extensions $\tilde{f}$ and $\tilde{g}$. In other words, the convolution is the bilinear mapping

$$\mathscr{B}_{[a,a+c)} \times \mathscr{B}_{[b,b+c)} \to \mathscr{B}_{[a+b,a+b+c)} \tag{46}$$

induced from the convolution of Laplace hyperfunctions under the representation (31).

$\mathscr{B}_{[0,c)}$ forms a commutative algebra under addition and convolution. If $c < \infty$, then it has divisors of zero, as the above counterexample shows. We do not know whether or not the convolution algebras $\mathscr{B}_{[0,\infty)}$ and

$$\mathscr{B}_+ = \bigcup_{a=-\infty}^{\infty} \mathscr{B}_{[a,\infty)} \tag{47}$$

have divisors of zero, although these algebras are the most natural frameworks for operational calculus.

Definition 5. A (Laplace) hyperfunction $f \in \mathscr{B}_+$ (resp. $\mathscr{B}_+^{\exp}$) is said to be *regular* if (40) holds for any $g \in \mathscr{B}_+$ (resp. $\mathscr{B}_+^{\exp}$).

Theorem 3. *A (Laplace) hyperfunction $f \in \mathscr{B}_+$ ($\mathscr{B}_+^{exp}$) is regular if one of the following conditions holds:*

(1) *The support $\operatorname{supp} f$ is one point.*

(2) *The support $\operatorname{supp} f$ is a compact set in $\mathbf{R}$ and the Laplace transform $\hat{f}(\lambda) = \hat{f}(\mu + i\nu)$ satisfies the condition*

$$\int_{-\infty}^{\infty} \frac{\log^+ |\hat{f}(i\nu)|}{1 + \nu^2} \, d\nu < \infty. \tag{48}$$

(3) *f is decomposed as*

$$f = f_0 + f_1, \tag{49}$$

where f_0 is regular and $\inf \operatorname{supp} f_0 < \inf \operatorname{supp} f_1$.

Proof. Suppose that $\inf \operatorname{supp} f = a$ and $\inf \operatorname{supp} g = b$. The minimum modulus theorem ([5], p. 51) in case (1) and the Ahlfors–Heins theorem ([5], p. 116) in case (2) assert that

$$\lim_{r \to \infty} \frac{\log |\hat{f}(r e^{i\theta})|}{r} = -a \cos \theta, \qquad |\theta| < \frac{\pi}{2}, \tag{50}$$

with few exceptions of θ and r. On the other hand, V. Bernstein's theorem implies that for any $\varepsilon > 0$

$$r^{-1} \log |\hat{g}(r e^{i\theta})| > -b \cos \theta - \varepsilon \tag{51}$$

for a set of positive upper linear density on each ray $\mathbf{R}_+ e^{i\theta}$ ([5], p. 190). Hence we have

$$\varlimsup_{r \to \infty} \frac{\log |\hat{f}\hat{g}(r e^{i\theta})|}{r} = -(a + b) \cos \theta.$$

The proof in case (3) is trivial.

Definition 6. A (Laplace) hyperfunction $f \in \mathscr{B}_+$ ($\mathscr{B}_+^{exp}$) is said to be *invertible* if it has the inverse $g \in \mathscr{B}_+$ ($\mathscr{B}_+^{exp}$) such that

$$f * g = \delta. \tag{52}$$

f is said to be *regularly invertible* if it has the inverse g such that

$$\inf \operatorname{supp} f + \inf \operatorname{supp} g = 0. \tag{53}$$

The invertibility of f is equivalent to the unique solvability of the equation

$$f * u = k$$

for any k. f is regularly invertible if and only if it is regular and invertible.

The invertibility of $f \in \mathcal{B}_+^{\mathrm{exp}}$ is easy to check; it has the inverse g if and only if the reciprocal $\hat{g}(\lambda) = 1/\hat{f}(\lambda)$ of the Laplace transform satisfies the conditions of Theorem 2.

In particular, if $\inf \operatorname{supp} f_1 > 0$, then $\delta - f_1$ is regularly invertible and $\sum_{n=0}^{\infty} \hat{f}_1(\lambda)^n$ is the Laplace transform of the inverse.

Obviously a necessary condition for invertibility of $f \in \mathcal{B}_+^{\mathrm{exp}}$ is that

$$\hat{f}(\lambda) \neq 0 \text{ in a neighborhood of } S_{-\pi/2,\pi/2}\infty \text{ in } \mathbf{O}. \tag{54}$$

Theorem 4. *A Laplace hyperfunction $f \in \mathcal{B}_+^{\mathrm{exp}}$ is regularly invertible if one of the following conditions holds*:

(1) *f satisfies (54) and $\operatorname{supp} f$ is one point.*
(2) *f satisfies (54) and condition (2) of Theorem 3.*
(3) *f is decomposed as (49), where f_0 is regularly invertible and $\inf \operatorname{supp} f_0 < \inf \operatorname{supp} f_1$.*

Proof. In cases (1) and (2) (50) holds without exceptions so that $\hat{g}(\lambda) = 1/\hat{f}(\lambda)$ belongs to $\mathcal{LB}_{[-a,\infty]}^{\mathrm{exp}}$.

In case (3) let g_0 be the inverse of f_0. Then $g_0 * f = \delta + g_0 * f_1$ is regularly invertible because $\inf \operatorname{supp} g_0 * f_1 > 0$. Hence $(\delta + g_0 * f_1)^{*-1} * g_0$ is the regular inverse of f.

Theorem 5. *A hyperfunction $f \in \mathcal{B}_+$ is (regularly) invertible if and only if so is any extension $\tilde{f} \in \mathcal{B}_+^{\mathrm{exp}}$ of f.*

Proof. If an extension $\tilde{f}$ has the (regular) inverse $\tilde{g}$, then $g = \rho(\tilde{g})$ is the (regular) inverse of f.

If f has the (regular) inverse g, then $r = \delta - \tilde{f} * \tilde{g}$ is in $\mathcal{B}_{\{\infty\}}^{\mathrm{exp}}$ for any extensions $\tilde{f}$ of f and $\tilde{g}$ of g, so that $\tilde{f}$ has the (regular) inverse $\tilde{g} * (\delta - r)^{*-1}$.

Kawai [13] has shown that a hyperfunction f with one point support is invertible in $\mathcal{B}_+$ if and only if it satisfies condition (54).

We do not know any examples which are invertible and not regular.

4. Ultradistributions as Elements in $\mathcal{B}_+ \cap \mathcal{M}$

The ultradistributions $\mathscr{D}^{*\prime}(\mathbf{R})$ are defined to be the continuous linear functionals on a non-quasi-analytic class $\mathscr{D}^*(\mathbf{R})$ of functions with compact support (see e.g. [15]). They are canonically embedded in the hyperfunctions $\mathscr{B}(\mathbf{R})$.

The Paley-Wiener theorem [16] says that the Laplace transform $\hat{f}(\mu + i\nu)$ of an ultradistribution $f \in \mathscr{D}^{*\prime}_{[a,b]}$ with compact support satisfies

$$|\hat{f}(i\nu)| \leq C \exp M(L\nu), \qquad \nu \in \mathbf{R}, \tag{55}$$

for a function $M(t) \geq 0$ such that

$$\int_{-\infty}^{\infty} \frac{M(\nu)}{1 + \nu^2} \, d\nu < \infty. \tag{56}$$

Hence f satisfies condition (48) of Theorems 3 and 4. Therefore ultra-distributions

$$\mathscr{D}^{*\prime}_+ = \bigcup_{a=-\infty}^{\infty} \mathscr{D}^{*\prime}_{[a,\infty)} \tag{57}$$

with support bounded from below are all regular in $\mathscr{B}_+$. In particular, the continuous functions $C([0, \infty))$ and

$$C_+ = \bigcup_{a=-\infty}^{\infty} C_{[a,\infty)} \tag{58}$$

are all regular, and hence are not divisors of zero in $\mathscr{B}_+$.

We may look upon the Mikusiński operators $\mathscr{M}$ also as the quotient field C_+/C_+. Thus $\mathscr{M}$ and the hyperfunctions $\mathscr{B}_+$ with support bounded from below are both embedded in the quotient algebra $\mathscr{B}_+/C_+$. The following theorem says that neither of them is included in the other.

Theorem 6.
(1) *All $f \in \mathscr{B}_+ \cap \mathscr{M}$ are regular in $\mathscr{B}_+$.*
(2) *If $f \in \mathscr{B}_+ \cap \mathscr{M}$ is not invertible in $\mathscr{B}_+$, then $f^{*-1} \in \mathscr{M}$ is not in $\mathscr{B}_+$.*

Proof. (1) Suppose that an $f \in \mathscr{B}_+$ is written $f = \varphi/\psi$ in $\mathscr{B}_+/C_+$ with $\varphi, \psi \in C_+$. Then we have for any $g \in \mathscr{B}_+$

$$\psi * f * g = \varphi * g.$$

Computing the inf supp of both sides in two ways, we obtain the regularity

$$\inf \operatorname{supp} f * g = \inf \operatorname{supp} f + \inf \operatorname{supp} g.$$

The proof of (2) is trivial.

The ultradistributions $\mathscr{D}^{*\prime}_+$ with support bounded from below are all included in the intersection $\mathscr{B}_+ \cap \mathscr{M}$ for the same reason as for the distributions $\mathscr{D}_+$. The following theorem gives a weak form of converse.

Theorem 7. *Let $f \in \mathcal{B}_{[a,b]}$ be a hyperfunction with compact support. If there is a non-zero continuous function φ with compact support such that $\varphi * f$ is continuous (or an ultradistribution), then the Laplace transform $\hat{f}(\mu + i\nu)$ satisfies condition (48).*

Conversely, if $\hat{f}(\mu + i\nu)$ satisfies condition (48), then for any $c > 0$ and non-quasi analytic class $\mathcal{D}^*$ there is a non-zero function $\varphi \in \mathcal{D}^*(\mathbf{R})$ with support in $[-c, c]$ such that $\varphi * f$ is in $\mathcal{D}^*(\mathbf{R})$.

Proof. Krein [20] has proved that an entire function $\hat{f}(\lambda)$ is of Nevanlinna class both in the half planes Re $\lambda > 0$ and Re $\lambda < 0$ if and only if it is of exponential type and satisfies (48). If $\hat{\varphi}(\lambda)$ and $\hat{\varphi}(\lambda)\hat{f}(\lambda)$ are of Nevanlinna class, as the Paley–Wiener theorem shows, then so is the quotient $\hat{f}(\lambda)$.

The converse is a slight modification of Theorem II of Beurling–Malliavin [4].

Burzyk [8] has shown that the conditions of the theorem also characterize the regular operators with compact support in the sense of Boehme [6], [23].

It is desirable to develop a general theory of ultradistributions based only on the estimate (48) independent of the choice of a non-quasi-analytic class $\mathcal{D}^*$.

5. The Vector-Valued Case

The theory of Laplace hyperfunctions can immediately be generalized to the case where Laplace hyperfunctions have values in a complex Banach space E.

If we apply it to the evolution equation

$$\begin{cases} \left(\dfrac{\partial}{\partial t} - A\right) u(t, x) = f(t, x), \\[2mm] u(0, x) = g(x), \end{cases} \tag{59}$$

where A is a closed linear operator in a Banach space F, then we find that the resolvent $q(t) \in \mathcal{B}_{[0,\infty]}^{\exp}(L(F, D(A)))$ (or $\mathcal{B}_{[0,\infty)}(L(F, D(A)))$) exists if and only if the resolvent $(\lambda - A)^{-1}$ in F exists in a neighborhood of $S_{-\pi/2,\pi/2}\infty$ in $\mathbf{O}$ and satisfies

$$\varlimsup_{r \to \infty} \frac{\log \|(r e^{i\theta} - A)^{-1}\|}{r} \le 0, \qquad |\theta| < \frac{\pi}{2}. \tag{60}$$

This is exactly Ōuchi's characterization of the generators of hyperfunction semi-groups [25].

The same method works for the second order equation

$$
\begin{cases}
\left(\dfrac{\partial^2}{\partial t^2} - A \right) u(t, x) = f(t, x), \\
u^{(i)}(0, x) = g_i(x), \qquad i = 0, 1.
\end{cases} \tag{61}
$$

If the resolvent $(\lambda - A)^{-1}$ in F exists in a neighborhood of $S_{-\pi,\pi}\infty$ in $\mathbf{O}$ and satisfies

$$
\varlimsup_{r \to \infty} \frac{\log \|(r^2 \, e^{i\theta} - A)^{-1}\|}{r} \leq 0, \qquad |\theta| < \pi, \tag{62}
$$

then a resolvent $q(t) \in \mathcal{B}_{[0,\infty]}^{\exp}(L(F, D(A)))$ exists.

Agmon [1] has shown that an even-order (resp. second-order) elliptic differential operator A with a coercive boundary condition satisfies (60) (resp. (61)) in $L^p(\Omega)$, $1 < p < \infty$, if its principal symbol satisfies

$$
\operatorname{Re} \sigma(A)(x, \xi) \leq 0 \tag{63}
$$

$$
(\text{resp. } \sigma(A)(x, \xi) \leq 0).
$$

References

[1] S. Agmon, On the eigenfunctions and on the eigenvalues of general elliptic boundary value problems, *Comm. Pure Appl. Math.* **15** (1962) 119–147.

[2] E. J. Berg, *Heaviside's Operational Calculus as Applied to Engineering and Physics*, McGraw-Hill, New York–London, 1936.

[3] L. Berg, Asymptotische Auffasung der Operatorenrechnung, *Studia Math.* **21** (1962) 215–229.

[4] A. Beurling and P. Malliavin, On the Fourier transforms of measures with compact support, *Acta Math.* **107** (1962) 291–309.

[5] R. P. Boas, Jr., *Entire Functions*, Academic Press, New York, 1954.

[6] T. K. Boehme, The support of Mikusiński operators, *Trans. Amer. Math. Soc.* **176** (1973) 319–334.

[7] E. Borel, *Leçons sur les Séries Divergentes*, 2e éd., Gauthier-Villars, Paris, 1928.

[8] J. Burzyk, A Paley–Wiener type theorem for operators and its applications, *Proc. Conf. on Generalized Functions, Convergence Structures and their Applications*, to appear.

[9] M. L. Cartwright, On integral functions of integral order, *Proc. London Math. Soc.* **33** (1931) 209–224.

[10] D. A. Ditkin, On the theory of operational calculus, *Doklady Akad. Nauk SSSR* **123** (1958) 395–396 (in Russian).

[11] G. Doetsch, *Theorie und Anwendung der Laplace-transformation*, Springer, Berlin, 1937.

[12] A. Kaneko, *An Introduction to Hyperfunctions*, Vol. 2, University of Tokyo Press, Tokyo, 1982 (in Japanese).

[13] T. Kawai, On the theory of Fourier hyperfunctions and its applications to partial differential equations with constant coefficients, *J. Fac. Sci., Univ. Tokyo, Sec. IA* **17** (1970) 467-517.

[14] H. Komatsu, An introduction to the theory of hyperfunctions, *Lecture Notes in Math.* **287** (1973) 3-40.

[15] H. Komatsu, Ultradistributions I, Structure theorems and a characterization, *J. Fac. Sci., Univ. Tokyo, Sec. IA* **20** (1973) 25-105.

[16] H. Komatsu, Ultradistributions II, The kernel theorem and ultradistributions with support in a submanifold, *J. Fac. Sci., Univ. Tokyo, Sec. IA* **24** (1977) 607-628.

[17] H. Komatsu, Ultradistributions III, Vector valued ultradistributions and the theory of kernels, *J. Fac. Sci., Univ. Tokyo, Sec. IA* **29** (1982) 653-718.

[18] H. Komatsu, Laplace transforms of hyperfunctions—A new foundation of the Heaviside calculus, *J. Fac. Sci., Univ. Tokyo, Sec. IA* **34** (1987) 805-820.

[19] H. Komatsu, Laplace transforms of hyperfunctions—Another foundation of the Heaviside operational calculus, *Proc. Conf. on Generalized Functions, Convergence Structures and their Applications*, to appear.

[20] M. G. Krein, A contribution to the theory of entire functions of exponential type, *Izvestiya Akad. Nauk SSSR* **11** (1947) 309-326 (in Russian).

[21] A. J. Macintyre, Laplace's transformation and integral functions, *Proc. London Math. Soc.* (2) **45** (1938) 1-20.

[22] J. Mikusiński, *Rachunek Operatorow*, Warszawa, 1953.

[23] J. Mikusiński and T. K. Boehme, *Operational Calculus*, 2nd Ed. Vol. II, Pergamon Press, Oxford, PWN, Warszawa, 1987.

[24] M. Morimoto, Analytic functionals with non-compact carrier, *Tokyo J. Math.* **1** (1978) 77-103.

[25] S. Ōuchi, On abstract Cauchy problem in the sense of hyperfunctions, *Lecture Notes in Math.* **287** (1973), 135-152.

[26] G. Pólya, Untersuchungen über Lücken und Singularitäten von Potenzreihen, *Math. Z.* **29** (1929) 549-640.

[27] Y. Saburi, Fundamental properties of modified Fourier hyperfunctions, *Tokyo J. Math.* **8** (1985) 231-273.

[28] M. Sato, Theory of hyperfunctions, *Sūgaku* **10** (1958) 1-27 (in Japanese).

[29] M. Sato, Theory of hyperfunctions, I, *J. Fac. Sci., Univ. Tokyo, Sec. I* **8** (1959) 139-193.

[30] L. Schwartz, Transformation de Laplace des distributions, *Medd. Lunds Math. Sem.* Suppl. (1952) 196-206.

[31] J. Sebastião e Silva, Les fonctions analytiques comme ultra-distributions dans le calcul opérationnel, *Math. Ann.* **136** (1958) 58-96.

[32] H. Triebel, *Interpolation Theory, Function Spaces, Differential Operators*, North-Holland, Amsterdam-New York-Oxford, 1978.

[33] K. Yosida, *Operational Calculus, A Theory of Hyperfunctions*, Springer, New York-Berlin-Heidelberg-Tokyo, 1984.

On a Conjectural Equation of Certain Kinds of Surfaces

Michio Kuga[*]

Department of Mathematics
State University of New York at Stony Brook
Stony Brook, New York

Let $\mathfrak{h} = \{z \in \mathbf{C} \mid z = x + iy, y > 0\}$ be the upper half plane; k be a total real number field with $[k : \mathbf{Q}] = m < \infty$; B be a quaternion algebra whose center is k. Let S be the set of all ∞-places of k; S_0 be the subset of all those places of k which split in B/k. Taking a maximal order $\mathfrak{O}$ of B, we put

$$\Gamma(1) = \Gamma(1 : \mathfrak{O}) = \{\gamma \in \mathfrak{O}^\times, \nu(\gamma) = 1\},$$

where $\nu : B^\times \to k^\times$ is the reduced norm map of B.

There is a Q-linear algebraic group G which satisfies $G(Q) \cong B_1^\times = \{g \in B^\times \mid \nu(g) = 1\}$. For that G, $G(\mathbf{R}) = SL(2, \mathbf{R})^{|S_0|} \times SU(2)^{|S-S_0|}$. So, identifying $B_1^\times$ with $G(Q)$, through $G(Q) \subset Q(R)$, via the projection of $G(\mathbf{R})$ to the first factor $SL(2, \mathbf{R})^{|S_0|}$, the group $B_1^\times$ acts on the product of the upper half planes $\mathfrak{h}^{|S_0|}$. Therefore the subgroup $\Gamma(1)$ operates on $\mathfrak{h}^{|S_0|}$, which is properly discontinuous; and we have a projective algebraic variety $U(1) = \Gamma(1) \backslash \mathfrak{h}^{|S_0|}$. In the following we assume that B is a division algebra and that $|S_0| = 2$. Then $U(1)$ is a compact algebraic surface. If Γ is a discrete subgroup

[*] This investigation has been supported by the Paul and Gabriella Rosenbaum Foundation.

in $SL(2, \mathbf{R})^{|S_0|}/\pm 1$, commensurable with $\Gamma(1)/\pm 1$ without the torsion element, $U = \Gamma \backslash \mathfrak{h}^2$ is a compact smooth projective algebraic surface. The Hodge diamond of U was determined by Matsushima and Shimura, and

$$\begin{pmatrix} & & h^{(2,2)} & & \\ & h^{(2,1)} & & h^{(1,2)} & \\ h^{(2,0)} & & h^{(1,1)} & & h^{(0,2)} \\ & h^{(1,0)} & & h^{(0,1)} & \\ & & h^{(0,0)} & & \end{pmatrix} = \begin{pmatrix} & & 1 & & \\ & 0 & & 0 & \\ p_g & & 2+p_g & & p_g \\ & 0 & & 0 & \\ & & 1 & & \end{pmatrix},$$

where p_g is the geometric genus of U. Therefore the arithmetic genus p_a of U, and the Euler characteristic $E(U)$ of U are

$$p_a = p_g + 1$$

$$E(U) = 4p_g + 4 = 4p_a.$$

On the other hand, the Euler characteristic E is computable by the Gauss–Bonnet formula, and we have

$$E = \frac{1}{4\pi^2} v(U),$$

where $v(U)$ is the total (4-dimensional) volume of $U = \Gamma \backslash \mathfrak{h}^2$ by the invariant measure

$$\omega_1 \wedge \omega_2 = \frac{dx_1 \wedge dy_1}{y_1^2} \wedge \frac{dx_2 \wedge dy_2}{y_2^2},$$

with $z_1 = x_1 + iy_1$, $z_2 = x_2 + iy_2$ being variables of $\mathfrak{h}^2$. And this $v(U)$ is determinable by the residue of the ζ-function $\zeta_B(s)$ of the quaternion algebra B (Shimizu's formula). Also the $\zeta_B(s)$ is expressible through the $\zeta_k(s)$ by Hey's formula, and finally we know the residue of $\zeta_k(s)$. Lastly we have

$$E(U) = \mu E(U(1)) = \frac{\mu^{2D^{3/2} \zeta_k(2)}}{2^{m-2} \pi^{2m}} \prod_{\mathfrak{p} | \mathfrak{D}_B} (N\mathfrak{p} - 1)$$

(Shimizu's formula), where

$$\mu = \frac{[\Gamma(1)/\pm 1 : \Gamma(1)/(\pm 1) \cap \Gamma]}{[\Gamma : (\Gamma(1)/(\pm 1) \cap \Gamma]},$$

D_k = the discriminant of k,

$\zeta_k(2)$ = the value of ζ-function at $s = 2$ of k,

$\mathfrak{D}_B$ = the product of the finite places of k which do not split in B/k,

$N\mathfrak{p} = N_{k/Q}\mathfrak{p} = [\mathfrak{D}_k : \mathfrak{p}]$,

$\mathfrak{D}_k$ = the integers of k.

The values of $\zeta_k(s)$ have been investigated by Siegel and by Shintani, etc. When k/Q is abelian, since then $\zeta_k(s)$ is expressible by the Dirichlet's L-function, and for the total real field the values of such functions at $s = 2$ are computable in finite form, the value of $E(U)$ is given finally as

$$E(U) = \mu \frac{1}{6} \frac{1}{2^{m-2}} \left| \prod_{x=1} B_x \right| \prod_{\mathfrak{p}|\mathfrak{D}_B} (N\mathfrak{p} - 1)$$

where

x are Dirichlet's characters corresponding to k,

$B_x = \left(\sum_{\mu=1}^{f_x - 1} \mu^2 x(\mu) \right) \frac{1}{f_x}$, the generalized Bernoulli's sum, $B_{x,2}$,

f_x = the conductor of x.

In particular for a real quadratic field k, we have

$$E(U) = \mu \tfrac{1}{6} |B_x| \prod_{\mathfrak{p}|\mathfrak{D}_B} (N\mathfrak{p} - 1).$$

For various k, various B, the values of $p_a(U(1))$, $E(U(1))$ are usually rather large, but in very rare instances they have small values. In fact Ira Shaval and Kisao Takeuchi and his students studied surfaces of this nature with the geometric genus $p_g \leq 1$.

Let $k = Q(\sqrt{d})$, ($d > 0$ the discriminant) be a real quadratic field; let B/k be a totally indefinite division quaternion algebra over k with discriminant $\mathfrak{D}_B = \mathfrak{p}_1 \mathfrak{p}_2 \cdots \mathfrak{p}_{2v}$. A surface $U(1) = \Gamma(1)\backslash\mathfrak{h}^2$, the quotient of $\mathfrak{h}^2$ by $\Gamma(1) = \{\gamma \in \mathfrak{D}^\times$, reduced norm of $\gamma = 1\}$, where $\mathfrak{D}$ is the maximal order of B, is also denoted by $U(d, \mathfrak{D})$ or by $U(d, \mathfrak{p}_1, \mathfrak{p}_2, \ldots, \mathfrak{p}_{2v})$. For a rational prime p, the prime ideal $\mathfrak{p}$ in $k = Q(\sqrt{d})$ which divides (p) is denoted by $\mathfrak{p}_p$. If p ramifies in k/Q, $\mathfrak{p}_p^2 = p$; if p is inert in k/Q $\mathfrak{p}_p' = p$; if p decomposes in k/Q, one prime ideal divisor of p is denoted by $\mathfrak{p}_p$, and another by $\mathfrak{p}_p'$, i.e. $\mathfrak{p}_p \mathfrak{p}_p' = p$.

For example $U(8, \mathfrak{p}_2, \mathfrak{p}_5)$, $U(12, \mathfrak{p}_2, \mathfrak{p}_{13})$, $U(12, \mathfrak{p}_2, \mathfrak{p}_{13}')$ are smooth surfaces with $p_g = 0$; and $U(8, \mathfrak{p}_3, \mathfrak{p}_7)$ is a smooth surface with $p_g = 1$. $U(17, \mathfrak{p}_2, \mathfrak{p}_2')$ is not a smooth surface, but it has a smooth covering of order 6 with $p_g = 0$ and a smooth covering of order 12 with $p_g = 1$. If $[k:Q] > 2$ there appears a tendency of the geometric genus of $U(1)$ to become bigger and bigger. By using an estimate of discriminants by Odlyzko, to have all of the surfaces of this type with $p_g = 0$, we need only to check all of the fields k with $[k : \mathbf{Q}] \leq 9$. Although there is no rigorous proof of it, we can conjecture that, when the value of p_g is given, the total number of such surfaces with given p_g will be finite.

For the smoothness of U, i.e., the existence (or not) of the torsion element in Γ, is almost determinable by the existence of subcyclotomic fields in B. Actually, Shaval and Takeuchi used that method. Here we discuss surfaces of this kind with $p_g = 0$. If $p_g = 0$ then $p_a = 1$, $E(U) = c_2 = 4p_a = 4$. Hence by the Riemann-Roch theorem,

$$p_a = \frac{c_1^2 + c_2}{12},$$

and therefore $c_1^2 = 12 - 4 = 8$. We will write the space of automorphic forms of the dimension m by S_m, i.e.,

$$S_m = \{\psi \text{ holomorphic on } \mathfrak{h}^2 \,|\, \psi(\gamma z) = \psi(z)(c_1 z_1 + d_1)^m (c_2 z_2 + d_2)^m\}$$

where

$$\gamma = \left(\begin{pmatrix} a_1 & b_1 \\ c_1 & d_1 \end{pmatrix} \begin{pmatrix} a_2 & b_2 \\ c_2 & d_2 \end{pmatrix} \right) \Big/ \pm 1 \in SL(2, \mathbf{R})^2 / \pm 1,$$

$$z = (z_1 z_2) \in \mathfrak{h}^2.$$

Then S_m defines a linear system (S_m), and by the Riemann-Roch theorem,

$$\dim(S_m) = \dim S_m = (m - 1)^2, \qquad \text{for } m > 3.$$

In particular,

$$\dim S_3 = 4,$$

$$\dim S_4 = 9,$$

$$\dim S_{10} = 81.$$

Let

$$S_3 = \{\psi_0, \psi_1, \psi_2, \psi_3\}$$

$$S_4 = \{\phi_0, \phi_1, \phi_2, \phi_4, \phi_5, \phi_6, \phi_7, \phi_8\}.$$

Our aim is to determine a (conjectural) equation of the image of U in the projective 3-space $P^3(\mathbf{C})$, by this linear system (S_3). Namely, the possible equation of $f(U)$ of

$$U \overset{f}{\to} P^3(\mathbf{C})$$
$$\underset{\uplus}{}$$
$$P \mapsto (\psi_0(P), \psi_1(P), \psi_2(P), \psi_3(P))$$

should be determined. More precisely, let

$$C = \{P \in U \,|\, \psi(P) = 0, \forall \psi \in S_3\}.$$

C might be a finite set or a curve in U, but can never be $C = U$. And the above mapping actually is defined at

$$U - C \to P^3(\mathbf{C})$$

$$P \to (\psi_0(P), \psi_1(P), \psi_2(P), \psi_3(P)).$$

The image of this map is denoted by $f(U)$. Since $(S_3) = (3/2)K$, and $(S_3)^2 = (3/2)K^2 = (9/4)K^2 = (9/4)8 = 18$, $f(U)$ satisfies an equation of order 18. We wish to find it. Let us consider the monomials

$$\psi_i \psi_j \phi_k \qquad (i, j = 0 \sim 3, \ k = 0 \sim 8).$$

There are 90 such monomials; these all belong to the 81-dimensional S_{10}. So there are 9 linearly independent relations

$$\sum_{ij, k} a_{i,j,k}^{(v)} \psi_i \psi_j \phi_k = 0 \qquad (v = 1 \sim 9).$$

Or more precisely, there are 9 linearly independent numerical vectors

$$\mathfrak{A}^{(v)} = (\cdots a_{ij,k}^{(v)} \cdots)_{i,j=0 \sim 4, k=0 \sim 9} \in \mathbf{C}^{90}, \qquad (v = 1 \sim 9)$$

such that

$$\sum_{i,j=0}^{4} \sum_{k=0}^{8} a_{i,j,k}^{(v)} \psi_i \psi_k \phi_k = 0 \qquad (v = 1 \sim 9). \tag{1}$$

If we put here the quadratic form

$$\sum_{i,j=0}^{4} a_{i,j,k}^{(v)} X_i X_j \tag{2}$$

of 4 variables by $Q_k^{(v)}(X) = Q_k^{(v)}(X_0 X_1 X_2 X_3)$, then (1) will be written as

$$\sum_{k=0}^{8} Q_k^{(v)}(\psi_0, \psi_1, \psi_2, \psi_3)\phi_k = 0, \qquad (v = 1 \sim 9). \tag{3}$$

At a generic point of U $(\phi_0(P), \ldots, \phi_8(P)) \neq 0$, so (3) implies

$$\det(Q_k^{(v)}(\psi_0, \psi_1, \psi_2, \psi_3)) = 0. \tag{4}$$

(4) means $f(U)$ satisfies the equation

$$\det(Q_k^{(v)}(X)) = 0 \tag{5}$$

of degree 18 (or constant 0). There is no proof however that (5) is not constant 0. That might need more precise forms $Q_k^{(v)}$. For each one of those examples of surfaces with $p_g = 0$ above, we would like to find out those $Q_k^{(v)}(X)$. But for that purpose, we might need to construct explicitly third- and fourth-dimensional automorphic forms. We cannot suggest how this might be done.

For surfaces with $p_g = 1$, we have no candidates for these equations; but we can look at similar considerations (Satake and the author's for $K3$ surfaces). Of course we have no interpretation of that for the moment. Consider a real 6-dimensional $H^2(U, \mathbf{R})$ with the quadratic form f of cup product.

We put $\mathfrak{A}(V, \mathbf{Q})$ the $\mathbf{Q}$-subspace of $H^2(U, \mathbf{R})$ spanned by all algebraic cycles (divisors). (We identified H^2 with H_2 through Poincaré duality.) The two forms

$$\omega_1 = \frac{-1}{2\pi} \frac{dx_1 \wedge dy_1}{y_1^2}, \qquad \omega_2 = \frac{-1}{2\pi} \frac{dx_2 \wedge dy_2}{y_1^2}$$

belong to algebraic cycles. We take an even-dimensional subspace $\not{h}$ of $H^2(U, \mathbf{Q})$ containing ω_1, and ω_2. $\not{h}$ is either $\mathbf{Q}\omega_1 + \mathbf{Q}\omega_2$ or $\mathfrak{A}'(V, \mathbf{Q})$ which is 4-dimensional.

We put

$$F = \not{h}^{\perp f}$$

the orthogonal complement of $\not{h}$ with respect to f, and we consider the quadratic form space $(F, f|F)$, where $f|F$ is the restriction of f on F. Consider the odd part $C^-(F, f|F)$ of the Clifford algebraic $C(F, f|F)$.

The unique complex structure J on $C^-(F, f|F) \otimes \mathbf{R}$ with the following properties (1), (2) makes this $C^-(F, f|F) \otimes \mathbf{R}$ a complex linear space E.

(1) J coincides with the left multiplication of some elements of C^+ on C^-

(2)   The projection with natural isomorphism

$$\mathbf{C} \cong H^{(2,0)}(U) \to (H^{(2,0)} + H^{(0,2)}) \cap H^2(U, R) \subset C^-(F, f \,|\, F) \otimes \mathbf{R}$$
$$\|$$
$$E$$

defined by $\omega \mapsto (\omega + \bar{\omega})/2$ is **C**-linear.

E is a 4- or 2-dimensional **C**-linear space according to $\mu = \mathbf{Q}\omega_1 + \mathbf{Q}_{v2}$ or $\mu = \mathfrak{A}'(V) = 4$-dimensional.

Let L denote the lattice

$$H^2(U, Z) \cap F.$$

Then, E/L is the wanted complex torus. This is actually an abelian variety. If $\dim_\mathbf{Q} \mathfrak{A}'(V, \mathbf{Q}) \geq \beta$ (this value is always ≥ 2), E/L is isogenous to the product of elliptic curves (even μ is taken as $\mathbf{Q}\omega_1 + \mathbf{Q}\omega_2$). If $\dim \mathfrak{A}'(V, \mathbf{Q}) = 4$ (maximal possible value), each of these elliptic curves is isogenous to the elliptic curve $H^{(2,0)}(U)/\text{projection of } H^2(U, \mathbf{Z})$ (constructed by Shioda); therefore it is expressed as $S_2/\text{lattice}$. From that we can see that the endomorphism ring contains the Hecke ring (the trace of which Shimizu calculated), and therefore the elliptic curve has a complex multiplication.

A mapping

$$C = \Delta\backslash\mathfrak{h} \xrightarrow{\varepsilon} U = \Gamma\backslash\mathfrak{h} \times \mathfrak{h}$$

of a curve C, obtained from $\mathfrak{h}$ divided by a Fuchsian group Δ to the surface U, is called an Eichler-map, if it is induced from

$$\mathfrak{h} \to \mathfrak{h} \times \mathfrak{h}$$

$$z \to (\psi(z), \phi(z)),$$

for a pair ψ, ϕ of fractional linear transformations. The image $\varepsilon(C)$ of an Eichler-map is called an Eichler curve. If U has a smooth Eichler curve $\varepsilon(C)$, we can show that $\omega_1, \omega_2, E(C)$ are linearly independent. But no such examples have been found.

I would like to dedicate this paper to the advent of Doctor Mikio Sato's 60th birthday. In this paper I investigated the surfaces which Dr. Sato and I discussed when Dr. Sato visited me at Stony Brook.

Vanishing Cycles and Second Microlocalization

Yves Laurent

Université de Paris-Sud
Département de Mathématiques
Orsay Cedex
France

The vanishing cycles of a holonomic $\mathscr{D}$-module on a hypersurface is a holonomic $\mathscr{D}$-module which has been defined by Malgrange [11] and Kashiwara [5]. When the $\mathscr{D}$-module is regular holonomic this corresponds to the classical vanishing cycles of Deligne via the Riemann–Hilbert transform. Here we give a new construction of the vanishing cycles of a holonomic $\mathscr{D}$-module using the technique of second microlocalization.

More precisely, given a complex manifold X and a smooth hypersurface Y in it, we define a sheaf of rings $\hat{\mathscr{D}}_{\Lambda}^{2}$ on the conormal bundle $\Lambda = T_{Y}^{*}X$ and we prove that if $\mathcal{M}$ is a holonomic $\mathscr{D}_{X}$-module (or more generally if $\mathcal{M}$ is a coherent $\mathscr{D}_{X}$-module with a b-function), then the module $\hat{\mathcal{M}}^{2}$ obtained from $\mathcal{M}$ by extension of the scalars to $\hat{\mathscr{D}}_{\Lambda}^{2}$ induces on Y a coherent $\mathscr{D}_{Y}$-module $\Phi(\mathcal{M})$ which is isomorphic to the vanishing cycles of $\mathcal{M}$.

This method gives a functorial construction which extends directly to the complexes of $\mathscr{D}$-modules and to the derived category. Moreover the same construction gives the microlocalized module of $\Phi(\mathcal{M})$.

Algebraic Analysis, Volume I

1. Formal Microdifferential Operators

If Y is a submanifold of a complex manifold X and $\mathscr{F}$ a sheaf on X we will denote by $\mu_Y(\mathscr{F})$ the microlocalization of $\mathscr{F}$ along Y (see [13], [7], [8] for the definition). It is a complex of sheaves on $T^*_Y X$ and identifying Y to the null section of $T^*_Y X$ we have

$$\mu_Y(\mathscr{F})|_Y = \mathbf{R}\Gamma_Y(\mathscr{F}).$$

The first example is given when $\mathscr{F}$ is the sheaf $\mathcal{O}_X$ of holomorphic functions on X, then $\mu_Y(\mathscr{F})$ is concentrated in degree d where d is the codimension of Y and we set $\mathscr{C}^{\mathbf{R}}_{Y|X} = \mu_Y(\mathcal{O}_X)[d]$.

If n is the dimension of X and $\Omega^{(0,n)}_{X \times X}$ is the sheaf of holomorphic differential forms on $X \times X$ of degree 0 in the first variables and of degree maximum in the second, the sheaf $\mathscr{E}^{\mathbf{R}}_X$ is $\mu_\Delta(\Omega^{(0,n)}_{X \times X})[n]$ where Δ is the diagonal of $X \times X$. It is a sheaf of rings on $T^*_\Delta X \times X \simeq T^*X$ and $\mathscr{C}^{\mathbf{R}}_{Y|X}$ is a $\mathscr{E}^{\mathbf{R}}_X$-module. From $\mathscr{E}^{\mathbf{R}}_X$ one defines the sheaves $\mathscr{E}^\infty_X$ and $\mathscr{E}_X$ on T^*X ([13] and [8]).

It is proved in [2] (see also [3]) that the sections of $\mathscr{E}^{\mathbf{R}}_X$ can be represented by symbols analogous to the symbols of $\mathscr{E}_X$ defined in [13].

Let $(x_1, \ldots, x_n)$ be a local system of coordinates of an open subset U of X and (x, ξ) be the corresponding coordinate system of T^*U, let V be an open subset of T^*U whose fibers are conic and proper for the real structure. A holomorphic function $f(x, \xi)$ on V is said to be sub-exponential if

$$\forall \varepsilon > 0, \quad \exists C_\varepsilon > 0, \quad \forall (x, \xi) \in V, \quad |f(x, \xi)| < C_\varepsilon \, e^{\varepsilon |\xi|},$$

and it is said to be exponentially decreasing on V if

$$\exists \delta > 0, \quad \exists C > 0, \quad \forall (x, \xi) \in V, \quad |f(x, \xi)| < C e^{-\delta |\xi|}.$$

Let us denote by $\mathscr{S}^\infty(V)$ the set of sub-exponential functions on V, by $\mathscr{S}^{-\infty}(V)$ the set of exponentially decreasing functions on V and $\mathscr{S}(V) = \varinjlim \mathscr{S}^\infty(V')/\mathscr{S}^{-\infty}(V')$ where the inductive limit is taken on all real conic open subsets V' of V such that $V' \cap \{|\xi| = 1\}$ is relatively compact in V.

In the same way we will denote by $\mathscr{S}^m(V)$ the subset of $\mathscr{S}^\infty(V)$ of functions $f(x, \xi)$ such that $|f(x, \xi)| < C|\xi|^m$ for some constant C and $\mathscr{S}_m(V) = \varinjlim \mathscr{S}^m(V')/\mathscr{S}^{-\infty}(V')$.

Then there is a one-to-one correspondence between $\mathscr{S}(V)$ and the set $\Gamma(V, \mathscr{E}^{\mathbf{R}}_X)$ of the sections of $\mathscr{E}^{\mathbf{R}}_X$ on V.

The sheaf $\mathscr{E}_X^{\mathbf{R},f}$ has been defined by Andronikof in [1]. It is a subsheaf of $\mathscr{E}_X^{\mathbf{R}}$ in one-to-one correspondence with $\bigcup_m \mathscr{S}_m(V)$.

We say that a section of $\mathscr{E}_X^{\mathbf{R},f}$ is of order at most m if its restrictions to conic proper open sets V are in correspondence with $\mathscr{S}_m(V)$.

We denote by $\mathscr{E}_{X,m}^{\mathbf{R}}$ the subsheaf of $\mathscr{E}_X^{\mathbf{R},f}$ of sections of order at most m and we define the sheaf $\hat{\mathscr{E}}_X^{\mathbf{R}}$ as the formal completion of the sheaf $\mathscr{E}_X^{\mathbf{R},f}$ for this filtration, that is, $\hat{\mathscr{E}}_X^{\mathbf{R}} = \bigcup \mathscr{E}_{X,m}^{\mathbf{R}}$ with $\hat{\mathscr{E}}_{X,m}^{\mathbf{R}} = \varprojlim_{j<m} \mathscr{E}_{X,m}^{\mathbf{R}} / \mathscr{E}_{X,j}^{\mathbf{R}}$.

On the null section of T^*X we have $\hat{\mathscr{E}}_{X|X}^{\mathbf{R}} = \mathscr{E}_{X|X}^{\mathbf{R}} = \mathscr{D}_X$ ($\mathscr{D}_X$ is the sheaf of differential operators with finite order on X).

Let us now define the set $\hat{\mathscr{S}}^m(V)$ as the set of formal series $u = \sum_{k\le m} u_k (k \in \mathbf{Z})$ where u_k belongs to $\mathscr{S}^k(V)$ for each k and by $\hat{\mathscr{S}}_m^{-\infty}(V)$ the subset of the series such that for each $j \ge 1$ $\sum_{l=0}^{j} u_{m-l}$ belongs to $\mathscr{S}^{m-j}(V)$. Finally we set $\hat{\mathscr{S}}_m(V) = \varprojlim \hat{\mathscr{S}}^m(V')/\hat{\mathscr{S}}_m^{-\infty}(V')$. From the definition we easily obtain

Proposition 1.1. *The set of sections of $\hat{\mathscr{E}}_{X,m}^{\mathbf{R}}$ on a proper conic open subset V of T^*U is in one-to-one correspondence with $\hat{\mathscr{S}}_m(V)$.*

It has been proved in [13] that $\mathscr{E}_X^{\mathbf{R}}$ is faithfully flat on $\mathscr{E}_X$ and as remarked in [1] the same proof still works for $\mathscr{E}_X^{\mathbf{R},f}$. The point is that the division theorem (Lemma 6.2.1 of [13]) is still true for $\mathscr{E}_X^{\mathbf{R},f}$. In fact this theorem is compatible with the filtration $(\mathscr{E}_{X,m}^{\mathbf{R}})$, so it is still true for $\hat{\mathscr{E}}_X^{\mathbf{R}}$, and with the same proof we obtain that $\hat{\mathscr{E}}_X^{\mathbf{R}}$ is *faithfully flat on $\mathscr{E}_X$*.

Let Y be a smooth submanifold of X. The sheaf $\mathscr{C}_{Y|X}$ is the quotient of $\mathscr{E}_X$ by the ideal of $\mathscr{E}_X$ generated by the vector fields on X tangent to Y. We have $\mathscr{C}_{Y|X}^{\mathbf{R}} = \mathscr{E}_X^{\mathbf{R}} \otimes_{\mathscr{E}} \mathscr{C}_{Y|X}$ and we will denote $\hat{\mathscr{C}}_{Y|X}^{\mathbf{R}} = \hat{\mathscr{E}}_X^{\mathbf{R}} \otimes_{\mathscr{E}} \mathscr{C}_{Y|X}$.

2. Sheaves of 2-Microdifferential Operators

The 2-microdifferential operators have been defined in [9] for the smooth involutive submanifolds of T^*X. Here we will consider the special case of a Lagrangian variety Λ which is of the form T_Y^*X, that is the conormal bundle to a submanifold Y of X.

The sheaf $\mathscr{E}_\Lambda^{2\mathbf{R}}$ was defined in [9] and [6] in the same way as $\mathscr{E}_X^{\mathbf{R}}$ replacing X by Λ and $\mathscr{O}_X$ by $\mathscr{C}_{Y|X}^\infty$:

$$\mathscr{E}_\Lambda^{2\mathbf{R}} = \mu_\Lambda(\mathscr{C}_{Y\times Y|X\times X}^\infty \otimes \Omega_{X\times X}^{(0,n)})[n]$$

(the tensor product $\otimes$ is taken over $\mathscr{O}_{X\times X}$).

It is proved in [9] that $\mu_\Lambda(\mathscr{C}_{Y\times Y|X\times X}^\infty \otimes \Omega_{X\times X}^{(0,n)})$ is concentrated in degree n, hence $\mathscr{E}_\Lambda^{2\mathbf{R}}$ is a sheaf on $T^*\Lambda$. As $\mathscr{E}_{X\times X}^\infty$ is faithfully flat on $\mathscr{E}_{X\times X}$ and $\hat{\mathscr{E}}_{X\times X}^{\mathbf{R}}$ is faithfully flat on $\mathscr{E}_{X\times X}$ the same result is still true if we replace $\mathscr{C}_{Y\times Y|X\times X}^\infty$ by $\mathscr{C}_{Y\times Y|X\times X}$ or by $\hat{\mathscr{C}}_{Y\times Y|X\times X}^{\mathbf{R}}$ and we set

$$\hat{\mathscr{E}}_\Lambda^{2\mathbf{R}} = \mu_\Lambda(\hat{\mathscr{C}}_{Y\times Y|X\times X}^{\mathbf{R}} \otimes \Omega_{X\times X}^{(0,n)})[n].$$

Remark. In what follows we will be mostly interested by the restriction of $\hat{\mathscr{E}}_\Lambda^{2\mathbf{R}}$ to the null section Λ of $T^*\Lambda$, that is, to the sheaf $\hat{\mathscr{E}}_\Lambda^{2\mathbf{R}}|_\Lambda = \mathscr{H}_\Lambda^n(\hat{\mathscr{C}}_{Y\times Y|X\times X}^{\mathbf{R}} \otimes \Omega_{X\times X}^{(0,n)})$, which will be denoted $\hat{\mathscr{D}}_\Lambda^{2\infty}$.

As $\mathscr{E}_X^\infty$ is obtained from $\mathscr{E}_X^{\mathbf{R}}$, we defined in [9] the sheaf $\mathscr{E}_\Lambda^{2\infty}$ on $T^*\Lambda$ from $\mathscr{E}_\Lambda^{2\mathbf{R}}$ by

$$\mathscr{E}_\Lambda^{2\infty}|_{\dot{T}^*\Lambda} = \gamma^{-1}\gamma_*(\mathscr{E}_\Lambda^{2\mathbf{R}}|_{\dot{T}^*\Lambda}) \quad \text{and} \quad \mathscr{E}_\Lambda^{2\infty}|_\Lambda = \mathscr{E}_{\Lambda|\Lambda}^{2\mathbf{R}}$$

where $\dot{T}^*\Lambda = T^*\Lambda\backslash\Lambda$ and γ is the canonical projection $\dot{T}^*\Lambda \to \mathbf{P}^*\Lambda$. In the same way we can define $\hat{\mathscr{E}}_\Lambda^{2\infty}$ from $\hat{\mathscr{E}}_\Lambda^{2\mathbf{R}}$.

Proposition 2.1.

(i) $\hat{\mathscr{E}}_\Lambda^{2\mathbf{R}}$ *is a sheaf of rings on* $T^*\Lambda$ *and* $\hat{\mathscr{E}}_\Lambda^{2\infty}$ *is a subring of* $\hat{\mathscr{E}}_\Lambda^{2\mathbf{R}}$.

(ii) $\pi^{-1}(\mathscr{E}_{X|\Lambda})$ *is a subring of* $\hat{\mathscr{E}}_\Lambda^{2\infty}$ *(with* $\pi: T^*\Lambda \to \Lambda$*).*

(iii) *The sheaf* $\hat{\mathscr{C}}_{Y|X}^{\mathbf{R}}$ *is a left* $\hat{\mathscr{D}}_\Lambda^{2\infty}$*-module.*

(The proof is just the same as the corresponding result in [9].) We can define a symbolic calculus for $\hat{\mathscr{E}}_\Lambda^{2\infty}$ just as we did for $\mathscr{E}_\Lambda^{2\infty}$ in [9]. Let us consider a local system of coordinates $(x_1, \ldots, x_p, t_1, \ldots, t_q)$ of X such that $Y = \{t = 0\}$ and (x, t, ξ, τ) the corresponding coordinate system of T^*X, then Λ has coordinates (x, τ) and $T^*\Lambda$ has coordinates (x, τ, x^*, τ^*). Let Ω be an open subset of $T^*\Lambda$ which is real conic in (τ, x^*), complex conic in (x^*, τ^*) and does not intersect $\{\tau = 0\}$.

For $m \in \mathbf{Z}$ we define $\hat{\mathscr{P}}_m^2(\Omega)$ as the set of formal series $P = \sum P_{i,k}(x, \tau, x^*, \tau^*)$ for $(i, k) \in \mathbf{Z} \times \mathbf{Z}$ such that:

(i)   For each $(i, k) \in \mathbf{Z}^2$, $P_{i,k}(x, \tau, x^*, \tau^*)$ is a holomorphic function on Ω homogeneous of degree i in (x^*, τ^*) and $P_{i,k} \equiv 0$ if $k > m$.

(ii) $\forall k \in \mathbf{Z}, \forall \varepsilon > 0, \exists C_{k,\varepsilon} > 0, \forall i \geq 0, \forall(x, \tau, x^*, \tau^*) \in \Omega$,

$$|P_{i,k}(x, \tau, x^*, \tau^*)| < C_{k,\varepsilon}\varepsilon^i \frac{1}{i!}(1 + |\tau|)^k(|x^*| + |\tau\tau^*|)^i.$$

(iii) $\forall k \in \mathbf{Z}, \exists C_k > 0, \forall i < 0, \forall(x, \tau, x^*, \tau^*) \in \Omega$,

$$|P_{i,k}(x, \tau, x^*, \tau^*)| < C_k^{1-i}(-i)!(1 + |\tau|)^k(|x^*| + |\tau\tau^*|)^i.$$

Then $\hat{\mathcal{S}}^2(\Omega) = \bigcup \hat{\mathcal{S}}^2_m(\Omega)$ and $\hat{\mathcal{S}}^2_{-\infty}(\Omega)$ is the subset of $\hat{\mathcal{S}}^2(\Omega)$ of $P = \sum P_{i,k}$ such that $Q = \sum Q_{i,k}$ defined by $Q_{i,k} = \sum_{l \geq k} P_{i,l}$ still belongs to $\hat{\mathcal{S}}^2(\Omega)$.

Theorem 2.2. *Let* $\Theta = (x_0, \tau_0, x_0^*, \tau_0^*)$ *be a point of* $T^*\Lambda$ *with* $\tau_0 \neq 0$. *There is a one-to-one correspondence between the germs of* $\hat{\mathcal{E}}^{2\infty}_\Lambda$ *at* Θ *and* $\varinjlim \hat{\mathcal{S}}^2(\Omega)/\hat{\mathcal{S}}^2_{-\infty}(\Omega)$, *where the inductive limit is taken on the open neighbourhoods of* Θ *in* $T^*\Lambda$ *which are real conic in* (τ, x^*) *and complex conic in* (x^*, τ^*).

(The proof of this theorem is exactly the same as the proof of the theorem 2.3.1 in [9] replacing the symbols of sections of $\mathcal{C}^\infty_{Y \times Y | X \times X}$ obtained from [13] by the symbols of $\hat{\mathcal{C}}^R_{Y \times Y | X \times X}$ from Proposition 1.1 here above.)

If Θ is in Λ, that is, if $x_0^* = \tau_0^* = 0$, the theorem is still true and the proof is easier (in this case $\hat{\mathcal{D}}^{2\infty}_\Lambda = \mathcal{H}^n_\Lambda(\hat{\mathcal{C}}^R_{Y \times Y | X \times X} \otimes \Omega^{(0,n)}_{X \times X})$ is calculated simply with Čech cohomology), the set $\hat{\mathcal{S}}^2(\Omega)$ is much simpler: $P_{i,k} = 0$ if $i < 0$ and if $i \geq 0$, $P_{i,k}$ is a homogeneous polynomial of degree i in (x^*, τ^*).

We will denote by $\hat{\mathcal{E}}^2_\Lambda$ (resp. $\hat{\mathcal{D}}^2_\Lambda$) the subsheaf of $\hat{\mathcal{E}}^{2\infty}_\Lambda$ (resp. $\hat{\mathcal{D}}^{2\infty}_\Lambda$) of the sections whose symbols $P = \sum P_{i,k}$ in $\hat{\mathcal{S}}^2(\Omega)$ satisfy:

$(ii)'$ $\quad \forall k \in \mathbf{Z}, \exists i_0, \forall i > i_0 \ P_{i,k} \equiv 0$.

In [9] we considered another sheaf of 2-microdifferential operators, $\mathcal{E}^2_\Lambda(\infty, \infty)$, which appears as the subsheaf of $\mathcal{E}^2_\Lambda$ of the sections whose symbols in $\hat{\mathcal{S}}^2(\Omega)$ satisfy

(iv) $\quad \forall(i, k) \ P_{i,k}(x, \tau, x^*, \tau^*)$ is homogeneous of degree $i + k$ in (τ, x^*).

The principal symbol $\sigma^{(\infty, \infty)}(P)$ of a section P of $\mathcal{E}^2_\Lambda(\infty, \infty)$ of symbol $\sum P_{i,k}$ is the function P_{i_0, k_0}, where k_0 is the lowest integer such that $P_{i,k} \equiv 0$ if $k > k_0$, and i_0 is the lowest integer such that $P_{i,k_0} \equiv 0$ if $i > i_0$.

Let us now describe how the injective morphism of $\pi^{-1}(\mathcal{E}_{X|\Lambda})$ into $\mathcal{E}^2_\Lambda(\infty, \infty)$ is given by symbols.

A section of $\mathcal{E}_X$ near Λ has a symbol $P = \sum_{j \leq m} P_j(x, t, \xi, \tau)$, where P_j is a holomorphic function near Λ homogeneous of degree j in (ξ, τ). P_j has a Taylor expansion along Λ:

$$P_j(x, t, \xi, \tau) = \sum P_{j,\alpha,\beta}(x, \tau) t^\alpha \xi^\beta.$$

Let us define the following functions on $T^*\Lambda$:

$$P_{i,k}(x, \tau, x^*, \tau^*) = \sum_{|\alpha| + |\beta| = i} P_{i+k,\alpha,\beta}(x, \tau)(-\tau^*)^\alpha (x^*)^\beta.$$

As proved in [9] (Prop. 1.4.9), $\sum P_{i,k}(x, \tau, x^*, \tau^*)$ is a symbol for P considered as a section of $\mathcal{E}^2_\Lambda(\infty, \infty)$.

The filtration $\hat{\mathscr{C}}_X^{\mathbf{R}} = \bigcup \hat{\mathscr{C}}_{X,m}^{\mathbf{R}}$ defines a filtration on $\hat{\mathscr{C}}_{Y|X}^{\mathbf{R}}$. This filtration on $\hat{\mathscr{C}}_{Y \times Y|X \times X}^{\mathbf{R}}$ induces a filtration $V. \hat{\mathscr{C}}_\Lambda^{2\mathbf{R}}$ on $\hat{\mathscr{C}}_\Lambda^{2\mathbf{R}}$, hence a filtration $V. \hat{\mathscr{C}}_\Lambda^2$. In fact $V_m \hat{\mathscr{C}}_\Lambda^2$ is the set of sections of $\hat{\mathscr{C}}_\Lambda^2$ whose symbols are in $\hat{\mathscr{P}}_m^2(\Omega)$.

The filtration $V. \hat{\mathscr{C}}_\Lambda^2$ induces in turn filtrations $V. \mathscr{C}_\Lambda^2(\infty, \infty)$ and $V. \mathscr{C}_X$ on $\mathscr{C}_\Lambda^2(\infty, \infty)$ and $\mathscr{C}_{X|\Lambda}$. The filtration $V. \mathscr{C}_X$ is the filtration which is denoted by $F_\infty \mathscr{C}_X$ in [10] and its restriction to the null section of T^*X is the filtration of $\mathscr{D}_{X|Y}$ which is denoted $F\mathscr{D}_X$ in [5] and $V. \mathscr{D}_X$ in [12].

Let us recall that the graduate $\mathrm{gr}_V \mathscr{D}_X$ (resp. $\mathrm{gr}_V \mathscr{C}_X$) is isomorphic to the sheaf $\mathscr{D}_{[\Lambda]}$ (resp. $\mathscr{D}_{(\Lambda)}$) of differential operators on Λ with polynomial (resp. homogeneous) coefficients in the fibers of Λ. A good V-filtration of a coherent $\mathscr{C}_X$-module $\mathscr{M}$ is the filtration induced on $\mathscr{M}$ by the filtration $(V. \mathscr{C}_X)^m$ for a morphism $(\mathscr{C}_X)^m \to \mathscr{M}$. The associated graduate is a coherent $\mathscr{D}_{(\Lambda)}$-module. The Euler vector field θ of the homogeneous manifold Λ is in $\mathscr{D}_{(\Lambda)}$ so it acts on $\mathrm{gr}_V \mathscr{M}$. A b-function for the module $\mathscr{M}$ is a polynomial b such that $b(\theta + k)$ annihilates $\mathrm{gr}_V^k(\mathscr{M})$ for each k.

It is proved that every holonomic $\mathscr{C}_X$- (or $\mathscr{D}_X$-) module admits a b-function (see [5], [12], [10]).

3. Equivalence Theorem

Let X be a complex manifold and Y be a smooth hypersurface of X. As before we will denote by $\Lambda = T_Y^* X$ the conormal bundle to Y.

Let us consider a local coordinate system $(x_1, \ldots, x_p, t)$ of X such that $Y = \{t = 0\}$, and (x, τ, x^*, τ^*) the corresponding coordinates of $T^*\Lambda$. We remark that the differential operator tD_t has the symbol $\tau\tau^*$ when considered as an element of $\mathscr{C}_\Lambda^2(\infty, \infty)$, it is of order 0 for the V-filtration, and its image in $\mathrm{gr}_V \mathscr{D}_X$ is θ, the Euler vector field of Λ.

Theorem 3.1.   *Let $\Theta = (x_0, \tau_0, x_0^*, \tau_0^*)$ be a point of $T^*\Lambda$ with $\tau_0 \neq 0$. Let P be a 2-microdifferential operator of $V_0 \mathscr{C}_\Lambda^2(\infty, \infty)$ which is of the form $b(tD_t) + Q$ where b is a polynomial of degree m and Q belongs to $V_{-1} \mathscr{C}_\Lambda^2(\infty, \infty)$. Then there is an isomorphism of $\hat{\mathscr{C}}_\Lambda^2$-modules near Θ:*

$$\hat{\mathscr{C}}_\Lambda^2 / \hat{\mathscr{C}}_\Lambda^2 P \overset{\sim}{\to} (\hat{\mathscr{C}}_\Lambda^2 / \hat{\mathscr{C}}_\Lambda^2 t)^m.$$

Proof.   It is very similar to the proof of Theorem 5.2.1 ch. II in [13] or Theorem 4.2.1 in [6].

First we use the theorem 2.7.2 of [9] to write $P = E \cdot \tilde{P}$ where E is invertible in $\mathscr{C}_\Lambda^2(\infty, \infty)$ and $\tilde{P}$ is a polynomial of degree m in t that is of the form $t^m + \sum_{j=0}^{m-1} \tilde{P}_j(x, D_x, D_t) \cdot t^j$ where $\tilde{P}_j$ does not contain t (i.e., $\tilde{P}_j$ commutes with D_t). As D_t is invertible (we suppose $\tau \neq 0$) $\tilde{P}$ can be rewritten as $(tD_t)^m + \sum_{j=0}^{m-1} P_j(x, D_x, D_t) \cdot (tD_t)^j$.

Moreover this theorem is compatible with the V-filtration, so $\tilde{P}$ is of the form $b(tD_t) + P'$ with P' in $V_{-1}\mathscr{E}_\Lambda^2(\infty, \infty)$, and then $P_j = \lambda_j + P_j'$ with λ_j a complex number and P_j' in $V_{-1}\mathscr{E}_\Lambda^2(\infty, \infty)$.

The module $\mathcal{M} = \hat{\mathscr{E}}_\Lambda^2 / \hat{\mathscr{E}}_\Lambda^2 P$ is isomorphic to $(\hat{\mathscr{E}}_\Lambda^2)^m / (\hat{\mathscr{E}}_\Lambda^2)^m (tD_t - A)$, where A is the (m, m)-matrix

$$A = \begin{pmatrix} 0 & 1 & 0 & \cdots & & 0 \\ 0 & 0 & 1 & \cdots & & 0 \\ \cdots & \cdots & \cdots & \cdots & & \cdots \\ 0 & 0 & \cdots & & 0 & 1 \\ P_0 & P_1 & & \cdots & & P_{m-1} \end{pmatrix}.$$

Let us remark that if B is an operator of $\mathscr{E}_\Lambda^2(\infty, \infty)$ which is independent of t its symbol is of the form $\sum B_{i,j}(x, x^*, \tau)$, where $B_{i,j}$ is homogeneous of degree i in x^* and j in (x, τ), hence it may be written as $\sum_{k \geq 0} \tau^{m-k} \sum_i \tilde{B}_{i,k}(x, x^*)$, and the estimates on a symbol of $\mathscr{E}_\Lambda^2(\infty, \infty)$ just say that for each k, $\sum_i \tilde{B}_{i,k}(x, x^*)$ is the symbol of a microdifferential operator B_k on $T^* Y$.

In the same way, the matrix A has a symbol of the form $\sum_{k \geq 0} \tau^{-k} A_k(x, x^*)$, where A_k is the symbol of an (m, m)-matrix of microdifferential operators (of finite order) on Y and A_0 is a constant matrix.

To prove the theorem we will show that there exists an invertible matrix R of operators of $\hat{\mathscr{E}}_\Lambda^2$ such that $R^{-1} \cdot (t - A \cdot D_t^{-1}) \cdot R = t \cdot I_m$, where I_m is the identity matrix of size m. This equation is equivalent to $[t, R] = A \cdot D_t^{-1} \cdot R$. We search R as an operator whose symbol is independent of τ^* and then the equation becomes $\partial R / \partial \tau = A\tau^{-1} R$.

Let R_0 be the solution of $\partial R_0 / \partial \tau = A_0 \tau^{-1} R_0$, which is equal to the identity matrix at τ_0. As A_0 is a constant matrix, the coefficients of the matrix R_0 are linear combinations of functions of the form $\sum_{i \leq k-1} (\log \tau)^i \tau^\alpha$. A solution of the problem will be given by $R = R_0(1 + \sum_{k \geq 1} R_k)$, where the R_k matrices are defined recursively by

$$\frac{\partial}{\partial \tau} R_k = R_0^{-1} A_k \tau^{-k-1} R_0 + \sum_{j=1}^{k-1} R_0^{-1} A_{j-k} \tau^{k-j-1} R_0 R_j, \qquad R_k|_{\tau = \tau_0} = 0.$$

For each k, R_k is a matrix of symbols of microdifferential operators with holomorphic parameter τ and a growth in τ of the type $N - k$ (with N the order of R_0^{-1} plus the order of R_0). So the series $\sum R_k$ define a symbol in $\hat{\mathscr{S}}^2(\Omega)$ and it is clear that $R = R_0(1 + \sum_{k \geq 1} R_k)$ is the symbol of a solution of the problem.

Let us come back to the general case of a smooth hypersurface Y of a complex manifold X with conormal $\Lambda = T_Y^* X$ and projection $p: \Lambda \to Y$. We denote by $j: (T^*X) \times_X Y \to T^*X$ and $\tilde{\omega}: (T^*X) \times_X Y \to T^*Y$ the canonical morphisms.

Let us recall that $\mathscr{E}_{Y \to X}$ is a sheaf on $(T^*X) \times_X Y$ which has been defined in [13] (where it was denoted $\mathscr{P}_{Y \to X}^f$). If $\mathscr{I}_Y$ is the ideal of definition of Y in $\mathcal{O}_X$ then $\mathscr{E}_{Y \to X}$ is isomorphic to $j^{-1}[\mathscr{E}_X / \mathscr{E}_X \mathscr{I}_Y)$ and it is a $(\tilde{\omega}^{-1} \mathscr{E}_Y, j^{-1} \mathscr{E}_X)$-bimodule.

We define the sheaf $\hat{\mathscr{D}}_{Y \leftarrow \Lambda}^2$ on Λ by

$$\hat{\mathscr{D}}_{Y \leftarrow \Lambda}^2 = (\mathscr{E}_{Y \to X}|_\Lambda) \otimes_{(\mathscr{E}_X|_\Lambda)} \hat{\mathscr{D}}_\Lambda^2.$$

As $(\tilde{\omega}^{-1} \mathscr{E}_Y)|_\Lambda = p^{-1} \mathscr{D}_Y$, $\hat{\mathscr{D}}_{Y \leftarrow \Lambda}^2$ is a $(p^{-1} \mathscr{D}_Y, \hat{\mathscr{D}}_\Lambda^2)$-bimodule, and as $\hat{\mathscr{D}}_\Lambda^2$ is flat over $\mathscr{E}_X|_\Lambda$ we have $\hat{\mathscr{D}}_{Y \leftarrow \Lambda}^2 = \hat{\mathscr{D}}_\Lambda^2 / \hat{\mathscr{D}}_\Lambda^2 \mathscr{I}_Y$.

If $\mathcal{M}$ is a coherent $\mathscr{E}_X$-module let us denote $\hat{\mathcal{M}}^2 = \hat{\mathscr{D}}_\Lambda^2 \otimes_{\mathscr{E}_X|_\Lambda} (\mathcal{M}|_\Lambda)$. We will now calculate the complex of $p^{-1} \mathscr{D}_Y$-module induced by $\hat{\mathcal{M}}^2$, that is, $\hat{\mathscr{D}}_{Y \leftarrow \Lambda}^2 \otimes_{\hat{\mathscr{D}}_\Lambda^2}^{\mathbf{L}} \hat{\mathcal{M}}^2$:

Theorem 3.2.   *Let $\mathcal{M}$ be a coherent $\mathscr{E}_X$-module defined in the neighbourhood of Λ. We suppose that $\mathcal{M}$ admits a b-function. Then*

(i) $\forall k \neq 1 \; \mathscr{T}\!or_k^{\hat{\mathscr{D}}_\Lambda^2}(\hat{\mathscr{D}}_{Y \leftarrow \Lambda}^2, \hat{\mathcal{M}}^2) = 0$ *on* $\Lambda \setminus Y$.

(ii) $\check{\Phi}(\mathcal{M}) = \mathscr{T}\!or_1^{\hat{\mathscr{D}}_\Lambda^2}(\hat{\mathscr{D}}_{Y \leftarrow \Lambda}^2, \hat{\mathcal{M}}^2)$ *is a local system of coherent $\mathscr{D}_Y$-modules on* $\dot{\Lambda} = \Lambda \setminus Y$, *i.e., $\check{\Phi}(\mathcal{M})$ is locally on $\dot{\Lambda}$ of the form $p^{-1}(\mathcal{N})$ for a coherent $\mathscr{D}_Y$-module $\mathcal{N}$.*

(iii) *If $\mathcal{M}$ and $\mathcal{N}$ are two coherent $\mathscr{E}_X$-modules which admit b-functions, there is a canonical isomorphism on* $\Lambda \setminus Y$:

$$\mathbf{R}\mathscr{H}\!om_{\hat{\mathscr{D}}_\Lambda^2}(\hat{\mathcal{M}}^2, \hat{\mathcal{N}}^2) \xrightarrow{\sim} \mathbf{R}\mathscr{H}\!om_{p^{-1}\mathscr{D}_Y}(\check{\Phi}(\mathcal{M}), \check{\Phi}(\mathcal{N})).$$

Proof.   The problem is local so we may choose an equation t of Y, and then $\hat{\mathscr{D}}_{Y \leftarrow \Lambda}^2 = \hat{\mathscr{D}}_\Lambda^2 / \hat{\mathscr{D}}_\Lambda^2 t$ and the complex $\hat{\mathscr{D}}_{Y \leftarrow \Lambda}^2 \otimes^{\mathbf{L}} \hat{\mathcal{M}}^2$ is quasi-isomorphic to $\hat{\mathcal{M}}^2 \xrightarrow{\cdot t} \hat{\mathcal{M}}^2$.

To prove the theorem we will first suppose that $\mathcal{M}$ is of the form $\mathscr{E}_X / \mathscr{E}_X P$. In this case, P satisfies the hypothesis of the Theorem 3.1 and $\hat{\mathcal{M}}^2$ is isomorphic to $(\hat{\mathscr{D}}_\Lambda^2 / \hat{\mathscr{D}}_\Lambda^2 t)^m$. We have then to prove that $\hat{\mathscr{D}}_\Lambda^2 / \hat{\mathscr{D}}_\Lambda^2 t \xrightarrow{\cdot t} \hat{\mathscr{D}}_\Lambda^2 / \hat{\mathscr{D}}_\Lambda^2 t$ is surjective and has a kernel equal to $p^{-1} \mathscr{D}_Y$; this is clear from the Theorem 2.2.

In the same way the part (*iii*) of the theorem is quite clear when P is equal to t and so for any P.

The theorem for general $\mathcal{M}$ is deduced from the case of one operator in the classical way (see the chapter 3 of [4] for example).

Remarks. (1) If $\mathcal{M}$ is a coherent $\mathcal{D}_X$-module in the neighbourhood of Y, the theorem applies with $\hat{\mathcal{M}}^2 = \hat{\mathcal{D}}_\Lambda^2 \otimes_{p^{-1}(\mathcal{D}_X|_Y)} p^{-1}(\mathcal{M}|_Y)$.

(2) We have stated the theorem with $\hat{\mathcal{D}}_\Lambda^2$ for simplicity but as Theorem 3.1 is true with $\hat{\mathcal{E}}_\Lambda^2$ we can define $\hat{\mathcal{E}}_{Y \leftarrow \Lambda}^2$ in the same way and state the theorem on $T^*\Lambda$ instead of Λ. We will obtain a $\tilde{p}^{-1}\mathcal{E}_Y$ coherent module $\Phi^\mu(\mathcal{M})$ which is the microlocalization of $\tilde{\Phi}(\mathcal{M})$ ($\tilde{p}$ is here the projection $(T^*Y) \times_Y \dot{\Lambda} \to T^*Y$).

The characteristic variety of $\tilde{\Phi}(\mathcal{M})$ is thus the image under $(T^*Y) \times_Y \dot{\Lambda} \to T^*\Lambda$ of the support of $\hat{\mathcal{E}}_\Lambda^2 \otimes \pi^{-1}(\mathcal{M}|_\Lambda)$ (with $\pi: T^*\Lambda \to \Lambda$). This support is equal to the support of $\mathcal{E}_\Lambda^2(\infty, \infty) \otimes \pi^{-1}(\mathcal{M}|_\Lambda)$ that is the variety which we called $\mathrm{Ch}_\Lambda^{(\infty)}(\mathcal{M})$ in [10]. When $\mathcal{M}$ admits a b-function this variety is contained in $(T^*Y) \times_Y \dot{\Lambda}$.

If $\mathcal{M}$ is a holonomic module, it admits a b-function and we know [10] that $\mathrm{Ch}_\Lambda^{(\infty)}(\mathcal{M})$ is Lagrangian, then $\tilde{\Phi}(\mathcal{M})$ is a *holonomic $\mathcal{D}_Y$-module.*

If we apply the part (*iii*) of the theorem to $\mathcal{N} = \mathcal{C}_{Y|X}$ we obtain

Corollary 3.3. *Let $\mathcal{M}$ be a coherent $\mathcal{E}_X$-module which admits a b-function. Then there is a canonical isomorphism*

$$\mathbf{R}\mathcal{H}om_{\mathcal{E}_X}(\mathcal{M}, \hat{\mathcal{C}}_{Y|X}^{\mathbf{R}}) \stackrel{\sim}{\to} \mathbf{R}\mathcal{H}om_{p^{-1}\mathcal{D}_Y}(\tilde{\Phi}(\mathcal{M}), p^{-1}\mathcal{O}_Y).$$

4.   Vanishing Cycles of a *D*-Module

The space $\dot{\Lambda} = \dot{T}_Y^*X$ is locally isomorphic to $Y \times \mathbf{C}^*$; if we fix a base point on $\dot{\Lambda}$, the local system $\tilde{\Phi}(\mathcal{M})$ is given by a coherent $\mathcal{D}_Y$-module $\Phi(\mathcal{M})$ with the action of the monodromy T.

The space of vanishing cycles of a coherent $\mathcal{D}_X$-module $\mathcal{M}$ admitting a b-function is a coherent $\mathcal{D}_Y$-module which has been defined in [11], [5] and [12]; here we will denote it by $\Phi'(\mathcal{M})$. The monodromy of $\Phi'(\mathcal{M})$ is by definition $T' = \exp[-2i\pi(\theta + 1)]$, where θ is the Euler vector field of Λ.

We now fix a function $f: X \to \mathbf{C}$ such that $Y = f^{-1}(0)$ and denote by t the coordinate of $\mathbf{C}$.

Theorem 4.1. *The $\mathcal{D}_Y$-module $\Phi(\mathcal{M})$ is isomorphic to the vanishing cycles $\Phi'(\mathcal{M})$ of $\mathcal{M}$, the action of the monodromy T on $\Phi(\mathcal{M})$ corresponding to the action of T' on $\Phi'(\mathcal{M})$.*

Proof. Let G be $\{\alpha \in \mathbf{C} \mid -1 \leq \alpha < 0\}$, where $<$ is relative to the lexicographic order on $\mathbf{C} \simeq \mathbf{R} \times \mathbf{R}$. From [5] we know that there exists a unique good filtration $V.\mathcal{M}$ for which the associated b-function has all its roots in G. Then $\Phi'(\mathcal{M})$ is by definition $\mathrm{gr}_V^0(\mathcal{M})$.

We define a new filtration of $\hat{\mathcal{D}}_\Lambda^2$ by $V_k^+ \hat{\mathcal{D}}_\Lambda^2 = \bigcap_{j>k} V_j \hat{\mathcal{D}}_\Lambda^2$ and a filtration of $\hat{\mathcal{M}}^2$ by $V_k^+(\hat{\mathcal{M}}^2) = \sum_{i+j=k} V_i^+ \hat{\mathcal{D}}_\Lambda^2 \cdot V_j(\mathcal{M})$.

For each $\alpha \in G$ and each $p \in \mathbf{N}$ we define $\Phi'_{\alpha,p}(\mathcal{M})$ as the kernel of $(T' - e^{2i\pi\alpha})^p$ in $\Phi'(\mathcal{M})$ and $\Phi'_\alpha(\mathcal{M}) = \bigcup \Phi'_{\alpha,p}(\mathcal{M})$. As $\Phi'(\mathcal{M})$ is annihilated by $b(\theta)$, the increasing sequence $\Phi'_{\alpha,p}(\mathcal{M})$ is stationary for each α and $\Phi'_\alpha(\mathcal{M})$ is nonzero for a finite number of α. Moreover $\Phi'(\mathcal{M})$ is the direct sum $\oplus \Phi'_\alpha(\mathcal{M})$.

For each $\alpha \in G$ and each $p \in \mathbf{N}$ we define $\Phi_{\alpha,p}(\mathcal{M})$ as the kernel of $(T - e^{2i\pi\alpha})^p$ in $\Phi(\mathcal{M})$ and $\Phi_\alpha(\mathcal{M}) = \bigcup \Phi_{\alpha,p}(\mathcal{M})$. As before, $\Phi(\mathcal{M})$ is the direct sum $\oplus \Phi_\alpha(\mathcal{M})$. In order to prove the theorem we will show that for each (α, p), $\Phi_{\alpha,p}(\mathcal{M})/\Phi_{\alpha,p-1}(\mathcal{M})$ is canonically isomorphic to $\Phi'_{\alpha,p}(\mathcal{M})/\Phi'_{\alpha,p-1}(\mathcal{M})$.

Lemma 4.2. *For each α in G we have*

(1) *$D_t^{-\alpha} \Phi_\alpha(\mathcal{M})$ is a subspace of $V_0^+ \mathcal{M}$, and the morphism $D_t^{-\alpha} \Phi_\alpha(\mathcal{M}) \to V_0^+ \mathcal{M}/V_{-1}^+ \mathcal{M}$ is injective.*

(2) *The image of $D_t^{-\alpha} \Phi_{\alpha,1}(\mathcal{M})$ in $V_0^+ \mathcal{M}/V_{-1}^+ \mathcal{M}$ is in $\mathrm{gr}^0 \mathcal{M}$ and this defines an isomorphism of $\mathcal{D}_Y$-modules $\varphi_{\alpha,1} : \Phi_{\alpha,1}(\mathcal{M}) \to \Phi'_{\alpha,1}(\mathcal{M})$.*

(3) *For each $p \geq 1$, the image of $(T - e^{2i\pi\alpha})^{p-1} \Phi_{\alpha,p}(\mathcal{M})$ under $\varphi_{\alpha,1}$ is in $(T' - e^{2i\pi\alpha})^{p-1} \Phi'_{\alpha,k}(\mathcal{M})$, and these two spaces are isomorphic.*

Proof. As $\Phi(\mathcal{M})$ is the direct sum of the $\Phi_\alpha(\mathcal{M})$, we may suppose in the proof that α is the unique root of b. Each section of $\mathcal{M}$ is then annihilated by an operator of the form $P = (tD_t - \alpha)^M + Q$ with $Q \in V_{-1}\mathcal{D}_X$, and we have a surjective homomorphism of $\mathcal{D}_X$-modules $\mathcal{L} \to \mathcal{M}$, where $\mathcal{L}$ is a direct sum of some $\mathcal{D}_X/\mathcal{D}_X P$ with P as above. Applying this to the kernel of $\mathcal{L} \to \mathcal{M}$, we obtain an exact sequence $\mathcal{L}' \to \mathcal{L} \to \mathcal{M} \to 0$ with $\mathcal{L}'$ of the same type as $\mathcal{L}$.

We have now to prove the lemma for $\mathcal{M} = \mathcal{D}_X/\mathcal{D}_X P$ with $P = (tD_t - \alpha)^m + Q$. We can write $\Phi(\mathcal{M})$ explicitly with the matrix R of theorem

3.1, and we see that the elements of $\Phi(\mathcal{M})$ are represented in $\hat{\mathcal{M}}^2 = \hat{\mathcal{D}}^2_X / \hat{\mathcal{D}}^2_X P$ by operators of the form $\tau^\alpha \sum_{k<m} a_k ((\log \tau)^k + Q)$ with a_k in $\mathcal{D}_Y$ and $Q \in V^+_{-1} \hat{\mathcal{D}}^2_\Lambda$ (τ is a symbol for D_t). It is now easy to verify the lemma.

As the kernel of $(T - e^{2i\pi\alpha})^{p-1} : \Phi_{\alpha,p}(\mathcal{M}) \to \Phi_{\alpha,1}(\mathcal{M})$ is $\Phi_{\alpha,p-1}(\mathcal{M})$, the lemma gives an isomorphism between $\Phi_{\alpha,p}(\mathcal{M})/\Phi_{\alpha,p-1}(\mathcal{M})$ and $\Phi'_{\alpha,p}(\mathcal{M})/\Phi'_{\alpha,p-1}(\mathcal{M})$.

References

[1] E. Andronikof, Microlocalisation Tempérée des Distributions et des Fonctions Holomorphes, Thèse à l'Université Paris-Nord (1987).

[2] T. Aoki, Invertibility for microdifferential operators of infinite order, *Publ. R.I.M.S., Kyoto* **18** 2 (1982) 1-29.

[3] L. Boutet de Monvel, Opérateurs pseudo-différentiels d'ordre infini, *Ann. Inst. Fourier, Grenoble* **22** (1972) 229-268.

[4] M. Kashiwara, Systems of microdifferential operators, *Progress in Math.* **34**, Birkhäuser (1983).

[5] M. Kashiwara, *Vanishing Cycles Sheaves and Holonomic Systems of Differential Equations*, Lecture Notes in Math No. 1016, Springer, 1983.

[6] M. Kashiwara and T. Kawai, *Second Microlocalization and Asymptotic Expansions*, Lecture Notes in Physics No. 126, Springer, 1980, pp. 21-76.

[7] M. Kashiwara and P. Schapira, Microhyperbolic systems, *Acta Math.* **142** (1979) 1-55.

[8] M. Kashiwara and P. Schapira, Microlocal study of sheaves, *Astérisque* **128** (1985).

[9] Y. Laurent, Théorie de la deuxième microlocalisation dans le domaine complexe, *Progress in Math.* **53**, Birkhäuser (1985).

[10] Y. Laurent, Polygône de Newton et b-fonctions pour les modules microdifférentiels, *Ann. E.N.S.*, 4ᵉ série, **20** (1987) 391-441.

[11] B. Malgrange, Polynômes Bernstein-Sato et cohomologie évanescente, *Astérisque* **101-102** (1983).

[12] C. Sabbah, $\mathcal{D}$-modules et cycles évanescents, Prépubl. Ecole Polyt., Paris, 1985.

[13] M. Sato, T. Kawai and M. Kashiwara, *Hyperfunctions and Pseudo-Differential Equations*, Lecture Notes in Math No. 287, Springer, 1983.

Extensions of Microfunction Sheaves up to the Boundary

J.-L. Lieutenant

Department of Mathematics
Liege, Belgium

1. Introduction

The sheaf $\mathscr{C}$ of microfunctions introduced by M. Sato allows us to describe the directional singularities of hyperfunctions in the cotangent bundle of the base space. Therefore, it was extensively used to solve many important problems in various fields of mathematics and particularly in partial differential equations. In order to extend such a useful tool for studying boundary value problems, several attempts were made to generalize this microlocalization up to the boundary of an open set (cf. [3], [4], [8], [10]). Most of these extensions mainly concern suitable subsheaves of the sheaf $\mathscr{B}$ of hyperfunctions constituted by sections which behave "fairly" at the boundary (e.g., mild hyperfunctions).

In the case of the boundary of an open *convex* set $\Omega \subset \mathbf{R}^n$, an answer to this question is presented here for general hyperfunctions. If we denote by F the closure of Ω in the radial compactification $\mathbf{D}^n$ of $\mathbf{R}^n$, by ι the natural imbedding $\Omega \to F$, by S_{n-1} and S_{n-1}^* the unit sphere and cosphere of $\mathbf{R}^n$ and

Algebraic Analysis, Volume I

by $S^*F \simeq F \times S_{n-1}^*$ the cotangential sphere bundle along F with its natural projection $\pi : S^*F \to F$, we construct

(a)   a sheaf $\mathscr{C}''^b$, whose restriction to $S^*\Omega \simeq \Omega \times S_{n-1}^*$ coincides with $\mathscr{C}$ together with a surjective morphism $\pi^{-1}\iota_*\mathscr{B} \to \mathscr{C}''^b$.

(b)   a sheaf $\mathscr{C}'^b$ concentrated over $\partial_{\mathbf{D}^n}\Omega \times S_{n-1}^*$ (i.e. vanishing over $S^*\Omega$) together with a surjective morphism $\tau^{-1}\iota_*\mathscr{A} \to \mathscr{C}''^b$, where $\mathscr{A}$ stands for the sheaf of real analytic functions over Ω.

It turns out that $\mathscr{C}''^b$ and $\mathscr{C}'^b$ respectively describe the directional singularities of hyperfunctions and real analytic functions near boundary points $x \in \partial_{\mathbf{D}^n}\Omega$.

The purpose of the present contribution is mainly explanatory and we shall only provide the main frame without proofs. For further details, the interested reader should consult [6], where the whole theory is exposed modulo slight changes of notations. In fact, for practical reasons, we have replaced the notations $\mathscr{C}^b$ and $\varinjlim_v \mathscr{H}_{S^*F}^n(\pi_v'^{-1}\iota_{v^*}\mathcal{O})$ there respectively by $\mathscr{C}'^b$ and $\mathscr{C}''^b$ here and the notations τ', π', $\tau_.'$ and $\pi_.'$ there respectively by τ, π, $\tau_.$ and $\pi_.$ here.

2.   Tapering Domains and Sheaves over *SF*

For the sake of simplicity, we shall suppose that the open set Ω of $\mathbf{R}^n$ contains the origin. We shall also consider positive numbers v strictly less than the Euclidean distance from 0 to the exterior of Ω. Then, we call a (linearly) *tapering domain* any neighborhood of Ω of the form

$$\Omega_v = \langle\!\langle \Omega \cup iC_v \rangle\!\rangle,$$

where $\langle\!\langle \ \rangle\!\rangle$ denotes the convex hull of a subset of $\mathbf{C}^n$, and C_v is any open convex neighborhood of 0 in $\mathbf{R}^n$ with diameter less than v (generally one takes $C_v = \{y \in \mathbf{R}^n : \sum_{j=1}^n |y_j| < v\}$). By F_v, we shall mean the closure of Ω_v in $\mathbf{D}^n + i\mathbf{R}^n$. Such intricate complexifications of Ω may look strange but it turns out to be indispensable to get some cohomology vanishing results necessary for our purpose (cf. [7]).

Definitions.   If ω is an open subset of F, let us then call a *profile of base* ω any open set $\Lambda \subset \omega + i\mathbf{R}^n$ of the form $\bigcup_{x \in \omega} [\{x\} + i\Lambda_x]$, where the *fibers* Λ_x are open convex cones with apex at the origin. By a *v-tuboid of profile* Λ, we shall mean an open subset V of Λ such that the following condition holds:

$$\forall K \subset\subset \Lambda, \exists \rho_0 > 0 \text{ s.t.} \{x + i\rho y \in F_v : x + iy \in K, \rho \in]0, \rho_0]\} \subset V.$$

We shall denote by $\mathcal{T}_v(\Lambda)$ the family (directed down by inclusion) of v-tuboids of profile Λ.

Let us now consider the sheaves $\mathcal{A}^0$, $\mathcal{A}'$, $\mathcal{A}''$ over $SF \simeq F \times S_{n-1}$ defined by their stalks at an arbitrary point $(x, \xi) \in SF$:

$$\mathcal{A}^i_{(x,\xi)} = \varinjlim_{v \downarrow 0} \varinjlim_{\substack{\omega \ni x \\ \Gamma \ni \xi}} \varinjlim_{V \in \mathcal{T}_v(\Lambda^i_{\omega}, \Gamma)} \mathcal{O}[V \cap \Omega_v] \qquad i = 0, ', ''$$

where ω runs through the neighborhoods of x in F, Γ through a family of open conic neighborhoods of ξ, and $\Lambda^i_{\omega,\Gamma}$ denotes one of the following profiles:

$$\Lambda^0_{\omega,\Gamma} = \omega + i\mathbf{R}^n \qquad (\Gamma \text{ is constantly equal to } \mathbf{R}^n)$$

$$\Lambda'_{\omega,\Gamma} = [(\omega \cap \Omega) + i\mathbf{R}^n] \cup [(\omega \cap \partial\Omega) + i\Gamma],$$

$$\Lambda''_{\omega,\Gamma} = \omega + i\Gamma.$$

If we consider the plane of the sheet as the plane of pure imaginary points of $\mathbf{C}^n$ and represent Ω by the left directed arrow, we get the following pictures for such tuboids:

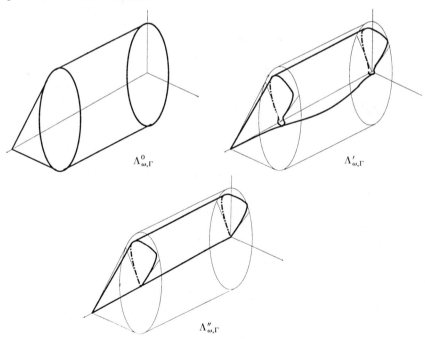

Such definitions directly provide injective morphisms $\mathcal{A}^0 \to \mathcal{A}'$, $\mathcal{A}^0 \to \mathcal{A}''$ and $\mathcal{A}' \to \mathcal{A}''$ and we may thus define the quotient sheaves $\mathcal{O}' = \mathcal{A}'/\mathcal{A}^0$, $\mathcal{O}'' = \mathcal{A}''/\mathcal{A}^0$ and $\mathcal{A}''/\mathcal{A}'$. It turns out that the last quotient $\mathcal{A}''/\mathcal{A}'$ is actually isomorphic to $\alpha_* \mathcal{Q}$, where $\mathcal{Q}$ denotes the sheaf $\mathcal{H}^1_{S\Omega}(\tau^{-1}\mathcal{O})$ introduced in [9] and α the natural imbedding $S\Omega \to SF$. Hence we get

Theorem 1. *One has a commutative diagram of short exact sequences*

$$
\begin{array}{ccccccccc}
& & 0 & & 0 & & 0 & & \\
& & \downarrow & & \downarrow & & \downarrow & & \\
0 & \to & \mathcal{A}^0 & \to & \mathcal{A}' & \to & \mathcal{O}' & \to & 0 \\
& & \| & & \downarrow & & \downarrow & & \\
0 & \to & \mathcal{A}^0 & \to & \mathcal{A}'' & \to & \mathcal{O}'' & \to & 0 \\
& & \downarrow & & \downarrow & & \downarrow & & \\
& & 0 & \to & \mathcal{A}''/\mathcal{A}' & = & \alpha_* \mathcal{Q} & \to & 0 \\
& & & & \downarrow & & \downarrow & & \\
& & & & 0 & & 0 & &
\end{array}
$$

3. Passage from Sphere to Cosphere

The rows and columns of the above diagram have of course intrinsic meaning in the derived category of complexes of sheaves over SF. Adapting Proposition 1.4.1 of [9], it is possible to show that this category is actually equivalent to the derived category of complexes of sheaves over S^*F through the microlocalization functor $\mathbf{F}$. This functor is constructed by means of the natural two projections $\pi_.: DF \to SF$ and $\tau_.: DF \to S^*F$, where DF denotes the set $\{(x, \xi, \eta) \in F \times S_{n-1} \times S^*_{n-1}: \langle \xi, \eta \rangle \leq 0\}$. Its value on a single sheaf $\mathcal{F}$ over SF (identified with the trivial complex $\cdots \to 0 \to \mathcal{F} \to 0 \to \cdots$ concentrated in degree 0) is given by

$$(\mathbf{F}\mathcal{F})^k = (R^k\tau_{.*})\pi_.^{-1}\mathcal{F}^a, \qquad \forall k \in \mathbf{N}$$

where $R^k\tau_{.*}$ means of course taking the kth derived image through $\tau_.$, and a is the antipodal map $(x, \xi) \mapsto (x, -\xi)$.

The key point of the theory consists in computing the image of the above diagram through the microlocalization functor $\mathbf{F}$. This can be done by looking first for an intrinsic geometric definition of the sheaves $\mathcal{A}^i$ ($i = {}^0, ', ''$) and by then performing explicit calculations of the derived functors. These computations may be carried over in a frame similar to the one of global microlocalization over $\mathbf{R}^n$ (cf. [5]). Nevertheless, the presence of the boundary brings several important difficulties which are mainly due to the non-closedness of the maps $\iota : \Omega \to F$, $\alpha : S\Omega \to SF$, $\beta : S^*\Omega \to S^*F$ and $\gamma : D\Omega \to DF$ (where $D\Omega$ is the restriction of DF above Ω). Such difficulties may be overcome by remarking that F, SF, S^*F and DF constitute b-spaces for their open subsets Ω, $S\Omega$, $S^*\Omega$ and $D\Omega$ respectively.

Definition. By a *b-space*, we mean a topological space E together with one of its open subsets E_0 such that one has $E = E_0 \cup \partial E_0$ and one can find for any $x \in E$ a basis of neighborhoods of x whose intersections with E_0 are connected. Such spaces are endowed with the interesting property that the decomposition in connected components of an open set of E is basically equivalent to the partition of its intersection with E_0. This new concept turns out to be particularly efficacious to recover some of the standard techniques of computing derived functors that are failing in this situation.

Beyond these ones, a more important difficulty breaks out when proving the concentration of $\mathbf{F}\mathcal{O}''$ in degree $n-1$ only. It can be bypassed by use of the following

Theorem 2. *The inductive limits*

$$\varinjlim_{v \downarrow 0} H^k_{\Omega + i\Gamma}(\Omega_v, \mathcal{O}) = 0 \qquad \forall k \neq n$$

vanish for any closed convex cone $\Gamma \subset \mathbf{R}^n$ which does not contain any straight line.

The preceding result is clearly of the Edge-of-the-Wedge type. Its proof (cf. [7]) is deeply connected with the linear tapering of the Ω_v near $\partial\Omega$ and is also based on the following new decomposition of the Cauchy kernel:

$$\frac{1}{(\zeta_1 - z_1) \ldots (\zeta_n - z_n)} = \sum_{l=1}^{n} \frac{1}{\varphi(z, \zeta) \prod_{l \neq m \leq n} (\zeta_m - z_m)},$$

where we have set $\varphi(z, \zeta) = \sum_{k=1}^{n} (\zeta_k - z_k)$.
Finally, we may define the sheaves

$$\mathcal{C}'^b = \mathbf{F}^{(n-1)}(\mathcal{O}'^a) \quad \text{and} \quad \mathcal{C}''^b = \mathbf{F}^{(n-1)}(\mathcal{O}''^a),$$

and prove that they give rise to the following

Theorem 3. (*a*) *The restrictions of $\mathcal{C}'^b$ and $\mathcal{C}''^b$ to $S^*\Omega$ verify*

$$\mathcal{C}'^b_{|S^*\Omega} = 0 \quad \text{and} \quad \mathcal{C}''^b_{|S^*\Omega} = \mathcal{C}.$$

(b) For any $(x_0, \eta_0) \in S^*F$, *the stalks* $\mathscr{C}^i_{(x_0, \eta_0)}$ $(i = ', '')$ *are respectively isomorphic to*

$$\frac{\mathcal{O}^i(\omega_m \times \bigcap_{j=1}^n \gamma_m^{(j)})}{\delta \bigoplus_{k=1}^n \mathcal{O}^i(\omega_m \times \bigcap_{j \neq k} \gamma_m^{(j)})} \qquad i = ', '',$$

where the sets ω_m *form a decreasing basis of neighborhoods of* x_0 *in F, where* δ *is the Cech coboundary operator (alternate sum of restrictions) and where we have set* $\gamma_m^{(j)} = \{\xi \in S_{n-1} : \langle \xi, \eta_m^{(j)} \rangle < 0\}$ *with sequences* $\eta_m^{(j)} \in S^*_{n-1}$ $(j = 1, \ldots, n)$ *such that* $\lim_{m \to \infty} \eta_m^{(j)} = \eta_0$ *holds for every j and such that the unions* $\bigcup_{j=1}^n \gamma_m^{(j)}$ *constitute a decreasing basis of neighborhoods of the closed hemisphere* $\{\xi \in S_{n-1} : \langle \xi, \eta_0 \rangle \leq 0\}$.

4. Microlocalization

Under the microlocalization functor **F**, the commutative square diagram of Theorem 1 is transformed in distinguished triangles of the derived category of sheaves over S^*F. Unfortunately, $\mathbf{F}(\mathscr{A}'')$ is not concentrated in one degree only and we have to compute further cohomology sheaves to get usable results. This can be done, namely by projecting the objects on the base space SF, and we consequently obtain

   Theorem 4.   *The following three commutative diagrams of sheaves over F, SF and* S^*F,

$$0 \to (\iota_*\mathcal{O})_{|F} \to (\iota_*\mathscr{A}) \to \pi_*\mathscr{C}'^b \to 0$$
$$\| \qquad\qquad \downarrow \qquad\qquad \downarrow$$
$$0 \to (\iota_*\mathcal{O})_{|F} \to (\iota_*\mathscr{B}) \to \pi_*\mathscr{C}''^b \to 0,$$

$$0 \to \mathscr{A}' \to \tau^{-1}\iota_*\mathscr{A} \to \pi_*\tau^{-1}\mathscr{C}'^b \to 0$$
$$\downarrow \qquad\qquad \downarrow \qquad\qquad \downarrow$$
$$0 \to \mathscr{A}'' \to \tau^{-1}\iota_*\mathscr{B} \to \pi_*\tau^{-1}\mathscr{C}''^b \to 0,$$

$$0 \to \ker \sigma' \to \pi^{-1}\iota_*\mathscr{A} \overset{\sigma'}{\to} \mathscr{C}'^b \to 0$$
$$\downarrow \qquad\qquad \downarrow \qquad\qquad \downarrow$$
$$0 \to \ker \sigma'' \to \pi^{-1}\iota_*\mathscr{B} \overset{\sigma''}{\to} \mathscr{C}''^b \to 0,$$

have exact rows. Moreover, decompositions similar to the ones appearing in Theorem 3.b allow us to compute explicit representations of the above morphisms in the stalks.

   Hence, we may introduce the following

Definitions. Denoting by σ' [resp. σ''] the two morphisms $\pi^{-1}\iota_*\mathcal{A} \to \mathscr{C}'^b$ [resp. $\pi^{-1}\iota_*\mathcal{B} \to \mathscr{C}''^b$] given by Theorem 4, we become able to define the *boundary wave-front sets* of real analytic functions and hyperfunctions by

$$\mathrm{WF}^b(f) = \mathrm{supp}_{C'^b}(\sigma'f) \qquad \forall f \in \mathcal{A}$$

$$\mathrm{WF}^b(f) = \mathrm{supp}_{C'^b}(\sigma''f) \qquad \forall f \in \mathcal{B}.$$

The meaning of such notions is then explicitly described by

Theorem 5. (a) *If $U \subset SF$ is an open set with conically convex fibers, the following are equivalent for every $g \in \mathscr{C}'^b(\tau'U \times S^*_{n-1})$ [resp. $\mathscr{C}''^b(\tau'U \times S^*_{n-1})$]*

$$\mathrm{supp}(g) \subset \{(x, \eta) : (x, \xi) \in U \Rightarrow \langle \xi, \eta \rangle \geq 0\}$$
$$\exists f \in \mathcal{A}'(U) \text{ [resp. } \mathcal{A}''(U)] \text{ such that } g = \sigma'(f) \text{ [resp. } g = \sigma''(f)].$$

(b) *Given $(x, \eta) \in \partial_{\mathbf{D}^n}\Omega \times S^*_{n-1}$ and f a section of $\iota_*\mathcal{A}$ [resp. $\iota_*\mathcal{B}$] near x, one has*

$$(x, \eta) \notin \mathrm{WF}^b(f) \Leftrightarrow f = \sum f_j$$

for some finite decomposition of f in functions $f_j \in \mathcal{A}'(\omega \times \gamma_j)$ [resp. $\mathcal{A}''(\omega \times \gamma_j)$] with ω an open neighborhood of x in F and γ_j conically convex open subsets of $\{\xi \in S_{n-1} : \langle \xi, \eta \rangle < 0\}$.

5. Some Applications and Remarks

Taking the compactness of F into account, we may directly extend assertion b) of Theorem 5 as a global one over F:

Proposition 6. *Any $f \in (\iota_*\mathcal{A})(F)$ [resp. $(\iota_*\mathcal{B})(F)$] whose WF^b does not meet $F \times \{\eta_0\}$ is finitely decomposable over F as a sum of sections f_j of $\mathcal{A}'$ [resp. $\mathcal{A}''$] over $F \times \gamma_j$ with conically convex sets $\gamma_j \subset \{\xi \in S_{n-1} : \langle \xi, \eta \rangle < 0\}$.*

A second straightforward application of this theory constitutes of course the concrete version of Theorem 2 which generalizes directly Bogolyubov's theorem and was also proved recently by Zarinov (cf. [11]).

Proposition 7. (a) *For any conically convex open subset $\gamma \subset S_{n-1}$ and any sections $f_\pm$ of $\mathcal{A}'$ [resp. $\mathcal{A}''$] over $F \times (\pm\gamma)$ such that f_+ and f_- coincide as* analytic functions [*resp. as hyperfunctions*] over Ω, there exists $v > 0$ and $f \in \mathcal{O}(\Omega_v)$ such that $f = f_{\pm}$ where such equalities make sense.

(b) For any conically convex open subsets γ, $\gamma' \subset S_{n-1}$ and any sections g, h of $\mathscr{A}'$ [*resp. $\mathscr{A}''$*] over $F \times \gamma$ and $F \times \gamma'$ respectively which agree over Ω as analytic functions [*resp. hyperfunctions*], there exists $f \in \mathscr{A}'(F \times \mathrm{conv}(\gamma \cup \gamma'))$ [*resp. $\mathscr{A}''(F \times \mathrm{conv}(\gamma \cup \gamma'))$*] such that one has $g = f$ and $h = f$ where such equalities make sense (*here "conv" means the trace on S_{n-1} of the convex hull of $\{r\xi : r > 0, \xi \in \gamma \cup \gamma'\}$*).

Let us now consider some applications of this theory to partial differential equations. We shall therefore denote indifferently by $P(D)$ the linear partial differential operator $\sum_{|\alpha| \le m} c_\alpha D_x^\alpha$ over Ω with constant coefficients $c_\alpha \in \mathbf{C}$ or its natural extension to the complex domain: $\sum c_\alpha D_z^\alpha$, where we have set $D_{z_j} = (D_{x_j} - i D_{y_j})/2$. The characteristic variety of P will be denoted by

$$\mathrm{Char}(P) = \{\xi \in \mathbf{C}^n \backslash \{0\} : P^0(\zeta) = 0\},$$

where P^0 denotes the principal symbol $\sum_{|\alpha| = m} c_\alpha \zeta^\alpha$.

One checks easily that P induces endomorphisms of $\mathcal{O}'$, $\mathcal{O}''$, $\pi^{-1} \iota_* \mathscr{A}$ and $\pi^{-1} \iota_* \mathscr{B}$. Moreover, taking the images of $P : \mathcal{O}' \to \mathcal{O}'$ and $P : \mathcal{O}'' \to \mathcal{O}''$ under F, we also obtain endomorphisms P' of $\mathscr{C}'^b$ and P'' of $\mathscr{C}''^b$. Writing the morphisms σ' and σ'' explicitly in the stalks and using the representations given in Theorem 3, one may prove easily that P' [resp. P''] is surjective in the stalks at points $(x_0, \eta_0) \in S^* F$, where x_0 admits a basis of neighborhoods $\{\omega\}$ in F verifying $P \mathscr{A}(\omega \cap \Omega) = \mathscr{A}(\omega \cap \Omega)$ [resp. $P \mathscr{B}(\omega \cap \Omega) = \mathscr{B}(\omega \cap \Omega)$].

Assuming moreover that η_0 is not characteristic for P, using elementary geometric considerations and the Malgrange-Ehrenpreis principle and taking the analyticity propagation results of [1] into account, we may even prove

Proposition 8. *The endomorphisms P' and P'' are bijective in the stalks at points $(x_0, \eta_0) \in S^* F$ such that $P^0(\eta_0) \ne 0$.*

From this we may deduce directly that the fundamental principle of microlocalization also holds for boundary wave front sets:

Sato's principle at the boundary. *If u and f are simultaneously real analytic functions or hyperfunctions verifying $Pu = f$ over $\omega \cap \Omega$, where ω is an open*

neighborhood of $x_0 \in F$ in $\mathbf{D}^n$, the following inclusions hold:

$$\mathrm{WF}^b(f) \subset \mathrm{WF}^b(u) \subset \mathrm{WF}^b(f) \cup \{(x, \eta) \in S^*F: P^0(\eta) = 0\}.$$

Let us finally consider an open subset ω of F such that $\omega \cap \Omega$ is convex and Γ an open convex cone of $\mathbf{R}^n$ whose polar set $\{\eta \in S^*_{n-1}: \langle \eta, \xi \rangle \geq 0, \xi \in \Gamma\}$ does not meet $\mathrm{Char}(P)$. Applying the Malgrange–Ehrenpreis principle and the results of Bony-Schapira once again, one directly gets that the equation $Pu = f$ is solvable in the spaces $\mathscr{A}'[\omega \times (\Gamma \cap S^{n-1})]$ and $\mathscr{A}''[\omega \times (\Gamma \cap S^{n-1})]$. As a corollary of this we directly recover the classical result concerning the analyticity of solutions of $Pu = f$ with f analytic and P elliptic.

Remarks. (a) This theory is certainly related to Schapira's corresponding one (cf. [10] and Schapira's contribution to the present volume).

(b) Concerning applications of this theory to P.D.E., much of the work remains to be done.

(c) Among open questions intrinsically connected with $\mathscr{C}'^b$ and $\mathscr{C}''^b$, one of the most interesting ones consists in determining whether these sheaves are flabby, smooth, fine . . . .

References

[1] J.-M. Bony and A. P. Schapira, Existence et prolongement des solutions holomorphes des équations aux dérivées partielles. *Inventiones Mathematicae* **17** (1972) 95–105.
[2] A. Kaneko, On the global existence of real analytic solutions of linear partial differential equations on unbounded domains, *J. Fac. Sci. Univ. Tokyo, Sec. I.A.* **32** (1985) 319–372.
[3] K. Kataoka, Microlocal theory of boundary value problems. I, *J. Fac. Sci. Univ. Tokyo Sect. I.A.* **27** (1980) 355–390; II, **28** (1981) 31–56.
[4] H. Komatsu, Boundary values for solutions of elliptic equations. *Proc. Int. Conf. Func. Anal. Rel. Topics*, Univ. Tokyo Press (1970) 107–121.
[5] J.-L. Lieutenant, Fronts d'onde à l'infini des fonctions analytiques réelles, *Ann. Inst. Fourier* **34** (1984) 164–193.
[6] J.-L. Lieutenant, Microlocalization at the boundary of a convex set, *J. Fac. Sci. Univ. Tokyo, Sec. I.A.* **33** (1986) 83–130.
[7] J.-L. Lieutenant, Vanishing cohomology over convex domains tapering at their real boundary, to appear.
[8] T. Oaku, Boundary value problems for systems of linear partial differential equations and propagation of micro-analyticity, *J. Fac. Sci. Tokyo, Sec. I.A.* **33** (1986) 175–232.
[9] M. Sato, T. Kawai, and M. Kashiwara, Microfunctions and pseudo-differential equations, *Lect. Notes in Math.* Vol. 287, Springer Verlag, Berlin-Heidelberg-New York, 1973, pp. 265–529.
[10] P. Schapira, Front d'onde analytique au bord I, *C.R. Acad. Sci. Paris* **302** (1986) 383–386.
[11] V. Zarinov, A generalization of Bogolyubov's "Edge of the Wedge" theorem, *A.M.S. Transl. of Soviet Math. Dokl.* **24** (1981) 425–429.

Extension of Holonomic $\mathscr{D}$-modules

B. Malgrange

Institut Fourier
St. Martin d'Hères
France

Introduction

Let X be a complex analytic manifold, and Y a closed hypersurface of X. We denote, as usual by $\mathcal{O}_X$ (resp. $\mathscr{D}_X$) the sheaf of holomorphic functions (resp. holomorphic differential operators) on X; we denote also by $\mathcal{O}_X[Y]$ (resp. $\mathscr{D}_X[Y]$) the sheaf of meromorphic functions with poles on Y (resp. $\mathcal{O}_X[Y] \otimes_{\mathcal{O}_X} \mathscr{D}_X$).

Let M be a holonomic left $\mathscr{D}_X$-module. In [K 1], Kashiwara proves the existence of a b-function for (M, Y); from that result, he deduces that $M[Y] = \mathscr{D}_X[Y] \otimes_{\mathscr{D}_X} M$ is also a holonomic $\mathscr{D}_X$-module. The purpose of this paper is to develop further this type of argument to classify the holonomic $\mathscr{D}_X$-modules N such that $N[Y] \simeq M[Y]$.

The result obtained can be considered as a generalization of a theorem by MacPherson and Vilonen on extension of perverse sheaves [M–V 2]; see also [M–V 1] and [V 1]. Actually, their result is equivalent to the special case where the holonomic modules to be considered are *regular*. This follows from the following double (and nontrivial!) translation:

Algebraic Analysis, Volume I

(*i*)   Translation "perverse sheaves" ↔ "holonomic regular modules" by the so-called "Riemann–Hilbert correspondence" [K-2], [Me], [B-B-D].

(*ii*) Translation "vanishing cycles" ↔ "*b*-function;" for this point, see [Be], [K 3] and, for a special case, [Ma 1].

The version of the second translation that I use here is closely related to [K 3]: from this paper, I have taken the basic idea of enlarging the class of $\mathscr{D}$-modules to be considered and to resolve them locally by "elementary" ones. The difference with [K 3] is that, here, one works with "usual" *b*-functions, and no regularity conditions.

The point of view used here was developed more or less independently in [Ma 2], [S], [Sa]; as far as I know, very close results are also known to Beilinson. I hope that he will publish his results some time.

M. Sato was the first to introduce and to study *b*-functions. Therefore, it is natural to publish these results in a volume dedicated to him. This is also a great pleasure to me.

1.   Specializable Modules

We will assume that Y is a *smooth* hypersurface of X. (The general case could be reduced to this one by the following trick: let E be the line bundle on X defined by Y, and σ its canonical section whose divisor is Y. Then, replace M by $\sigma_* M$ and Y by the zero section of E. I leave the details to the reader.)

Denote by $\mathscr{I} \subset \mathcal{O}_X$ the ideal of Y, and by $V^{\cdot}$ the Kashiwara filtration of $\mathscr{D}_X \mid Y : a \in V^i(\mathscr{D}_X)$ iff, $\forall k \geq 0$, one has $a\mathscr{I}^k \subset \mathscr{I}^{k+i}$; then $\mathrm{gr}_V \mathscr{D}_X$ is canonically isomorphic to the sheaf (on Y) of differential operators on $T_Y X$, polynomial in the fibers.

Let M be a coherent (left) $\mathscr{D}_X$-module. A decreasing filtration $\{M^i\}_{i \in \mathbf{Z}}$ on $M \mid Y$ is a *V*-filtration iff one has $V^i \mathscr{D}_X M^j \subset M^{i+j}$ for all i, j and $U M^i = M$. The filtration is called *good* iff it satisfies the following further conditions:

(*i*)   The M^i are $V^0 \mathscr{D}_X$-coherent.

(*ii*) On every compact, one has $V^1 \mathscr{D}_X M^i = M^{i+1}$ for $i \gg 0$, and $V^{-1} \mathscr{D}_X M^i = M^{i-1}$ for $i \ll 0$.

Locally, any coherent M admits a good *V*-filtration (take a surjection $\mathscr{D}^p \to M \to 0$, and take on M the filtration quotient).

Definition 1.1. A coherent $\mathscr{D}_X$-module M is said to be "specializable on Y" if, for a good V-filtration of M, the action of the Euler vector $\theta \in \Gamma(Y, \mathrm{gr}_V \mathscr{D}_X)$ is locally finite, e.g., for any $y \in Y$ and any $m \in (\mathrm{gr}\, M)_y$ there exists $b \in \mathbf{C}[s]$ such that $b(\theta)m = 0$.

One verifies easily that this definition is independent of the (local) good filtration which has been chosen.

Fix now a lifting τ of the canonical map $\mathbf{C} \to \mathbf{C}/\mathbf{Z}$, and denote by A the image of τ. For instance, one could take $A = \{-1 < \mathrm{Re}\, \alpha \leq 0\}$, but it does not matter. We will however suppose that $0 \in A$ (this is required for some of the results of Section 2). The next construction is an extension of a well-known result in the theory of O.D.E. with regular singularities.

Theorem 1.2. *If M is specializable on Y, then $M \,|\, Y$ admits one and only one good V-filtration with the following property: The eigenvalues of θ, acting on $\mathrm{gr}^i M$, belong to $A + i$.*

For the proof, see [K 3] or [S]. From now on, we will call "canonical" this filtration, and denote by $\mathrm{Sp}\, M$ the corresponding $\mathrm{gr}\, M$. It is easy to verify that $\mathrm{Sp}\, M$ does not depend on τ if one forgets about the degrees.

Theorem 1.3. *Let $0 \to M' \to M \to M'' \to 0$ be an exact sequence of coherent $\mathscr{D}_X$-modules. Then*

(*i*) *M is specializable on Y iff M' and M'' have the same property.*

(*ii*) *The canonical V-filtration of M' (resp. M'') is induced by (resp. quotient of) the canonical V-filtration of M.*

The only nontrivial point is the fact that the filtration induced by $V(M)$ on M' is good. This is done by an argument of the Artin–Rees type; *cf.* [S].

Examples 1.4.

(*i*) A holonomic module is specializable; this is proved in [K 3] (one can prove that "specializable" is equivalent to the existence of a b-function in the sense of [K 3]; I leave this question to the reader).

(*ii*) Suppose that X is a disc in $\mathbf{C}$ and $Y = \{0\}$. Take M holonomic on X. It is known that the formal completion $\hat{M}$ of M at 0 can be decomposed: $\hat{M} = \hat{M}' \oplus \hat{M}''$, with $\hat{M}'$ regular and $\hat{M}''$ purely irregular ($=$ it has no regular submodule). Then, one has $\widehat{\mathrm{Sp}\, M} = \hat{M}'$.

Sketch of the proof: if M is regular, then Sp $M \simeq M$ (this is essentially Fuchs's theory). On the other hand, if $\hat{M}$ is purely irregular, one can write $\hat{M} = \hat{\mathscr{D}} / \hat{\mathscr{D}}_p$, with $p = 1 + q$, $q \in V^1 \hat{\mathscr{D}}$; then, one has $\mathrm{gr}_V\, p = 1$ and Sp $\hat{M} = 0$.

(iii) Take again for ξ an arbitrary complex manifold, and let M be a coherent $\mathscr{D}_X$-module with support on Y; then M is specializable on Y and the eigenvalues of θ are the (strictly) negative integers.

In fact, let $(y, t) = (y_1, \ldots, y_{n-1}, t)$ be local coordinates on X such that $Y = \{t = 0\}$. Any $m \in M \mid Y$ verifies $t^i m = 0$ for some i; taking $\tilde{\theta} = t\, \partial / \partial t$ as lifting of θ, one has

$$\frac{\partial^i}{\partial t^i} t^i m = (\tilde{\theta} + 1) \cdots (\tilde{\theta} + i) m = 0.$$

(iv) If Y is non-characteristic for M, then M is specializable on Y. One reduces the proof to the case where $M = \mathscr{D} / \mathscr{D}_p$, with Y non-characteristic for p; and, in that case, the result is easy.

2. Direct and Inverse Images

In this section, I recall very briefly some results which are essentially in [Ma 2] or [S]; I will need a part of them in the next section. I shall use the following notations: I denote by $Y \rightarrow_i X \leftarrow^j X - Y$ the canonical embeddings, and I set also $Y \rightarrow^i T_Y X \leftarrow^j T_Y X - Y$. For M a (left) $\mathscr{D}_X$-module, or more generally an object of $D^b(\mathscr{D}_X)$, I set $i^! M = \mathscr{D}_{Y \rightarrow X} \otimes_{\mathscr{D}_X}^L M[-1]$, and $i^* M = Di^! DM$, where D is the dualizing functor (for the notations not defined here, see for instance [B], Chapter VI). For $N \in \mathrm{ob}\, D^b(\mathscr{D}_Y)$, I set $i_! B = i_* N = \mathscr{D}_{X \leftarrow Y} \otimes_{\mathscr{D}_Y} N$.

For the open embedding j, I use the "moderate" images, defined in the following way: for $M \in \mathrm{ob}\, D^b(\mathscr{D}_X)$, I write $j^! M = j^* M = M[Y]$ as $\mathscr{D}_X[Y]$-modules. For $N \in \mathrm{ob}\, D^b(\mathscr{D}_X[Y])$, I denote by $j_* N$ the object of $D^b(\mathscr{D}_X)$ obtained by restriction of the scalars from $\mathscr{D}_X[Y]$ to $\mathscr{D}_X$; finally, I set $j_! N = Dj_* DN$.

To analyze these notions for specializable modules, it is convenient to look first at the special case $X = Y \times D$, $D \subset \mathbf{C}$ a disc $|t| < r$; of course one identifies Y with $Y \times \{0\}$.

Now, take for M a specializable $\mathscr{D}_X$-module. Then the following properties are true.

2.1. One has a functorial isomorphism $i^!M = [\mathrm{Sp}_{\cdot}^{-1}M \overset{t}{\to} \mathrm{Sp}^0 M]$ (the dot under the term of degree zero). For, one has an exact sequence of bi-$(\mathscr{D}_Y, \mathscr{D}_X)$-modules $0 \to \mathscr{D}_X \to {}^t\mathscr{D}_X \to \mathscr{D}_{Y \to X} \to 0$ (left multiplication by t); therefore one has $i^!M = [\underset{\cdot}{M} \to {}^t M]$. Now, consider the maps of $\mathscr{D}_Y$-complexes

$$[\underset{\cdot}{M} \overset{t}{\to} M] \to [\underset{\cdot}{M}/M^0 \overset{t}{\to} M/M^1] \leftarrow [\underset{\cdot}{M}^{-1}/M^0 \overset{t}{\to} M^0/M^1].$$

To prove the result, it suffices to prove that these maps are quasi-isomorphisms; this is a consequence of the following lemma.

Lemma 2.2. *For $k \neq -1$ (resp. $k \geq 0$), the map $t: \mathrm{Sp}^k M \to \mathrm{Sp}^{k+1} M$ (resp. $t: M^k \to M^{k+1}$) is bijective.*

The first assertion follows from the fact that $t\partial_t$ (resp. $\partial_t t$) is bijective on $\mathrm{Sp}^k M$ for $k \neq 0$ (resp. $k \neq -1$). The second follows from the first one, and the fact that $t: M^k \to M^{k+1}$ is bijective for $k \gg 0$ (property *(ii)* of good V-filtrations).

2.3. The preceding result shows that t is bijective on M iff it is so on $\mathrm{Sp}\, M$; this is equivalent to say that one has $M \overset{\sim}{\to} j_* j^* M$ iff one has $\mathrm{Sp}\, M \overset{\sim}{\to} \bar{j}_* \bar{j}^* \mathrm{Sp}\, M$.

A little bit more careful analysis shows actually the following result: if M is specializable, then $j_* j^* M$ is specializable, and one has an isomorphism $\bar{j}_* \bar{j}^* \mathrm{Sp}\, M = \mathrm{Sp}\, j_* j^* M$. (Actually, $j_* j^* M$ is isomorphic to $\mathscr{D}_X \otimes_{V^0 \mathscr{D}_X} t^{-1} M^0$, with the obvious structure of $V^0 \mathscr{D}_X$-module on $t^{-1} M^0$.)

2.4. If M is specializable, then, for each k, $\underline{H}^k DM$ is specializable and one has an isomorphism $\mathrm{Sp}\, \underline{H}^k DM = \underline{H}^k D \,\mathrm{Sp}\, M$; this is proved in [S] by resolving locally M by "elementary modules," in the sense explained there.

From this follows that $j_! j^! M$ is specializable in each degree; actually one proves that $\underline{H}^k j_! j^! M = 0$ for $k \neq 0$ and that $M'' = \underline{H}^0 j_! j^! M$ satisfies "$\partial_t : \mathrm{Sp}^0 M'' \to \mathrm{Sp}^{-1} M''$ is bijective." (It is sufficient to prove the similar assertions for $\mathrm{Sp}\, M$ instead of M; to do this, one takes local resolutions of $\mathrm{Sp}\, M$ by direct sums of molecules of the type $\mathrm{gr}_V \mathscr{D}_X / \mathrm{gr}_V \mathscr{D}_X : b(\theta)$.)

Since the natural map $M'' \to M$ has its kernel and cokernel with support on Y, one has $\mathrm{Sp}^k M'' \overset{\sim}{\to} \mathrm{Sp}^k M$ for $k \geq 0$. From all that one deduces the isomorphism $\bar{j}_! \bar{j}^! \mathrm{Sp}\, M = \mathrm{Sp}\, j_! j^! M$.

2.5. One has $i^*M = [\mathrm{Sp}^0 M \xrightarrow{\partial_t} \mathrm{Sp}_{\cdot}^{-1} M]$; actually, one has

$$i^*M = i^*[j_!j^!M \to M] \qquad \text{since } i^*j_!j^!M = 0$$

$$= i^![j_!j^!M \to M] \qquad \text{since the complex } [\cdots] \text{ has cohomology with support on } Y \text{ (see, for instance } [B]).$$

Then, the results follow easily from 2.1 and 2.4.

2.6. The preceding results can be essentially summarized by saying that, in the case where $X = Y \times D$, the functors $i^!$, i^*, $j_!j^!$ and j_*j^* commute with Sp.

Now, suppose that Y is an arbitrary smooth hypersurface of X. It is easy to deduce from these results that $j_!j^!$ and j_*j^* commute with Sp. It seems to me likely that this is true also for $i^!$ and i^*, but I have no proof of it (of course, the result is true in each degree, but the problem is to define suitable morphisms).

There is however one case where this result can be proved, e.g., the case where one has a smooth morphism $X \xrightarrow{f} D$, (D a disc in $\mathbf{C}$), and $Y = f^{-1}(0)$ (one can reduce this case to the case of a product by the usual trick: embed X in $X \times D$ by the graph map, and replace Y by $X \times \{0\}$).

2.7. In the case $X = Y \times D$, or, more generally, in the case $X \xrightarrow{f} D$, it is an easy application of 2.5 to interpret $\mathrm{Sp}^0 M$ and $\mathrm{Sp}^{-1} M$ respectively as "moderate nearby cycles" and "moderate vanishing cycles," the formulas for "can" and "var" being given, as usual, by

$$\mathrm{can} = \partial_t, \qquad \mathrm{var} = h(t\partial_t)t, \qquad h(\xi) = \frac{e^{2\pi i \xi} - 1}{\xi}.$$

As this is an obvious generalization of the regular case, and as this last one is written in many places, I omit this point.

3. Extensions

Again, Y is an arbitrary smooth closed hypersurface of X. By definition, a coherent $\mathscr{D}_X[Y]$-module $\tilde{M}$ is specializable iff $j_*\tilde{M}$ is specializable; as already mentioned, one can prove that this is equivalent to the existence of a functional equation for mf^s, $m \in \tilde{M}$, of the type considered in [K 3] and other papers. We are interested to classify the coherent $\mathscr{D}_X$-modules M provided with an isomorphism $j^*M \simeq \tilde{M}$.

Since $j_*\tilde{M}$ is specializable, and since the kernel and the cokernel of $M \to j_*j^*M = j_*\tilde{M}$ have their support on Y, M is specializable; furthermore, the $\mathrm{gr}_V\mathscr{D}_X$-module $\mathrm{Sp}\, M$ is *monodromic* in the sense of [V 2], e.g., it is coherent and the action of θ on it is locally finite.

Now, denote

(i)   by $\mathrm{Sp}(X, Y)$ the category of coherent $\mathscr{D}_X$-modules, specializable on Y.

(ii)   by $\widetilde{\mathrm{Sp}}(X, Y)$ the category of triples $(\tilde{M}, \bar{M}, \sigma)$ where $\tilde{M}$ is a specializable $\mathscr{D}_X[Y]$-module, $\bar{M}$ a monodromic $\mathrm{gr}_V\mathscr{D}_X$-module, and σ an isomorphism $\mathrm{Sp}\, j_*\tilde{M} \overset{\sim}{\to} \bar{j}_*\bar{j}^*\bar{M}$.

(iii) by φ the functor $\mathrm{Sp}(X, Y) \to \widetilde{\mathrm{Sp}}(X, Y)$ which associate to M: j^*M, $\mathrm{Sp}\, M$ and the obvious σ. Then the result is

Theorem 3.1. *The functor φ is an equivalence.*

This theorem reduces the problem of the extension of $\tilde{M}$ to the monodromic case. It can be stated in several ways, using vanishing cycles, can, var, *etc.* For these questions, I refer to [V 2].

3.1. φ is Fully Faithful

$\mathrm{Sp}(X, Y)$ is an abelian category; it does not have enough injectives, but in any case we have the functors Hom and Ext^1, and their local versions $\underline{\mathrm{Hom}}$ and $\underline{\mathrm{Ext}}^1$. For M and $N \in \mathrm{ob}\,\mathrm{Sp}(X, Y)$, we have therefore morphisms

$$\underline{\mathrm{Hom}}(\varphi): \underline{\mathrm{Hom}}(M, N) \to \underline{\mathrm{Hom}}(\varphi(M), \varphi(N))$$

$$\underline{\mathrm{Ext}}^1(\varphi): \underline{\mathrm{Ext}}^1(M, N) \to \underline{\mathrm{Ext}}^1(\varphi(M), \varphi(N)).$$

Lemma 3.2. *These maps are bijective in the following cases:*

(i)   *$j_!j^!M \to M$ is an isomorphism*;

(ii)   *M has support on Y.*

(i) is an easy consequence of the following fact, applied to $k = 0, 1$: the natural map $\mathrm{Ext}^k(j_!j^!M, N) \to \mathrm{Ext}^k(j^!M, j^!N)$ is an isomorphism. (It suffices to prove this fact for N replaced by $\mathscr{D}_X$; in this case, it is just the definition of $j_!$).

To prove (ii), note first that there exists L, coherent on $\mathscr{D}_Y$ such that $M = i_!L$; as the verification is local it is therefore sufficient to study the case

$L = \mathscr{D}_Y$ and $M = \mathscr{D}_{X \leftarrow Y}$; we can also suppose that $X = Y \times D$, D a disc $|t| < r$, and $Y = Y \times \{0\}$. Then we have a resolution

$$0 \to \mathscr{D}_X \xrightarrow{t} \mathscr{D}_X \to \mathscr{D}_X \to \mathscr{D}_{X \leftarrow Y} \to 0$$

(right multiplication by t); therefore, $R \underline{\mathrm{Hom}}(\mathscr{D}_{X \leftarrow Y}, N)$ is the complex $0 \to N \xrightarrow{t} N \to 0$ (left multiplication). The arguments of (2.1) show that this complex is quasi-isomorphic to $0 \to \mathrm{Sp}\, N \xrightarrow{t} \mathrm{Sp}\, N \to 0$. I leave to the reader the verification that the isomorphisms obtained here are precisely $\underline{\mathrm{Hom}}(\varphi)$ and $\underline{\mathrm{Ext}}^1(\varphi)$. This proves the lemma.

(In principle, one could also have used the duality map $\underline{\mathrm{Ext}}^k(i_! L, N) = i_* \underline{\mathrm{Ext}}^k(L, i^! N)$ and apply (2.1). But, due to the rather complicated definition of the duality map, the commutativity of diagram to be proved seems to me less obvious.)

Now, to prove that φ is fully faithful, it is sufficient to prove that $\underline{\mathrm{Hom}}(\varphi)$ is an isomorphism. To prove this result, we factorize the natural map $j_! j^! M \to M$ into exact sequences: $0 \to P \to j_! j^! M \to M' \to 0$, $0 \to M' \to M \to Q \to 0$, with P and Q having support on Y. One has an exact sequence

$$0 \to \underline{\mathrm{Hom}}(M', N) \to \underline{\mathrm{Hom}}(j_! j^! M, N) \to \underline{\mathrm{Hom}}(P, N) \to \cdots$$

and the similar exact sequence when one applies φ; the last two terms are isomorphic, therefore the first one is so. Then, one applies the same arguments to the second exact sequence (here, one needs "$\underline{\mathrm{Ext}}^1$"). This proves the result.

3.2. φ is Essentially Surjective

This is done essentially by the same argument: given $(\tilde{M}, \bar{M})$ with $\mathrm{Sp}\, j_* \tilde{M} = \bar{j}_* \bar{j}^* \bar{M}$, we factorize $\bar{j}_! \bar{j}^! \bar{M} \to \bar{M}$ into $0 \to \bar{P} \to \bar{j}_! \bar{j}^! \bar{M} \to \bar{M}' \to 0$ and $0 \to \bar{M}' \to \bar{M} \to \bar{Q} \to 0$. The pair $(\tilde{M}, \bar{j}_! \bar{j}^! \bar{M})$ is represented by $j_! \tilde{M}$ and $(0, \bar{P})$ and $(0, \bar{Q})$ are represented by P and Q with support on Y. Using (3.2) one gets a map $P \to j_! \tilde{M}$ whose cokernel M' represent $(\tilde{M}, \bar{M}')$; then, using again (3.2), one gets an extension $Q \to M'[1]$ which gives the required M. This ends the proof of the theorem.

References

[B] A. Borel *et al.*, *Algebraic D-modules*, Academic Press, 1987.

[B–B–D] A. Beilinson, J. Bernstein, and P. Deligne, *Faisceaux pervers*, *Astérisque* **100** (1982).

[Be] A. Beilinson, Letter to P. Deligne (March 1981), unpublished.

[K 1]   M. Kashiwara, On the holonomic systems of differential equations, II, *Inv. Math.* **49** (1978) 121–135.

[K 2]   M. Kashiwara, The Riemann Hilbert problem for holonomic systems, *RIMS* 437, Kyoto University, 1983.

[K 3]   M. Kashiwara, Vanishing cycle sheaves and holonomic systems of differential equations, *Springer Lecture Notes* **1016** (1983) 134–142.

[M-V 1]   R. MacPherson and K. Vilonen, Une construction élémentaire des faisceaux pervers, *C. R. Acad. Sci. Sér. I Math.* **299** (1984), 443–446.

[M-V 2]   R. MacPherson and K. Vilonen, Elementary construction of perverse sheaves, *Inv. Math.* **84** (1986) 403–435.

[Ma 1]   B. Malgrange, Polynôme de Bernstein–Sato et cohomologie évanescente, *Astérisque* **101–102** (1983) 243–267.

[Ma 2]   B. Malgrange, Letter to P. Deligne (Janvier 84), unpublished.

[Me]   Z. Mebkhout, Une équivalence de catégories, et une autre équivalence de catégories, *Comp. Math.* **51** (1984) 55–69.

[S]   C. Sabbah, $\mathscr{D}$-modules et cycles évanescents, *Proceedings of the Conference held at La Rabida, Spain 1984*; Hermann, Paris, 1987.

[Sa]   M. Saito, Modules de Hodge polarisables, *RIMS* 553, Kyoto University, 1986.

[V 1]   J. L. Verdier, Extension of a perverse sheaf over a closed subspace, *Astérisque* **130** (1985) 210–217.

[V 2]   J. L. Verdier, Prolongement de faisceaux pervers monodromiques, *Astérisque* **130** (1985) 218–236.

On the Irregularity of the $\mathscr{D}_X$ Holonomic Modules

Z. Mebkhout

U.E.R. de Mathématiques
Université de Paris
Paris, France

0. Introduction

We would like to introduce the reader to the irregularity of the $\mathscr{D}_X$ holonomic modules. A key result of the last decade is the equivalence of categories between the triangulated category of holonomic regular complexes and the triangulated category of constructible complexes on a smooth complex algebraic or analytic variety [K 1], [Me 1]. The proof of this result depended heavily on Hironaka's theorem on the resolution of the singularities.

But the category of the regular $\mathscr{D}_X$ modules is smaller than the category of the irregular $\mathscr{D}_X$ modules. It turns out that it is much easier and simpler to deduce the properties of the regularity from the properties of the irregularity.

The studies of the irregularity are motivated (for us) first by Malgrange's work [M 1], [M 2] on the singular irregular points of the differential equations in the complex domain, and second by Deligne's Principle: The properties of the wild ramification of the l-adic sheaves in characteristic

Algebraic Analysis, Volume I

$p > 0$ are similar to the properties of the irregularity of the $\mathcal{D}_X$ modules in characteristic zero. The tame ramification corresponds to the regularity.

We illustrate in Sections 1 and 2 Deligne's Principle, and in Section 3 we show how to avoid Hironaka's theorem in the theory of the $\mathcal{D}_X$ modules over the complex number field. Since the analogue of Hironaka theorem is not (yet) available in characteristic $p > 0$ this could help to prove the properties of the cohomology of algebraic varieties over a field of positive characteristic. The reader is asked to see the references for complete proofs.

1. The Semi-Continuity Theorem

Let $(X, \mathcal{O}_X)$ be a smooth algebraic variety over a field of characteristic zero or a complex analytic manifold. Let $\mathcal{D}_X$ be the sheaf of differential operators of finite order with coefficients in $\mathcal{O}_X$. If $\dim X = 1$ any $\mathcal{D}_X$-holonomic module $\mathcal{M}$ has only isolated singularities. If $o \in X$ the Fuchs irregularity of $\mathcal{M}$ at o is a non-negative natural number algebraically defined and is denoted $Ir_o(\mathcal{M})$. If X is a Riemann surface let $\mathcal{O}_{,o}$ be the local (analytic) ring at o and $\hat{\mathcal{O}}_{,o}$ its completion. For a left $\mathcal{D}_X$ holonomic module $\mathcal{M}$, the Malgrange comparison theorem [M 1] says that $\mathrm{Ext}^i_{\mathcal{D}_{,o}}(\mathcal{M}_{,o}, \mathcal{Q}_{,o}) = 0$ if $i \neq 0$ and $Ir_o(\mathcal{M}) = \dim_{\mathbb{C}} \hom_{\mathcal{D}_{,o}}(\mathcal{M}_{,o}, \mathcal{Q}_{,o})$ where $\mathcal{Q}_{,o} := \hat{\mathcal{O}}_{,o}/\mathcal{O}_{,o}$.

The semi-continuity theorem for the irregularity is the analogue of the semi-continuity theorem of the Swan conductor due to Deligne (cf. [La]). It describes the behaviour of the irregularity in a family of differential equations. Let $f : X \to S$ be a smooth morphism of relative dimension one between two smooth complexes algebraic or analytic varieties and Z a divisor in X such that the restriction of f to Z is finite. The localisation of a holonomic module $\mathcal{M}$ along Z is the $\mathcal{D}_X$ module $\mathcal{M}(*Z) := \varinjlim_k \hom_{\mathcal{O}_X}(\mathcal{I}_Z^k, \mathcal{M})$ where $\mathcal{I}_Z$ is a defining ideal of Z. In the algebraic case it is isomorphic to $i_* i^{-1}\mathcal{M}$ where i is the canonical inclusion of $U := X - Z$ in X. As a consequence of the Bernstein–Sato polynomial existence the localisation of a holonomic module is holonomic. For each $s \in S$ we denote by X_s and Z_s the fibers of f over s. Suppose that $\mathcal{M}$ is smooth over $U := X - Z$ of rank r. The restriction $\mathcal{M}_s := \mathcal{O}_{X_s} \otimes_{\mathcal{O}_X} \mathcal{M}$ of $\mathcal{M}$ to X_s is a $\mathcal{D}_{X_s}$ holonomic module smooth over $X_s - Z_s$. Since X_s is a curve for each s, $o \in Z_s$, $Ir_o(\mathcal{M}_s)$ is a non-negative natural number. Consider the functions Ψ, $\Phi : S \to N$, $\Psi(s) := \sum_{o \to s} Ir_o(\mathcal{M}_s)$, $\Phi(s) := \Psi(s) + r \# Z_s$ where $\# Z_s$ is the number of points of Z_s.

We have the following theorem (cf. [Me 2]):

Theorem 1.1.   *In the previous situation we have the following.*

(α)   *The functions Ψ and Φ are constructible;*

(β)   *If* dim $S = 1$ *the jump of the function Φ is a non-negative natural number;*

(γ)   *The function Φ is lower semi-continuous and if it is locally constant the characteristic variety of $\mathcal{M}(*Z)$ is contained in $T_X^* X \cup T_Z^* X$.*

We note $T_Z^* X$ is the closure in the cotangent bundle of X of the conormal bundle of the smooth part of Z. In fact we can make part (γ) more precise [Me 3]:

Theorem 1.2.   *If $r \neq 0$ the function Φ is locally constant if and only if the morphism $Z \to S$ is étale.*

In the l-adic case the jump of the function Φ when dim $S = 1$ is equal to the dimension of the vanishing space of the l-adic sheaf [La]. In the complex case if $DR(\mathcal{M}) := R \hom_{\mathcal{D}_X}(\mathcal{O}_X, \mathcal{M})$ is the transcendental De Rham complex of $\mathcal{M}$, the jump of the function Φ when dim $S = 1$ in general is not equal to the dimension of the vanishing space $R^1\Phi_f(DR(\mathcal{M}))$ [Me 2]. So we must look to a vanishing theory for the $\mathcal{D}_X$ modules in characteristic zero analogous to the l-adic theory.

2.   The Irregular Vanishing Cycles

2.1.

We first recall the definition and the properties of Fuchs–Malgrange–Kashiwara V-filtration of a holonomic $\mathcal{D}_X$ module [M 3], [K 2], [M 4], [S.M].

Let $f: X \to \mathbf{C}$ be a smooth complex function of the complex smooth variety X, Y the fiber of f over 0 and $T_Y X$ the normal bundle of Y in X. Let $\mathcal{I}_Y$ be the reduced ideal of Y and $\mathcal{I}_Y^p = \mathcal{O}_X$ if p is a non-positive natural number. For each $k \in \mathbf{Z}$ put $V_k(\mathcal{D}_X) := \{P \in \mathcal{D}_X, P(\mathcal{I}_Y^p) \subset \mathcal{I}_Y^{p-k}$ for every p in $\mathbf{Z}\}$. The filtration $V(\mathcal{D}_X)$ is a good increasing filtration. The sheaf $\mathrm{gr}^V(\mathcal{D}_X) := \oplus_k V_{k+1}(\mathcal{D}_X)/V_k(\mathcal{D}_X)$ can be identified with $\mathcal{D}_{T_Y X}$ in the algebraic case and with the subsheaf of $\mathcal{D}_{T_Y X}$ of differential operators with polynomial coefficients along the fiber of $T_Y X \to Y$ in the analytic case. In $\mathrm{gr}_0^V(\mathcal{D}_X)$ we have the Euler vector field Eu, we denote by Eu an element of $V_0(\mathcal{D}_X)$

such that its class is Eu. If u is a local section of a holonomic $\mathcal{D}_X$ module $\mathcal{M}$ the Bernstein–Sato functional equation says that there is a non-zero polynomial $B \in \mathbf{C}[s]$ (if u is not zero) such that $B(Eu)u \in V_{-1}(\mathcal{D}_X)u$. The monic generator of the ideal of such B is denoted b_u and its set of zeros is noted $\mathrm{ord}_Y(u)$. Fix a total order $\leq$ in $\mathbf{C}$ compatible with the natural order in $\mathbf{R}$. For a $\mathcal{D}_X$ module $\mathcal{M}$ put for $x \in X$, $\alpha \in \mathbf{C}$, $V_\alpha(\mathcal{M})_x :=$ $\{u \in \mathcal{M}_x, \mathrm{ord}_Y(u) \subset \{s \in \mathbf{C}, -\alpha - 1 \leq s\}\}$ and $V_\alpha(\mathcal{M}) := \bigcup_x V_\alpha(\mathcal{M})_x$. We have (cf. [S.M]):

Proposition 2.1.1. *For each $\alpha \in \mathbf{C}$ and any holonomic $\mathcal{D}_X$ module $\mathcal{M}$, $V_\alpha(\mathcal{M})$ is a $V_0(\mathcal{D}_X)$ coherent module. The filtration $V_\alpha(\mathcal{M})$, $\alpha \in \mathbf{C}$, is a good V filtration indexed (locally) by a finite number of lattices of type $\alpha + \mathbf{Z}$.*

Put $\mathrm{gr}_\alpha^V(\mathcal{M}) := V_\alpha(\mathcal{M})/\bigcup_{\beta < \alpha} V_\beta(\mathcal{M})$, which is a $\mathrm{gr}_0^V(\mathcal{D}_X)$ coherent module such that the action of $Eu + \alpha + 1$ is (locally) nilpotent so it is in fact D_Y coherent. We have (cf. [S.M]):

Theorem 2.1.2. *For a $\mathcal{D}_X$ holonomic module $\mathcal{M}$, $\mathrm{gr}_\alpha^V(\mathcal{M})$ is a $\mathcal{D}_Y$ holonomic module for each α in $\mathbf{C}$.*

For each α there is an action, the monodromy action, which is $\exp(-2\pi\sqrt{-1}Eu)$ on $\mathrm{gr}_\alpha^V(\mathcal{M})$.

Definition 2.1.3. Put

$$\Psi_f^m(\mathcal{M}) := \bigoplus_{-1 \leq \alpha < 0} \mathrm{gr}_\alpha^V(\mathcal{M}) \quad \text{and} \quad \Phi_f^m(\mathcal{M}) := \bigoplus_{-1 < \alpha \leq 0} \mathrm{gr}_\alpha^V(\mathcal{M}).$$

$\Psi_f^m(\mathcal{M})$ and $\Phi_f^m(\mathcal{M})$ are $\mathcal{D}_Y$ holonomic modules with the monodromy action. If t is a local coordinate for Y we have a morphism $\mathrm{Can} : \Psi_f^m(\mathcal{M}) \to \Phi_f^m(\mathcal{M})$ which is the identity on $\mathrm{gr}_\alpha^V(\mathcal{M})$ for $-1 < \alpha < 0$ and $-\partial_t : \mathrm{gr}_1^V(\mathcal{M}) \to \mathrm{gr}_0^V(\mathcal{M})$ for $\alpha = -1$.

2.2.

Let t be a coordinate on $\mathbf{C}$. A differential form ω, a finite sum of $a_q t^q \, dt/t (a_q \in \mathbf{C}, q \in \mathbf{Q} q < 0)$, defines a connexion on the trivial bundle of a finite covering of C. Remember that for any $\mathcal{D}_\mathbf{C}^-$ holonomic module $\mathcal{N}$ there is only a finite set $V(\mathcal{N})$ of non-zero forms ω which appear in the formal decomposition at the origins of $\mathcal{N}$ (cf. [B]). The following theorem is due to Deligne [D 2] in the algebraic case. One finds a proof in [L.M] which works also in the local analytic case.

Theorem 2.2.1. *With the previous notations let $\mathcal{M}$ be a $\mathcal{D}_X$ holonomic module. There is a finite set $V(\mathcal{M})$ of forms ω such that for any $\mathcal{D}_C$ holonomic module $\mathcal{N}$ with $V(\mathcal{M}) \cap -V(\mathcal{N}) \neq \varnothing$, then $\mathrm{gr}_\alpha^V(\mathcal{M} \otimes_{\mathcal{O}_X} f^* \mathcal{N}(* \theta)) = 0$ for any α in $\mathbf{C}$.*

For a $\mathcal{D}_X$ holonomic module $\mathcal{M}$ there is a finite covering $\mathbf{C}(\sqrt{m})$ of $\mathbf{C}$ such that every ω in $V(\mathcal{M}) - \omega$ defines a bundle of rank one with connexion $\mathcal{L}^\omega$ on $\mathbf{C}(\sqrt{m})$. Let

$$
\begin{array}{ccc}
X(\sqrt{m}) & \longrightarrow & X \\
\downarrow{\scriptstyle f} & & \downarrow{\scriptstyle f} \\
\mathbf{C}(\sqrt{m}) & \longrightarrow & \mathbf{C}
\end{array}
$$

be the natural fiber product and $\tilde{\mathcal{M}}$ the inverse image of $\mathcal{M}$ on $X(\sqrt{m})$.

Definition 2.2.2. Let $\mathcal{M}$ be a $\mathcal{D}_X$ holonomic module and ω in $V(\mathcal{M})$; put $\quad \Psi_f^\omega(\mathcal{M}) := \Psi_f^m(\tilde{\mathcal{M}} \otimes_{\mathcal{O}_X} L^\omega(* \mathcal{O})), \quad \Phi_f^\omega(\mathcal{M}) := \Psi_f^\omega(\mathcal{M}), \quad \Psi_f^s(\mathcal{M}) := \bigoplus_{\omega \in V(\mathcal{M})} \Psi_f^\omega(\mathcal{M}), \quad \Phi_f^s(\mathcal{M}) := \bigoplus_{\omega \in V(\mathcal{M})} \Phi_f^\omega(\mathcal{M}), \quad \Psi_f(\mathcal{M}) := \Psi_f^m(\mathcal{M}) \oplus \Psi_f^s(\mathcal{M}),$ and $\Phi_f(\mathcal{M}) := \Phi_f^m(\mathcal{M}) \oplus \Phi_f^s(\mathcal{M}).$

The functors $\mathcal{M} \to \Psi_f(\mathcal{M})$ and $\mathcal{M} \to \Phi_f(\mathcal{M})$ are exact from the category of $\mathcal{D}_X$ holonomic modules to the category of $\mathcal{D}_Y$ holonomic modules with the monodromy action. We get a vanishing cycle theory analogous to the l-adic theory.

Let $f: X \to \mathbf{C}$ be a smooth complex function on a smooth complex surface X, and Z a divisor in X such that the restriction of f to Z is finite. Let $\mathcal{M}$ be a $\mathcal{D}_X$ holonomic module smooth over $U := X - Z$; then $\Phi_f(\mathcal{M})$ is a $\mathcal{D}_Y$ holonomic module supported by $Y \cap Z$ where $Y := f^{-1}(0)$. We have the following theorem analogous to [La]:

Theorem 2.2.3. *The jump of the function Φ (see Section 1) of the previous situation at the origin is the sum of the multiplicities of $\Phi_f(\mathcal{M})$ at the points of $Y \cap Z$.*

3. The Irregularity Sheaf of a $\mathscr{D}_X$ Holonomic Module and the Comparison Theorem

3.1. The Irregularity Sheaf

Let X be a complex smooth variety (algebraic or analytic as always), Z a divisor and i the transcendental inclusion of $U := X - Z$ in X. If $\mathscr{M}$ is a bounded complex of $\mathscr{D}_X$ modules with holonomic cohomology we put

$$\mathscr{M}(*Z) := \varinjlim_k \hom_{\mathcal{O}_X}(I_Z^k, \mathscr{M}),$$

which is a bounded complex of $\mathscr{D}_X$ modules with holonomic cohomology. We have a natural morphism between the transcendental De Rham complexes:

$$\mathbf{DR}(\mathscr{M}(*Z)) \to \mathbf{R}i_* i^{-1}\mathbf{DR}(\mathscr{M}).$$

Definition 3.1.1. We call the cone of the previous morphism the *irregularity complex* of $\mathscr{M}$ along Z and we denote it $\mathbf{IR}_Z(\mathscr{M})$. It is a constructible complex on Z.

By construction we get an exact functor $\mathbf{IR}_Z$ of triangulated categories:

$$\mathbf{IR}_Z : D_h^b(\mathscr{D}_X) \to D_c^b(\mathbf{C}_Z).$$

We have the positivity theorem which generalizes Malgrange's theorem:

Theorem 3.1.2. [Me 3]. *If $\mathscr{M}$ is a $\mathscr{D}_X$ holonomic module $\mathbf{IR}_Z(\mathscr{M})$ is a perverse sheaf on Z.*

By the positivity theorem the functor $\mathbf{IR}_Z$ sends the category of $\mathscr{D}_X$ holonomic modules into the category of perverse sheaves on Z.

Recall that a constructible complex $\mathscr{F}$ is called a perverse sheaf on Z if (i) its cohomology is concentrated in $[0, \dim Z]$; (ii) the dimension of the support of $h^i(\mathscr{F})$ is $\leq \dim Z - i$ where $h^i(\mathscr{F})$ is the i-th cohomology sheaf of $\mathscr{F}$, (iii) the dual $F^v := \mathbf{R}\hom_{\mathbf{C}_Z}(\mathscr{F}, \mathscr{H}_Z)[-2\dim Z]$ of $\mathscr{F}$ has properties (i) and (ii). The category of perverse sheaves $\mathrm{Perv}(\mathbf{C}_Z)$ is an abelian subcategory of $D_c^b(\mathbf{C}_Z)$ [B.B.D]. The positivity theorem has many consequences, for example

Corollary 3.1.3. [Me 3]. (α) *If* $0 \to \mathcal{M}_1 \to \mathcal{M} \to \mathcal{M}_2 \to 0$ *is an exact sequence of* $\mathcal{D}_X$ *holonomic modules* $0 \to \mathbf{IR}_Z(\mathcal{M}_1) \to \mathbf{IR}_Z(\mathcal{M}) \to \mathbf{R}_Z(\mathcal{M}_2) \to 0$ *is an exact sequence of perverse sheaves.*

(β) *If* $\mathcal{M}$ *is* $\mathcal{D}_X$ *holonomic complex* ($\varepsilon D_h^b(\mathcal{D}_X)$) *the irregularity along* Z *of the i-th (ordinary) cohomology sheaf of* $\mathcal{M}$ *is isomorphic to the i-th (perverse) cohomology sheaf of* $\mathbf{IR}_Z(\mathcal{M})$.

(γ) *If* $\mathcal{M}^*$ *is the dual module of a holonomic module* $\mathcal{M}$, *then* $\mathbf{IR}_Z(\mathcal{M}^*)$ *vanishes if and only if* $\mathbf{IR}_Z(\mathcal{M})$ *vanishes.*

As suggested by the one-dimensional case it is possible to define an exact functor ir_Z from the category of the $\mathcal{D}_X$ holonomic modules to the category of $\mathcal{D}_X$ holonomic modules with support in $f^{-1}(0)$ such that

$$\mathbf{DR}(ir_Z(\mathcal{M})) \approx \mathbf{IR}_Z(\mathcal{M})[-1]$$

for any holonomic module $\mathcal{M}$. The tool is the vanishing cycles theory of Section 2. The construction is purely algebraic and should have an l-adic analogue cf. [L.M].

3.2. The Comparison Theorem

Let $Z \subset Y \subset X$ be subvarieties of X, possibly with singularities, such that $Y - Z$ is smooth and Z is locally defined by one equation. If $\mathcal{M}$ is a holonomic module such that its support is contained in Y then $\mathbf{IR}_Z(\mathcal{M})$ is a perverse sheaf on Z.

Theorem 3.2.1. *Suppose that* $\mathcal{M}$ *is a* $\mathcal{D}_X$ *holonomic module with support contained in* Y *and smooth over* $Y - Z$. *Then the sheaf* $\mathbf{IR}_Z(\mathcal{M})$ *vanishes if and only if the codimension in* Z *of its support is at least one.*

The proof of Theorem 3.2.1 [Me 4] does not use Hironaka's theorem on the resolution of singularities. It contains as particular cases Grothendieck's comparison theorem [G] and Deligne's comparison theorem [D 1].

Let us denote by $Mhr(\mathcal{D}_X)$ the category of $\mathcal{D}_X$ holonomic modules such that their inverse image on any smooth curve over X in the algebraic case, and any complex disc in the analytic case, has only regular singularities (even at infinity in the algebraic case). Let us denote by $D_{hr}^b(\mathcal{D}_X)$ the category of bounded complex of $\mathcal{D}_X$ modules with cohomology in $Mhr(\mathcal{D}_X)$. As a consequence of 3.2.1 we get ([Me 4]) that a $\mathcal{D}_X$ holonomic module is in $Mhr(\mathcal{D}_X)$ if and only if its irregularity sheaf along any divisor (even at infinity in the algebraic case) vanishes. We denote by $\mathbf{DR}_r$ the restriction of the functor $\mathbf{DR}$ to the category $D_{hr}^b(\mathcal{D}_X)$. We get by "dévissage" [Me 1]:

Theorem 3.2.2. *The functor $\mathbf{DR}_r$ is an exact fully faithful functor of triangulated categories between $D_{hr}^b(\mathcal{D}_X)$ and $D_c^b(\mathbf{C}_X)$.*

Let $f: X \to \mathbf{C}$ be a complex function and $\mathcal{M}$ a $\mathcal{D}_X$ holonomic module. Then the vanishing complex $\mathbf{R}\Psi_f(DR(\mathcal{M}))$ of its De Rham complex is a perverse sheaf on $f^{-1}(0)$ with the monodromy action T. Suppose f is smooth. Then $\Psi_f^m(\mathcal{M})$ is a holonomic module on $f^{-1}(0)$ with the monodromy action.

Theorem 3.2.3. *If $\mathcal{M}$ is in $Mhr(\mathcal{D}_X)$, then $\Psi_f^m(\mathcal{M})$ is in $Mhr(\mathcal{D}_{f^{-1}(0)})$ and we have a canonical isomorphism of perverse sheaves with a monodromy action*

$$\mathbf{DR}(\Psi_f^m(\mathcal{M})) \simeq \mathbf{R}\Psi_f(DR(\mathcal{M})).$$

As a consequence of Theorem 3.2.3 and of the Monodromy theorem [L] we get by the Malgrange method [M 3] the rationality of the zeros of the Bernstein–Sato polynomial of a germ of a holomorphic function.

It is also possible to prove Riemann's Existence Theorem (cf. [D 1]) in its modern form, that is to say the functor $\mathbf{DR}_r$ is essentially surjective without Hironaka's theorem. But its proof and motivation are of a different nature than the comparison theorem [Me 4]. Grothendieck's motivation [G] for the comparison theorem between the De Rham cohomologies is that the De Rham cohomology is a good (or a Weil) cohomology for algebraic varieties over a field of characteristic zero. This motivation is completely independent of Riemann's Existence Theorem.

References

[B.B.D]   A. A. Beilonson, J. Bernstein, and P. Deligne, Faisceaux pervers *Astérisque* **100** (1983).

[B]   D. Bertrand, Travaux récents due les points singuliers des équations différentielles linéaires, *Séminaire Bourbaki* n° 538, juin 1979.

[D 1]   P. Deligne, Equations différentielles à points singuliers réguliers, *Springer Lecture Notes in Math*, 163, 1970.

[D2]   P. Deligne, Lettre à Malgrange datée du 20 décembre 1983.

[G]   A. Grothendieck, On the De Rham cohomology of algebraic varieties, *I.H.E.S.* **29** (1966) 93–103.

[K 1]   M. Kashiwara, The Riemann Hilbert for holonomic systems, *Publ. R.I.M.S., Kyoto, Univ.* **20** (1984) 319–365.

[K 2]   M. Kashiwara, Vanishing cycle sheaves and holonomic systems of differential equations, *Springer Lecture Notes in Math.* 1016, 1983, 134–142.

[La]   G. Laumon, Semi continuité du conducteur de Swan (d'aprés Deligne), *Astérisque* **82-83** (1981) 173–219.

[L.M]   Y. Laurent and Z. Mebkhout, La théorie des cycles évanescents irréguliers (to appear).

[L]   D. T. Lê, The geometry of the Monodromy theorem, *C. P. Ramanujam, A Tribute*, Studies in Math, n° 8, Tata Institute, Bombay, 1978, pp. 157–173.

[M 1]   B. Malgrange, Remarques sur les points singuliers des équations différentielles, *C. R. Acad. Sci., Paris* **273**-23 (1971) 1136–1137.

[M 2]   B. Malgrange, Sur les points singuliers des équations différentielles, *Ens. Math,* **20** (1974) 147–176.

[M 3]   B. Malgrange, Polynôme de Bernstein Sato et cohomologie évanescentes, *Astérisque* **101-102** (1983) 243–267.

[M 4]   B. Malgrange, Extension of $\mathscr{D}$ holonomic modules, this volume.

[Me 1]   Z. Mebkhout, Une équivalence de catégories et une autre équivalence de catégories, *Comp., Math.* **51** (1984) 51–88.

[Me 2]   Z. Mebkhout, Le théorème de semi-continuité de l'irrégularité des équations différentielles, *Astérisque* **130** (1985) 365–417.

[Me 3]   Z. Mebkhout, La positivité de l'irrégularité des $\mathscr{D}_X$ modules holonomes, (to appear) cf. *C. R. Acad. Sci., Paris* **303**, série I (1986) 803–806.

[Me 4]   Z. Mebkhout, Le théorème de Comparaison entre cohomologies de De Rham et le théorème d'existence de Riemann, (to appear) cf. *C. R. Acad. Sci. Paris*, October 1987.

[S.M]   C. Sabbah and Z. Mebkhout, $\mathscr{D}_X$ modules et cycles évanescents. In conférences des Plans-sur-Bex (mars 1984), Travaux en cours Hermann (to appear).

[S.G.A.7] *Séminaire de Géometrie Algébrique du Bois Marie 1967-69*, dirigé par Grothendieck, Deligne, Katz, *Springer Lecture Notes in Math*, 288, 340, 1972–1973.

An Error Analysis of Quadrature Formulas Obtained by Variable Transformation

Masatake Mori

Institute of Information Sciences and Electronics
University of Tsukuba
Tsukuba, Ibaraki, Japan

1. Introduction

When we evaluate an integral numerically using some quadrature formula in laboratory practice, we seldom carry out its error estimation. The reason is that, since the error formulas found in a common textbook are usually based on the mean-value theorem, they have an unknown parameter which makes it impossible to evaluate them specifically. For example, the error of the n-point Gauss–Legendre rule is given [2] by

$$\int_{-1}^{1} f(x)\, dx - \sum_{j=1}^{n} w_j f(x_j) = \frac{2^{2n+1}(n!)^4}{(2n+1)[(2n)!]^3} f^{(2n)}(\xi), \qquad -1 < \xi < 1$$

(1.1)

in which ξ is unknown. Although it may seem to be possible to estimate $\max_{-1 < x < 1} |f^{(2n)}(x)|$ instead of $f^{(2n)}(\xi)$, it is not the case because to formally differentiate $f(x)$ n times is quite a cumbersome task and, even if it is successful, estimation of $\max_{-1 < x < 1} |f^{(2n)}(x)|$ usually gives a very conservative result.

Almost all of the conventional quadrature formulas of high-order accuracy can be regarded as designed to integrate analytic functions as precisely as possible. For example, the 8-point Gauss–Legendre rule integrates polynomials of degree up to 15 exactly, as seen from (1.1), and a function which can be approximated favorably by a polynomial of degree 15 may be regarded as an analytic function from the standpoint of practical analysis.

When the integrand is an analytic function several useful methods to estimate the quadrature error are known [5]. In a preceding paper [3] we developed a method of error estimation based on a contour integral representation of the error in the complex plane combined with a contour map of a function which characterizes the quadrature error. It may be regarded as an application of the theory of hyperfunctions [1] to numerical analysis. In the present paper we apply the method to quadrature formulas obtained by variable transformation.

First we consider the integral

$$I = \int_a^b f(x)\,dx, \tag{1.2}$$

where the interval (a, b) of integration is finite and $f(z)$ is assumed to be analytic in a domain D in which the real-line segment $a \le x \le b$ is contained. Suppose that C is a simple contour which lies in D and runs around the line segment $a \le x \le b$ in the positive sense. Then for $a \le x \le b$ we have the Cauchy's formula

$$f(x) = \frac{1}{2\pi i} \oint_C \frac{1}{z - x} f(z)\,dz. \tag{1.3}$$

Substituting (1.3) into (1.2) and interchanging the order of integration we have

$$I = \frac{1}{2\pi i} \oint_C \Psi(z) f(z)\,dz, \tag{1.4}$$

$$\Psi(z) = \int_a^b \frac{1}{z - x}\,dx = \log \frac{z - a}{z - b}. \tag{1.5}$$

Next we consider a quadrature formula

$$I_n = \sum_{j=1}^n A_j f(a_j) \tag{1.6}$$

which approximates (1.2). a_j and A_j are the points and the weights of the formula, respectively. We assume that every point a_j is contained in the interior of C. Then again substituting (1.3) into (1.6) we have

$$I_n = \frac{1}{2\pi i} \oint_C \Psi_n(z)f(z)\, dz, \tag{1.7}$$

$$\Psi_n(z) = \sum_{j=1}^{n} \frac{A_j}{z - a_j}. \tag{1.8}$$

Therefore the error of the quadrature formula can be written

$$\Delta I_n = I - I_n$$

$$= \frac{1}{2\pi i} \oint_C \Phi_n(z)f(z)\, dz, \tag{1.9}$$

$$\Phi_n(z) = \Psi(z) - \Psi_n(z). \tag{1.10}$$

$\Phi_n(z)$ is a function which depends only on the interval of integration and on the quadrature formula, and we call it the *characteristic function* of the error of the quadrature formula. Thus we see that, as long as we deal with the integral of an analytic function, the error (1.9) can be regarded as a hyperfunction and $-\Phi_n(z)/(2\pi i)$ is nothing but its defining function.

The analysis stated so far is not only of theoretical interest but also of practical interest. In fact, in the case of numerical integration of an analytic function, it is often possible to get a specific and reasonable error estimate for the given integrand. To this end we gave contour maps of the characteristic functions of the error for various kinds of quadrature formulas [3].

For example, the characteristic function of the error of the n-point Gauss–Legendre rule over $(-1, 1)$ is given in terms of a continued fraction expansion

$$\Phi_n(z) = \frac{\gamma_1 \gamma_2 \cdots \gamma_{n+1}}{P_n(z)\{P_n(z)R_{n+1}(z) - \gamma_{n+1}P_{n-1}(z)\}},$$

$$\gamma_k = 1 - \frac{1}{k}, \qquad \gamma_1 = 2, \tag{1.11}$$

where $P_n(z)$ is the Legendre polynomial of degree n. $R_k(z)$ is a function defined recursively in terms of a continued fraction expansion

$$R_k(z) = \alpha_k z + \beta_k - \frac{\gamma_{k+1}}{R_{k+1}(z)},$$

$$\alpha_k = 2 - \frac{1}{k}, \qquad \beta_k = 0. \tag{1.12}$$

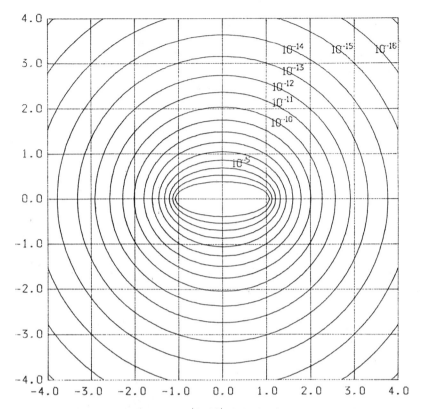

FIGURE 1   The contour map of $|\Phi_n(z)|$ of the 8-point Gauss–Legendre rule.

In Fig. 1 the contour map of $|\Phi_n(z)|$ for $n = 8$ is shown.

As an example consider the integral

$$I = \int_{-1}^{1} \frac{dx}{(x-1)(x^2+1)}. \tag{1.13}$$

Suppose that it is evaluated by means of the 8-point Gauss–Legendre rule. Then, since $\lim_{z\to\infty}|z\Phi_n(z)| = 0$ and by the residue theorem we have

$$\Delta I_n = -\mathrm{Res}(2)\Phi_n(2) - \mathrm{Res}(i)\Phi_n(i) - \mathrm{Res}(-i)\Phi_n(-i). \tag{1.14}$$

From Fig. 1 we see that $|\Phi_n(2)| \ll |\Phi_n(\pm i)| \approx 2 \times 10^{-6}$, and hence $|\Delta I_n| \approx 0.9 \times 10^{-6}$. It is evident that this is a good estimate of the actual error. On the other hand, if we try to estimate the error using (1.1) we immediately find that it is practically impossible. When the integrand is not a rational function the saddle-point method may be used as an alternative [3].

2. Quadrature Formulas Obtained by Variable Transformation and the DE-Rule

If the integrand $f(x)$ of (1.2) has a singularity at one or both of the endpoints $x = a$ and b, we usually are not able to get a good result with high accuracy by a quadrature formula designed for integrands without singularities at the endpoints like the Gauss–Legendre rule. In such a case we can successfully use a quadrature formula obtained by a variable transformation which maps both of the endpoints to infinity.

The basic theorem on which we rely is an optimum property of the simple trapezoidal rule applied to the integral over $(-\infty, \infty)$. Suppose that $g(u)$ is an analytic function which has no singularity on the real axis except infinity, and consider the integral

$$I = \int_{-\infty}^{\infty} g(u) \, du. \tag{2.1}$$

We apply the trapezoidal rule with an equal mesh size h to (2.1),

$$I_h = h \sum_{n=-\infty}^{\infty} g(nh). \tag{2.2}$$

Then, under the assumption that the average number of the points a_n per unit length is the same, the trapezoidal rule (2.2) is optimal among formulas of the form

$$I_A = \sum_{n=-\infty}^{\infty} A_n g(a_n), \tag{2.3}$$

in the sense that the decay rate of the characteristic function is asymptotically maximum as the imaginary part of the independent variable becomes large [3].

Now consider again the integral

$$I = \int_{a}^{b} f(x) \, dx. \tag{2.4}$$

We assume that $f(x)$ may have an integrable singularity at the end-points $x = a$ and $x = b$. We apply a variable transformation

$$x = \phi(u) \tag{2.5}$$

to (2.4), where $\phi(u)$ is an analytic increasing function without singularities on the real axis except infinity satisfying

$$a = \phi(-\infty), \qquad b = \phi(+\infty). \tag{2.6}$$

Then we have

$$I = \int_{-\infty}^{\infty} f(\phi(u))\phi'(u) \, du, \tag{2.7}$$

in which $f(\phi(u))\phi'(u)$ has no singularities on the real axis except infinity. Here we recall the optimality of the trapezoidal rule and apply it to (2.7) to obtain a good quadrature formula:

$$I_h = h \sum_{n=-\infty}^{\infty} f(\phi(nh))\phi'(nh). \tag{2.8}$$

At this stage we still have a free choice of $\phi(u)$. We showed that (2.5) is optimal in a certain sense when the integrand of (2.7) after the transformation decays in a double exponential way like

$$|f(\phi(u))\phi'(u)| \sim c_1 \exp(-c_2 \exp(c_3|u|)), \qquad u \to \pm\infty, \tag{2.9}$$

where c_1, c_2 and c_3 are some constants [4, 5]. We call the quadrature formula obtained by such a variable transformation the *double exponential formula*, abbreviated as the DE-rule. Sugihara gave an analysis of the optimal property of the DE-rule in a different way introducing some functional space for the integrands [6].

We give here some typical examples of the DE-rule. First consider the integral

$$I = \int_{-1}^{1} f(x) \, dx, \tag{2.10}$$

where $f(x)$ may have integrable algebraic or logarithmic singularities at the endpoints. Then

$$x = \tanh\left(\frac{\pi}{2} \sinh u\right) \tag{2.11}$$

gives a DE-rule. Although the constant $\pi/2$ is chosen for some reason [4], another choice of this constant will not give a serious effect on the efficiency of the formula.

Next consider the integral

$$I = \int_{0}^{\infty} f(x) \, dx, \tag{2.12}$$

where $f(x)$ may have an integrable algebraic or logarithmic singularity at $x = 0$ and is assumed to decay as an algebraic function as $x \to \infty$. Then the transformation

$$x = \exp(\pi \sinh u) \tag{2.13}$$

gives a DE-rule. Note that if $f(x)$ in (2.12) has the form of (algebraic function) × (trigonometric function) the transformation by (2.13) does not work well. In such a case some extrapolation method often improves the convergence property [7].

If $f(x)$ in (2.12) behaves like

$$f(x) \sim f_1(x) \times \exp(-x), \qquad x \to +\infty, \tag{2.14}$$

where $f_1(x)$ is an algebraic function, then the transformation

$$x = \exp(u - \exp(-u)), \tag{2.15}$$

which gives a single exponential decay of the weight as $u \to +\infty$, results in a DE-rule.

Finally for the integral

$$I = \int_{-\infty}^{\infty} f(x) \, dx, \tag{2.16}$$

where $f(x)$ is an algebraic function which decays slowly as $x \to \pm\infty$, the transformation

$$x = \sinh\left(\frac{\pi}{2} \sinh u\right) \tag{2.17}$$

gives a DE-rule which integrates (2.16) efficiently.

3. Characteristic Function of the Error of the DE-Rule

In this section we give the characteristic function of the error of the DE-rule. In order to do so we need the characteristic function of the error of the trapezoidal rule (2.2). If $g(u)$ has no singularity on the real axis, then the error of (2.2) is given [3] by

$$\Delta I_h = I - I_h$$

$$= \frac{1}{2\pi i} \int_{\hat{C}} \hat{\Phi}_h(w) g(w) \, dw, \tag{3.1}$$

$$\hat{\Phi}_h(w) = \begin{cases} \dfrac{-2\pi i}{1-\exp(-2\pi i w/h)}; & \text{Im } w > 0 \\[2ex] \dfrac{+2\pi i}{1-\exp(+2\pi i w/h)}; & \text{Im } w < 0. \end{cases} \tag{3.2}$$

The path $\hat{C}$ of integration consists of two infinite curves one of which runs above the real axis towards the left and the other runs below the real axis towards the right. It is also assumed that $g(w)$ has no singularity between these two curves. If $|\text{Im } w| \gg h/(2\pi)$, then (3.2) may be approximated as

$$\frac{1}{2\pi i}\hat{\Phi}_h(w) \approx \begin{cases} +\exp(-2\pi i w/h); & \text{Im } w > 0 \\ -\exp(+2\pi i w/h); & \text{Im } w < 0, \end{cases} \tag{3.3}$$

and hence, if the mesh size h is small enough, we see that the contour map of (3.3) or of

$$\frac{1}{2\pi i}|\hat{\Phi}_h(w)| = 10^{-m}, \qquad m = 1, 2, 3, \ldots \tag{3.4}$$

consists approximately of lines

$$\text{Im } w = \pm v_m, \qquad v_m = \frac{1}{2\pi} hm \log 10, \qquad m = 1, 2, 3, \ldots \tag{3.5}$$

lying parallel to the real axis at an equal interval with each other. The constant $1/(2\pi i)$ is multiplied in order to make the real axis lie also at the equal distance $h \log 10/(2\pi)$ from the nearest contour lines.

If we substitute

$$g(w) = f(\phi(w))\phi'(w) \tag{3.6}$$

into (3.1) we have the error representation of the quadrature formula (2.8)

$$\Delta I_h = \frac{1}{2\pi i} \int_{\hat{C}} \hat{\Phi}_h(w) f(\phi(w))\phi'(w) \, dw. \tag{3.7}$$

By the inverse transformation of (2.5),

$$w = \phi^{-1}(z), \tag{3.8}$$

(3.7) becomes

$$\Delta I_h = \frac{1}{2\pi i} \int_C \Phi_h(z) f(z) \, dz, \tag{3.9}$$

$$\Phi_h(z) = \Phi_h(\phi(w)) = \hat{\Phi}_h(w), \tag{3.10}$$

where the path C is the image of $\hat{C}$ by (3.8). Thus we obtained the characteristic function $\Phi_h(z)$ of the error of the formula (2.8) for the original integral (2.4).

Therefore the contour map of $|\Phi_h(z)|$ for the formula, i.e., the map of the curves satisfying

$$\frac{1}{2\pi}|\Phi_h(z)| = 10^{-m}, \qquad m = 1, 2, 3, \ldots, \tag{3.11}$$

is obtained by mapping the approximate lines (3.5) onto the z-plane by $z = \phi(w)$. The contour map of $|\Phi_h(z)|/(2\pi)$ for the DE-rule based on (2.11) is shown in Fig. 2.

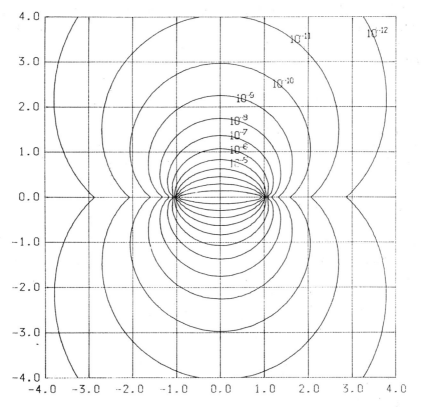

FIGURE 2   The contour map of $|\Phi_h(z)|/(2\pi)$, $h = 0.25$ of the DE-rule by (2.11). Although this is a map for $h = 0.25$, it can be used for other values of h, say 0.5 or 0.125. For example, for $h = 0.125$, 10^{-m} in the figure should read 10^{-2m} by (3.5).

There are two methods to plot the contour map using an XY-plotter. The first one is to divide the square or rectangular domain in the z-plane on which the map is going to be displayed into, say, a 100×100 mesh, to evaluate $|\Phi_h(z)|/(2\pi)$ at each mesh point and to invoke a subroutine to plot a contour map. The second one is to represent the line (3.4) for each m as

$$w = u + i\frac{hm \log 10}{2\pi}, \qquad -\infty < u < \infty \qquad (3.12)$$

and to trace out the image of the line by $z = \phi(w)$ from some large negative value up to some large positive value of u. Fig. 2 was plotted by means of the former method.

It should be noted that, in contrast to the interpolatory type formula like the Newton–Cotes or the Gauss–Legendre rule, the DE-rule does not integrate a constant function exactly. We call the error induced when integrating 1 the *intrinsic error* of the formula. The intrinsic error ε_I can be given as

$$\varepsilon_I = \frac{1}{2\pi i}\int_C \Phi_h(z)\,dz = \frac{1}{2\pi i}\int_{\hat{C}} \hat{\Phi}_h(w)\phi'(w)\,dw \qquad (3.13)$$

from (3.9) and (3.7), and can be evaluated specifically if we actually integrate 1 using the DE-rule.

4. A Generalized DE-Rule

In order to integrate a function efficiently a quadrature formula suitable for the integrand should be chosen. Moreover, if the computer time necessary for constructing a formula is much less than that necessary for evaluating the given integral, it would be worthwhile to design a quadrature formula suitable for the integrand. Suppose that an integral

$$I = \int_{-1}^{1} f(x)\,dx \qquad (4.1)$$

is given and that $f(x)$ requires much computer time to evaluate. Now, in order to evaluate (4.1), we will try to use a DE-rule based on a generalized DE-transformation

$$z = \tanh(A\,e^{aw} - B\,e^{-bw}), \qquad (4.2)$$

where a, b, A and B are positive constants. We decompose (4.2) into

$$z = \tanh \zeta \qquad (4.3)$$

and

$$\zeta = A\,e^{aw} - B\,e^{-bw}. \qquad (4.4)$$

From (4.4) we have

$$\xi = A\,e^{au} \cos av - B\,e^{-bu} \cos bv$$
$$\eta = A\,e^{au} \sin av + B\,e^{-bu} \sin bv, \qquad (4.5)$$

where $\zeta = \xi + i\eta$ and $w = u + iv$. We see that the image of the line $\operatorname{Im} w = v_m =$ constant corresponding to (3.5) in the ζ-plane approaches asymptotically to the half lines

$$\eta = (\tan av_m)\xi,\ \xi > 0 \qquad \text{as } u \to +\infty,$$
$$\eta = (\tan bv_m)\xi,\ \xi < 0 \qquad \text{as } u \to -\infty, \qquad (4.6)$$

as shown in Fig. 3. Furthermore it would be interesting to note that, as is easily seen from (4.6), (4.4) maps the two lines

$$\operatorname{Im} w = \pm \frac{\pi}{a+b} \qquad (4.7)$$

in the w-plane onto the two half lines in the ζ-plane,

$$\eta = \pm \left(\tan \frac{a\pi}{a+b} \right) \xi, \qquad |\eta| > \eta_{\min}, \qquad (4.8)$$

where

$$\eta_{\min} = \left(\sin \frac{a\pi}{a+b} \right) \left(\left(\frac{b}{a} \right)^{a/(a+b)} + \left(\frac{a}{b} \right)^{b/(a+b)} \right) A^{b/(a+b)} B^{a/(a+b)}, \qquad (4.9)$$

as shown in Fig. 3. In other words, (4.4) defines a one-to-one mapping of the ζ-plane with a boundary (4.8) onto the strip domain in the w-plane,

$$|\operatorname{Im} w| < \frac{\pi}{a+b}. \qquad (4.10)$$

If $a = b$, then the image of $\operatorname{Im} w = $ constant is symmetric with respect to the imaginary axis in the ζ-plane, and (4.9) becomes

$$\eta_{\min} = 2\sqrt{AB}. \qquad (4.11)$$

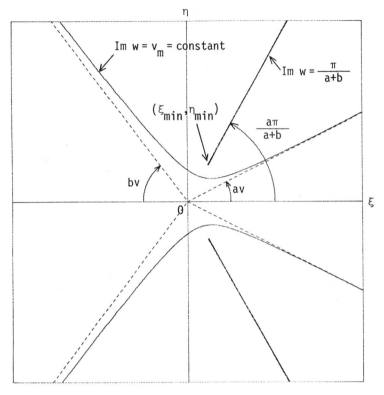

FIGURE 3 The image of Im $w = v_m$ = constant by $\zeta = A\,e^{aw} - B\,e^{-bw}$.

Next consider the mapping by $z = \tanh \zeta$. It maps the strip domain

$$|\text{Im } \zeta| = |\eta| < \frac{\pi}{2} \qquad (4.12)$$

onto the z-plane with a boundary,

$$\text{Im } z = y = 0, \qquad |\text{Im } z| = |x| > 1. \qquad (4.13)$$

Therefore, if we first map the line Im $w = v_m$ = constant onto the ζ-plane by (4.4) and then map the part of the image satisfying $|\text{Im } \zeta| = |\eta| < \pi/2$ onto the z-plane by (4.3), we obtain the contour map of the characteristic function of the error of the DE-rule based on (4.2).

If $\eta_{\min} \geq \pi/2$, the image of the half lines (4.8) in the ζ-plane corresponding to (4.7) in the w-plane does not appear in the z-plane. On the other hand, if $\eta_{\min} < \pi/2$ this image appears in the z-plane. As an example, the contour

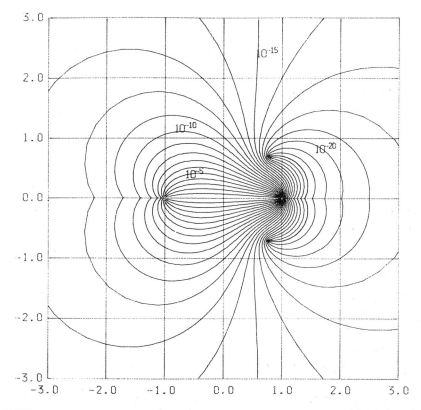

FIGURE 4 The contour map of $|\Phi_h(z)|/(2\pi)$ of the DE-rule defined by (4.2) with $a=\frac{1}{2}$, $b=1$, $A=B=\frac{1}{2}$, $h=0.25$.

map of $|\Phi_h(z)|/(2\pi)$ for the generalized DE-rule defined by (4.2) with $a=\frac{1}{2}$, $b=1$, $A=B=\frac{1}{2}$, $h=0.25$ is shown in Fig. 4. We see that an arc-shaped part of the image of (4.7) appears in the neighborhood of $z=1$.

It is evident from (3.10) that $|\Phi_h(z)|$ becomes minimum along this image. Therefore, when the integrand is a rational function, the error will be small if we choose the parameters a, b, A and B in (4.2) in such a way that the image be located in the neighborhood of those poles of the integrand which are expected to contribute significantly to the quadrature error. For example, the DE-rule whose contour map is shown in Fig. 4 will integrate favorably a function having poles in the neighborhood of the arc-shaped image appearing around $z=1$.

Finally we consider the effect of changing the parameters a, b, A and B on the accuracy of the formula when the integrand has endpoint algebraic singularities with different powers, e.g.,

$$I = \int_{-1}^{1} (1-x)^{\lambda}(1+x)^{\mu}\, dx, \qquad \lambda \neq \mu. \tag{4.14}$$

If we apply the transformation

$$x = \tanh(A\, e^{aw} - B\, e^{-bw}) \tag{4.15}$$

to (4.14) we have, for large $|w|$,

$$(1-x)^{\lambda}(1+x)^{\mu}\, dx \sim \begin{cases} \exp(-2(1+\lambda)A\, e^{aw})\, dw, & w \to +\infty, \\ \exp(-2(1+\mu)B\, e^{-bw})\, dw, & w \to -\infty. \end{cases} \tag{4.16}$$

Thus the decay of the integrand at large $|w|$ is essentially double exponential, and hence the influence of λ and μ on the efficiency of the formula is small. The situation is the same for the endpoint logarithmic singularity. A more important point to which more attention should be paid is to avoid the cancellation of the significant digits which may arise in the computation of the integrand of (4.14) near the endpoints $x = \pm 1$ when λ and μ are close to -1.

We conclude that the choice of the parameters a, b, A and B gives less influence on the error due to the algebraic or logarithmic singularities at the endpoints than that due to the location of other singular points such as poles. Therefore, if you want to integrate a class of functions which require much computer time to evaluate and have singularities at some fixed points in the complex plane, it is advisable to design a suitable DE-rule for this class of functions. However, in a general-purpose subroutine package, it is not a wise policy from a practical point of view to choose suitable parameters for a quadrature formula corresponding to each integrand, and it is recommended for the DE-rule in a general-purpose subroutine package to fix the parameters as (2.11), i.e., $a = b = 1$ and $A = B = \pi/4$.

References

[1] M. Sato, Theory of Hyperfunctions I, *J. Fac. of Sci., Univ. of Tokyo* **8** (1959), 139–193.
[2] M. Abramowitz and I. A. Stegun, eds. *Handbook of Mathematical Functions*, NBS, Dover, 1964, 1965.
[3] H. Takahasi and M. Mori, Error Estimation in the Numerical Integration of Analytic Functions, *Report of the Computer Centre, Univ. of Tokyo* **3** (1970) 41–108.
[4] H. Takahasi and M. Mori, Double Exponential Formulas for Numerical Integration, *Publ. RIMS, Kyoto Univ.* **9** (1974) 721–741.

[5] P. J. Davis and P. Rabinowitz, *Methods of Numerical Integration*, 2nd ed., Academic Press (1984).
[6] M. Sugihara, On the Optimality of the DE transformation (in Japanese), *Kokyuroku of RIMS, Kyoto Univ.*, No. 585 (1986) 150–175.
[7] M. Sugihara, Methods of Numerical Integration of Oscillatory Functions by the DE-formula with the Richardson Extrapolation, *J. Comput. Appl. Math.* **17** (1987) 47–68.

Analytic Functionals on the Complex Light Cone and Their Fourier–Borel Transformations

Mitsuo Morimoto and Ryoko Wada

Department of Mathematics
Sophia University
Tokyo, Japan

Introduction

We constructed a theory of analytic functions and functionals on the *complex sphere*

$$\tilde{S} = \{z \in \mathbf{C}^{d+1}; z^2 = 1\}, \tag{0.1}$$

where $z^2 = z \cdot z$, $z \cdot \zeta = z_1\zeta_1 + z_2\zeta_2 + \cdots + z_{d+1}\zeta_{d+1}$. The complex sphere $\tilde{S}$ is the complexification of the *real sphere*

$$S = \{x \in \mathbf{R}^{d+1}; \|x\| = 1\}, \tag{0.2}$$

where $\|x\| = (x^2)^{1/2}$ is the Euclidean norm in $\mathbf{R}^{d+1}$. See Morimoto [7, 8], Wada [9, 10], Wada–Morimoto [11].

A similar theory of analytic functions and functionals is possible if we replace the complex sphere $\tilde{S}$ and the real sphere S by the *complex light*

Algebraic Analysis, Volume I

cone M and its compact subset Σ, where

$$M = \{z \in \mathbf{C}^{d+1}; \, z^2 = 0\}. \tag{0.3}$$

$$\Sigma = \{z = x + iy \in M : \|x\| = \|y\| = 1\}. \tag{0.4}$$

The compact set Σ is isomorphic to the unit cotangent bundle of the real sphere S and we will call it tentatively the *spherical sphere*. See Ii [3] and Wada [9], where Σ is denoted by N.

We will state in this note the main results of these theories and announce some new results. We will not describe how to define the linear topology of function spaces. The proof of new results will be published elsewhere.

It is well known that on $\mathbf{C}^{d+1}$, the Fourier–Borel transformation P_λ establishes the linear topological isomorphisms (see for example Martineau [5]):

$$\mathrm{Exp}'(\mathbf{C}^{d+1}) \overset{\sim}{\to} \mathcal{O}(\mathbf{C}^{d+1}), \tag{0.5}$$

$$\mathcal{O}'(\mathbf{C}^{d+1}) \overset{\sim}{\to} \mathrm{Exp}(\mathbf{C}^{d+1}), \tag{0.6}$$

where $\mathcal{O}(\mathbf{C}^{d+1})$ (resp. $\mathrm{Exp}(\mathbf{C}^{d+1})$) denotes the space of entire functions (resp. entire functions of exponential type) and $\mathcal{O}'(\mathbf{C}^{d+1})$ (resp. $\mathrm{Exp}'(\mathbf{C}^{d+1})$) denotes the space dual to $\mathcal{O}(\mathbf{C}^{d+1})$ (resp. $\mathrm{Exp}(\mathbf{C}^{d+1})$). We may say that the Fourier–Borel transformation P_λ on $\mathbf{C}^{d+1}$ establishes the duality between $\mathrm{Exp}'(\mathbf{C}^{d+1})$ and $\mathcal{O}'(\mathbf{C}^{d+1})$ via (0.5) and (0.6). We will show that similar phenomena also appear on the complex sphere $\tilde{S}$ (Theorem 4.3) and on the complex light cone M (Theorem 4.5).

We will suppose $d \geq 2$ in this paper. If $d = 1$, the situation becomes one-dimensional and very simple (see Morimoto [6]).

1. Spaces of Complex Harmonic Functions

Let $P_{n,d}(\mathbf{C}^{d+1})$ (resp. $H_{n,d}(\mathbf{C}^{d+1})$) be the space of homogeneous polynomials (resp. homogeneous harmonic polynomials) of degree n in dimension $d+1$. $P_d(\mathbf{C}^{d+1}) = \sum_{n=0}^{\infty} P_{n,d}(\mathbf{C}^{d+1})$ (resp. $H_d(\mathbf{C}^{d+1}) = \sum_{n=0}^{\infty} H_{n,d}(\mathbf{C}^{d+1})$) denotes the space of polynomials (resp. harmonic polynomials) in dimension $d+1$. We consider the spaces $P_d(\mathbf{C}^{d+1})$, $H_d(\mathbf{C}^{d+1})$, etc. as function spaces over $\mathbf{C}^{d+1}$. We put

$$N(n, d) = \dim H_{n,d}(\mathbf{C}^{d+1}) = \frac{(2n+d-1)(n+d-2)!}{n!(d-1)!}. \tag{1.1}$$

Let us recall some properties of $N(n, d)$:

$$N(n, d) = O(n^{d-1}), \tag{1.2}$$

$$\sum_{n=0}^{\infty} N(n, d) t^n = \frac{1+t}{(1-t)^d} \qquad \text{for } |t| < 1. \tag{1.3}$$

On $\mathbf{C}^{d+1}$, we introduce three kinds of norms:

$$\|z\| = [\|x\|^2 + \|y\|^2]^{1/2} \qquad \text{(Euclidean norm)}, \tag{1.4}$$

$$L(z) = [\|x\|^2 + \|y\|^2 + 2\{\|x\|^2\|y\|^2 - (x \cdot y)^2\}^{1/2}]^{1/2} \qquad \text{(Lie norm)}, \tag{1.5}$$

$$L^*(z) = \sup\{|\zeta \cdot z|; L(\zeta) \le 1\}$$

$$= \frac{1}{\sqrt{2}} [\|x\|^2 + \|y\|^2 + \{(\|x\|^2 - \|y\|^2)^2 + 4(x \cdot y)^2\}^{1/2}]^{1/2}$$

$$\text{(dual Lie norm)}, \tag{1.6}$$

where $z = x + iy$, $x, y \in \mathbf{R}^{d+1}$.

For r with $0 < r \le \infty$ (resp. $0 \le r < \infty$),

$$\tilde{B}(r) = \{z \in \mathbf{C}^{d+1}; L(z) < r\} \tag{1.7}$$

(resp. $\tilde{B}[r] = \{z \in \mathbf{C}^{d+1}; L(z) \le r\}$) is the open (resp. closed) Lie ball of radius r. $\mathcal{O}(\tilde{B}(r))$ denotes the space of holomorphic functions on $\tilde{B}(r)$. We will consider the space $\mathcal{O}_\Delta(\tilde{B}(r))$ of complex harmonic functions, where

$$\mathcal{O}_\Delta(\tilde{B}(r)) = \{f \in \mathcal{O}(\tilde{B}(r)); \Delta_z f = 0\}, \tag{1.8}$$

where

$$\Delta_z = \frac{\partial^2}{\partial z_1^2} + \frac{\partial^2}{\partial z_2^2} + \cdots + \frac{\partial^2}{\partial z_{d+1}^2}. \tag{1.9}$$

The Lie norm $L(z)$ has the following property:

Lemma 1.1 (Drużkowski [1]). *For every homogeneous polynomial f on $\mathbf{C}^{d+1}$, we have*

$$\sup\{|f(z)|; L(z) < 1\} = \sup\{|f(s)|; s \in S\}. \tag{1.10}$$

Relying on Lemma 1.1, we can prove

Proposition 1.2. *For any $f \in \mathcal{O}_\Delta(\tilde{B}(r))$, we define the n-th harmonic component f_n of f as follows:*

$$f_n(z) = \frac{1}{2\pi i} \int_{|t|=\rho} \frac{f(tz)}{t^{n+1}} \, dt \qquad \text{for every } \rho \text{ with } 0 < \rho < r. \qquad (1.11)$$

Then $f_n \in H_{n,d}(\mathbf{C}^{d+1})$, $n = 0, 1, 2, \ldots$,

$$f(z) = \sum_{n=0}^{\infty} f_n(z) \qquad (1.12)$$

converges on every compact subset of $\tilde{B}(r)$, and

$$\varlimsup_{n\to\infty} \|f_n\|_S^{1/n} \leq \frac{1}{r}, \qquad (1.13)$$

where $\| \ \|_S$ denotes the L^2-norm on S (see Section 2).

Conversely if we have a sequence of $f_n \in H_{n,d}(\mathbf{C}^{d+1})$ which satisfies the condition (1.13), then the right side of (1.12) converges uniformly on every compact subset of $\tilde{B}(r)$ and defines a function $f \in \mathcal{O}_\Delta(\tilde{B}(r))$.

We define

$$\mathcal{O}_\Delta(\tilde{B}[r]) = \operatorname{ind}\lim_{r'>r} \mathcal{O}_\Delta(\tilde{B}(r')) \qquad \text{for } 0 \leq r < \infty. \qquad (1.14)$$

Let N be a norm in $\mathbf{C}^{d+1}$. We define

$$\operatorname{Exp}_\Delta(\mathbf{C}^{d+1}; (A:N)) = \{f \in \mathcal{O}_\Delta(\mathbf{C}^{d+1}); \text{ for any } \varepsilon > 0 \text{ there exists } C \geq 0$$

$$\text{s.t. } |f(z)| \leq C e^{(A+\varepsilon)N(z)}\} \qquad \text{for } 0 \leq A < \infty, \qquad (1.15)$$

$$\operatorname{Exp}_\Delta(\mathbf{C}^{d+1}; [A:N]) = \{f \in \mathcal{O}_\Delta(\mathbf{C}^{d+1}); \text{ there exist } 0 \leq B < A \text{ and } C \geq 0$$

$$\text{s.t. } |f(z)| \leq C e^{BN(z)}\} \qquad \text{for } 0 < A \leq \infty. \qquad (1.16)$$

$\mathcal{O}_\Delta(\tilde{B}(r))$ and $\operatorname{Exp}_\Delta(\mathbf{C}^{d+1}; (A:N))$ are FS spaces, while $\mathcal{O}_\Delta(\tilde{B}[r])$ and $\operatorname{Exp}_\Delta(\mathbf{C}^{d+1}; [A:N])$ are DFS spaces. We denote

$$\operatorname{Exp}_\Delta(\mathbf{C}^{d+1}) = \operatorname{Exp}_\Delta(\mathbf{C}^{d+1}; [\infty:N]).$$

Because we have

$$\operatorname{Exp}_\Delta(\mathbf{C}^{d+1}; (A:L)) = \operatorname{Exp}_\Delta(\mathbf{C}^{d+1}; (2A:L^*)), \qquad (1.17)$$

$$\operatorname{Exp}_\Delta(\mathbf{C}^{d+1}; [A:L]) = \operatorname{Exp}_\Delta(\mathbf{C}^{d+1}; [2A:L^*]), \qquad (1.17')$$

we can state a theorem in Morimoto [8] as follows:

Theorem 1.3.

$$f \in \mathrm{Exp}_\Delta(\mathbf{C}^{d+1}; (A:L^*)) \ \textit{iff} \ \varlimsup_{n\to\infty}(n!\|f_n\|_S)^{1/n} \le \frac{A}{2}, \qquad 0 \le A < \infty, \qquad (1.18)$$

$$f \in \mathrm{Exp}_\Delta(\mathbf{C}^{d+1}; [A:L^*]) \ \textit{iff} \ \varlimsup_{n\to\infty}(n!\|f_n\|_S)^{1/n} < \frac{A}{2}, \qquad 0 < A \le \infty, \qquad (1.18')$$

$$f \in \mathcal{O}_\Delta(\tilde{B}[r]) \ \textit{iff} \ \varlimsup_{n\to\infty}\|f_n\|_S^{1/n} < \frac{1}{r}, \qquad\qquad 0 \le r < \infty, \qquad (1.19)$$

$$f \in \mathcal{O}_\Delta(\tilde{B}(r)) \ \textit{iff} \ \varlimsup_{n\to\infty}\|f_n\|_S^{1/n} \le \frac{1}{r}, \qquad\qquad 0 < r \le \infty. \qquad (1.19')$$

The precise meaning of $(1.19')$ is Proposition 1.2. (1.18), $(1.18')$ and (1.19) should be understood similarly.

To describe our theorems in the following sections, we need some more function spaces. For $\lambda \in \mathbf{C}$, we define

$$\mathcal{O}_\lambda(\tilde{B}(r)) = \{f \in \mathcal{O}(\tilde{B}(r)); (\Delta_z + \lambda^2)f = 0\}. \qquad (1.20)$$

Note that $\mathcal{O}_0(\tilde{B}(r)) = \mathcal{O}_\Delta(\tilde{B}(r))$ and that $\mathcal{O}_\lambda(\tilde{B}(r))$ is a closed linear subspace of $\mathcal{O}(\tilde{B}(r))$. $\mathcal{O}_\lambda(\tilde{B}[r])$, $\mathrm{Exp}_\lambda(\mathbf{C}^{d+1}: (A:N))$ and $\mathrm{Exp}_\lambda(\mathbf{C}^{d+1}; [A:N])$ have similar meanings.

2. Holomorphic Functions on the Complex Sphere

Let us denote by $P_{n,d}(\tilde{S})$, $H_{n,d}(\tilde{S})$, etc. ($P_{n,d}(S)$, $H_{n,d}(S)$, etc., resp.) the restriction of $P_{n,d}(\mathbf{C}^{d+1})$, $H_{n,d}(\mathbf{C}^{d+1})$, etc., on the complex sphere $\tilde{S}$ (on the real sphere S, resp.).

$H_{n,d}(S)$ is called the *space of spherical harmonics* of degree n in dimension $d+1$. It is known that the restriction mappings

$$H_{n,d}(\mathbf{C}^{d+1}) \xrightarrow{\approx} H_{n,d}(\tilde{S}) \xrightarrow{\approx} H_{n,d}(S) \qquad (2.1)$$

are linear isomorphisms and

$$P_{n,d}(\tilde{S}) = \sum_{k=0}^{n} H_{k,d}(\tilde{S}), \qquad P_{n,d}(S) = \sum_{k=0}^{n} H_{k,d}(S). \qquad (2.2)$$

The orthogonal group $O(d+1)$ acts transitively on S,

$$S = O(d+1)/O(d).$$

and there exists a unique $O(d+1)$-invariant measure dS on S such that $\int_S dS = 1$. We have

$$P_d(S) = H_d(S) = \sum_{n=0}^{\infty} H_{n,d}(S), \qquad (2.3)$$

where the last term is an orthogonal sum with respect to the L^2 inner product $(\ ,\ )_S$ of $L^2(S)$, where $(f, g)_S = \langle f, \bar{g} \rangle_S$ and

$$\langle f, g \rangle_S = \int_S f(s)g(s)\, dS(s). \qquad (2.4)$$

The Legendre polynomial $P_{n,d}(t)$ has a reproducing property for $f_n \in H_{n,d}(S)$, namely, the following is valid for any $\rho > 0$ and $z \in \mathbf{C}^{d+1}$:

$$f_n(z) = N(n, d) \int_S f_n(\rho s)\rho^{-n}(z^2)^{n/2} P_{n,d}\left(s \cdot \frac{z}{(z^2)^{1/2}}\right) dS(s). \qquad (2.5)$$

The n-th harmonic component f_n of $f \in \mathcal{O}_\Delta(\tilde{B}(r))$ is given by the integral on the real sphere S:

$$f_n(z) = N(n, d) \int_S f(\rho s)\rho^{-n}(z^2)^{n/2} P_{n,d}\left(s \cdot \frac{z}{(z^2)^{1/2}}\right) dS(s), \qquad (2.6)$$

for every ρ with $0 < \rho < r$. Summing up the formula (2.6), we can get the *Poisson formula* for a harmonic function $f \in \mathcal{O}_\Delta(\tilde{B}(r))$:

$$f(z) = \int_S \frac{(1 - (z/\rho)^2)f(\rho s)}{(1 + (z/\rho)^2 - 2s \cdot (z/\rho))^{(d+1)/2}}\, dS(s), \qquad (2.7)$$

for some ρ, $0 < \rho < r$ and $z \in \tilde{B}(\rho)$.

Let us define

$$\tilde{S}(r) = \tilde{S} \cap \tilde{B}(r) \qquad \text{for } 1 < r \le \infty, \qquad (2.8)$$

$$\tilde{S}[r] = \tilde{S} \cap \tilde{B}[r] \qquad \text{for } 1 \le r < \infty. \qquad (2.9)$$

$\tilde{S}(r)$ is an open complex neighborhood of the real sphere $S = S[1]$ for $r > 1$.

For $r > 1$, $\mathcal{O}(\tilde{S}(r))$ denotes the space of holomorphic functions on the open subset $\tilde{S}(r)$ of the complex sphere $\tilde{S}$. By the Oka–Cartan Theorem B, we know the restriction mapping $\mathcal{O}(\tilde{B}(r)) \to \mathcal{O}(\tilde{S}(r))$ is surjective. For $r \ge 1$, we define

$$\mathcal{O}(\tilde{S}[r]) = \operatorname{ind}\lim_{r' > r} \mathcal{O}(\tilde{S}(r')). \qquad (2.10)$$

Let us define the space of holomorphic functions of exponential type on $\tilde{S}$. For a norm N on $\mathbf{C}^{d+1}$, we define

$$\mathrm{Exp}(\tilde{S}; (A:N)) = \{f \in \mathcal{O}(\tilde{S}); \text{ for any } \varepsilon > 0 \text{ there exists } C \geq 0$$

$$\text{s.t. } |f(z)| \leq C\, e^{(A+\varepsilon)N(z)} \text{ for } z \in \tilde{S}\} \qquad \text{for } 0 \leq A < \infty,$$

$$(2.11)$$

$$\mathrm{Exp}(\tilde{S}; [A:N]) = \{f \in \mathcal{O}(\tilde{S}); \text{ there exist } 0 \leq B < A \text{ and } C \geq 0$$

$$\text{s.t. } |f(z)| \leq C\, e^{BN(z)} \text{ for } z \in \tilde{S}\} \qquad \text{for } 0 < A \leq \infty. \quad (2.12)$$

$\mathcal{O}(\tilde{S}(r))$ and $\mathrm{Exp}(\tilde{S}: (A:N))$ are FS spaces while $\mathcal{O}(\tilde{S}[r])$ and $\mathrm{Exp}(\tilde{S}: [A:N])$ are DFS spaces. We denote

$$\mathrm{Exp}(\tilde{S}) = \mathrm{Exp}(\tilde{S}; [\infty : N]).$$

Note that for $z = x + iy \in \tilde{S}$ we have $\|z\| = (2\|x\|^2 - 1)^{1/2}$, $L(z) = \|x\| + (\|x\|^2 - 1)^{1/2}$ and $L^*(z) = \|x\|$. Therefore we have

$$\mathrm{Exp}(\tilde{S}; (A:L)) = \mathrm{Exp}(\tilde{S}; (2A:L^*))$$

and

$$\mathrm{Exp}(\tilde{S}; [A:L]) = \mathrm{Exp}(\tilde{S}; [2A:L^*]).$$

Theorem 2.1. *Suppose $\lambda \in \mathbf{C}$ satisfies the following condition*

$$(\lambda/2)^{-n-(d-1)/2} J_{n+(d-1)/2}(\lambda) \neq 0 \qquad \text{for all } n = 0, 1, 2, 3, \ldots, \quad (2.13)$$

where $J_{n+(d-1)/2}$ is the Bessel function. Then the following restriction mappings α_λ are linear topological isomorphisms:

$$\alpha_\lambda : \mathrm{Exp}_\lambda(\mathbf{C}^{d+1}; (A:L^*)) \stackrel{\approx}{\to} \mathrm{Exp}(\tilde{S}; (A:L^*)), \qquad 0 < A \leq \infty, \qquad (2.14)$$

$$\alpha_\lambda : \mathrm{Exp}_\lambda(\mathbf{C}^{d+1}; [A:L^*]) \stackrel{\approx}{\to} \mathrm{Exp}(\tilde{S}; [A:L^*]), \qquad 0 \leq A < \infty, \qquad (2.14')$$

$$\alpha_\lambda : \mathcal{O}_\lambda(\tilde{B}[r]) \stackrel{\approx}{\to} \mathcal{O}(\tilde{S}[r]), \qquad 1 \leq r < \infty, \qquad (2.15)$$

$$\alpha_\lambda : \mathcal{O}_\lambda(\tilde{B}(r)) \stackrel{\approx}{\to} \mathcal{O}(\tilde{S}(r)), \qquad 1 < r \leq \infty. \qquad (2.15')$$

Wada [10] proved (2.15) and (2.15') and (2.14') with $r = \infty$. Note that $\lambda = 0$ satisfies the condition (2.13).

By (2.14), (2.14'), (2.15) and (2.15') with $\lambda = 0$, we will identify the spaces $\mathcal{O}(\tilde{S}(r))$, $\mathrm{Exp}(\tilde{S}; (A:L^*))$, etc. with $\mathcal{O}_\Delta(\tilde{B}(r))$, $\mathrm{Exp}_\Delta(\mathbf{C}^{d+1}, (A:L^*))$, etc., and we can speak of the n-th harmonic component f_n of $f \in \mathcal{O}(\tilde{S}(r))$,

$\text{Exp}(\tilde{S}; (A:L^*))$, etc. Therefore the spaces $\mathcal{O}(\tilde{S}(r))$, $\text{Exp}(\tilde{S}, (A:L^*))$, etc., can be characterized by the growth condition of $\|f_n\|_S$ by Theorem 1.3.

The n-th harmonic component f_n of $f \in \mathcal{O}(\tilde{S}(r))$ is represented directly by the integral formula (2.6), because S is a subset of $\tilde{S}(r)$.

The proof of Theorem 2.1 and 2.1bis relies on the fact that the integral path S in (2.6) can be shifted to S_r or S_r^* for some $r > 1$, where

$$S_r = \{z \in \tilde{S}; L(z) = r\} \text{ and} \qquad\qquad (2.16)$$

$$S_r^* = \{z \in \tilde{S}; L^*(z) = r\} = S_{r'}, \qquad \text{with } r' = r + (r^2-1)^{1/2}. \quad (2.16')$$

The orthogonal group $O(d+1)$ acts on S_r^* transitively,

$$S_r^* = O(d+1)/O(d-1),$$

and there is a unique $O(d+1)$-invariant measure dS_r^* on S_r^* such that $\int_{S_r^*} dS_r^* = 1$.

Lemma 2.2. *Let $r > 1$. The n-th harmonic component f_n of $f \in \mathcal{O}(\tilde{S})$ is given by the following formula:*

$$f_n(z) = \frac{N(n,d)}{C_n(r)} \int_{S_r^*} f(\sigma)(z^2)^{n/2} P_{n,d}\left(\bar{\sigma} \cdot \frac{z}{(z^2)^{1/2}}\right) dS_r^*(\sigma), \qquad (2.17)$$

$z \in \mathbf{C}^{d+1}$, *where*

$$C_n(r) = P_{n,d}(2r^2-1). \qquad\qquad (2.18)$$

As polynomials form a dense subspace of $\mathcal{O}(\tilde{S})$, $\text{Exp}(\tilde{S})$ is dense in $\mathcal{O}(\tilde{S})$ and we can consider $\mathcal{O}'(\tilde{S}) \subset \text{Exp}'(\tilde{S})$, where $\mathcal{O}'(\tilde{S})$ and $\text{Exp}'(\tilde{S})$ are dual spaces of $\mathcal{O}(\tilde{S})$ and $\text{Exp}(\tilde{S})$ respectively. $\langle\ ,\ \rangle_{(S)}$ will denote the canonical inner product of the dual space and the original space.

We will define the n-th harmonic component T_n of a functional $T \in \text{Exp}'(\tilde{S})$ as follows:

$$T_n(z) = N(n,d)\left\langle T_s, (z^2)^{1/2} P_{n,d}\left(s \cdot \frac{z}{(z^2)^{1/2}}\right)\right\rangle_{(S)} \qquad \text{for } z \in \mathbf{C}^{d+1}, \quad (2.19)$$

$s \in S$ being a dummy variable. $T_n(z)$ belongs to the space $H_{n,d}(\mathbf{C}^{d+1})$ and the spaces of analytic functionals $\mathcal{O}'(\tilde{S})$, $\text{Exp}'(\tilde{S})$, etc., can be characterized by the growth conditions of $\|T_n\|_S$.

Theorem 2.3. *Suppose $T \in \mathrm{Exp}'(\tilde{S})$. Let us denote by $T_n(z)$ the n-th harmonic component of T. Then we have*

$$T \in \mathrm{Exp}'(\tilde{S}; [A:L^*]) \text{ iff } \varlimsup_{n \to \infty} \left(\frac{1}{n!} \|T_n\|_S\right)^{1/n} \le \frac{2}{A}, \qquad 0 < A \le \infty, \tag{2.20}$$

$$T \in \mathrm{Exp}'(\tilde{S}; (A:L^*)) \text{ iff } \varlimsup_{n \to \infty} \left(\frac{1}{n!} \|T_n\|_S\right)^{1/n} < \frac{2}{A}, \qquad 0 \le A < \infty, \tag{2.20'}$$

$$T \in \mathcal{O}'(\tilde{S}(r)) \text{ iff } \varlimsup_{n \to \infty} \|T_n\|_S^{1/n} < r, \qquad\qquad 1 < r \le \infty, \tag{2.21}$$

$$T \in \mathcal{O}'(\tilde{S}[r]) \text{ iff } \varlimsup_{n \to \infty} \|T_n\|_S^{1/n} \le r, \qquad\qquad 1 \le r < \infty. \tag{2.21'}$$

(2.21), (2.21′) and (2.20) with $A = \infty$ were proved in Morimoto [8]. (2.20) and (2.20′) can be proved similarly.

We have the following formula:

Theorem 2.4. *Suppose $f \in \mathrm{Exp}(\tilde{S})$ (resp. $\mathcal{O}(\tilde{S}(r))$, etc.) and $T \in \mathrm{Exp}'(\tilde{S})$ (resp. $\mathcal{O}'(\tilde{S}(r))$, etc.). Then we have*

$$\langle T, f \rangle_{(S)} = \sum_{n=0}^{\infty} \langle T, f_n \rangle_{(S)} = \sum_{n=0}^{\infty} \langle T_n, f \rangle_S$$

$$= \sum_{n=0}^{\infty} \langle T_n, f_n \rangle_S. \tag{2.22}$$

3. Holomorphic Functions on the Complex Light Cone

Let us denote by $P_{n,d}(M)$, $H_{n,d}(M)$, etc., $(P_{n,d}(\Sigma)$, $H_{n,d}(\Sigma)$, etc., resp.) the restriction of $P_{n,d}(\mathbf{C}^{d+1})$, $H_{n,d}(\mathbf{C}^{d+1})$, etc., on the complex light cone M (on the spherical sphere Σ). As in the spherical case, we know the restriction mappings

$$H_{n,d}(\mathbf{C}^{d+1}) \overset{\approx}{\to} H_{n,d}(M) \overset{\approx}{\to} H_{n,d}(\Sigma) \tag{3.1}$$

are linear isomorphisms and

$$P_{n,d}(M) = H_{n,d}(M), \qquad P_{n,d}(\Sigma) = H_{n,d}(\Sigma). \tag{3.2}$$

As $\|x\| = \|y\|$ and $x \cdot y = 0$ for $z = x + iy \in M$, we have for $z \in M$,

$$\|z\| = \sqrt{2}\,\|x\| = \sqrt{2}\,\|y\|,$$

$$L(z) = 2\|x\| = 2\|y\| = \sqrt{2}\,\|z\|,$$

and

$$L^*(z) = \|x\| = \|y\| = \|z\|/\sqrt{2}.$$

Therefore the spherical sphere Σ is represented as follows:

$$\Sigma = \{z \in M;\ \|z\| = \sqrt{2}\} = \{z \in M;\ L(z) = 2\}$$

$$= \{z \in M;\ L^*(z) = 1\}. \tag{3.3}$$

Note that M is spanned by Σ: $M = \{t\sigma;\ t \in \mathbf{C},\ \sigma \in \Sigma\}$. The orthogonal group $O(d+1)$ acts canonically on Σ,

$$\Sigma = O(d+1)/O(d-1),$$

and there exists a unique $O(d+1)$-invariant measure $d\Sigma$ on Σ such that $\int_\Sigma d\Sigma = 1$. We have

$$P_d(\Sigma) = H_d(\Sigma) = \sum_{n=0}^{\infty} H_{n,d}(\Sigma), \tag{3.4}$$

where the last term is an orthogonal sum with respect to the bilinear form

$$\langle f, g \rangle_\Sigma = \int_\Sigma f(\bar{\sigma})g(\sigma)\,d\Sigma(\sigma). \tag{3.5}$$

The polynomial $(z' \cdot z)^n$ has a reproducing property for $f_n \in H_{n,d}(\mathbf{C}^{d+1})$ on the spherical sphere Σ, namely we have

$$f_n(z) = N(n, d) \int_\Sigma f_n(\rho\sigma/2)\left(\bar{\sigma} \cdot \frac{z}{\rho}\right)^n d\Sigma(\sigma) \tag{3.6}$$

for $z \in \mathbf{C}^{d+1}$ and $\rho > 0$.

The n-th harmonic component f_n of $f \in \mathcal{O}_\Delta(\tilde{B}(r))$ is given by the integral on the spherical sphere Σ:

$$f_n(z) = N(n, d) \int_\Sigma f(\rho\sigma/2)\left(\bar{\sigma} \cdot \frac{z}{\rho}\right)^n d\Sigma(\sigma) \tag{3.7}$$

for $z \in \mathbf{C}^{d+1}$, where $0 < \rho < r$ and the right-hand side is independent of ρ. Summing up the formula (3.7) using (1.3), we can get the *Poisson formula* for harmonic functions:

$$f(z) = \int_{\Sigma} f(\rho\sigma/2) \frac{1 + \bar{\sigma} \cdot (z/\rho)}{(1 - \bar{\sigma} \cdot (z/\rho))^d} \, d\Sigma(\sigma) \tag{3.8}$$

for $z \in \tilde{B}(\rho)$ with $0 < \rho < r$.

The proof of the formulas (3.7) and (3.8) relies on the following close relationship between two bilinear forms $\langle \ , \ \rangle_S$ and $\langle \ , \ \rangle_{\Sigma}$.

Lemma 3.1 (Wada–Morimoto [11]). *For f_n and $g_n \in H_{n,d}(\mathbf{C}^{d+1})$, we have*

$$\langle f_n, g_n \rangle_S = 2^{-2n} \frac{n! \, N(n, d) \Gamma((d+1)/2)}{\Gamma(n + (d+1)/2)} \langle f_n, g_n \rangle_{\Sigma}. \tag{3.9}$$

Thanks to (1.2) we have

Corollary. *For a sequence $f_n \in H_{n,d}(\mathbf{C}^{d+1})$, $n = 0, 1, 2, \ldots$, we have*

$$\overline{\lim_{n \to \infty}} \|f_n\|_S^{1/n} = \tfrac{1}{2} \overline{\lim_{n \to \infty}} \|f_n\|_{\Sigma}^{1/n}, \tag{3.10}$$

where $\| \ \|_{\Sigma}$ denotes the L^2-norm on Σ:

$$\|f\|_{\Sigma} = \int_{\Sigma} |f(\sigma)|^2 \, d\Sigma(\sigma). \tag{3.11}$$

Let us define the truncated cone:

$$M(r) = M \cap \tilde{B}(r) \qquad \text{for } 0 < r \le \infty, \tag{3.12}$$

$$M[r] = M \cap \tilde{B}[r] \qquad \text{for } 0 \le r < \infty. \tag{3.13}$$

The complex light cone M is a complex cone which has a singularity at $z = 0$. For $r > 0$, we define $\mathcal{O}(M(r))$ to be the restriction image of $\mathcal{O}(\tilde{B}(r))$ onto $M(r)$. We will call an element of $\mathcal{O}(M(r))$ a *holomorphic* function on $m(r)$. ($\mathcal{O}(M)$ was denoted by Holo(M) in Ii [3] and Wada [9].) By the definition, the restriction mapping $\mathcal{O}(\tilde{B}(r)) \to \mathcal{O}(M(r))$ is surjective. For $r \ge 0$, we define

$$\mathcal{O}(M[r]) = \text{ind} \lim_{r' > r} \mathcal{O}(M(r')). \tag{3.14}$$

Let us define the space of holomorphic functions of exponential type on M. For a norm N on $\mathbf{C}^{d+1}$, we define

$\text{Exp}(M; (A:N)) = \{f \in \mathcal{O}(M); \text{ for any } \varepsilon > 0 \text{ there exists } C \geq 0$

$$\text{s.t. } |f(z)| \leq C \, e^{(A+\varepsilon)N(z)} \text{ for } z \in M\} \qquad \text{for } 0 \leq A < \infty, \tag{3.15}$$

$\text{Exp}(M; [A:N]) = \{f \in \mathcal{O}(M); \text{ there exist } B < A \text{ and } C \geq 0$

$$\text{s.t. } |f(z)| \leq C \, e^{BN(z)} \text{ for } z \in M\} \qquad \text{for } 0 < A \leq \infty. \tag{3.16}$$

$\mathcal{O}(M(r))$ and $\text{Exp}(M; (A:N))$ are FS spaces while $\mathcal{O}(M[r])$ and $\text{Exp}(M:[A:N])$ are DFS spaces. We will denote

$$\text{Exp}(M) = \text{Exp}(M; [\infty:N]).$$

Note that we have

$$\text{Exp}(M; (A:L)) = \text{Exp}(M; (2A:L^*))$$

and

$$\text{Exp}(M; [A:L]) = \text{Exp}(M; [2A:L^*]).$$

Theorem 3.2. *For any* $\lambda \in \mathbf{C}$, *the following restriction mappings* α_λ *are linear topological isomorphisms*:

$$\alpha_\lambda : \text{Exp}_\lambda(\mathbf{C}^{d+1}; (A:L^*)) \tilde{\to} \text{Exp}(M; (A:L^*)), \qquad 0 \leq A < \infty, \tag{3.17}$$

$$\alpha_\lambda : \text{Exp}_\lambda(\mathbf{C}^{d+1}; [A:L^*]) \tilde{\to} \text{Exp}(M; [A:L^*]), \qquad 0 < A \leq \infty, \tag{3.17'}$$

$$\alpha_\lambda : \mathcal{O}_\lambda(\tilde{B}[r]) \tilde{\to} \mathcal{O}(M[r]), \qquad 1 \leq r < \infty, \tag{3.18}$$

$$\alpha_\lambda : \mathcal{O}_\lambda(\tilde{B}(r)) \tilde{\to} \mathcal{O}(M(r)), \qquad 1 < r \leq \infty. \tag{3.18'}$$

(3.17) and (3.17') were proved by Wada [9] using a result in Ii [3], while (3.18) and (3.18') were proved by Wada–Morimoto [11].

By Theorem 3.2, we will identify the spaces $\mathcal{O}(M(r))$, $\text{Exp}(M)$, etc., with the spaces $\mathcal{O}_\lambda(\tilde{B}(r))$, $\text{Exp}_\lambda(\mathbf{C}^{d+1})$, etc., and we can speak of the n-th harmonic component f_n of $f \in \mathcal{O}(M(r))$, $\text{Exp}(M)$, etc. Then f_n can be represented directly by $f \in \mathcal{O}(M(r))$, $\text{Exp}(M)$, etc., with the integral formula (3.7).

Theorem 3.3.

$$f \in \text{Exp}(M; (A:L^*)) \text{ iff } \varlimsup_{n \to \infty} (n! \|f_n\|_\Sigma)^{1/n} \leq A, \qquad 0 \leq A < \infty, \tag{3.19}$$

$$f \in \text{Exp}(M; [A:L^*]) \text{ iff } \varlimsup_{n \to \infty} (n! \|f_n\|_\Sigma)^{1/n} < A, \qquad 0 < A \leq \infty, \tag{3.19'}$$

$$f \in \mathcal{O}(M[r]) \text{ iff } \varlimsup_{n \to \infty} \|f_n\|_{\Sigma}^{1/n} < \frac{2}{r}, \qquad\qquad 0 \leq r < \infty, \qquad (3.20)$$

$$f \in \mathcal{O}(M(r)) \text{ iff } \varlimsup_{n \to \infty} \|f_n\|_{\Sigma}^{1/n} \leq \frac{2}{r}, \qquad\qquad 0 < r \leq \infty. \qquad (3.20')$$

As polynomials form a dense subspace of $\mathcal{O}(M)$, $\mathrm{Exp}(M)$ is dense in $\mathcal{O}(M)$ and we can consider $\mathcal{O}'(M) \subset \mathrm{Exp}'(M)$, where $\mathcal{O}'(M)$ and $\mathrm{Exp}'(M)$ are dual spaces of $\mathcal{O}(M)$ and $\mathrm{Exp}(M)$ respectively. $\langle \ , \ \rangle_{(\Sigma)}$ will denote the canonical inner product of duality for these spaces.

We will define the n-th harmonic component T_n of a functional $T \in \mathrm{Exp}'(M)$ as follows:

$$T_n(z) = N(n, d) \left\langle T_\sigma, \left(z \cdot \frac{\sigma}{2} \right)^n \right\rangle_{(\Sigma)}. \qquad (3.21)$$

$T_n(z)$ belongs to the space $H_{n,d}(\mathbf{C}^{d+1})$, and the spaces of analytic functionals $\mathcal{O}'(M)$, $\mathrm{Exp}'(M)$, etc., can be characterized by the growth conditions of $\|T_n\|_{\Sigma}$.

Theorem 3.4. *Suppose $T \in \mathrm{Exp}'(M)$. Let us denote by $T_n(z)$ the n-th harmonic component of T.*

$$T \in \mathrm{Exp}'(M; [A:L^*]) \text{ iff } \varlimsup \left(\frac{1}{n!} \|T_n\|_{\Sigma} \right)^{1/n} < \frac{1}{A}, \qquad 0 < A \leq \infty, \qquad (3.22)$$

$$T \in \mathrm{Exp}'(M; (A:L^*)) \text{ iff } \varlimsup \left(\frac{1}{n!} \|T_n\|_{\Sigma} \right)^{1/n} \leq \frac{1}{A}, \qquad 0 \leq A < \infty, \qquad (3.22')$$

$$t \in \mathcal{O}'(M(r)) \text{ iff } \varlimsup_{n \to \infty} \|T_n\|_{\Sigma}^{1/n} < \frac{r}{2}, \qquad\qquad 0 < r \leq \infty, \qquad (3.23)$$

$$T \in \mathcal{O}'(M[r]) \text{ iff } \varlimsup_{n \to \infty} \|T_n\|_{\Sigma}^{1/n} \leq \frac{r}{2}, \qquad\qquad 0 \leq r < \infty. \qquad (3.23')$$

Suppose $f \in \mathrm{Exp}(M)$ and $T \in \mathrm{Exp}'(M)$. Then we have

$$\langle T, f \rangle_{(\Sigma)} = \sum_{n=0}^{\infty} \langle T, f_n \rangle_{(\Sigma)}$$

$$= \sum_{n=0}^{\infty} \left\langle T_z, N(n, d) \int_{\Sigma} f\left(\frac{\rho\sigma}{2} \right) \left(\bar{\sigma} \cdot \frac{z}{\rho} \right)^n d\Sigma(\sigma) \right\rangle_{(\Sigma)}$$

$$= \sum_{n=0}^{\infty} N(n,d) \int_{\Sigma} \left\langle T_z, \left(\bar{\sigma} \cdot \frac{z}{\rho} \right)^n \right\rangle_{(\Sigma)} f\left(\frac{\rho\sigma}{2} \right) d\Sigma(\sigma)$$

$$= \sum_{n=0}^{\infty} \int_{\Sigma} T_n\left(\frac{2\bar{\sigma}}{\rho} \right) f\left(\frac{\rho\sigma}{2} \right) d\Sigma(\sigma)$$

$$= \sum_{n=0}^{\infty} \int_{\Sigma} T_n\left(\frac{2\bar{\sigma}}{\rho} \right) f_n\left(\frac{\rho\sigma}{2} \right) d\Sigma(\sigma)$$

$$= \sum_{n=0}^{\infty} \langle T_n, f_n \rangle_{\Sigma}.$$

Therefore we have

Theorem 3.5. Suppose $f \in \mathrm{Exp}(M)$ (resp. $\mathcal{O}(M(r))$, etc.) and $T \in \mathrm{Exp}'(M)$ (resp. $\mathcal{O}'(M(r))$, etc.). Then we have

$$\langle T, f \rangle_{(\Sigma)} = \sum_{n=0}^{\infty} \langle T, f_n \rangle_{(\Sigma)} = \sum_{n=0}^{\infty} \langle T_n, f \rangle_{\Sigma} = \sum_{n=0}^{\infty} \langle T_n, f_n \rangle_{\Sigma}. \qquad (3.24)$$

Remark. For $0 \le \rho \le r$, let us consider the set

$$S_{\rho,r} = \{ z \in \mathbf{C}^{d+1}; z^2 = \rho^2, L(z) = r \}. \qquad (3.25)$$

If $\rho = r$, $S_{\rho,r}$ collapses to the real sphere of radius ρ: $S_{\rho,\rho} = \rho S$. If $\rho < r$, $S_{\rho,r}$ are all isomorphic to the spherical sphere $\Sigma = S_{0,1}$ as $O(d+1)$-spaces. In particular $S_r = S_{1,r} \cong \Sigma$. In fact,

$$S_{\rho,r} = \{ g\omega_{\rho,r}; g \in O(d+1) \} \cong O(d+1)/O(d-1), \qquad (3.26)$$

where $\omega_{\rho,r} = ((r^2+\rho^2)/(2r), 0, 0, \ldots, 0) + i(0, (r^2-\rho^2)/(2r), 0, \ldots, 0)$. If we change ρ and r suitably, we get a diffeomorphism $S_r \cong \Sigma$.

4. The Fourier–Borel Transformation

We mentioned in the Introduction that the Fourier–Borel transformation on $\mathbf{C}^{d+1}$ establishes the duality between $\mathrm{Exp}'(\mathbf{C}^{d+1})$ and $\mathcal{O}'(\mathbf{C}^{d+1})$. Martineau [5] gave a more precise version:

Theorem 4.1. *The Fourier–Borel transformation on* $\mathbf{C}^{d+1}$

$$P_{\lambda} : T_z \to \langle T_z, e^{i\lambda z \cdot \zeta} \rangle$$

establishes the following linear isomorphisms:

$$P_\lambda : \mathrm{Exp}'(\mathbf{C}^{d+1}; [|\lambda|r : L^*]) \cong \mathcal{O}(\tilde{B}(r)), \qquad 0 < r \le \infty, \qquad (4.1)$$

$$P_\lambda : \mathrm{Exp}'(\mathbf{C}^{d+1}; (|\lambda|r : L^*)) \cong \mathcal{O}(\tilde{B}[r]), \qquad 0 \le r < \infty, \qquad (4.1')$$

$$P_\lambda : \mathcal{O}'(\tilde{B}(r)) \cong \mathrm{Exp}(\mathbf{C}^{d+1}; [|\lambda|r : L^*]), \qquad 0 < r \le \infty, \qquad (4.2)$$

$$P_\lambda : \mathcal{O}'(\tilde{B}[r]) \cong \mathrm{Exp}(\mathbf{C}^{d+1}; (|\lambda|r : L^*)), \qquad 0 \le r < \infty. \qquad (4.2')$$

We will show that the same kind of duality holds on the complex sphere $\tilde{S}$ and the complex light cone M.

4.1. Spherical Case

Suppose an analytic functional $T \in \mathrm{Exp}'(\tilde{S})$ is given. For $\lambda \ne 0$, we define the *Fourier–Borel transform* G_λ of T as follows:

$$G_\lambda(z) = \langle T_\zeta, e^{i\lambda z \cdot \zeta} \rangle_{(S)} \qquad \text{for } z \in \mathbf{C}^{d+1}. \qquad (4.3)$$

The transformation $P_\lambda : T \to G_\lambda$ is called the *Fourier–Borel transformation* on the complex sphere $\tilde{S}$.

Theorem 4.2. *The Fourier–Borel transformation P_λ establishes topological linear isomorphisms:*

$$P_\lambda : \mathrm{Exp}'(\tilde{S}; [|\lambda|r : L^*]) \cong \mathcal{O}_\lambda(\tilde{B}(r)), \qquad 0 < r \le \infty, \qquad (4.4)$$

$$P_\lambda : \mathrm{Exp}'(\tilde{S}; (|\lambda|r : L^*)) \cong \mathcal{O}_\lambda(\tilde{B}[r]), \qquad 0 \le r < \infty, \qquad (4.4')$$

$$P_\lambda : \mathcal{O}'(\tilde{S}(r)) \cong \mathrm{Exp}_\lambda(\mathbf{C}^{d+1}; [|\lambda|r : L^*]), \qquad 1 < r \le \infty, \qquad (4.5)$$

$$P_\lambda : \mathcal{O}'(\tilde{S}[r]) \cong \mathrm{Exp}_\lambda(\mathbf{C}^{d+1}; (|\lambda|r : L^*)), \qquad 1 \le r < \infty. \qquad (4.5')$$

This kind of theorem was first proved by Hashizume *et al.* [2]. The case where $r = \infty$ was treated by Morimoto [8]. The isomorphisms (4.4) and (4.4') were proved by Wada–Morimoto [11], while (4.5) and (4.5') were proved by Wada [9].

Let α_λ be the restriction mapping defined in Theorem 2.1. We call the composite transformation $\alpha_\lambda \circ P_\lambda$ the *spherical Fourier–Borel transformation*. Combining Theorem 2.1 and Theorem 4.2, we have

Theorem 4.3. *Suppose λ satisfies* (2.13). *Then the spherical Fourier–Borel transformation defines the following linear topological isomorphisms*:

$$\alpha_\lambda \circ P_\lambda : \mathrm{Exp}'(\tilde{S}; [|\lambda|r:L^*]) \xrightarrow{\sim} \mathcal{O}(\tilde{S}(r)), \qquad 1 < r \le \infty, \qquad (4.6)$$

$$\alpha_\lambda \circ P_\lambda : \mathrm{Exp}'(\tilde{S}; (|\lambda|r:L^*)) \xrightarrow{\sim} \mathcal{O}(\tilde{S}[r]), \qquad 1 \le r < \infty, \qquad (4.6')$$

$$\alpha_\lambda \circ P_\lambda : \mathcal{O}'(\tilde{S}(r)) \xrightarrow{\sim} \mathrm{Exp}(\tilde{S}; [|\lambda|r:L^*]), \qquad 1 < r \le \infty, \qquad (4.7)$$

$$\alpha_\lambda \circ P_\lambda : \mathcal{O}'(\tilde{S}[r]) \xrightarrow{\sim} \mathrm{Exp}(\tilde{S}; (|\lambda|r:L^*)), \qquad 1 < r \le \infty. \qquad (4.7')$$

Therefore we can say that the spherical Fourier–Borel transformation establishes the duality between $\mathrm{Exp}'(\tilde{S}; [|\lambda|r:L^*])$ and $\mathcal{O}'(\tilde{S}(r))$, or between $\mathrm{Exp}'(\tilde{S}; (|\lambda|r:L^*))$ and $\mathcal{O}'(\tilde{S}[r])$.

4.2. Conical Case

Suppose an analytic functional $T \in \mathrm{Exp}'(M)$ is given. For $\lambda \ne 0$, we define similarly the *Fourier–Borel transform* G_λ of T as follows:

$$G_\lambda(z) = \langle T_\zeta, e^{i\lambda z \cdot \zeta} \rangle_{(\Sigma)} \qquad \text{for } z \in \mathbf{C}^{d+1}. \qquad (4.8)$$

The transformation $P_\lambda : T \to G_\lambda$ is called the *Fourier–Borel transformation* on the complex light cone M.

Theorem 4.4. *The Fourier–Borel transformation P_λ establishes topological linear isomorphisms*:

$$P_\lambda : \mathrm{Exp}'(M; [|\lambda|r:L^*]) \xrightarrow{\sim} \mathcal{O}_\Delta(\tilde{B}(r)), \qquad 0 < r \le \infty, \qquad (4.9)$$

$$P_\lambda : \mathrm{Exp}'(M; (|\lambda|r:L^*)) \xrightarrow{\sim} \mathcal{O}_\Delta(\tilde{B}[r]), \qquad 0 \le r < \infty, \qquad (4.9')$$

$$P_\lambda : \mathcal{O}'(M(r)) \xrightarrow{\sim} \mathrm{Exp}_\Delta(\mathbf{C}^{d+1}; [|\lambda|r:L^*]), \qquad 0 < r \le \infty, \qquad (4.10)$$

$$P_\lambda : \mathcal{O}'(M[r]) \xrightarrow{\sim} \mathrm{Exp}_\Delta(\mathbf{C}^{d+1}; (|\lambda|r:L^*)), \qquad 0 \le r < \infty. \qquad (4.10')$$

This kind of theorem was first proved by Kowata–Okamoto [4] in a different formulation.

We call the composite transformation $\alpha_0 \circ P_\lambda$ the *conical Fourier–Borel transformation*. Combining Theorem 3.2 and Theorem 4.4, we have

Theorem 4.5. *The conical Fourier–Borel transformation defines the following linear topological isomorphism*:

$$\alpha_0 \circ P_\lambda : \mathrm{Exp}'(M; [|\lambda|r:L^*]) \xrightarrow{\sim} \mathcal{O}(M(r)), \qquad 0 < r \le \infty, \qquad (4.11)$$

$$\alpha_0 \circ P_\lambda : \text{Exp}'(M; (|\lambda|r:L^*)) \xrightarrow{\sim} \mathcal{O}(M[r]), \qquad 0 \le r < \infty, \qquad (4.11')$$

$$\alpha_0 \circ P_\lambda : \mathcal{O}'(M(r)) \xrightarrow{\sim} \text{Exp}(M; [|\lambda|r:L^*]), \qquad 0 < r \le \infty, \qquad (4.12)$$

$$\alpha_0 \circ P_\lambda : \mathcal{O}'(M[r]) \xrightarrow{\sim} \text{Exp}(M; (|\lambda|r:L^*)), \qquad 0 \le r < \infty. \qquad (4.12')$$

This theorem establishes the duality of $\text{Exp}'(M; [|\lambda|r: L^*])$ and $\mathcal{O}'(M(r))$, or between $\text{Exp}'(M; (|\lambda|r:L^*))$ and $\mathcal{O}'(M[r])$ via the conical Fourier-Borel transformation.

References

[1] L. Drużkowski, Effective formula for the crossnorm in complexified unitary spaces, *Zeszyty Nauk. Uniw. Jagiéllon., Prace Mat.* **16** (1974) 47-53.

[2] M. Hashizume, A. Kowata, K. Minemura, and K. Okamoto, An integral representation of an eigenfunction of the Laplacian in the Euclidean space, *Hiroshima Math. J.* **2** (1972) 535-545.

[3] K. Ii, On a Bargmann-type transform and a Hilbert space of holomorphic functions, *Tôhoku Math. J.* **38** (1986) 57-69.

[4] A. Kowata and K. Okamoto, Harmonic functions and the Borel-Weil theorem, *Hiroshima Math. J.* **4** (1974) 89-97.

[5] A. Martineau, Equations différentielles d'ordre infini, *Bull. Soc. Math. France* **95** (1967) 109-154.

[6] M. Morimoto, A generalization of the Fourier-Borel transformation for the analytic functionals with non-convex carrier, *Tokyo J. Math.* **2** (1979) 301-322.

[7] M. Morimoto, Hyperfunctions on the Sphere, *Sophia Kokyuroku in Mathematics* **12**, Sophia Univ. Dept. Math., 1982 (in Japanese).

[8] M. Morimoto, Analytic functionals on the sphere and their Fourier-Borel transformations, *Complex Analysis, Banach Center Publications* **11**, PWN-Polish Scientific Publishers, Warsaw, 1983, 223-250.

[9] R. Wada, On the Fourier-Borel transformations of analytic functionals on the complex sphere, *Tôhoku Math. J.* **38** (1986) 417-432.

[10] R. Wada, Holomorphic functions on the complex sphere.

[11] R. Wada and M. Morimoto, A uniqueness set for the differential operator $\Delta_z + \lambda^2$, *Tokyo J. Math.* **10** (1987) 93-105.

A p-adic Theory of Hyperfunctions, II

Yasuo Morita

Mathematical Institute
Tohoku University
Aoba, Sendai
Japan

In our previous paper [7], we have constructed a p-adic theory of hyperfunctions of one variable by elementary methods. We continue the research in this paper, and construct a p-adic theory of hyperfunctions of several variables by using relative cohomologies of rigid analytic spaces.

Let k be a complete nonarchimedean field, let $\mathscr{X} = (X, \mathscr{O}_X, \mathfrak{T})$ be the standard rigid analytic structure on the affine n-space $X = k^n$, and let K be a compact subset of k^n with respect to the natural topology of k^n. Let $H_K^p(X, \mathscr{O}_X)$ be the p-th relative cohomology group of the sheaf $\mathscr{O}_X$ with supports in K. Then we show that (i) for any admissible open subset U of X, the cohomology group $H_{K \cap U}^p(U, \mathscr{O}_X)$ vanishes for $p \neq n$, (ii) for any open subset V of K, V can be written as $K \cap U$ with an admissible open subset U of X, and $\mathscr{B}(V) = H_{K \cap U}^n(U, \mathscr{O}_X)$ is independent of a special choice of U, (iii) the functor $\mathscr{B} : V \mapsto \mathscr{B}(V)$ defines a flabby sheaf on K, and (iv) $\mathscr{B}(K)$ gives the dual space of the space $\mathscr{A}(K)$ of locally analytic functions on K with respect to the natural locally convex topologies of $\mathscr{A}(K)$ and $\mathscr{B}(K)$.

Algebraic Analysis, Volume I

In Section 1, we study the general theory of relative cohomologies of rigid analytic spaces. We study the relation between the usual topology of K and the Grothendieck topology of X in Section 2. In Section 3, we prove a lemma on the relative cohomologies on a polydisk. We obtain the duality in Section 4, and construct the sheaf $\mathscr{B}$ of hyperfunctions on K in Section 5.

As for the theory of rigid analytic spaces, we use the terminology of Bosch-Günter-Remmert [3].

1. Cohomologies of Rigid Analytic Spaces

Let k be a complete field with a nontrivial nonarchimedean valuation $|\ |$. Let $\mathscr{X} = (X, \mathscr{O}_X, \mathfrak{T})$ be a rigid analytic space. Hence $\mathfrak{T}$ is a Grothendieck topology on X, and $\mathscr{O}_X$ is the structure sheaf. For example, the n-dimensional affine space k^n has a natural rigid analytic structure (cf., e.g., Bosch-Günter-Remmert [3], Chap. 9, pp. 362–363, Example 2).

Let $\mathscr{F}$ be a sheaf of $\mathscr{O}_X$-modules on $\mathfrak{T}$. Then we say $\mathscr{F}$ is *flabby* if for any admissible covering $\mathscr{U}$ of any admissible open subset U of X, the covering cohomology $H^p(\mathscr{U}, \mathscr{F})$ vanishes for any $p \geq 1$ (cf. M. Artin [1], Chap. II). We say that $\mathscr{F}$ is *strongly flabby* if, for any admissible coverings $\mathscr{U}$ and $\mathscr{V}$ of admissible subsets U and V of X such that $U \supset V$, the relative covering cohomology $H^p(\mathscr{U}, \mathscr{V}, \mathscr{F})$ vanishes for any $p \geq 1$. It follows from the long exact sequence

$$0 \to \Gamma(\mathscr{U}, \mathscr{V}, \mathscr{F}) \to \Gamma(\mathscr{U}, \mathscr{F}) \to \Gamma(\mathscr{V}, \mathscr{F}) \to \cdots$$

$$\cdots \to H^p(\mathscr{U}, \mathscr{V}, \mathscr{F}) \to H^p(\mathscr{U}, \mathscr{F}) \to H^p(\mathscr{V}, \mathscr{F}) \to \cdots$$

of covering cohomologies that $\mathscr{F}$ is strongly flabby if and only if $\mathscr{F}$ is flabby and $H^1(\mathscr{U}, \mathscr{V}, \mathscr{F})$ vanishes for any $\mathscr{U}$ and $\mathscr{V}$. Further, any injective sheaf is strongly flabby in our sense.

Let $\mathscr{F}$ be a sheaf of $\mathscr{O}_X$-modules on $\mathfrak{T}$. Then we denote by $\mathscr{C}(\mathscr{F})$ the presheaf on $\mathfrak{T}$ defined by

$$U \mapsto \mathscr{C}(\mathscr{F})(U) = \prod_{x \in X} \mathscr{F}_x,$$

where $\mathscr{F}_x$ ($x \in X$) denotes the germ of $\mathscr{F}$ at x. Then this sheaf $\mathscr{C}(\mathscr{F})$ is induced from a flabby sheaf on the discrete topological space X. Since any flabby sheaf on a topological space is strongly flabby in our sense, $\mathscr{C}(\mathscr{F})$

is strongly flabby with respect to the discrete topology. Since $\mathfrak{T}$ is weaker than the discrete topology, $\mathscr{C}(\mathscr{F})$ is a strongly flabby sheaf on $\mathfrak{T}$.

Put $\mathscr{C}^{-1}(\mathscr{F}) = \mathscr{F}$, $\mathscr{C}^0(\mathscr{F}) = \mathscr{C}(\mathscr{F})$, and define $\mathscr{C}^i(\mathscr{F})$ $(i \geq 0)$ inductively by $\mathscr{C}^i(\mathscr{F}) = \mathscr{C}(\mathrm{Coker}\,(\mathscr{C}^{i-2}(\mathscr{F}) \to \mathscr{C}^{i-1}(\mathscr{F})))$. Then we have a resolution

$$\mathscr{C}^{\cdot}(\mathscr{F}): 0 \to \mathscr{F} \to \mathscr{C}^0(\mathscr{F}) \to \mathscr{C}^1(\mathscr{F}) \to \mathscr{C}^2(\mathscr{F}) \to \cdots$$

of $\mathscr{F}$ by strongly flabby sheaves. Hereafter we call $\mathscr{C}^{\cdot}(\mathscr{F})$ the *canonical flabby resolution* of $\mathscr{F}$.

Let $\mathscr{F}$ be a sheaf of $\mathscr{O}_X$-modules, let $V \subset U \subset X$ be admissible open subsets of X, and let $S = U - V$. Let $\Gamma(U, \mathscr{F})$ denote the set of all sections of $\mathscr{F}$ on U. For any $f \in \Gamma(U, \mathscr{F})$, let $\mathrm{supp}(f)$ denote the support of f. Since the functor $\Gamma_S(U, \ ): \mathscr{F} \mapsto \Gamma_S(U, \mathscr{F}) = \{f \in \Gamma(U, \mathscr{F}); \ \mathrm{supp}(f) \subset S\}$ is left exact, we denote the derived functors of $\Gamma_S(U, \mathscr{F})$ by $H_S^p(U, \mathscr{F})$ $(p = 0, 1, 2, \ldots)$.

Now we apply Lemma 4.1 of M. Artin [1], p. 39 to our case, and see that, if $\mathscr{F}$ is a strongly flabby sheaf, then the relative cohomology $H_S^p(U, \mathscr{F})$ vanishes for any $p \geq 1$, and the canonical flabby resolution $\mathscr{C}^{\cdot}(\mathscr{F})$ can be used to calculate the relative cohomologies $H_S^p(U, \mathscr{F})$ (cf. the proof of Corollary 4.4 of *ibid.* p. 41). Since $\mathscr{C}^i(\mathscr{F})$ is strongly flabby,

$$0 \to \Gamma(\mathscr{U}, \mathscr{V}, \mathscr{C}^i(\mathscr{F})) \to \Gamma(\mathscr{U}, \mathscr{C}^i(\mathscr{F})) \to \Gamma(\mathscr{V}, \mathscr{C}^i(\mathscr{F})) \to 0$$

is exact for any $\mathscr{U}$, $\mathscr{V}$, and i. Here we have $H^p(\Gamma(\mathscr{U}, \mathscr{V}, \mathscr{C}^{\cdot}(\mathscr{F}))) \simeq H^p(\Gamma_S(U, \mathscr{C}^{\cdot}(\mathscr{F}))) \simeq H_S^p(U, \mathscr{F})$, $H^p(\Gamma(\mathscr{U}, \mathscr{C}^{\cdot}(\mathscr{F}))) \simeq H^p(\Gamma(U, \mathscr{C}^{\cdot}(\mathscr{F}))) \simeq H^p(U, \mathscr{F})$, $H^p(\Gamma(\mathscr{V}, \mathscr{C}^{\cdot}(\mathscr{F}))) \simeq H^p(\Gamma(U-S, \mathscr{C}^{\cdot}(\mathscr{F}))) \simeq H^p(U-S, \mathscr{F})$, where $U = \bigcup_{W \in \mathscr{U}} W$ and $U - S = \bigcup_{W \in \mathscr{V}} W$. Hence, by taking the cohomologies, we obtain a long exact sequence

$$0 \to \Gamma_S(U, \mathscr{F}) \to \Gamma(U, \mathscr{F}) \to \Gamma(U-S, \mathscr{F}) \to \cdots$$

$$\to H_S^p(U, \mathscr{F}) \to H^p(U, \mathscr{F}) \to H^p(U-S, \mathscr{F}) \to \cdots,$$

where U, V, $S = U - V$, and $\mathscr{F}$ are as before. In particular, if $\mathscr{F}$ is strongly flabby, then the restriction map $\Gamma(U, \mathscr{F}) \to \Gamma(V, \mathscr{F})$ is surjective for any admissible open subsets $V \subset U$.

Let S be a subset of X such that $X - S$ is admissible open in X. Let $\mathscr{F}$ be a sheaf of $\mathscr{O}_X$-modules, and let U be an admissible open subset of X. Then the functor $U \mapsto \Gamma_{S \cap U}(U, \mathscr{F})$ defines a sheaf $\Gamma_S(\mathscr{F})$ on $\mathfrak{T}$. Since this functor $\mathscr{F} \mapsto \Gamma_S(\mathscr{F})$ is left exact, we denote the derived functors by $\mathscr{H}_S^p(\mathscr{F})$ $(p = 0, 1, 2, \ldots)$. It is easy to see that $\mathscr{H}_S^p(\mathscr{F})$ is the sheaficication of the presheaf $U \mapsto H_{S \cap U}^p(U, \mathscr{F})$.

Let U be an admissible open subset of X, and let $S \supset Z$ be subsets of X such that $X \supset X - Z \supset X - Z$ are admissible open. Then, by the standard arguments, we obtain a spectral sequence

$$E_2^{p,q} = H_{Z \cap U}^p(U, \mathcal{H}_S^q(\mathcal{F})) \Rightarrow H_{Z \cap U}^n(U, \mathcal{F}).$$

It follows from the degeneracy of this spectral sequence that (i) if $\mathcal{H}_S^q(\mathcal{F}) = 0$ for $q = 0, 1, \ldots, m-1$, then $\Gamma(U, \mathcal{H}_S^m(\mathcal{F})) = H_{S \cap U}^m(U, \mathcal{F})$, and ($ii$) if $\mathcal{H}_S^q(\mathcal{F}) = 0$ for any $q \neq m$, then $H_{Z \cap U}^n(U, \mathcal{F})$ vanishes for any $n < m$, and $H_{Z \cap U}^n(U, \mathcal{F}) = H_{Z \cap U}^{n-m}(U, \mathcal{H}_S^m(\mathcal{F}))$ otherwise.

Similarly, we have a spectral sequence

$$E_2^{p,q} = H^p(\mathcal{U}, \mathcal{V}, \mathcal{H}_*^q(\mathcal{F})) \Rightarrow H_S^n(U, \mathcal{F}),$$

where $U, V, S, \mathcal{U}, \mathcal{V}, \mathcal{F}$ are as before, and $\mathcal{H}_*^q(\mathcal{F})$ denotes the correspondence $T \to H^q(U_T, \mathcal{F})$ defined on the nerves of the covering $\mathcal{U}$. In particular, $H_S^p(U, \mathcal{F}) = H^p(\mathcal{U}, \mathcal{V}, \mathcal{F})$ holds if the cohomology $H^q(V_{i_0} \cap V_{i_1} \cap \cdots \cap V_{i_p}, \mathcal{F})$ vanishes for any $V_{i_0}, V_{i_1}, \ldots, V_{i_p} \in \mathcal{U}$, and for any $q \geq 1$.

Lemma 1. *Let U be an admissible open subset of X, let S be a subset of U such that $U - S$ is admissible open in U, and let $\mathcal{F}$ be a sheaf of $\mathcal{O}_X$-modules on $\mathfrak{T}$. Then we have*

$$H_S^p(X, \mathcal{F}) = H_S^p(U, \mathcal{F}) \qquad p = 0, 1, 2, \ldots.$$

Proof. Let $\mathcal{C}^\cdot(\mathcal{F}) = \{0 \to \mathcal{F} \to \mathcal{C}^1(\mathcal{F}) \to \mathcal{C}^2(\mathcal{F}) \to \cdots\}$ be the canonical flabby resolution of $\mathcal{F}$. Then the restriction map $\Gamma_S(X, \mathcal{C}^i(\mathcal{F})) \to \Gamma_S(U, \mathcal{C}^i(\mathcal{F}))$ is surjective, because $\mathcal{C}^i(\mathcal{F})$ is strongly flabby, and U is admissible open in X. Since $S \subset U$, the kernel of the restriction map is $\{s \in \Gamma_S(X, \mathcal{C}^i(\mathcal{F})); \text{supp}(s) \cap U = \varnothing\} = \{0\}$. Hence $\Gamma_S(X, \mathcal{C}^i(\mathcal{F})) \simeq \Gamma_S(U, \mathcal{C}^i(\mathcal{F}))$. Hence, taking the cohomologies of $\mathcal{C}^\cdot(\mathcal{F})$, we obtain the lemma.   Q.E.D.

2. Lemmas on the Grothendieck Topology on k^n

Let k and $X = k^n$ be as in Section 1, and let $(X, \mathcal{O}_X, \mathfrak{T})$ be the standard rigid analytic structure on k^n. We define a metric $d(\ , \ )$ on X by $d(x, y) = \max\{|x_j - y_j|; 1 \leq j \leq n\}$ for any $x = (x_1, x_2, \ldots, x_n)$ and $y = (y_1, y_2, \ldots, y_n)$.

Let $\mathbf{R}$ be the ordered set of real numbers, let $\{-, 0, +\}$ be the ordered set with $-<0<+$, and let $\mathbf{R}\times\{-,0,+\}$ be the product with respect to the lexicographic order (cf. Morita [7], p. 3). For any $a=(a_1, a_2, \ldots, a_n)\in X$ and $r\in \mathbf{R}\times\{-,0,+\}$, we denote by $B(a, r)=\{z\in X; d(a, z)\leq r\}$ the product of balls $\{z\in k; |z-a_j|\leq r\}$ of diameter r. Let $\{r_m\}_{m=1}^{\infty}$ be a strictly decreasing sequence of positive numbers such that $r_m\in |k|$ and $r_m\to 0\ (m\to 0)$.

Let K be a compact subset of X, and let $U=X-K=\{x\in X; x\notin K\}$. Then U is an open subset of X. Let

$$K_m=\{z\in X; d(z, K)<r_m\}.$$

Then $\{K_m\}_{m=1}^{\infty}$ is a strictly decreasing sequence of open subsets of X such that each K_m is covered by a finite number of mutually disjoint open balls $B(a_i, (r_m)^-)\ (a_i\in K, i=1, 2, \ldots, l_m)$ of diameter r_m, and $\bigcap_{m\geq 1} K_m=K$ holds. Let

$$K_m^c=\bigcup_{1\leq i\leq l_m} B(a_i, r_m).$$

Then $K_m^c\supset K_m\supset K_{m+1}^c$. Furthermore, if we put $U_m=X-K_m$, then U_m is an open subset of X, and $\bigcup_{m\geq 1} U_m=U=X-K$ holds.

Put

$$D(a_i, r_m, j)=\{x\in k^n; |x_j-a_{ij}|\geq r_m\}\qquad (j=1, 2, \ldots, n),$$

$$X_m=\{x\in k^n; d(x, 0)\leq (r_m)^{-1}\}.$$

Then $\{X_m\}_{m=1}^{\infty}$ is an admissible open covering of k^n, and the rigid analytic structure of k^n is obtained by gluing the affine structures of these X_m.

Lemma 2.   *Each U_m is an admissible open subset of X, and $\{U_m\}_{m=1}^{\infty}$ is an admissible open covering of U.*

Proof.   Let X_m and the $D(a_i, r_m, j)$ be as above. Then U_m is covered by the $D(a_i, r_m, j)$, and the intersections $X_m\cap D(a_i, r_m, j)$ are affinoid subdomains of X_m. It follows from [3], p. 343, Corollary 4 that $X_m\cap U_m$ is an admissible open subset of X_m. Since the strong G-topology satisfies the condition (G_1), U_m is an admissible open subset of X.

Since the strong G-topology satisfies the condition (G_2), it is sufficient to prove that $\{X_M\cap U_m\}_{m=1}^{\infty}$ is an admissible covering of $X_M\cap U$ for any $M\geq 1$. Let $\varphi: Y\to X_M$ be an affinoid morphism with $\varphi(Y)\subset X_M\cap U$. Then, by the maximum principle (cf. *ibid.* p. 237, Proposition 4), for any point $x=(x_1, x_2, \ldots, x_n)$ of K, and for any $1\leq j\leq n$, there exists a point y_0 of Y

such that the inequality

$$|\varphi_j(y) - x_j| \geq |\varphi_j(y_0) - x_j| > 0$$

holds for any $y \in Y$, where $\varphi = (\varphi_1, \varphi_2, \ldots, \varphi_n)$.

Put $\rho_j(x) = |\varphi_j(y_0) - x_j|$. Then $\rho_j(x)$ is a continuous function on K. Since K is compact, $\rho_j(x)$ have a minimum on K. We assume that r_m is smaller than this minimum for any $j = 1, 2, \ldots, n$. Then $\varphi(Y)$ is contained in some $X_M \cap U_m$. Hence, by *ibid.* p. 342, Proposition 2, (ii), $\{X_M \cap U_m\}_{m=1}^{\infty}$ is an admissible covering. Q.E.D.

The following lemma relates the induced topology on a compact subset $K \subset X$ and the induced $\mathfrak{T}$-topology on K.

Lemma 3. *Let $X = k^n$, K, $U = X - K$, etc., be as before. Let $\{V_i\}_{i \in I}$ be an open covering of K. Put $U_i = U \coprod V_i$. Then each U_i is an admissible open subset of X, and $\{U_i\}_{i \in I}$ is an admissible covering of X.*

Proof. Since $U_i = (X - K) \cup V_i = X - (K - V_i)$, and since $K - V_i$ is compact, it follows from Lemma 2 that U_i is an admissible open subset of X.

Since each U_i has the form which was studied in Lemma 2, we put $U_{im} = \{x \in X; d(x, K - V_i) \geq r_m\}$. Then $\{U_{im}\}_{m=1}^{\infty}$ is an admissible open covering of U_i, and $\{U_{im}\}_{i \in I, 1 \leq m < \infty}$ is an open covering of X. Since K is compact, there exists a finite subset $\{U_{im}\}_{1 \leq i \leq N, 1 \leq m \leq l}$ of $\{U_{im}\}_{i \in I, 1 \leq m < \infty}$ such that $K \subset \bigcup_{1 \leq i \leq N, 1 \leq m \leq l} U_{im}$. It is obvious that this $\{U_{im}\}_{1 \leq i \leq N, 1 \leq m \leq l}$ is a refinement of $\{U_i\}$.

Let $\{X_m\}_{m=1}^{\infty}$ be as before. Since the strong $\mathfrak{T}$-topology satisfies the condition (G_2), it is sufficient to prove that $\{X_M \cap U_{im}\}_{1 \leq i \leq N, 1 \leq m \leq l}$ is an admissible covering of X_M for any positive integer M. Since this is a finite covering, and since each $X_M \cap U_{im}$ is a union of a finite number of affinoids (cf. the proof of Lemma 2), this is an admissible covering. Q.E.D.

3. The Key Lemma

Let $\mathscr{X} = (X, \mathscr{O}_X, \mathfrak{T})$ be the standard rigid analytic structure on $X = k^n$. For any $a \in k$ and $r \in \mathbf{R} = \{-, 0, +\}$, let

$$B(a, r) = \{x \in k; |x - a| \leq r\}.$$

Let $D_1, D_2, \ldots, D_n$ be affinoid subsets of k, and let $a_i \in D_i$ and $0 \neq r_i \in |k|$
$(i = 1, \ldots, n)$. Let

$$S = B(a_1, (r_1)^-) \times B(a_2, (r_2)^-) \times \cdots \times B(a_n, (r_n)^-)$$

be the product of the open balls $B(a_i, (r_i)^-)$. Let U be an affinoid subset
of X containing $B(a_1, r_1) \times B(a_2, r_2) \times \cdots \times B(a_n, r_n)$. Then $U - S$ is a
nonempty admissible open subset of U. Let $U_0 = U$, and let

$$U_i = \{x \in U; \, |x_i - a_i| \geq r_i\}$$

$(i = 1, 2, \ldots, n)$. Clearly $\mathcal{U} = \{U_1, U_2, \ldots, U_n\}$ is an affinoid covering of
$U - S$, and $\mathcal{U}^0 = \{U_0, U_1, U_2, \ldots, U_n\}$ is an affinoid covering of U. Now
we have the following lemma.

Lemma 4 (Key Lemma).   *Let $(X, \mathcal{O}_X, \mathfrak{X})$, $U, S, \mathcal{U}^0$, and $\mathcal{U}$ be as above.
Then the relative cohomology group $H_S^p(U, \mathcal{O}_X)$ vanishes for $p \neq n$, and
$H_S^n(U, \mathcal{O}_X)$ is isomorphic to*

$$\mathcal{O}_X(U_1 \cap U_2 \cap \cdots \cap U_n) \Big/ \sum_{i=1}^n \mathcal{O}_X(U_1 \cap U_2 \cap \cdots \cap U_{i-1} \cap U_{i+1} \cap \cdots \cap U_n).$$

Proof.   Since $U - S$ is an open subset of U, it follows from the theorem
of identity that $\Gamma_S(U, \mathcal{O}_X) = H_S^0(U, \mathcal{O}_X)$ vanishes. Since U is an affinoid
space, the cohomology $H^p(U, \mathcal{O}_X)$ vanishes for any $p \geq 1$. It follows from
the exact sequence

$$0 \to \Gamma_S(U, \mathcal{O}_X) \to \Gamma(U, \mathcal{O}_X) \to \Gamma(U - S, \mathcal{O}_X) \to \cdots$$

$$\to H_S^p(U, \mathcal{O}_X) \to H^p(U, \mathcal{O}_X) \to H^p(U - S, \mathcal{O}_X) \to \cdots$$

that there exist canonical isomorphisms

$$H_S^1(U, \mathcal{O}_X) \simeq \Gamma(U - S, \mathcal{O}_X)/\Gamma(U, \mathcal{O}_X)$$

$$H_S^p(U, \mathcal{O}_X) \simeq H^p(U - S, \mathcal{O}_X) \qquad (p \geq 2).$$

Since each U_i is an affinoid, the cohomology $H^p(U_i, \mathcal{O}_X)$ vanishes for any
$p \geq 1$. Since $\mathcal{U}$ is an admissible open covering of $U - S$ such that $H^p(U_i, \mathcal{O}_X)$
vanishes for any $U_i \in \mathcal{U}$ and for any $p \geq 1$, we have $H^{p-1}(U - S, \mathcal{O}_X) \simeq$
$H^{p-1}(\mathcal{U}, \mathcal{O}_X)$ for any p. Since the covering $\mathcal{U}$ contains only n elements,
$H_S^p(U, \mathcal{O}_X) \simeq H^{p-1}(\mathcal{U}, \mathcal{O}_X)$ vanishes for $p \geq n + 1$.

Let $C^q(\mathcal{U}, \mathcal{O}_X)$ be the module of $\mathcal{O}_X$-valued alternating q-cochains on $\mathcal{U}$.

Then we have

$$C^{n-2}(\mathcal{U}, \mathcal{O}_X) = \bigoplus_{1 < i < n} \mathcal{O}_X(U_1 \cap U_2 \cap \cdots \cap U_{i-1} \cap U_{i+1} \cap \cdots \cap U_n),$$

$$C^{n-1}(\mathcal{U}, \mathcal{O}_X) = \mathcal{O}_X(U_1 \cap U_2 \cap \cdots \cap U_n), \qquad C^n(\mathcal{U}, \mathcal{O}_X) = 0.$$

It follows that $Z^{n-1}(\mathcal{U}, \mathcal{O}_X) = \mathcal{O}_X(U_1 \cap U_2 \cap \cdots \cap U_n)$, and $B^{n-1}(\mathcal{U}, \mathcal{O}_X) = \sum_{1 \leq i \leq n} \mathcal{O}_X(U_1 \cap \cdots \cap U_{i-1} \cap U_{i+1} \cap \cdots \cap U_n)$, where $\mathcal{O}_X(U_1 \cap \cdots \cap U_{i-1} \cap U_{i+1} \cap \cdots \cap U_n)$ are considered as submodules of $\mathcal{O}_X(U_1 \cap U_2 \cap \cdots \cap U_n)$ by the natural restriction maps. Hence we have calculated $H_S^n(U, \mathcal{O}_X)$. In particular, the lemma is proved for $n = 1$.

Now we assume that $n \geq 2$ and $n - 1 \geq p \geq 2$, and prove that $H_S^p(U, \mathcal{O}_X) \simeq H^{p-1}(\mathcal{U}, \mathcal{O}_X)$ vanishes. Let $q = p - 1$, and let $\varphi = \bigoplus \varphi(i_0, i_1, \ldots, i_q)$ be an element of $C^q(\mathcal{U}, \mathcal{O}_X)$. Hence each $\varphi(i_0, i_1, \ldots, i_q)$ is a holomorphic mapping of the intersection $U(i_0, i_1, \ldots, i_q)$ of $U_{i_0}, U_{i_1}, \ldots, U_{i_q}$ with values in k, and $\varphi(i_0, i_1, \ldots, i_q)$ is alternating in $(i_0, i_1, \ldots, i_n)$. Furthermore, if φ is a cocycle, then we have

$$\sum_{j=0}^{q+1} (-1)^j \varphi(i_0, i_1, \ldots, i_{j-1}, i_{j+1}, \ldots, i_{q+1}) = 0.$$

We restrict each $\varphi(i_0, i_1, \ldots, i_q) \in \mathcal{O}_X(U(i_0, i_1, \ldots, i_q))$ to the product of the circles $\{x \in k; |x_j - a_j| = r_j\}$ $(j = 1, 2, \ldots, n)$, and expand it into the Laurent series. Then this Laurent expansion has a negative power of $x_i - a_i$ only for $i = i_0, i_1, \ldots, i_q$. Let $\varphi_j^1(i_0, i_1, \ldots, i_q)$ denote the sum of the terms of this Laurent expansion which have negative powers of $x_j - a_j$, and let $\varphi_j^1 = (\varphi_j^1(i_0, i_1, \ldots, i_q))$. Then this cochain φ_j^1 satisfies the cocycle condition, because it is the singular part of the cocycle φ with respect to $x_j - a_j$. Put

$$\varphi_j^2(i_0, i_1, \ldots, i_q) = \varphi(i_0, i_1, \ldots, i_q) - \varphi_j^1(i_0, i_1, \ldots, i_q).$$

Then $\varphi_j^2 = (\varphi_j^2(i_0, i_1, \ldots, i_q))$ also satisfies the cocycle condition. Further, the Laurent expansion of $\varphi_j^2(i_0, i_1, \ldots, i_q)$ has a negative power of $x_i - a_i$ only if $i = i_0, i_1, \ldots, i_q$, and $i \neq j$. Hence $\varphi_j^2 = (\varphi_j^2(i_0, i_1, \ldots, i_q))$ is a q-cocycle of the covering $\{U_1, U_2, \ldots, U_{j-1}, U, U_{j+1}, \ldots, U_n\}$. Since this covers the affinoid space U, the cohomology $H^q(\{U_1, U_1, \ldots, U_{j-1}, U, U_{j+1}, \ldots, U_n\}, \mathcal{O}_X)$ vanishes for $q \geq 1$.

Since we have assumed $q \geq 1$, φ_j^2 can be written as a coboundary of a $(q-1)$-cochain on $\{U_1, U_2, \ldots, U_{j-1}, U, U_{j+1}, \ldots, U_n\}$. Since this induces a coboundary on $\{U_1, U_2, \ldots, U_n\}$, by subtracting φ_j^2 if necessary, we may assume that $\varphi = \varphi_j^1$. We repeat this replacement for $j = 1, 2, \ldots, n$, and may assume that the Laurent expansion of $\varphi(i_0, i_1, \ldots, i_q)$ has the form

$$\sum_{m_1, m_2, \ldots, m_n = -\infty}^{-1} c(m_1, m_2, \ldots, m_n)(x_1 - a_1)^{m_1} \ldots (x_n - a_n)^{m_n}.$$

Now the cocycle condition can be written as

$$\varphi(i_0, i_1, \ldots, i_{j-1}, i_{j+1}, \ldots, i_{q+1})$$
$$= \sum_{0 \le l \le q+1, l \ne j} (-1)^{l+1} \varphi(i_0, i_1, \ldots, i_{l-1}, i_{l+1}, \ldots, i_{q+1}).$$

Here the left-hand side is an element of $\mathcal{O}_X(U(i_0, \ldots, i_{j-1}, i_{j+1}, \ldots, i_q))$. Hence no negative power of $x_j - a_j$ appears. On the other hand, by our assumption, the right-hand side has only terms with negative powers of $x_j - a_j$. Hence $\varphi(i_0, i_1, \ldots, i_q) = 0$, and hence $H^q(\mathcal{U}, \mathcal{O}_X)$ vanishes in this case.

Now we assume that $n \ge 2$ and $p = 1$. Let φ be an element of $\Gamma(U - S, \mathcal{O}_X)$. Then φ is an analytic function on $U - S$. Hence we can expand it into a Laurent series of the form

$$\varphi(x) = \sum_{m = -\infty}^{+\infty} c_m(x_1, x_2, \ldots, x_{n-1})(x_n - a_n)^m,$$

where the coefficients $c_m(x_1, x_2, \ldots, x_{n-1})$ are elements of $\mathcal{O}_X(D_1 \times D_2 \times \cdots \times D_{n-1})$. Since $\varphi(x)$ is an element of $\Gamma(U - S, \mathcal{O}_X)$, $c_m(x_1, x_2, \ldots, x_{n-1})$ vanishes on $D_1 \times D_2 \times \cdots \times D_{n-1} - B(a_1, (r_1)^-) \times B(a_2, (r_2)^-) \times \cdots \times B(a_{n-1}, (r_{n-1})^-)$ if m is negative. It follows from the theorem of identity that these coefficients vanish identically on $D_1 \times D_2 \times \cdots \times D_{n-1}$. Hence $\varphi(x)$ is an analytic function on $(U - S) \cup D_1 \times D_2 \times \cdots \times D_{n-1} \times \{x_n; |x_n| \le r_n\} = U$. Hence the cohomology $H^1_S(U, \mathcal{O}_X) \simeq \Gamma(U - S, \mathcal{O}_X)/\Gamma(U, \mathcal{O}_X)$ vanishes also in this case.   Q.E.D.

Corollary. *Let $X, \mathcal{O}_X, U, S$, etc., be as in Lemma 4. Then the n-th relative cohomology group $H^n_S(U, \mathcal{O}_X)$ can be identified with the set consisting of all Laurent series*

$$\varphi(x) = \sum_{m_1, \ldots, m_n = -\infty}^{-1} c(m_1, \ldots, m_n)(x_1 - a_1)^{m_1} \cdots (x_n - a_n)^{m_n}$$

$(c(m_1, \ldots, m_n) \in k)$ which converge on the complement of S.

Proof. Since we have calculated $H^n_S(U, \mathcal{O}_X)$ in Lemma 4, it is easy to obtain this corollary by expanding $\varphi(z) \in \mathcal{O}_X(U_1 \cap \cdots \cap U_n)$ into the

Laurent series on the product of the circles $\{x_i \in k; |x_i - a_i| = r_i\}$ $(i = 1, \ldots, n)$.   Q.E.D.

4.   Projective Limits and the Duality

Let $X = k^n$ and let $(X, \mathcal{O}_X, \mathfrak{T})$ be as before. Let U be an admissible open subset of X, and let S_1 and S_2 be subsets of U such that $S_1 \cap S_2 = \varnothing$, and $U - S_i$ $(i = 1, 2)$ are admissible open in U. Let $\mathcal{F}$ be a sheaf of $\mathcal{O}_X$-modules on $\mathfrak{T}$, and let $\mathcal{C}^{\cdot}(\mathcal{F})$ be the canonical flabby resolution of $\mathcal{F}$. Then

$$0 \to \Gamma_{S_1}(U, \mathcal{C}^{\cdot}(\mathcal{F})) \to \Gamma_{S_1 \coprod S_2}(U, \mathcal{C}^{\cdot}(\mathcal{F})) \to \Gamma_{S_2}(U, \mathcal{C}^{\cdot}(\mathcal{F})) \to 0$$

is exact. Hence, taking the cohomologies, we obtain the following long exact sequence:

$$0 \to \Gamma_{S_1}(U, \mathcal{F}) \to \Gamma_{S_1 \coprod S_2}(U, \mathcal{F}) \to \Gamma_{S_2}(U, \mathcal{F}) \to \cdots$$

$$\to H^p_{S_1}(U, \mathcal{F}) \to H^p_{S_1 \coprod S_2}(U, \mathcal{F}) \to H^p_{S_2}(U, \mathcal{F}) \to \cdots.$$

Hence the following proposition follows from Lemma 4.

Proposition 1. *Let $(X, \mathcal{O}_X, \mathfrak{T})$ and $U = D_1 \times D_2 \times \cdots \times D_n$ be as in Lemma 4. Let*

$$S_i = \prod_{1 \leq j \leq n} \{x \in k; |x_j - a_{ij}| \leq (r_i)^-\} \qquad (1 \leq i \leq M)$$

$(a_{ij} \in k, 0 \neq r_i \in |k|)$ be mutually disjoint open balls. Put $S = S_1 \coprod S_2 \coprod \cdots \coprod S_M$. Then the relative cohomology $H^p_S(U, \mathcal{F})$ vanishes for $p \neq n$, and $H^n_S(U, \mathcal{O}_X) \simeq \bigoplus_{1 \leq i \leq M} H^n_{S_i}(U, \mathcal{O}_X)$.

Let K be a compact subset of $X = k^n$. Let $\{r_m\}$, $K_m = \coprod_{i=1}^{l_m} B(a_i, (r_m)^-)$, K_m^c $(m = 1, 2, 3, \ldots)$ be as before. Let U be an affinoid subset of X containing $K_1^c \supset K$. Let $\mathcal{B}(a_i, (r_m)^-)$ denote the space of series $\sum c_M (x - a_i)^M$ $(c_M \in k)$ which converge outside $B(a_i, (r_m)^-)$, where M runs over all n-tuples $(m_1, m_2, \ldots, m_n)$ of negative integers, and $(x - a_i)^M = \prod_{1 \leq j \leq n} (x_j - a_{ij})^{m_j}$. Then, by Proposition 1, Lemma 1 and the corollary to Lemma 4, $H^p_{K_m}(U, \mathcal{O}_X) = 0$ for any $p \neq n$, and $H^n_{K_m}(U, \mathcal{O}_X) \simeq \bigoplus_{1 \leq i \leq l_m} \mathcal{B}(a_i, (r_m)^-)$. We define a norm $\| \ \|_{i,m}$ on $\mathcal{B}(a_i, (r_m)^-)$ by

$$\|\sum c_M (x - a_i)^M\|_{i,m} = \sup_M |c_M| r_m^{m_1 + m_2 + \cdots + m_n}.$$

Then $\mathscr{B}(a_i, (r_m)^-)$ becomes a Banach space with this norm. Further, $H^n_{K_m}(U, \mathcal{O}_X) \simeq \bigoplus_{1 \le i \le l_m} \mathscr{B}(a_i, (r_m)^-)$ becomes a Banach space with the norm $\| \ \|_m = \| \ \|_{1,m} + \| \ \|_{2,m} + \cdots + \| \ \|_{l_m}$.

Let $D(a_i, r_m, j)$ be as in Section 2, let $\mathcal{U}_m = \{U \cap D(a_i, r_m, j)\}_{1 \le i \le l_m, 1 \le j \le n}$, and let $\mathcal{U}^0_m = \mathcal{U}_m \cup \{U\}$. Let $\mathscr{V}^0_m = \bigcup_{1 \le h < m} \mathcal{U}_h$, $\mathscr{V}_m = \bigcup_{1 \le h < m} \mathcal{U}_h$, $\mathscr{V}^0 = \bigcup_{1 \le m < \infty} \mathscr{V}^0_m$, $\mathscr{V} = \bigcup_{1 \le m < \infty} \mathscr{V}_m$. Then $\mathscr{V}^0_m$ and $\mathscr{V}_m$ are affinoid coverings of U and $U - K_m$, respectively, and $\mathscr{V}^0$ and $\mathscr{V}$ are admissible coverings of U and $U - K$, respectively. Since $H^p(V, \mathcal{O}_X)$ vanishes for any $V \in \mathscr{V}^0$ and for any $p \ge 1$, it follows from the results of Section 1 that there exist canonical isomorphisms

$$H^p_{K_m}(U, \mathcal{O}_X) \simeq H^p(\mathscr{V}^0_m, \mathscr{V}_m, \mathcal{O}_X),$$

$$H^p_K(U, \mathcal{O}_X) \simeq H^p(\mathscr{V}^0, \mathscr{V}, \mathcal{O}_X),$$

for any $p \ge 0$ and for any $m \ge 1$.

Let $C^p(\mathscr{V}^0_m, \mathscr{V}_m, \mathcal{O}_X)$ and $C^p(\mathscr{V}^0, \mathscr{V}, \mathcal{O}_X)$ be the modules of all $\mathcal{O}_X$-valued alternating p-cochains on $\mathscr{V}^0_m$ and $\mathscr{V}^0$ which vanish on $\mathscr{V}_m$ and $\mathscr{V}$, respectively. Since $\mathscr{V}^0_m \supset \mathscr{V}_m$, $\mathscr{V}^0_{m-1} \supset \mathscr{V}_{m-1}$, $\mathscr{V}^0_m \supset \mathscr{V}^0_{m-1}$, $\mathscr{V}_m \supset \mathscr{V}_{m-1}$, we have a homomorphism

$$\rho^p_m : C^p(\mathscr{V}^0_m, \mathscr{V}_m, \mathcal{O}_X) \to C^p(\mathscr{V}^0_{m-1}, \mathscr{V}_{m-1}, \mathcal{O}_X).$$

We see that this map induces a continuous map

$$\rho^p_m : H^p_{K_m}(U, \mathcal{O}_X) \to H^p_{K_{m-1}}(U, \mathcal{O}_X).$$

By the definition of $\mathscr{V}^0$ and $\mathscr{V}$, $C^p(\mathscr{V}^0, \mathscr{V}, \mathcal{O}_X) = \text{proj} \lim_{m \to \infty} C^p(\mathscr{V}^0_m, \mathscr{V}_m, \mathcal{O}_X)$, and $H^p(\mathscr{V}^0, \mathscr{V}, \mathcal{O}_X)$ is the p-th cohomology group of the complex $\{\text{proj} \lim_{m \to \infty} C^{\cdot}(\mathscr{V}^0_m, \mathscr{V}_m, \mathcal{O}_X)\}$. Hence we have a canonical isomorphism

$$H^p_K(U, \mathcal{O}_X) \simeq H^p\left(\text{proj} \lim_{m \to \infty} C^{\cdot}(\mathscr{V}^0_m, \mathscr{V}_m, \mathcal{O}_X)\right).$$

Since $\mathscr{V}^0_m \supset \mathscr{V}^0_{m-1}$ and $\mathscr{V}_m \supset \mathscr{V}_{m-1}$, the restriction map $\rho^p_m : C^p(\mathscr{V}^0_m, \mathscr{V}_m, \mathcal{O}_X) \to C^p(\mathscr{V}^0_{m-1}, \mathscr{V}_{m-1}, \mathcal{O}_X)$ is surjective for any p. Hence ρ^p_m induces a surjective map $B^p(\mathscr{V}^0_m, \mathscr{V}_m, \mathcal{O}_X) \to B^p(\mathscr{V}^0_{m-1}, \mathscr{V}_{m-1}, \mathcal{O}_X)$ of coboundaries. Since the cohomology $H^p(\mathscr{V}^0_m, \mathscr{V}_m, \mathcal{O}_X) \simeq H^p_{K_m}(U, \mathcal{O}_X)$ vanishes for any m and for any $p \ne n$, ρ^p_m induces a surjective map $Z^p(\mathscr{V}^0_m, \mathscr{V}_m, \mathcal{O}_X) \to Z^p(\mathscr{V}^0_{m-1}, \mathscr{V}_{m-1}, \mathcal{O}_X)$ of cocyles for any $p \ne n$.

It follows from Hilfssatz 2.6 of Kiehl [5] that

$$H_K^p(U, \mathcal{O}_X) \simeq H^p(\mathcal{V}^0, \mathcal{V}, \mathcal{O}_X) \simeq H^p\left(\operatorname*{proj\,lim}_{m \to \infty} C^{\cdot}(\mathcal{V}_m^0, \mathcal{V}_m, \mathcal{O}_X)\right)$$

$$\simeq \operatorname*{proj\,lim}_{m \to \infty} H^p(C^{\cdot}(\mathcal{V}_m^0, \mathcal{V}_m, \mathcal{O}_X))$$

$$\simeq \operatorname*{proj\,lim}_{m \to \infty} H^p(\mathcal{V}_m^0, \mathcal{V}_m, \mathcal{O}_X) = 0,$$

for any $p \neq n+1$.

If $p = n+1$, then we have a canonical isomorphism

$$H_K^{n+1}(U, \mathcal{O}_X) \simeq H^n(U - K, \mathcal{O}_X)$$

$$\simeq H^n(\mathcal{V}, \mathcal{O}_X) \simeq H^n\left(\operatorname*{proj\,lim}_{m \to \infty} C^{\cdot}(\mathcal{V}_m, \mathcal{O}_X)\right),$$

where $C^{\cdot}(\mathcal{V}_m, \mathcal{O}_X)$ denotes the complex of $\mathcal{O}_X$-valued alternating cochains on $\mathcal{V}_m$. Since $\mathcal{V}_m \supset \mathcal{V}_{m-1}$, $\rho_m^p : C^p(\mathcal{V}_m, \mathcal{O}_X) \to C^p(\mathcal{V}_{m-1}, \mathcal{O}_X)$ is surjective. Hence ρ_m^p induces a surjective map $B^p(\mathcal{V}_m, \mathcal{O}_X) \to B^p(\mathcal{V}_{m-1}, \mathcal{O}_X)$ for any p. On the other hand, the module of $(n-1)$-cocycles of this complex $\{C^{\cdot}(\mathcal{V}_m, \mathcal{O}_X)\}$ is isomorphic to $\mathcal{O}_X(U - K_m)$. Hence the homomorphism $Z^{n-1}(\mathcal{V}_m, \mathcal{O}_X) \to Z^{n-1}(\mathcal{V}_{m-1}, \mathcal{O}_X)$ of cocycles can be identified with the restriction map $\mathcal{O}_X(U - K_m) \to \mathcal{O}_X(U - K_{m-1})$. Since this map has a dense image, it follows from a standard argument that

$$H^n\left(\operatorname*{proj\,lim}_{m \to \infty} C^{\cdot}(\mathcal{V}_m, \mathcal{O}_X)\right) = \operatorname*{proj\,lim}_{m \to \infty} H^n(C^{\cdot}(\mathcal{V}_m, \mathcal{O}_X)) = 0.$$

Therefore we have proved the following theorem.

Theorem 1.   *Let $(X, \mathcal{O}_X, \mathfrak{T})$ be the standard rigid analytic structure on $X = k^n$, let K be a compact subset of X, and let U be an affinoid subset of X containing K. Then the relative cohomology $H_K^p(U, \mathcal{O}_X)$ vanishes for any $p \neq n$. Let $\{r_m\}_{m=1}^{\infty}$, $K_m = \coprod_{1 \leq i \leq l_m} B(a_i, (r_m)^-)$ be as in Section 2. Let $\mathcal{B}(a_i, (r_m)^-)$ be the space of series $\sum c_M (x - a_i)^M (c_M \in k)$ which converge outside $B(a_i, (r_m)^-)$. Then $H_K^n(U, \mathcal{O}_X) \simeq \operatorname{proj\,lim}_{m \to \infty} H_{K_m}^n(U, \mathcal{O}_X)$, and $H_{K_m}^n(U, \mathcal{O}_X) \simeq \bigoplus_{1 \leq i \leq l_m} \mathcal{B}(a_i, (r_m)^-)$.*

For any series $\psi(x) = \sum_M d_M (x - a_i)^M (d_M \in k)$, where M runs over all n-tuples $(m_1, m_2, \ldots, m_n)$ of non-negative integers, and $(x - a_i)^M =$

$\prod_{1 \le j \le n} (x_j - a_{ij})^{m_j}$, put

$$\|\psi(x)\|_{i,m} = \sup_M |d_M| (r_m)^{m_1 + m_2 + \cdots + m_n}.$$

Let

$$\mathscr{A}_b(B(a_i, r_m)) = \{\psi(x) = \sum_M d_M (x - a_i)^M; d_M \in k, \|\psi(x)\|_{i,m} < \infty\}.$$

Then $\mathscr{A}_b(B(a_i, r_m))$ becomes a Banach space over k with this norm $\| \ \|_{i,m}$. Let $\mathscr{A}_b(K_m)$ be the direct sum of these Banach spaces $\mathscr{A}_b(B(a_i, r_m))$ ($i = 1, 2, \ldots, l_m$). Then the restriction map gives a continuous homomorphism $\mathscr{A}_b(K_{m-1}) \to \mathscr{A}_b(K_m)$. Let $\mathscr{A}(K)$ be the locally convex inductive limit of these Banach spaces. Then $\mathscr{A}(K)$ can be naturally identified with the space of locally analytic functions on K.

Let

$$\langle \ , \ \rangle_{i,m} : \mathscr{A}_b(B(a_i, r_m)) \times \mathscr{B}(a_i, (r_m)^-) \to k$$

be the pairing defined by

$$\left\langle \sum_L a_L (x - a_i)^L, \sum_M b_M (x - a_i)^M \right\rangle_{i,m} = \sum a_L b_M,$$

where $L = (l_1, l_2, \ldots, l_n)$ (resp. $M = (m_1, m_2, \ldots, m_n)$) runs over all n-tuples of non-negative integers (resp. all n-tuples of negative integers), $a_L, b_M \in k$ satisfy the obvious conditions, and the sum on the right-hand side runs over all pairs (L, M) such that $l_1 + m_1 = -1$, $l_2 + m_2 = -1, \ldots, l_n + m_n = -1$.

Let $\mathscr{B}(K_m)$ be the direct sum of the Banach spaces $\mathscr{B}(a_i, (r_m)^-)$ ($i = 1, 2, \ldots, l_m$), and let $\mathscr{B}(K) = H^n_K(U, \mathcal{O}_X)$ be the locally convex projective limit of these Banach spaces $\mathscr{B}(K_m)$ (cf. Theorem 1). Let $\langle \ , \ \rangle_m : \mathscr{A}_b(K_m) \times \mathscr{B}(K_m) \to k$ be the sum of the above pairings $\langle \ , \ \rangle_{i,m}$ for $i = 1, 2, \ldots, l_m$. Then, taking the limit for $m \to \infty$, these pairings $\langle \ , \ \rangle_m$ define a continuous pairing $\langle \ , \ \rangle : \mathscr{A}(K) \times \mathscr{B}(K) \to k$. Further, we repeat the arguments in Morita [7] and Schikhof–Morita [12] and obtain the following theorem.

Theorem 2. *Let K be a compact subset of the rigid analytic space $X = k^n$, U an affinoid subset of X containing K, $\mathscr{A}(K)$ the space of locally analytic functions on K, and $\mathscr{B}(K)$ the space $H^n_K(U, \mathcal{O}_X)$ of relative cohomologies. Then*

   (i)   *$\mathscr{A}(K)$ is a complete Hausdorff k-vector space.*

  (ii)   *$\mathscr{B}(K)$ is a Fréchet k-vector space.*

 (iii)   *$\mathscr{A}(K)$ and $\mathscr{B}(K)$ become mutually dual linear topological k-vector spaces with the pairing $\langle \ , \ \rangle$.*

5.   The Main Results

Let k, $X = k^n$, $(X, \mathcal{O}_X, \mathfrak{T})$ be as in Section 3, and let K be a compact subset of X. Let $\mathcal{H}_K^m(\mathcal{O}_X)$ be the sheaf on $\mathfrak{T}$ defined in Section 1. Hence, if U is a sufficiently small admissible open subset of X, $\Gamma(U, \mathcal{H}_K^m(\mathcal{O}_X)) = H_{K \cap U}^m(U, \mathcal{O}_X)$ for any $m \geq 0$. If U is an affinoid subset of X, then it follows from Theorem 1 that $\Gamma(U, \mathcal{H}_K^m(\mathcal{O}_X))$ vanishes for any $m \neq n$, because $K \cap U$ is a compact subset of X. Since the set of all affinoid subsets of X forms a fundamental system of the topology of X, the sheaf $\mathcal{H}_K^m(\mathcal{O}_X)$ vanishes for any $m \neq n$. Therefore, by the results of Section 1, we have a canonical isomorphism

$$H_{K \cap U}^{m+n}(U, \mathcal{O}_X) \simeq H^m(U, \mathcal{H}_K^n(\mathcal{O}_X)),$$

for any $m \geq 0$, and for any admissible open subset U of X. In particular, the presheaf $U \mapsto H_{K \cap U}^n(U, \mathcal{O}_X) \simeq \Gamma(U, \mathcal{H}_K^n(\mathcal{O}_X))$ is a sheaf on the Grothendieck topology $\mathfrak{T}$, and the cohomology $H^p(U, \mathcal{H}_K^n(\mathcal{O}_X))$ of this sheaf vanishes for any $p \geq 1$.

Let V be an open subset of K. Let U be an admissible open subset of X such that $K \cap U = V$. For example, $U = (X - K) \coprod V = X - (K - V)$ satisfies this condition (cf. Lemma 2). Put

$$\mathcal{B}(V) = H_{K \cap U}^n(U, \mathcal{O}_X) \simeq \Gamma(U, \mathcal{H}_K^n(\mathcal{O}_X)).$$

By Lemma 1, $\mathcal{B}(V)$ is independent of a special choice of U.

Let $\{V_i\}_{i \in I}$ be an open covering of an open subset V of K, and let $U_i = (X - K) \coprod V_i$. Then $(X - K) \coprod V = X - (K - V)$ is an admissible open subset of X, and $\{U_i\}_{i \in I}$ is an admissible covering of $(X - K) \coprod V$. Since $\mathcal{H}_K^n(\mathcal{O}_X)$ is a sheaf on the Grothendieck topology $\mathfrak{T}$, $\varphi \in \Gamma(\{U_i\}_{i \in I}, \mathcal{H}_K^n(\mathcal{O}_X))$ is zero if and only if the restriction φ_i of φ to each U_i vanishes for any $i \in I$. It follows from the above isomorphism that $\varphi \in \mathcal{B}(V)$ vanishes if and only if the restriction φ_i of φ to each V_i vanishes for any $i \in I$. Similarly, a set $\{\varphi_i\}_{i \in I}$ of local sections $\varphi_i \in \mathcal{B}(V_i)$ comes from a global section $\varphi \in \mathcal{B}(V)$ if and only if they are compatible, i.e., $\varphi_i(z) = \varphi_j(z)$ holds for any $z \in V_i \cap V_j$ and for any $i, j \in I$, because the corresponding condition holds for $\{U_i\}_{i \in I}$. Hence the presheaf $\mathcal{B} : V \mapsto \mathcal{B}(V)$ is in fact a sheaf on K.

Let V be an open subset of K, and let $U = X - (K - V)$. Then $K \cap U = V$ and $K - V = K \cap (X - U)$ is compact. Since $\mathcal{H}_K^{n+1}(\mathcal{O}_X) = 0$, for any $x \in K$, there exists a closed ball U_x of the form $\{x \in X; |x_j - a_j| \leq r \ (j = 1, 2, \ldots, n)\}$

$(a \in X, r \in |k|)$ such that

$$H^1_{(K-V) \cap U_x}(U_x, \mathcal{H}^n_K(\mathcal{O}_X)) = \Gamma_{(K-V) \cap U_x}(U_x, \mathcal{H}^{n+1}_K(\mathcal{O}_X)) = 0.$$

Since K is compact, a finite number of such balls $U_1, U_2, \ldots, U_m$ cover K. Here we may assume that $U_i \cap U_j = \varnothing$ for $i \neq j$. Put $V_i = K \cap U_i$. Then the following sequence is exact:

$$\Gamma(U_i, \mathcal{H}^n_K(\mathcal{O}_X)) \to \Gamma(U_i - (K - V) \cap U_i, \mathcal{H}^n_K(\mathcal{O}_X))$$

$$\to H^1_{(K-V) \cap U_i}(U_i, \mathcal{H}^n_K(\mathcal{O}_X)) = 0.$$

Let φ be any element of $\mathcal{B}(V) = H^n_K(U, \mathcal{O}_X) = \Gamma(U, \mathcal{H}^n_K(\mathcal{O}_X))$. Let $\varphi_i \in \mathcal{B}(V \cap V_i) = \Gamma(U \cap U_i, \mathcal{H}^n_K(\mathcal{O}_X)) = \Gamma(U_i - (K - V) \cap U_i, \mathcal{H}^n_K(\mathcal{O}_X))$ be the restriction of φ to $V \cap V_i$. We put $\varphi_i = 0$ if $V \cap V_i = \varnothing$. Since the restriction map $\Gamma(U_i, \mathcal{H}^n_K(\mathcal{O}_X)) \to \Gamma(U_i - (K - V) \cap U_i, \mathcal{H}^n_K(\mathcal{O}_X))$ is surjective, we can extend each φ_i to an element ψ_i of $\Gamma(U_i, \mathcal{H}^n_K(\mathcal{O}_X))$.

Let $W = \bigcup_{1 \leq i \leq n} U_i$. Since $U_1, U_2, \ldots, U_n$ are mutually disjoint, W is an affinoid subset of X. Since $\{V_1, V_2, \ldots, V_m\}$ is a finite affinoid covering of W, it is admissible for $\mathfrak{T}$. Hence $\{\psi_i\}_{1 \leq i \leq m}$ defines an element ψ of $\Gamma(W, \mathcal{H}^n_K(\mathcal{O}_X))$. Since W is an admissible subset of X containing K, ψ is an element of $\mathcal{B}(K)$. Since the restriction of ψ to V coincides with the original φ, the restriction map $\mathcal{B}(K) \to \mathcal{B}(V)$ is surjective. Hence $\mathcal{B}$ is a flabby sheaf on K.

Therefore we have proved the following theorem.

Theorem 3. *Let $(X, \mathcal{O}_X, \mathfrak{T})$ be the standard rigid analytic structure on $X = k^n$, and let K be a compact subset of X. For any open subset V of K, take an admissible subset U of X such that $K \cap U = V$, and put*

$$\mathcal{B}(V) = H^n_{K \cap U}(U, \mathcal{O}_X).$$

Then $\mathcal{B}(V)$ is independent of a special choice of U, and the correspondence $V \mapsto \mathcal{B}(V)$ gives a flabby sheaf on K.

References

[1] M. Artin, *Grothendieck topologies*, Lecture notes of Harvard University, 1962.
[2] M. Artin, A. Grothendieck, and J. L. Verdier, *Théorie des Topos et Cohomologie Etale des Schémas* (SGA 4), Lecture Notes in Math. **269** (1972), **270** (1972), **305** (1973).
[3] S. Bosch, U. Günter and R. Remmert, *Non-archimedean Analysis*, Springer-Verlag, Berlin–Heidelberg–New York–Tokyo, 1984.

[4] D. Barsky, *Mesures p-adique et éléments analytiques*, Sémin. Théor. Nombres, 1973-1974, Univ. Bordeaux, Exposé 6, 1974.

[5] R. Kiehl, Theorem A und Theorem B in der nichtarchimedischen Funktionentheorie, *Inventiones Math.* **2** (1967) 256-273.

[6] Y. Morita, Analytic functions on an open subset of $\mathbf{P}^1(k)$, *J. Reine Angew. Math.* **311/312** (1979) 361-383.

[7] Y. Morita, A p-adic theory of hyperfunctions, I, *Publ. Res. Inst. Math. Sci.* **17** (1981) 1-24.

[8] Y. Morita, Analytic representations of SL_2 over a p-adic number field, I-III, *J. Fac. Sci. Univ. Tokyo, Sect. IA Math.* **28** (1982), 891-905 (with A. Murase); *Progress in Math.* **46** (1984), 282-297; *Advanced Studies in Pure Math.* **7** (1986), 185-222.

[9] A. C. M. Van Rooij, *Nonarchimedean functional analysis*, Monographs and Textbooks in Pure and Applied Math, **51**, Marcel Dekker, Inc. New York, Basel, 1978.

[10] M. Sato, Theory of hyperfunction, I-II, *J. Fac. Sci. Univ. Tokyo, Sec. IA* **8** (1959) 139-193; **8** (1960) 387-437.

[11] W. H. Schikhof, Locally convex spaces over nonspherically complete valued fields, I-II, *Tijdschrift van het Belgisch Wiskundig Genootschap, Ser. B, Fasc. I* **28** (1986).

[12] W. H. Schikhof and Y. Morita, Duality of projective limit spaces and inductive limit spaces over a nonspherically complete nonarchimedean field, *Tohoku Math. J.* **38** (1986) 387-397.

[13] J. Tate, Rigid analytic spaces, *Inventiones Math.* **12** (1971) 257-289.

[14] J. van Tiel, Espaces localement K-convexes, I-III, *Indag. Math.* **27** (1965) 249-258; 259-272; 273-289.